CALCULUS

A COMPLETE COURSE

SECOND EDITION

STUDENT'S SOLUTIONS MANUAL

VOLUME 2

CALCULUS
A COMPLETE COURSE
SECOND EDITION

Student's Solutions Manual
Volume 2

Ross L. Finney

Franklin D. Demana
The Ohio State University

Bert K. Waits
The Ohio State University

Daniel Kennedy
Baylor School

ADDISON-WESLEY

An imprint of Addison Wesley Longman, Inc.

Reading, Massachusetts • Menlo Park, California • New York • Harlow, England
Don Mills, Ontario • Sydney • Mexico City • Madrid • Amsterdam

Introduction

This publication provides complete, worked out solutions for Quick Review, odd numbered Exercises, and Lesson Explorations, including Cumulative Review and Appendices exercises.

Many of the analytic solutions could alternatively be done graphically. Teachers are encouraged to accept creative, graphical solutions that show understanding of the graphing utilities. In general, the solutions in this manual provide one possible method of solution.

Reproduced by Addison Wesley Longman from camera-ready copy prepared by Laurel Technical Services.

ISBN 0-201-66979-X

1 2 3 4 5 6 7 8 9 10 PHTH 02010099

CONTENTS

Chapter 11
Vectors and Analytic Geometry in Space

Chapter Opener

$\mathbf{F} = (7\mathbf{i} + 8\mathbf{j} + \mathbf{k}) + (6\mathbf{i} + 9\mathbf{j} + 2\mathbf{k})$
$= 13\mathbf{i} + 17\mathbf{j} + 3\mathbf{k}$
$W = \mathbf{F} \cdot \mathbf{D} = (13\mathbf{i} + 17\mathbf{j} + 3\mathbf{k}) \cdot (12\mathbf{i} + 16\mathbf{j} + 0\mathbf{k})$
$= (13)(12) + (17)(16) + (3)(0)$
$= 428$
The amount of work is 428 ft-lb.

■ Section 11.1 Cartesian (Rectangular) Coordinates and Vectors in Space (pp. 575–585)

Quick Review 11.1

1. $d = \sqrt{(x_2 - x_1)^2 + (y_2 - y_1)^2}$
$= \sqrt{[5 - (-2)]^2 + (2 - 3)^2}$
$= \sqrt{7^2 + (-1)^2}$
$= \sqrt{50}$
$= 5\sqrt{2}$

2. $M = \left(\frac{x_1 + x_2}{2}, \frac{y_1 + y_2}{2}\right)$
$= \left(\frac{-2 + 5}{2}, \frac{3 + 2}{2}\right)$
$= \left(\frac{3}{2}, \frac{5}{2}\right)$

3. $\sqrt{2^2 + 3^2} = \sqrt{13}$

4. $\frac{\mathbf{u}}{|\mathbf{u}|} = \frac{2\mathbf{i} + 3\mathbf{j}}{\sqrt{13}} = \frac{2}{\sqrt{13}}\mathbf{i} + \frac{3}{\sqrt{13}}\mathbf{j}$

5. Since the unit vector in the direction of **u** is
$\frac{2}{\sqrt{13}}\mathbf{i} + \frac{3}{\sqrt{13}}\mathbf{j}$,
the desired vector is
$-3\left(\frac{2}{\sqrt{13}}\mathbf{i} + \frac{3}{\sqrt{13}}\mathbf{j}\right) = -\frac{6}{\sqrt{13}}\mathbf{i} - \frac{9}{\sqrt{13}}\mathbf{j}$.

6. Ellipse, center (0, 0), vertices: (0, ±3), x-intercepts: (±2, 0)

7. Ellipse plus interior, center (0, 0), vertices: (0, ±3), x-intercepts: (±2, 0)

8. $(x - x_1)^2 + (y - y_1)^2 = r^2$
$(x - 2)^2 + [y - (-3)]^2 = 3^2$
$(x - 2)^2 + (y + 3)^2 = 9$

9. $x^2 + y^2 + 6x - 2y = -6$
$x^2 + 6x + y^2 - 2y = 6$
$(x^2 + 6x + 9) + (y^2 - 2y + 1) = -6 + 9 + 1$
$(x + 3)^2 + (y - 1)^2 = 4$
Center: $(-3, 1)$, radius $= 2$

10. $[3 - (-2)]\mathbf{i} + (-1 - 3)\mathbf{j} = 5\mathbf{i} - 4\mathbf{j}$

Section 11.1 Exercises

1. The line through the point (2, 3, 0) parallel to the z-axis

3. The x-axis

5. The circle $x^2 + y^2 = 4$ in the plane $z = -2$

7. The circle $y^2 + z^2 = 1$ in the yz-plane

9. Substitute 0 for z in $x^2 + y^2 + (z + 3)^2 = 25$.
$x^2 + y^2 + 3^2 = 25$
$x^2 + y^2 = 16$
The set of points can be described as the circle $x^2 + y^2 = 16$ in the plane $z = 0$.

11. a. The first quadrant of the xy-plane
b. The fourth quadrant of the xy-plane

13. a. The ball of radius 1 centered at the origin
b. All points at distance greater than 1 unit from the origin

15. a. The upper hemisphere of radius 1 centered at the origin
b. The solid upper hemisphere of radius 1 centered at the origin

17. a. $x = 3$
b. $y = -1$
c. $z = -2$

19. a. $z = 1$
b. $x = 3$
c. $y = -1$

21. a. $x^2 + (y - 2)^2 = 4, z = 0$
b. $(y - 2)^2 + z^2 = 4, x = 0$
c. $x^2 + z^2 = 4, y = 2$

23. a. $y = 3, z = -1$
b. $x = 1, z = -1$
c. $x = 1, y = 3$

25. $x^2 + y^2 + z^2 = 25, z = 3$

27. $0 \le z \le 1$

29. $z \le 0$

31. a. $(x - 1)^2 + (y - 1)^2 + (z - 1)^2 < 1$
b. $(x - 1)^2 + (y - 1)^2 + (z - 1)^2 > 1$

33. Length $= |2\mathbf{i} + \mathbf{j} - 2\mathbf{k}| = \sqrt{2^2 + 1^2 + (-2)^2} = 3$
Direction: $\frac{2}{3}\mathbf{i} + \frac{1}{3}\mathbf{j} - \frac{2}{3}\mathbf{k}$
$2\mathbf{i} + \mathbf{j} - 2\mathbf{k} = 3\left(\frac{2}{3}\mathbf{i} + \frac{1}{3}\mathbf{j} - \frac{2}{3}\mathbf{k}\right)$

35. Length $= |5\mathbf{k}| = \sqrt{25} = 5$
Direction: **k**
$5\mathbf{k} = 5(\mathbf{k})$

37. Length $= \left|\frac{1}{\sqrt{6}}\mathbf{i} - \frac{1}{\sqrt{6}}\mathbf{j} - \frac{1}{\sqrt{6}}\mathbf{k}\right| = \sqrt{3\left(\frac{1}{\sqrt{6}}\right)^2} = \sqrt{\frac{1}{2}}$
Direction: $\frac{1}{\sqrt{3}}\mathbf{i} - \frac{1}{\sqrt{3}}\mathbf{j} - \frac{1}{\sqrt{3}}\mathbf{k}$
$\frac{1}{\sqrt{6}}\mathbf{i} - \frac{1}{\sqrt{6}}\mathbf{j} - \frac{1}{\sqrt{6}}\mathbf{k} = \sqrt{\frac{1}{2}}\left(\frac{1}{\sqrt{3}}\mathbf{i} - \frac{1}{\sqrt{3}}\mathbf{j} - \frac{1}{\sqrt{3}}\mathbf{k}\right)$

39. a. $2\mathbf{i}$
b. $-\sqrt{3}\mathbf{k}$
c. $\frac{3}{10}\mathbf{j} + \frac{2}{5}\mathbf{k}$
d. $6\mathbf{i} - 2\mathbf{j} + 3\mathbf{k}$

41. $|\mathbf{A}| = \sqrt{12^2 + 5^2} = \sqrt{169} = 13$

$\frac{\mathbf{A}}{|\mathbf{A}|} = \frac{1}{13}\mathbf{A} = \frac{1}{13}(12\mathbf{i} - 5\mathbf{k})$

the desired vector is $\frac{7}{13}(12\mathbf{i} - 5\mathbf{k})$

43. a. the distance = the length = $|\overrightarrow{P_1P_2}| = |3\mathbf{i} + 4\mathbf{j} - 5\mathbf{k}|$
$= \sqrt{9 + 16 + 25} = 5\sqrt{2}$

b. $3\mathbf{i} + 4\mathbf{j} - 5\mathbf{k} = 5\sqrt{2}\left(\frac{3}{5\sqrt{2}}\mathbf{i} + \frac{4}{5\sqrt{2}}\mathbf{j} - \frac{1}{\sqrt{2}}\mathbf{k}\right)$

the direction is $\frac{3}{5\sqrt{2}}\mathbf{i} + \frac{4}{5\sqrt{2}}\mathbf{j} - \frac{1}{\sqrt{2}}\mathbf{k}$

c. the midpoint is $\left(\frac{-1+2}{2}, \frac{1+5}{2}, \frac{5+0}{2}\right) = \left(\frac{1}{2}, 3, \frac{5}{2}\right)$

45. a. the distance = the length = $|\overrightarrow{P_1P_2}| = |-\mathbf{i} - \mathbf{j} - \mathbf{k}|$
$= \sqrt{3}$

b. $-\mathbf{i} - \mathbf{j} - \mathbf{k} = \sqrt{3}\left(-\frac{1}{\sqrt{3}}\mathbf{i} - \frac{1}{\sqrt{3}}\mathbf{j} - \frac{1}{\sqrt{3}}\mathbf{k}\right)$

the direction is $-\frac{1}{\sqrt{3}}\mathbf{i} - \frac{1}{\sqrt{3}}\mathbf{j} - \frac{1}{\sqrt{3}}\mathbf{k}$

c. the midpoint is $\left(\frac{3+2}{2}, \frac{4+3}{2}, \frac{5+4}{2}\right) = \left(\frac{5}{2}, \frac{7}{2}, \frac{9}{2}\right)$

47. center $(-2, 0, 2)$, radius $\sqrt{8} = 2\sqrt{2}$

49. $(x-1)^2 + (y-2)^2 + (z-3)^2 = 14$

51. $x^2 + y^2 + z^2 + 4x - 4z = 0$
$(x^2 + 4x + 4) + y^2 + (z^2 - 4z + 4) = 4 + 4$
$(x+2)^2 + (y-0)^2 + (z-2)^2 = (\sqrt{8})^2$
the center is at $(-2, 0, 2)$ and the radius is $\sqrt{8}$

53. $2x^2 + 2y^2 + 2z^2 + x + y + z = 9$

$x^2 + \frac{1}{2}x + y^2 + \frac{1}{2}y + z^2 + \frac{1}{2}z = \frac{9}{2}$

$\left(x^2 + \frac{1}{2}x + \frac{1}{16}\right) + \left(y^2 + \frac{1}{2}y + \frac{1}{16}\right) + \left(z^2 + \frac{1}{2}z + \frac{1}{16}\right) = \frac{9}{2} + \frac{3}{16}$

$\left(x + \frac{1}{4}\right)^2 + \left(y + \frac{1}{4}\right)^2 + \left(z + \frac{1}{4}\right)^2 = \left(\frac{5\sqrt{3}}{4}\right)^2$

the center is at $\left(-\frac{1}{4}, -\frac{1}{4}, -\frac{1}{4}\right)$ and the radius is $\frac{5\sqrt{3}}{4}$

55. $\overrightarrow{AB} = (5-a)\mathbf{i} + (1-b)\mathbf{j} + (3-c)\mathbf{k} = \mathbf{i} + 4\mathbf{j} - 2\mathbf{k}$
$5 - a = 1$, $1 - b = 4$, and $3 - c = -2$
$a = 4$, $b = -3$, and $c = 5$
A is the point $(4, -3, 5)$

57. a. the distance between (x, y, z) and $(x, 0, 0)$ is $\sqrt{y^2 + z^2}$

b. the distance between (x, y, z) and $(0, y, 0)$ is $\sqrt{x^2 + z^2}$

c. the distance between (x, y, z) and $(0, 0, z)$ is $\sqrt{x^2 + y^2}$

59. Let $A = (a_1, a_2, a_3)$, $B = (b_1, b_2, b_3)$, $C = (c_1, c_2, c_3)$, and $D = (d_1, d_2, d_3)$.
The midpoints of the sides are:

$M_{AB} = \left(\frac{a_1+b_1}{2}, \frac{a_2+b_2}{2}, \frac{a_3+b_3}{2}\right)$,
$M_{BC} = \left(\frac{b_1+c_1}{2}, \frac{b_2+c_2}{2}, \frac{b_3+c_3}{2}\right)$,
$M_{CD} = \left(\frac{c_1+d_1}{2}, \frac{c_2+d_2}{2}, \frac{c_3+d_3}{2}\right)$,
$M_{AD} = \left(\frac{a_1+d_1}{2}, \frac{a_2+d_2}{2}, \frac{a_3+d_3}{2}\right)$.

The midpoint of the segment joining M_{AB} and M_{CD} and the midpoint of the segment joining M_{BC} and M_{AD} have coordinates

$\frac{a_i + b_i + c_i + d_i}{4}$ for $i = 1, 2, 3$. Thus, the midpoints bisect each other.

61. $\overrightarrow{P_1P_2} = \mathbf{i} + 2\mathbf{j} + 3\mathbf{k} = \overrightarrow{P_4P_3}$ and $\overrightarrow{P_2P_3} = 2\mathbf{i} + 3\mathbf{j} + 4\mathbf{k} = \overrightarrow{P_1P_4}$. The opposite sides are represented by equal vectors. Therefore, they are parallel and have the same length, and the figure is a parallelogram.

63. a. $x = -1: (-1)^2 + y^2 + z^2 = 9$
$y^2 + z^2 = 8$
$x = 0: 0^2 + y^2 + z^2 = 9$
$y^2 + z^2 = 9$
$x = 2: 2^2 + y^2 + z^2 = 9$
$y^2 + z^2 = 5$

b.

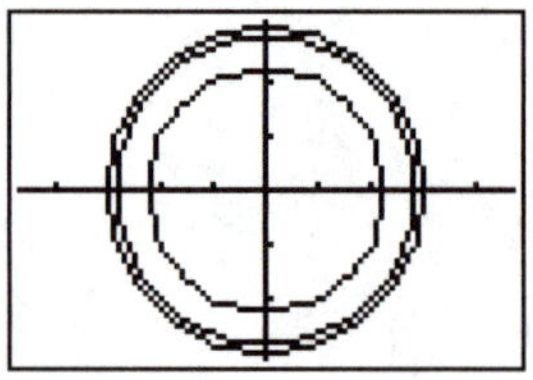

c. If $0 \le |c| \le 3$, then for $x = c$ graph $r = \sqrt{9 - c^2}$ in polar coordinate mode. If $|c| > 3$ there is no graph.

d. If $|x| \le 3$, then $A_x = \pi(9 - x^2)$.

e. It is the volume of the sphere.

65. The midpoint of AB is $M\left(\frac{3}{2}, 0, \frac{5}{2}\right)$ and

$\left(\frac{2}{3}\right)\overrightarrow{CM} = \frac{2}{3}\left[\left(\frac{3}{2} + 1\right)\mathbf{i} + (0 - 2)\mathbf{j} + \left(\frac{5}{2} + 1\right)\mathbf{k}\right]$
$= \frac{2}{3}\left(\frac{5}{2}\mathbf{i} - 2\mathbf{j} + \frac{7}{2}\mathbf{k}\right) = \frac{5}{3}\mathbf{i} - \frac{4}{3}\mathbf{j} + \frac{7}{3}\mathbf{k}.$

The terminal point of $\left(\frac{5}{3}\mathbf{i} - \frac{4}{3}\mathbf{j} + \frac{7}{3}\mathbf{k}\right) + \overrightarrow{OC}$
$= \left(\frac{5}{3}\mathbf{i} - \frac{4}{3}\mathbf{j} + \frac{7}{3}\mathbf{k}\right) + (-\mathbf{i} + 2\mathbf{j} - \mathbf{k}) = \frac{2}{3}\mathbf{i} + \frac{2}{3}\mathbf{j} + \frac{4}{3}\mathbf{k}$ is the point $\left(\frac{2}{3}, \frac{2}{3}, \frac{4}{3}\right)$ which is the location of the intersection of the medians.

■ Section 11.2 Dot Products (pp. 586–593)

Exploration 1 Proving a Theorem in Solid Geometry

1. Since the origin is the midpoint of line segment PQ, the coordinates of Q are $(-a, -b, -c)$.

2. The equation of a sphere is $x^2 + y^2 + z^2 = r^2$. Since (a, b, c) is on the sphere, $a^2 + b^2 + c^2 = r^2$, so $x^2 + y^2 + z^2 = a^2 + b^2 + c^2$.

3. $\overrightarrow{PX} = (x - a)\mathbf{i} + (y - b)\mathbf{j} + (z - c)\mathbf{k}$
$\overrightarrow{QX} = (x + a)\mathbf{i} + (y + b)\mathbf{j} + (z + c)\mathbf{k}$

4. $\overrightarrow{PX} \cdot \overrightarrow{QX}$
$= (x - a)(x + a) + (y - b)(y + b) + (z - c)(z + c)$
$= (x^2 - a^2) + (y^2 - b^2) + (z^2 - c^2)$
$= (x^2 + y^2 + z^2) - (a^2 + b^2 + c^2)$
$= 0$

5. Angle PXQ is a right angle. We have proved that if a triangle is inscribed in a sphere and one side of the triangle is a diameter of the sphere, then the triangle is a right triangle.

Quick Review 11.2

1. $\mathbf{A} \cdot \mathbf{B} = (1)(1) + (1)(-1) = 0$

2. $\mathbf{A} \cdot \mathbf{B} = 0 = |\mathbf{A}||\mathbf{B}| \cos \theta$
$0 = \cos \theta$
$\theta = \frac{\pi}{2}$

3. $\mathbf{A} \cdot \mathbf{i} = (1)(1) + (1)(0) = 1$

4. $m = \frac{-1}{1} = -1$
$y - y_1 = m(x - x_1)$
$y - 2 = -1(x - 3)$
$x + y = 5$

5. $m = -\frac{1}{-1} = 1$
$y - y_1 = m(x - x_1)$
$y - 2 = 1(x - 3)$
$x - y = 1$

6. $y = \sin x$

$\frac{dy}{dx} = \cos x$

$\frac{dy}{dx}\left(\frac{\pi}{6}\right) = \cos \frac{\pi}{6} = \frac{\sqrt{3}}{2}$

The vector $\mathbf{i} + \frac{\sqrt{3}}{2}\mathbf{j}$ is tangent to the curve at $x = \frac{\pi}{6}$, and its length is $\sqrt{1^2 + \left(\frac{\sqrt{3}}{2}\right)^2} = \frac{\sqrt{7}}{2}$, so the desired vectors are $\pm\frac{2}{\sqrt{7}}\left(\mathbf{i} + \frac{\sqrt{3}}{2}\mathbf{j}\right)$, or $\pm\left(\frac{2}{\sqrt{7}}\mathbf{i} + \sqrt{\frac{3}{7}}\mathbf{j}\right)$.

7. The normal unit vectors can be obtained by exchanging the coordinates of a unit tangent vector and changing the sign of one of the coordinates. The desired vectors are $\pm\left(\sqrt{\frac{3}{7}}\mathbf{i} - \frac{2}{\sqrt{7}}\mathbf{j}\right)$.

8. $x = 5 \cos t$, $y = 3 \sin t$

$\frac{dx}{dt} = -5 \sin t$, $\frac{dy}{dt} = 3 \cos t$

$\frac{dx}{dt}\left(\frac{3\pi}{4}\right) = -\frac{5}{\sqrt{2}}$, $\frac{dy}{dt}\left(\frac{3\pi}{4}\right) = -\frac{3}{\sqrt{2}}$

The vector $-\frac{5}{\sqrt{2}}\mathbf{i} - \frac{3}{\sqrt{2}}\mathbf{j}$ is tangent to the curve at $t = \frac{3\pi}{4}$, and its length is $\sqrt{\left(-\frac{5}{\sqrt{2}}\right)^2 + \left(-\frac{3}{\sqrt{2}}\right)^2} = \sqrt{17}$, so the desired vectors are $\pm\frac{1}{\sqrt{17}}\left(-\frac{5}{\sqrt{2}}\mathbf{i} - \frac{3}{\sqrt{2}}\mathbf{j}\right)$, or $\pm\left(\frac{5}{\sqrt{34}}\mathbf{i} + \frac{3}{\sqrt{34}}\mathbf{j}\right)$.

9. The normal unit vectors can be obtained by exchanging the coordinates of a unit tangent vector and changing the sign of one of the coordinates. The desired vectors are $\pm\left(\frac{3}{\sqrt{34}}\mathbf{i} - \frac{5}{\sqrt{34}}\mathbf{j}\right)$.

10. Assuming the distance is given in meters,
$W = \mathbf{F} \cdot \mathbf{D} = (50\text{ N}) \cdot (5\text{ m}) = 250\text{ N-m}$

Section 11.2 Exercises

NOTE: In Exercises 1–10 below we calculate $\text{proj}_{\mathbf{A}}\,\mathbf{B}$ as the vector $\left(\frac{|\mathbf{B}|\cos\theta}{|\mathbf{A}|}\right)\mathbf{A}$, so the scalar multiplier of **A** is the number in column 6 divided by the number in column 2.

	$\mathbf{A}\cdot\mathbf{B}$	$\lvert\mathbf{A}\rvert$	$\lvert\mathbf{B}\rvert$	$\cos\theta$	θ	$\lvert\mathbf{B}\rvert\cos\theta$	$\text{proj}_{\mathbf{A}}\,\mathbf{B}$
1.	-25	5	5	-1	π	-5	$-2\mathbf{i}+4\mathbf{j}-\sqrt{5}\mathbf{k}$
3.	25	15	5	$\frac{1}{3}$	$\cos^{-1}\left(\frac{1}{3}\right)\approx 1.23$ rad	$\frac{5}{3}$	$\frac{1}{9}(10\mathbf{i}+11\mathbf{j}-2\mathbf{k})$
5.	2	$\sqrt{34}$	$\sqrt{3}$	$\frac{2}{\sqrt{102}}$	$\cos^{-1}\left(\frac{2}{\sqrt{102}}\right)\approx 1.37$ rad	$\frac{2}{\sqrt{34}}$	$\frac{1}{17}(5\mathbf{j}-3\mathbf{k})$

7. $\mathbf{A}\cdot\mathbf{B}=3$ and $\mathbf{A}\cdot\mathbf{A}=2$

$$\mathbf{B}=\left(\frac{\mathbf{A}\cdot\mathbf{B}}{\mathbf{A}\cdot\mathbf{A}}\right)\mathbf{A}+\left(\mathbf{B}-\left(\frac{\mathbf{A}\cdot\mathbf{B}}{\mathbf{A}\cdot\mathbf{A}}\right)\mathbf{A}\right)=\frac{3}{2}\mathbf{A}+\left(\mathbf{B}-\frac{3}{2}\mathbf{A}\right)=\frac{3}{2}(\mathbf{i}+\mathbf{j})+\left[(3\mathbf{j}+4\mathbf{k})-\frac{3}{2}(\mathbf{i}+\mathbf{j})\right]=\left(\frac{3}{2}\mathbf{i}+\frac{3}{2}\mathbf{j}\right)+\left(-\frac{3}{2}\mathbf{i}+\frac{3}{2}\mathbf{j}+4\mathbf{k}\right)$$

9. $\mathbf{A}\cdot\mathbf{B}=28$ and $\mathbf{A}\cdot\mathbf{A}=6$

$$\mathbf{B}=\left(\frac{\mathbf{A}\cdot\mathbf{B}}{\mathbf{A}\cdot\mathbf{A}}\right)\mathbf{A}+\left(\mathbf{B}-\left(\frac{\mathbf{A}\cdot\mathbf{B}}{\mathbf{A}\cdot\mathbf{A}}\right)\mathbf{A}\right)=\frac{28}{6}\mathbf{A}+\left(\mathbf{B}-\frac{28}{6}\mathbf{A}\right)=\frac{14}{3}(\mathbf{i}+2\mathbf{j}-\mathbf{k})+\left[(8\mathbf{i}+4\mathbf{j}-12\mathbf{k})-\left(\frac{14}{3}\mathbf{i}+\frac{28}{3}\mathbf{j}-\frac{14}{3}\mathbf{k}\right)\right]=\left(\frac{14}{3}\mathbf{i}+\frac{28}{3}\mathbf{j}-\frac{14}{3}\mathbf{k}\right)+\left(\frac{10}{3}\mathbf{i}-\frac{16}{3}\mathbf{j}-\frac{22}{3}\mathbf{k}\right)$$

11. The sum of two vectors of equal length is *always* orthogonal to their difference, as we can see from the equation $(\mathbf{v}_1+\mathbf{v}_2)\cdot(\mathbf{v}_1-\mathbf{v}_2)$

$$=\mathbf{v}_1\cdot\mathbf{v}_1+\mathbf{v}_2\cdot\mathbf{v}_1-\mathbf{v}_1\cdot\mathbf{v}_2-\mathbf{v}_2\cdot\mathbf{v}_2=|\mathbf{v}_1|^2-|\mathbf{v}_2|^2=0$$

13. a. $\cos\alpha=\frac{\mathbf{i}\cdot\mathbf{v}}{|\mathbf{i}||\mathbf{v}|}=\frac{a}{|\mathbf{v}|}$, $\cos\beta=\frac{\mathbf{j}\cdot\mathbf{v}}{|\mathbf{j}||\mathbf{v}|}=\frac{b}{|\mathbf{v}|}$, $\cos\gamma=\frac{\mathbf{k}\cdot\mathbf{v}}{|\mathbf{k}||\mathbf{v}|}=\frac{c}{|\mathbf{v}|}$ and

$$\cos^2\alpha+\cos^2\beta+\cos^2\gamma=\left(\frac{a}{|\mathbf{v}|}\right)^2+\left(\frac{b}{|\mathbf{v}|}\right)^2+\left(\frac{c}{|\mathbf{v}|}\right)^2=\frac{a^2+b^2+c^2}{|\mathbf{v}||\mathbf{v}|}=\frac{|\mathbf{v}||\mathbf{v}|}{|\mathbf{v}||\mathbf{v}|}=1$$

b. If $|\mathbf{v}|=1$, then $\cos\alpha=\frac{a}{|\mathbf{v}|}=a$, $\cos\beta=\frac{b}{|\mathbf{v}|}=b$ and $\cos\gamma=\frac{c}{|\mathbf{v}|}=c$ are the direction cosines of **v**.

15. Use a cube whose edges represent **i**, **j**, and **k**. Then the diagonal of the cube and the diagonal of one face are $\mathbf{A}=\mathbf{i}+\mathbf{j}+\mathbf{k}$ and $\mathbf{B}=\mathbf{i}+\mathbf{k}$, respectively.

$$\theta=\cos^{-1}\left(\frac{\mathbf{A}\cdot\mathbf{B}}{|\mathbf{A}||\mathbf{B}|}\right)=\cos^{-1}\left(\frac{2}{\sqrt{2}\sqrt{3}}\right)=\cos^{-1}\sqrt{\frac{2}{3}}\approx 0.62\text{ rad}\approx 35.26^\circ$$

17. $\mathbf{v}\cdot\mathbf{u}_1=(a\mathbf{u}_1+b\mathbf{u}_2)\cdot\mathbf{u}_1=a\mathbf{u}_1\cdot\mathbf{u}_1+b\mathbf{u}_2\cdot\mathbf{u}_1=a|\mathbf{u}_1|^2+b(\mathbf{u}_2\cdot\mathbf{u}_1)=a(1)^2+b(0)=a$

19. a. $|\mathbf{D}|^2=\mathbf{D}\cdot\mathbf{D}=25\mathbf{A}\cdot\mathbf{A}+36\mathbf{B}\cdot\mathbf{B}+9\mathbf{C}\cdot\mathbf{C}=25+36+9=70$

$|\mathbf{D}|=\sqrt{70}$

b. $|\mathbf{D}|^2=25|\mathbf{A}|^2+36|\mathbf{B}|^2+9|\mathbf{C}|^2=(25)(4)+(36)(9)+(9)(16)=568$

$|\mathbf{D}|=\sqrt{568}$

21. $P(0,0,0)$, $Q(1,1,1)$ and $\mathbf{F}=5\mathbf{k}$

$\overrightarrow{PQ}=\mathbf{i}+\mathbf{j}+\mathbf{k}$ and $\mathbf{W}=\mathbf{F}\cdot\overrightarrow{PQ}=(5\mathbf{k})\cdot(\mathbf{i}+\mathbf{j}+\mathbf{k})=5$ N-m $=5$ J

23. $\mathbf{W}=|\mathbf{F}||\overrightarrow{PQ}|\cos\theta=(200)(20)(\cos 30^\circ)=2000\sqrt{3}=3464.10$ N-m $=3464.10$ J

25. $P(x_1,y_1)=P\left(x_1,\frac{c}{b}-\frac{a}{b}x_1\right)$ and $Q(x_2,y_2)=Q\left(x_2,\frac{c}{b}-\frac{a}{b}x_2\right)$ are any two points P and Q on the line with $b\neq 0$

$$\overrightarrow{PQ}=(x_2-x_1)\mathbf{i}+\frac{a}{b}(x_2-x_1)\mathbf{j}$$

$$\overrightarrow{PQ}\cdot\mathbf{v}=\left[(x_2-x_1)\mathbf{i}+\frac{a}{b}(x_2-x_1)\mathbf{j}\right]\cdot(a\mathbf{i}+b\mathbf{j})=a(x_2-x_1)+b\left(\frac{a}{b}\right)(x_2-x_1)=0$$

v is perpendicular to $\overrightarrow{PQ}$ for $b\neq 0$. If $b=0$, then $\mathbf{v}=a\mathbf{i}$ is perpendicular to the vertical line $ax=c$. Alternatively, the slope of **v** is $\frac{b}{a}$ and the slope of the line $ax+by=c$ is $-\frac{a}{b}$, so the slopes are negative reciprocals. The vector **v** and the line are perpendicular.

27. $\mathbf{v} = \mathbf{i} + 2\mathbf{j}$ is perpendicular to the line $x + 2y = c$.
$P(2, 1)$ is on the line,
so $2 + 2 = c$.
The line is $x + 2y = 4$.

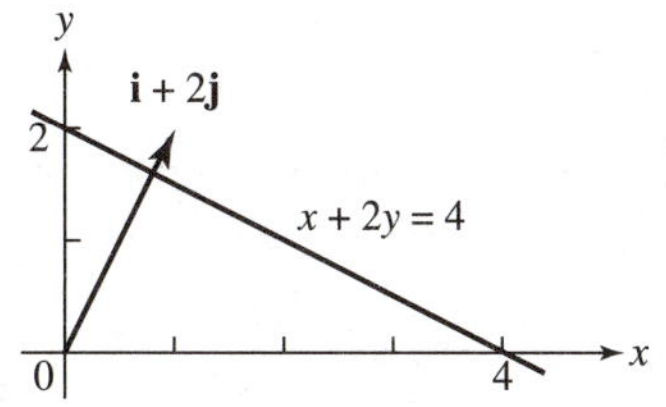

29. $\mathbf{v} = -2\mathbf{i} + \mathbf{j}$ is perpendicular to the line $-2x + y = c$.
$P(-2, -7)$ is on the line,
so $(-2)(-2) - 7 = c$.
The line is $-2x + y = -3$.

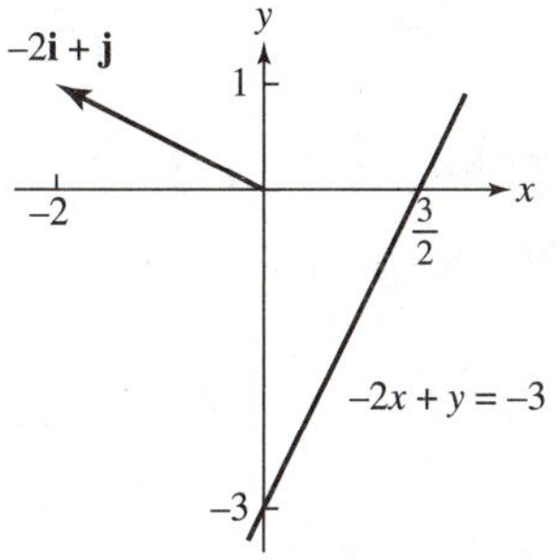

31. $\mathbf{v} = \mathbf{i} - \mathbf{j}$ is parallel to the line $-x - y = c$.
$P(-2, 1)$ is on the line,
so $-(-2) - 1 = c$.
The line is $-x - y = -1$.

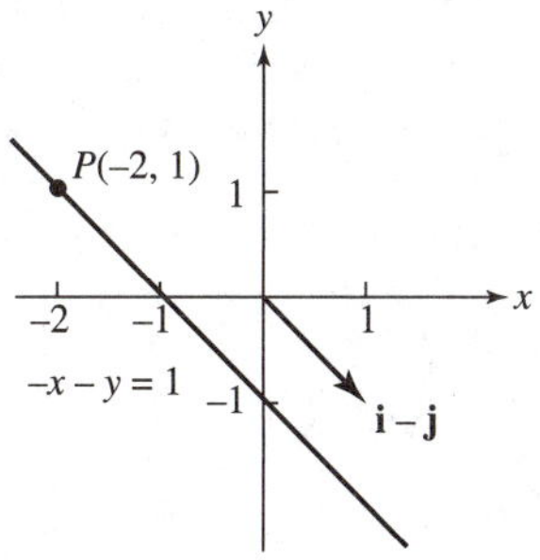

33. $\mathbf{v} = -\mathbf{i} - 2\mathbf{j}$ is parallel to the line $2x - y = c$
$P(1, 2)$ is on the line,
so $(2)(1) - 2 = c$.
The line is $2x - y = 0$.

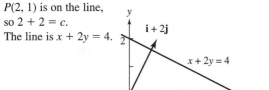

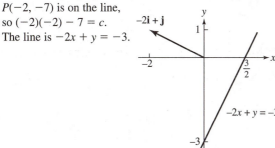

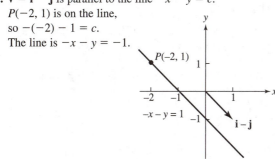

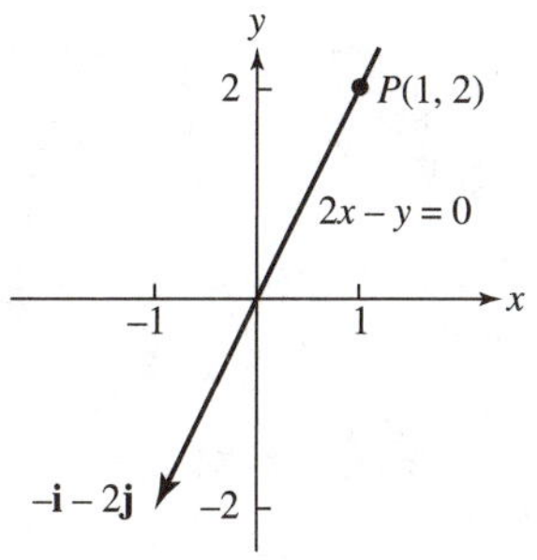

In Exercises 35–40, we use the fact that $\mathbf{n} = a\mathbf{i} + b\mathbf{j}$ is normal to the line $ax + by = c$.

35. $\mathbf{n}_1 = 3\mathbf{i} + \mathbf{j}$ and $\mathbf{n}_2 = 2\mathbf{i} - \mathbf{j}$, so

$$\theta = \cos^{-1}\left(\frac{\mathbf{n}_1 \cdot \mathbf{n}_2}{|\mathbf{n}_1||\mathbf{n}_2|}\right) = \cos^{-1}\left(\frac{6-1}{\sqrt{10}\sqrt{5}}\right) = \cos^{-1}\left(\frac{1}{\sqrt{2}}\right) = \frac{\pi}{4}$$

37. $\mathbf{n}_1 = \sqrt{3}\mathbf{i} - \mathbf{j}$ and $\mathbf{n}_2 = \mathbf{i} - \sqrt{3}\mathbf{j}$

$$\theta = \cos^{-1}\left(\frac{\mathbf{n}_1 \cdot \mathbf{n}_2}{|\mathbf{n}_1||\mathbf{n}_2|}\right) = \cos^{-1}\left(\frac{\sqrt{3}+\sqrt{3}}{\sqrt{4}\sqrt{4}}\right) = \cos^{-1}\left(\frac{\sqrt{3}}{2}\right) = \frac{\pi}{6}$$

39. $\mathbf{n}_1 = 3\mathbf{i} - 4\mathbf{j}$ and $\mathbf{n}_2 = \mathbf{i} - \mathbf{j}$

$$\theta = \cos^{-1}\left(\frac{\mathbf{n}_1 \cdot \mathbf{n}_2}{|\mathbf{n}_1||\mathbf{n}_2|}\right) = \cos^{-1}\left(\frac{3+4}{\sqrt{25}\sqrt{2}}\right) = \cos^{-1}\left(\frac{7}{5\sqrt{2}}\right)$$

≈ 0.14 rad

41. The angle between the corresponding normals is equal to the angle between the corresponding tangents. The points of intersection are $\left(-\frac{\sqrt{3}}{2}, \frac{3}{4}\right)$ and $\left(\frac{\sqrt{3}}{2}, \frac{3}{4}\right)$. At $\left(-\frac{\sqrt{3}}{2}, \frac{3}{4}\right)$ the tangent line for $f(x) = x^2$ is

$$y - \frac{3}{4} = f'\left(-\frac{\sqrt{3}}{2}\right)\left(x - \left(-\frac{\sqrt{3}}{2}\right)\right), \text{ or}$$

$$y = -\sqrt{3}\left(x + \frac{\sqrt{3}}{2}\right) + \frac{3}{4}, \text{ which is } y = -\sqrt{3}x - \frac{3}{4},$$

and the tangent line for $f(x) = \frac{3}{2} - x^2$ is

$$y - \frac{3}{4} = f'\left(-\frac{\sqrt{3}}{2}\right)\left(x - \left(-\frac{\sqrt{3}}{2}\right)\right), \text{ or}$$

$$y = \sqrt{3}\left(x + \frac{\sqrt{3}}{2}\right) + \frac{3}{4} \text{ which is } y = \sqrt{3}x + \frac{9}{4}.$$

The corresponding normals are $\mathbf{n}_1 = \sqrt{3}\mathbf{i} + \mathbf{j}$ and $\mathbf{n}_2 = -\sqrt{3}\mathbf{i} + \mathbf{j}$. The angle at $\left(-\frac{\sqrt{3}}{2}, \frac{3}{4}\right)$ is

$$\theta = \cos^{-1}\left(\frac{\mathbf{n}_1 \cdot \mathbf{n}_2}{|\mathbf{n}_1||\mathbf{n}_2|}\right) = \cos^{-1}\left(\frac{-3+1}{\sqrt{4}\sqrt{4}}\right)$$

$= \cos^{-1}\left(-\frac{1}{2}\right) = \frac{2\pi}{3}$, the acute angle between the lines is $\frac{\pi}{3}$.

At $\left(\frac{\sqrt{3}}{2}, \frac{3}{4}\right)$ the tangent line for $f(x) = x^2$ is $y = \sqrt{3}\left(x + \frac{\sqrt{3}}{2}\right) + \frac{3}{4}$, or $y = \sqrt{3}x + \frac{9}{4}$ and the tangent line for $f(x) = \frac{3}{2} - x^2$ is $y = -\sqrt{3}\left(x + \frac{\sqrt{3}}{2}\right) + \frac{3}{4}$ or $y = -\sqrt{3}x - \frac{3}{4}$. The corresponding normals are $\mathbf{n}_1 = -\sqrt{3}\mathbf{i} + \mathbf{j}$ and $\mathbf{n}_2 = \sqrt{3}\mathbf{i} + \mathbf{j}$. The angle at $\left(\frac{\sqrt{3}}{2}, \frac{3}{4}\right)$ is $\theta = \cos^{-1}\left(\frac{\mathbf{n}_1 \cdot \mathbf{n}_2}{|\mathbf{n}_1||\mathbf{n}_2|}\right) = \cos^{-1}\left(\frac{-3+1}{\sqrt{4}\sqrt{4}}\right)$

$= \cos^{-1}\left(-\frac{1}{2}\right) = \frac{2\pi}{3}$, the acute angle between the lines is $\frac{\pi}{3}$.

43. The curves intersect when $y = x^3 = (y^2)^3 = y^6$.

$y = 0$ or $y = 1$

The points of intersection are $(0, 0)$ and $(1, 1)$. At $(0, 0)$, the tangent line for $y = x^3$ is $y = 0$ and the tangent line for $y = \pm\sqrt{x}$ is $x = 0$. Therefore, the angle of intersection at $(0, 0)$ is $\frac{\pi}{2}$. At $(1, 1)$, the tangent line for $y = x^3$ is $y = 3x - 2$ and the tangent line for $y = \sqrt{x}$ is $y = \frac{1}{2}x + \frac{1}{2}$. The corresponding normal vectors are $\mathbf{n}_1 = -3\mathbf{i} + \mathbf{j}$ and $\mathbf{n}_2 = -\frac{1}{2}\mathbf{i} + \mathbf{j}$

$$\theta = \cos^{-1}\left(\frac{\mathbf{n}_1 \cdot \mathbf{n}_2}{|\mathbf{n}_1||\mathbf{n}_2|}\right) = \cos^{-1}\left(\frac{\frac{3}{2}+1}{\sqrt{10}\sqrt{\frac{5}{4}}}\right) = \cos^{-1}\left(\frac{1}{\sqrt{2}}\right) = \frac{\pi}{4},$$

the angle is $\frac{\pi}{4}$.

45. Let $\mathbf{A} = a_1\mathbf{i} + a_2\mathbf{j} + a_3\mathbf{k}$, $\mathbf{B} = b_1\mathbf{i} + b_2\mathbf{j} + b_3\mathbf{k}$, $\mathbf{C} = c_1\mathbf{i} + c_2\mathbf{j} + c_3\mathbf{k}$, and $\mathbf{D} = d_1\mathbf{i} + d_2\mathbf{j} + d_3\mathbf{k}$.

a.
$$\begin{aligned}\mathbf{A}\cdot\mathbf{B} &= a_1b_1 + a_2b_2 + a_3b_3\\ &= b_1a_1 + b_2a_2 + b_3a_3\\ &= \mathbf{B}\cdot\mathbf{A}\end{aligned}$$

b.
$$\begin{aligned}(c\mathbf{A})\cdot\mathbf{B} &= (ca_1\mathbf{i} + ca_2\mathbf{j} + ca_3\mathbf{k})\cdot(b_1\mathbf{i} + b_2\mathbf{j} + b_3\mathbf{k})\\ &= ca_1b_1 + ca_2b_2 + ca_3b_3\\ \mathbf{A}\cdot(c\mathbf{B}) &= (a_1\mathbf{i} + a_2\mathbf{j} + a_3\mathbf{k})\cdot(cb_1\mathbf{i} + cb_2\mathbf{j} + cb_3\mathbf{k})\\ &= a_1cb_1 + a_2cb_2 + a_3cb_3\\ &= ca_1b_1 + ca_2b_2 + ca_3b_3\\ c(\mathbf{A}\cdot\mathbf{B}) &= c(a_1b_1 + a_2b_2 + a_3b_3)\\ &= ca_1b_1 + ca_2b_2 + ca_3b_3\end{aligned}$$

Thus, $(c\mathbf{A})\cdot\mathbf{B} = \mathbf{A}\cdot(c\mathbf{B}) = c(\mathbf{A}\cdot\mathbf{B})$
$= ca_1b_1 + ca_2b_2 + ca_3b_3$

c.
$$\begin{aligned}\mathbf{A}\cdot(\mathbf{B}+\mathbf{C}) &= (a_1\mathbf{i} + a_2\mathbf{j} + a_3\mathbf{k})\cdot[(b_1 + c_1)\mathbf{i} + (b_2 + c_2)\mathbf{j} + (b_3 + c_3)\mathbf{k}]\\ &= a_1(b_1 + c_1) + a_2(b_2 + c_2) + a_3(b_3 + c_3)\\ &= a_1b_1 + a_1c_1 + a_2b_2 + a_2c_2 + a_3b_3 + a_3c_3\\ &= (a_1b_1 + a_2b_2 + a_3b_3) + (a_1c_1 + a_2c_2 + a_3c_3)\\ &= \mathbf{A}\cdot\mathbf{B} + \mathbf{A}\cdot\mathbf{C}\end{aligned}$$

d.
$$\begin{aligned}(\mathbf{A}+\mathbf{B})\cdot\mathbf{C} &= [(a_1 + b_1)\mathbf{i} + (a_2 + b_2)\mathbf{j} + (a_3 + b_3)\mathbf{k}]\cdot(c_1\mathbf{i} + c_2\mathbf{j} + c_3\mathbf{k})\\ &= (a_1 + b_1)c_1 + (a_2 + b_2)c_2 + (a_3 + b_3)c_3\\ &= a_1c_1 + b_1c_1 + a_2c_2 + b_2c_2 + a_3c_3 + b_3c_3\\ &= (a_1c_1 + a_2c_2 + a_3c_3) + (b_1c_1 + b_2c_2 + b_3c_3)\\ &= \mathbf{A}\cdot\mathbf{C} + \mathbf{B}\cdot\mathbf{C}\end{aligned}$$

e.
$$\begin{aligned}&(\mathbf{A}+\mathbf{B})\cdot(\mathbf{C}+\mathbf{D})\\ &= [(a_1 + b_1)\mathbf{i} + (a_2 + b_2)\mathbf{j} + (a_3 + b_3)\mathbf{k}]\cdot[(c_1 + d_1)\mathbf{i} + (c_2 + d_2)\mathbf{j} + (c_3 + d_3)\mathbf{k}]\\ &= (a_1 + b_1)(c_1 + d_1) + (a_2 + b_2)(c_2 + d_2) + (a_3 + b_3)(c_3 + d_3)\\ &= a_1c_1 + a_1d_1 + b_1c_1 + b_1d_1 + a_2c_2 + a_2d_2 + b_2c_2 + b_2d_2 + a_3c_3 + a_3d_3 + b_3c_3 + b_3d_3\\ &= (a_1c_1 + a_2c_2 + a_3c_3) + (a_1d_1 + a_2d_2 + a_3d_3) + (b_1c_1 + b_2c_2 + b_3c_3) + (b_1d_1 + b_2d_2 + b_3d_3)\\ &= \mathbf{A}\cdot\mathbf{C} + \mathbf{A}\cdot\mathbf{D} + \mathbf{B}\cdot\mathbf{C} + \mathbf{B}\cdot\mathbf{D}\end{aligned}$$

f.
$$\begin{aligned}\mathbf{A}\cdot\mathbf{A} &= (a_1\mathbf{i} + a_2\mathbf{j} + a_3\mathbf{k})\cdot(a_1\mathbf{i} + a_2\mathbf{j} + a_3\mathbf{k})\\ &= a_1^2 + a_2^2 + a_3^2\\ &= \left(\sqrt{a_1^2 + a_2^2 + a_3^2}\right)^2\\ &= |\mathbf{A}|^2\end{aligned}$$

47. $(x\mathbf{i} + y\mathbf{j})\cdot\mathbf{v} = |x\mathbf{i} + y\mathbf{j}||\mathbf{v}|\cos\theta \le 0$ when $\frac{\pi}{2} \le \theta \le \pi$.

This means (x, y) has to be a point whose position vector makes an angle with **v** that is a right angle or bigger.

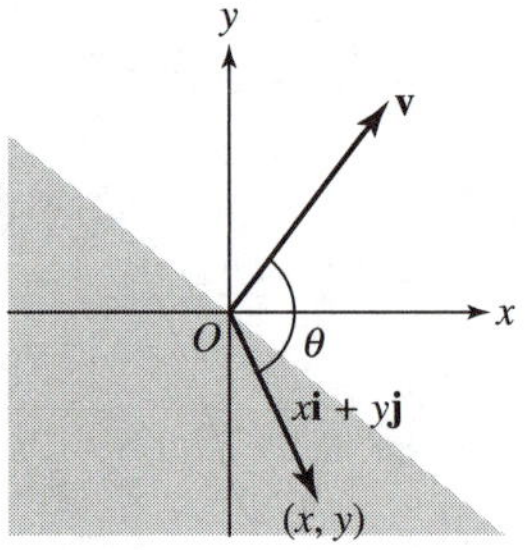

Section 11.3 Cross Products (pp. 593–600)

Quick Review 11.3

1. Slope of $AB = \frac{4-1}{3-1} = \frac{3}{2}$

Slope of $DC = \frac{6-3}{7-5} = \frac{3}{2}$

Slope of $AD = \frac{3-1}{5-1} = \frac{1}{2}$

Slope of $BC = \frac{6-4}{7-3} = \frac{1}{2}$

Opposite sides have the same slope, so they are parallel, and the figure is a parallelogram.

2. Let $\mathbf{u} = \overrightarrow{AB} = 2\mathbf{i} + 3\mathbf{j}$ and $\mathbf{v} = \overrightarrow{AD} = 4\mathbf{i} + 2\mathbf{j}$.

Then $|\mathbf{u}| = \sqrt{2^2 + 3^2} = \sqrt{13}$

and $|\mathbf{v}| = \sqrt{4^2 + 2^2} = 2\sqrt{5}$.

If θ is the angle between **u** and **v**, $\theta = \cos^{-1}\frac{\mathbf{u}\cdot\mathbf{v}}{|\mathbf{u}||\mathbf{v}|}$

$= \cos^{-1}\frac{(2)(4) + (3)(2)}{(\sqrt{13})(2\sqrt{5})} = \cos^{-1}\frac{7}{\sqrt{65}}$.

The area of the parallelogram is

$$|\mathbf{u}||\mathbf{v}|\sin\theta = (\sqrt{13})(2\sqrt{5})\sqrt{1 - \left(\frac{7}{\sqrt{65}}\right)^2} = 8.$$

3. Use the solution to Exercise 2.

$\angle A$ and $\angle C$: $\cos^{-1}\frac{7}{\sqrt{65}} \approx 29.74°$

$\angle B$ and $\angle D$: $180° - \cos^{-1}\frac{7}{\sqrt{65}} = \cos^{-1}\left(-\frac{7}{\sqrt{65}}\right)$
$= 150.26°$

4.
$$\begin{aligned}\sin 2x &= 0\\ 2x &= k\pi\\ x &= \frac{k\pi}{2}\end{aligned}$$

For $0 \le x \le 2\pi$, the solutions are $x = 0, \frac{\pi}{2}, \pi, \frac{3\pi}{2}, 2\pi$.

5.
$$\begin{aligned}\cos 2x &= 0\\ 2x &= (2k+1)\frac{\pi}{2}\\ x &= (2k+1)\frac{\pi}{4}\end{aligned}$$

For $0 \le x \le 2\pi$, the solutions are $x = \frac{\pi}{4}, \frac{3\pi}{4}, \frac{5\pi}{4}, \frac{7\pi}{4}$.

6. $\begin{vmatrix} 2 & 3 \\ -1 & 1 \end{vmatrix} = (2)(1) - (3)(-1) = 2 + 3 = 5$

7. $\begin{vmatrix} -3 & 4 \\ 0 & 1 \end{vmatrix} = (-3)(1) - (4)(0) = -3 - 0 = -3$

8. $\begin{vmatrix} 2 & 3 & -1 \\ -1 & 2 & 4 \\ 1 & 0 & 3 \end{vmatrix} = 2\begin{vmatrix} 2 & 4 \\ 0 & 3 \end{vmatrix} - 3\begin{vmatrix} -1 & 4 \\ 1 & 3 \end{vmatrix} + (-1)\begin{vmatrix} -1 & 2 \\ 1 & 0 \end{vmatrix}$

$= 2(6 - 0) - 3(-3 - 4) - (0 - 2)$
$= 35$

9. $\begin{vmatrix} -1 & 0 & 1 \\ 1 & 1 & 1 \\ 0 & 1 & -1 \end{vmatrix} = -1\begin{vmatrix} 1 & 1 \\ 1 & -1 \end{vmatrix} - 0\begin{vmatrix} 1 & 1 \\ 0 & -1 \end{vmatrix} + 1\begin{vmatrix} 1 & 1 \\ 0 & 1 \end{vmatrix}$

$= -(-1 - 1) - 0 + (1 - 0)$
$= 3$

10. $\begin{vmatrix} 2 & -1 & 3 \\ 4 & 0 & -2 \\ -6 & 1 & -1 \end{vmatrix} = 2\begin{vmatrix} 0 & -2 \\ 1 & -1 \end{vmatrix} - (-1)\begin{vmatrix} 4 & -2 \\ -6 & -1 \end{vmatrix} + 3\begin{vmatrix} 4 & 0 \\ -6 & 1 \end{vmatrix}$

$= 2(0 + 2) + (-4 - 12) + 3(4 - 0)$
$= 0$

Section 11.3 Exercises

1. $\mathbf{A} \times \mathbf{B} = \begin{vmatrix} \mathbf{i} & \mathbf{j} & \mathbf{k} \\ 2 & -2 & -1 \\ 1 & 0 & -1 \end{vmatrix}$

$= \begin{vmatrix} -2 & -1 \\ 0 & -1 \end{vmatrix}\mathbf{i} - \begin{vmatrix} 2 & -1 \\ 1 & -1 \end{vmatrix}\mathbf{j} + \begin{vmatrix} 2 & -2 \\ 1 & 0 \end{vmatrix}\mathbf{k}$

$= 2\mathbf{i} + \mathbf{j} + 2\mathbf{k} = 3\left(\frac{2}{3}\mathbf{i} + \frac{1}{3}\mathbf{j} + \frac{2}{3}\mathbf{k}\right)$

length $= 3$ and the direction is $\frac{2}{3}\mathbf{i} + \frac{1}{3}\mathbf{j} + \frac{2}{3}\mathbf{k}$

$\mathbf{B} \times \mathbf{A} = -(\mathbf{A} \times \mathbf{B}) = -3\left(\frac{2}{3}\mathbf{i} + \frac{1}{3}\mathbf{j} + \frac{2}{3}\mathbf{k}\right)$

length $= 3$ and the direction is $-\frac{2}{3}\mathbf{i} - \frac{1}{3}\mathbf{j} - \frac{2}{3}\mathbf{k}$

3. $\mathbf{A} \times \mathbf{B} = \begin{vmatrix} \mathbf{i} & \mathbf{j} & \mathbf{k} \\ 2 & -2 & 4 \\ -1 & 1 & -2 \end{vmatrix}$

$= \begin{vmatrix} -2 & 4 \\ 1 & -2 \end{vmatrix}\mathbf{i} - \begin{vmatrix} 2 & 4 \\ -1 & -2 \end{vmatrix}\mathbf{j} + \begin{vmatrix} 2 & -2 \\ -1 & 1 \end{vmatrix}\mathbf{k} = \mathbf{0}$

length $= 0$ and has no direction
$\mathbf{B} \times \mathbf{A} = -(\mathbf{A} \times \mathbf{B}) = \mathbf{0}$
length $= 0$ and has no direction

5. $\mathbf{A} \times \mathbf{B} = \begin{vmatrix} \mathbf{i} & \mathbf{j} & \mathbf{k} \\ 2 & 0 & 0 \\ 0 & -3 & 0 \end{vmatrix}$

$= \begin{vmatrix} 0 & 0 \\ -3 & 0 \end{vmatrix}\mathbf{i} - \begin{vmatrix} 2 & 0 \\ 0 & 0 \end{vmatrix}\mathbf{j} + \begin{vmatrix} 2 & 0 \\ 0 & -3 \end{vmatrix}\mathbf{k} = -6(\mathbf{k})$

length $= 6$ and the direction is $-\mathbf{k}$
$\mathbf{B} \times \mathbf{A} = -(\mathbf{A} \times \mathbf{B}) = 6(\mathbf{k})$
length $= 6$ and the direction is $\mathbf{k}$

7. $\mathbf{A} \times \mathbf{B} = \begin{vmatrix} \mathbf{i} & \mathbf{j} & \mathbf{k} \\ -8 & -2 & -4 \\ 2 & 2 & 1 \end{vmatrix}$

$= \begin{vmatrix} -2 & -4 \\ 2 & 1 \end{vmatrix}\mathbf{i} - \begin{vmatrix} -8 & -4 \\ 2 & 1 \end{vmatrix}\mathbf{j} + \begin{vmatrix} -8 & -2 \\ 2 & 2 \end{vmatrix}\mathbf{k}$

$= 6\mathbf{i} - 12\mathbf{k} = 6\sqrt{5}\left(\frac{1}{\sqrt{5}}\mathbf{i} - \frac{2}{\sqrt{5}}\mathbf{k}\right)$

length $= 6\sqrt{5}$ and the direction is $\frac{1}{\sqrt{5}}\mathbf{i} - \frac{2}{\sqrt{5}}\mathbf{k}$

$\mathbf{B} \times \mathbf{A} = -(\mathbf{A} \times \mathbf{B}) = -(6\mathbf{i} - 12\mathbf{k})$

length $= 6\sqrt{5}$ and the direction is $-\frac{1}{\sqrt{5}}\mathbf{i} + \frac{2}{\sqrt{5}}\mathbf{k}$

9. $\mathbf{A} \times \mathbf{B} = \mathbf{i} \times \mathbf{j} = \mathbf{k}$

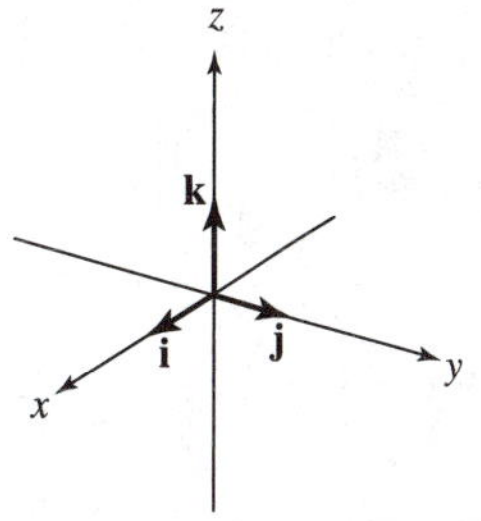

11. $\mathbf{A} \times \mathbf{B} = \begin{vmatrix} \mathbf{i} & \mathbf{j} & \mathbf{k} \\ 1 & 0 & -1 \\ 0 & 1 & 1 \end{vmatrix}$

$= \begin{vmatrix} 0 & -1 \\ 1 & 1 \end{vmatrix}\mathbf{i} - \begin{vmatrix} 1 & -1 \\ 0 & 1 \end{vmatrix}\mathbf{j} + \begin{vmatrix} 1 & 0 \\ 0 & 1 \end{vmatrix}\mathbf{k} = \mathbf{i} - \mathbf{j} + \mathbf{k}$

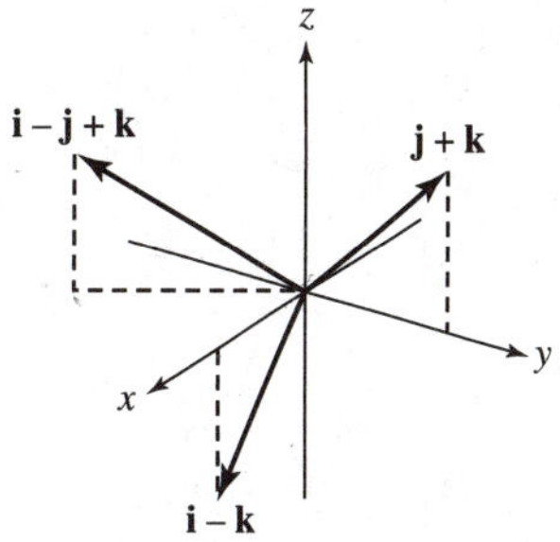

13. $\mathbf{A} \times \mathbf{B} = \begin{vmatrix} \mathbf{i} & \mathbf{j} & \mathbf{k} \\ 1 & 1 & 0 \\ 1 & -1 & 0 \end{vmatrix}$

$= \begin{vmatrix} 1 & 0 \\ -1 & 0 \end{vmatrix}\mathbf{i} - \begin{vmatrix} 1 & 0 \\ 1 & 0 \end{vmatrix}\mathbf{j} + \begin{vmatrix} 1 & 1 \\ 1 & -1 \end{vmatrix}\mathbf{k} = -2\mathbf{k}$

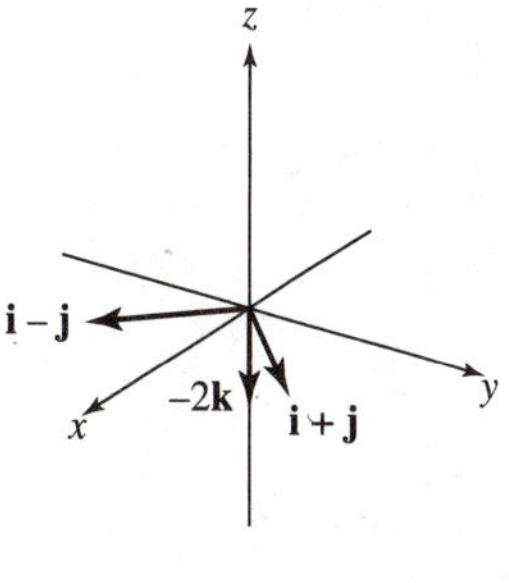

15. a. $\overrightarrow{PQ} \times \overrightarrow{PR} = \begin{vmatrix} \mathbf{i} & \mathbf{j} & \mathbf{k} \\ 1 & 1 & -3 \\ -1 & 3 & -1 \end{vmatrix}$

$= \begin{vmatrix} 1 & -3 \\ 3 & -1 \end{vmatrix}\mathbf{i} - \begin{vmatrix} 1 & -3 \\ -1 & -1 \end{vmatrix}\mathbf{j} + \begin{vmatrix} 1 & 1 \\ -1 & 3 \end{vmatrix}\mathbf{k}$

$= 8\mathbf{i} + 4\mathbf{j} + 4\mathbf{k}$

$\text{Area} = \frac{1}{2}|\overrightarrow{PQ} \times \overrightarrow{PR}| = \frac{1}{2}\sqrt{64 + 16 + 16} = 2\sqrt{6}$

b. $\mathbf{u} = \pm \frac{\overrightarrow{PQ} \times \overrightarrow{PR}}{|\overrightarrow{PQ} \times \overrightarrow{PR}|}$

$= \pm\frac{1}{4\sqrt{6}}(8\mathbf{i} + 4\mathbf{j} + 4\mathbf{k}) = \pm\frac{1}{\sqrt{6}}(2\mathbf{i} + \mathbf{j} + \mathbf{k})$

17. a. $\overrightarrow{PQ} \times \overrightarrow{PR} = \begin{vmatrix} \mathbf{i} & \mathbf{j} & \mathbf{k} \\ 1 & 1 & 1 \\ 1 & 1 & 0 \end{vmatrix}$

$= \begin{vmatrix} 1 & 1 \\ 1 & 0 \end{vmatrix}\mathbf{i} - \begin{vmatrix} 1 & 1 \\ 1 & 0 \end{vmatrix}\mathbf{j} + \begin{vmatrix} 1 & 1 \\ 1 & 1 \end{vmatrix}\mathbf{k}$

$= -\mathbf{i} + \mathbf{j}$

$\text{Area} = \frac{1}{2}|\overrightarrow{PQ} \times \overrightarrow{PR}| = \frac{1}{2}\sqrt{1 + 1} = \frac{\sqrt{2}}{2}$

b. $\mathbf{u} = \pm \frac{\overrightarrow{PQ} \times \overrightarrow{PR}}{|\overrightarrow{PQ} \times \overrightarrow{PR}|} = \pm\frac{1}{\sqrt{2}}(-\mathbf{i} + \mathbf{j}) = \pm\frac{1}{\sqrt{2}}(\mathbf{i} - \mathbf{j})$

19. a. $\mathbf{A} \cdot \mathbf{B} = 0 + (-1) + (-5) = -6$,
$\mathbf{A} \cdot \mathbf{C} = -75 + (-3) + (-3) = -81$,
$\mathbf{B} \cdot \mathbf{C} = 0 + 3 + 15 = 18$
None of the vectors are perpendicular.

b. $\mathbf{A} \times \mathbf{B} = \begin{vmatrix} \mathbf{i} & \mathbf{j} & \mathbf{k} \\ 5 & -1 & 1 \\ 0 & 1 & -5 \end{vmatrix} \neq \mathbf{0}$

$\mathbf{A} \times \mathbf{C} = \begin{vmatrix} \mathbf{i} & \mathbf{j} & \mathbf{k} \\ 5 & -1 & 1 \\ -15 & 3 & -3 \end{vmatrix} = \mathbf{0}$

$\mathbf{B} \times \mathbf{C} = \begin{vmatrix} \mathbf{i} & \mathbf{j} & \mathbf{k} \\ 0 & 1 & -5 \\ -15 & 3 & -3 \end{vmatrix} \neq \mathbf{0}$

A and **C** are parallel.

21. $|\overrightarrow{PQ} \times \mathbf{F}| = |\overrightarrow{PQ}||\mathbf{F}| \sin(60°)$

$= \frac{2}{3} \cdot 30 \cdot \frac{\sqrt{3}}{2}$ ft-lb

$= 10\sqrt{3}$ ft-lb

23. If $\mathbf{A} = a_1\mathbf{i} + a_2\mathbf{j} + a_3\mathbf{k}$, $\mathbf{B} = b_1\mathbf{i} + b_2\mathbf{j} + b_3\mathbf{k}$, and $\mathbf{C} = c_1\mathbf{i} + c_2\mathbf{j} + c_3\mathbf{k}$, then

$\mathbf{A} \cdot (\mathbf{B} \times \mathbf{C}) = \begin{vmatrix} a_1 & a_2 & a_3 \\ b_1 & b_2 & b_3 \\ c_1 & c_2 & c_3 \end{vmatrix}$,

$\mathbf{B} \cdot (\mathbf{C} \times \mathbf{A}) = \begin{vmatrix} b_1 & b_2 & b_3 \\ c_1 & c_2 & c_3 \\ a_1 & a_2 & a_3 \end{vmatrix}$ and

$\mathbf{C} \cdot (\mathbf{A} \times \mathbf{B}) = \begin{vmatrix} c_1 & c_2 & c_3 \\ a_1 & a_2 & a_3 \\ b_1 & b_2 & b_3 \end{vmatrix}$

which all have the same value, since the interchanging of two pairs of rows in a determinant does not change its value.

$(\mathbf{A} \times \mathbf{B}) \cdot \mathbf{C} = \begin{vmatrix} 2 & 0 & 0 \\ 0 & 2 & 0 \\ 0 & 0 & 2 \end{vmatrix} = 8$

The volume is $|8| = 8$.

25. $(\mathbf{A} \times \mathbf{B}) \cdot \mathbf{C} = \begin{vmatrix} 2 & 1 & 0 \\ 2 & -1 & 1 \\ 1 & 0 & 2 \end{vmatrix} = -7$

The volume is $|-7| = 7$.
(for details about verification, see Exercise 23)

27. a. true, $|\mathbf{A}| = \sqrt{a_1^2 + a_2^2 + a_3^2} = \sqrt{\mathbf{A} \cdot \mathbf{A}}$

b. not always true, $\mathbf{A} \cdot \mathbf{A} = |\mathbf{A}|^2$

c. true, $\mathbf{A} \times \mathbf{0} = \begin{vmatrix} \mathbf{i} & \mathbf{j} & \mathbf{k} \\ a_1 & a_2 & a_3 \\ 0 & 0 & 0 \end{vmatrix} = 0\mathbf{i} + 0\mathbf{j} + 0\mathbf{k} = \mathbf{0}$

d. true, $\mathbf{A} \times (-\mathbf{A}) = \begin{vmatrix} \mathbf{i} & \mathbf{j} & \mathbf{k} \\ a_1 & a_2 & a_3 \\ -a_1 & -a_2 & -a_3 \end{vmatrix}$

$= (-a_2a_3 + a_2a_3)\mathbf{i}$
$+ (-a_1a_3 + a_1a_3)\mathbf{j}$
$+ (-a_1a_2 + a_1a_2)\mathbf{k}$
$= \mathbf{0}$

e. not always true, $\mathbf{i} \times \mathbf{j} = \mathbf{k} \neq -\mathbf{k} = \mathbf{j} \times \mathbf{i}$, for example

f. true, this is one of the vector distributive laws.

g. true, $(\mathbf{A} \times \mathbf{B}) \cdot \mathbf{B} = \mathbf{A} \cdot (\mathbf{B} \times \mathbf{B}) = \mathbf{A} \cdot \mathbf{0} = 0$

h. true, this is the triple scalar product.

29. a. $\text{proj}_{\mathbf{B}} \mathbf{A} = \left(\frac{\mathbf{A} \cdot \mathbf{B}}{\mathbf{B} \cdot \mathbf{B}}\right)\mathbf{B}$

b. $\pm(\mathbf{A} \times \mathbf{B})$

c. $\pm(\mathbf{A} \times \mathbf{B}) \times \mathbf{C}$

d. $|(\mathbf{A} \times \mathbf{B}) \cdot \mathbf{C}|$

31. a. yes, $\mathbf{A} \times \mathbf{B}$ and $\mathbf{C}$ are both vectors

b. no, **A** is a vector but $\mathbf{B} \cdot \mathbf{C}$ is a scalar

c. yes, **A** and $\mathbf{B} \times \mathbf{C}$ are both vectors

d. no, **A** is a vector but $\mathbf{B} \cdot \mathbf{C}$ is a scalar

33. $\overrightarrow{AB} = -\mathbf{i} + \mathbf{j}$ and $\overrightarrow{AD} = -\mathbf{i} - \mathbf{j}$

$\overrightarrow{AB} \times \overrightarrow{AD} = \begin{vmatrix} \mathbf{i} & \mathbf{j} & \mathbf{k} \\ -1 & 1 & 0 \\ -1 & -1 & 0 \end{vmatrix}$

$= \begin{vmatrix} 1 & 0 \\ -1 & 0 \end{vmatrix}\mathbf{i} - \begin{vmatrix} -1 & 0 \\ -1 & 0 \end{vmatrix}\mathbf{j} + \begin{vmatrix} -1 & 1 \\ -1 & -1 \end{vmatrix}\mathbf{k} = 2\mathbf{k}$

$\text{area} = |\overrightarrow{AB} \times \overrightarrow{AD}| = 2$

35. $\overrightarrow{AB} = 3\mathbf{i} - 2\mathbf{j}$ and $\overrightarrow{AD} = 5\mathbf{i} + \mathbf{j}$

$$\overrightarrow{AB} \times \overrightarrow{AD} = \begin{vmatrix} \mathbf{i} & \mathbf{j} & \mathbf{k} \\ 3 & -2 & 0 \\ 5 & 1 & 0 \end{vmatrix}$$

$$= \begin{vmatrix} -2 & 0 \\ 1 & 0 \end{vmatrix}\mathbf{i} - \begin{vmatrix} 3 & 0 \\ 5 & 0 \end{vmatrix}\mathbf{j} + \begin{vmatrix} 3 & -2 \\ 5 & 1 \end{vmatrix}\mathbf{k} = 13\mathbf{k}$$

area $= |\overrightarrow{AB} \times \overrightarrow{AD}| = 13$

37. $\overrightarrow{AB} = -2\mathbf{i} + 3\mathbf{j}$ and $\overrightarrow{AC} = 3\mathbf{i} + \mathbf{j}$

$$\overrightarrow{AB} \times \overrightarrow{AC} = \begin{vmatrix} \mathbf{i} & \mathbf{j} & \mathbf{k} \\ -2 & 3 & 0 \\ 3 & 1 & 0 \end{vmatrix}$$

$$= \begin{vmatrix} 3 & 0 \\ 1 & 0 \end{vmatrix}\mathbf{i} - \begin{vmatrix} -2 & 0 \\ 3 & 0 \end{vmatrix}\mathbf{j} + \begin{vmatrix} -2 & 3 \\ 3 & 1 \end{vmatrix}\mathbf{k} = -11\mathbf{k}$$

area $= \frac{1}{2}|\overrightarrow{AB} \times \overrightarrow{AC}| = \frac{11}{2}$

39. $\overrightarrow{AB} = 6\mathbf{i} - 5\mathbf{j}$ and $\overrightarrow{AC} = 11\mathbf{i} - 5\mathbf{j}$

$$\overrightarrow{AB} \times \overrightarrow{AC} = \begin{vmatrix} \mathbf{i} & \mathbf{j} & \mathbf{k} \\ 6 & -5 & 0 \\ 11 & -5 & 0 \end{vmatrix}$$

$$= \begin{vmatrix} -5 & 0 \\ -5 & 0 \end{vmatrix}\mathbf{i} - \begin{vmatrix} 6 & 0 \\ 11 & 0 \end{vmatrix}\mathbf{j} + \begin{vmatrix} 6 & -5 \\ 11 & -5 \end{vmatrix}\mathbf{k} = 25\mathbf{k}$$

area $= \frac{1}{2}|\overrightarrow{AB} \times \overrightarrow{AC}| = \frac{25}{2}$

41. The vectors determining two of the sides are $\mathbf{A} = a_1\mathbf{i} + a_2\mathbf{j}$ and $\mathbf{B} = b_1\mathbf{i} + b_2\mathbf{j}$, so

$$\mathbf{A} \times \mathbf{B} = \begin{vmatrix} \mathbf{i} & \mathbf{j} & \mathbf{k} \\ a_1 & a_2 & 0 \\ b_1 & b_2 & 0 \end{vmatrix} = \begin{vmatrix} a_1 & a_2 \\ b_1 & b_2 \end{vmatrix}\mathbf{k}$$

and the triangle's area is $\frac{1}{2}|\mathbf{A} \times \mathbf{B}| = \pm\frac{1}{2}\begin{vmatrix} a_1 & a_2 \\ b_1 & b_2 \end{vmatrix}$.

The applicable sign is (+) if the acute angle from **A** to **B** runs counterclockwise in the *xy*-plane, and (−) if it runs clockwise, because the area must be a nonnegative number.

43. No. For example, $\mathbf{i} + \mathbf{j} \neq -\mathbf{i} + \mathbf{j}$, but $\mathbf{i} \times (\mathbf{i} + \mathbf{j}) = \mathbf{i} \times (-\mathbf{i} + \mathbf{j}) = \mathbf{k}$.

■ Section 11.4 Lines and Planes in Space (pp. 600–607)

Quick Review 11.4

1. a.
$$\begin{aligned} 2x - y + 3z &= -3 \\ 2(1) - 2 + 3(-1) &= -3 \\ -3 &= -3 \end{aligned}$$
Yes, the point (1, 2, −1) is a solution.

b.
$$\begin{aligned} 2x - y + 3z &= -3 \\ 2(3) - 1 + 3(2) &= -3 \\ 11 &\neq -3 \end{aligned}$$
No, the point (3, 1, 2) is not a solution.

c.
$$\begin{aligned} 2x - y + 3z &= -3 \\ 2(-1) - 3 + 3(-4) &= -3 \\ -17 &\neq -3 \end{aligned}$$
No, the point (−1, 3, −4) is not a solution.

d.
$$\begin{aligned} 2x - y + 3z &= -3 \\ 2(5) - 0 + 3(4) &= -3 \\ 22 &\neq -3 \end{aligned}$$
No, the point (5, 0, 4) is not a solution.

2. Assume (−2, −3) corresponds to $t = 0$ and (3, 1) corresponds to $t = 1$. Then $x = -2 + at$ and $y = -3 + bt$. Solving $3 = -2 + a$ and $1 = -3 + b$ gives $a = 5$ and $b = 4$, so one parametrization of the line is $x = -2 + 5t$, $y = -3 + 4t$, $-\infty < t < \infty$. (Other parametrizations are possible.)

3. Using the solution to Exercise 2, one parametrization of the line segment is $x = -2 + 5t$, $y = -3 + 4t$, $0 \le t \le 1$. (Other parametrizations are possible.)

4. Since $2x + 3y = 1$ can be written as $y = -\frac{2}{3}x + \frac{1}{3}$, the line has slope $-\frac{2}{3}$, so x increases 3 units when y decreases 2 units. Starting at (1, 2), one parametrization of the line is $x = 1 + 3t$, $y = 2 - 2t$, $-\infty < t < \infty$. (Other parametrizations are possible.)

5. $x - 2y = -1$
$2x + y = 2$

$$\begin{aligned} x - 2y &= -1 \\ 4x + 2y &= 4 \\ \hline 5x \quad &= 3 \\ x &= \frac{3}{5} \end{aligned}$$

Substitute $\frac{3}{5}$ for x in $2x + y = 2$.

$$2\left(\frac{3}{5}\right) + y = 2$$
$$y = \frac{4}{5}$$

The solution is $x = \frac{3}{5}$, $y = \frac{4}{5}$.

6. $3x + 2y = 1$
$4x - 3y = -1$

$$\begin{aligned} 9x + 6y &= 3 \\ 8x - 6y &= -2 \\ \hline 17x \quad &= 1 \\ x &= \frac{1}{17} \end{aligned}$$

Substitute $\frac{1}{17}$ for x in $3x + 2y = 1$.

$$3\left(\frac{1}{17}\right) + 2y = 1$$
$$2y = \frac{14}{17}$$
$$y = \frac{7}{17}$$

The solution is $x = \frac{1}{17}$, $y = \frac{7}{17}$.

7. $\mathbf{v} = -2\mathbf{i} + \mathbf{j} + 2\mathbf{k}$

$$|\mathbf{v}| = \sqrt{(-2)^2 + 1^2 + 2^2} = \sqrt{9} = 3$$

The unit vector is $\frac{\mathbf{v}}{|\mathbf{v}|} = -\frac{2}{3}\mathbf{i} + \frac{1}{3}\mathbf{j} + \frac{2}{3}\mathbf{k}$.

8. $$\mathbf{u} = \mathbf{v}_1 \times \mathbf{v}_2 = \begin{vmatrix} \mathbf{i} & \mathbf{j} & \mathbf{k} \\ 3 & -2 & 1 \\ -1 & 1 & -2 \end{vmatrix} = 3\mathbf{i} + 5\mathbf{j} + \mathbf{k}$$

$$\mathbf{w} = \mathbf{v}_2 \times \mathbf{v}_1 = \begin{vmatrix} \mathbf{i} & \mathbf{j} & \mathbf{k} \\ -1 & 1 & -2 \\ 3 & -2 & 1 \end{vmatrix} = -3\mathbf{i} - 5\mathbf{j} - \mathbf{k}$$

9. $$\begin{aligned} \mathbf{u} \cdot \mathbf{v}_1 &= (3\mathbf{i} + 5\mathbf{j} + \mathbf{k}) \cdot (3\mathbf{i} - 2\mathbf{j} + \mathbf{k}) \\ &= 9 - 10 + 1 = 0 \end{aligned}$$
$$\begin{aligned} \mathbf{w} \cdot \mathbf{v}_1 &= (-3\mathbf{i} - 5\mathbf{j} + \mathbf{k}) \cdot (3\mathbf{i} - 2\mathbf{j} + \mathbf{k}) \\ &= -9 + 10 - 1 = 0 \end{aligned}$$

10. $\mathbf{u} \cdot \mathbf{v}_2 = (3\mathbf{i} + 5\mathbf{j} + \mathbf{k}) \cdot (-\mathbf{i} + \mathbf{j} - 2\mathbf{k})$
$= -3 + 5 - 2 = 0$
$\mathbf{w} \cdot \mathbf{v}_2 = (-3\mathbf{i} - 5\mathbf{j} - \mathbf{k}) \cdot (-\mathbf{i} + \mathbf{j} - 2\mathbf{k})$
$= 3 - 5 + 2 = 0$

Section 11.4 Exercises

1. The direction is $\mathbf{i} + \mathbf{j} + \mathbf{k}$ and a point is $P(3, -4, -1)$, so $x = 3 + t$, $y = -4 + t$, $z = -1 + t$.

3. The direction is $\overrightarrow{PQ} = 5\mathbf{i} + 5\mathbf{j} - 5\mathbf{k}$ and a point is $P(-2, 0, 3)$, so $x = -2 + 5t$, $y = 5t$, $z = 3 - 5t$.

5. The direction is $2\mathbf{i} - \mathbf{j} + 3\mathbf{k}$ and a point is $P(3, -2, 1)$, so $x = 3 + 2t$, $y = -2 - t$, $z = 1 + 3t$.

7. The direction is $3\mathbf{i} + 7\mathbf{j} - 5\mathbf{k}$ and a point is $P(2, 4, 5)$, so $x = 2 + 3t$, $y = 4 + 7t$, $z = 5 - 5t$.

9. The direction is $\mathbf{A} \times \mathbf{B} = \begin{vmatrix} \mathbf{i} & \mathbf{j} & \mathbf{k} \\ 1 & 2 & 3 \\ 3 & 4 & 5 \end{vmatrix}$
$= -2\mathbf{i} + 4\mathbf{j} - 2\mathbf{k}$
and a point is $P(2, 3, 0)$, so $x = 2 - 2t$, $y = 3 + 4t$, $z = -2t$.

11. The direction is $\overrightarrow{PQ} = \mathbf{i} + \mathbf{j} + \frac{3}{2}\mathbf{k}$ and a point is $P(0, 0, 0)$, so $x = t$, $y = t$, $z = \frac{3}{2}t$, where $0 \le t \le 1$.

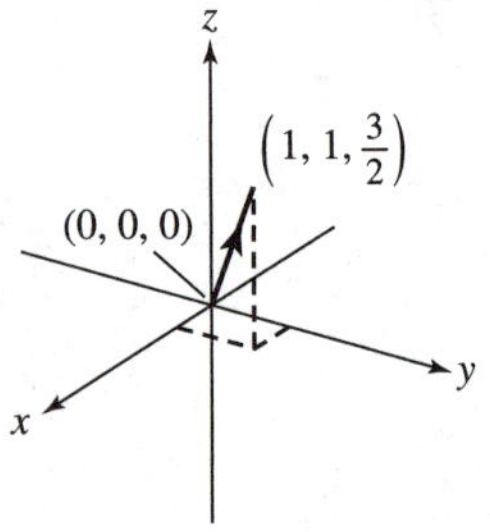

13. The direction is $\overrightarrow{PQ} = -2\mathbf{j}$ and a point is $P(0, 1, 1)$, so $x = 0$, $y = 1 - 2t$, $z = 1$, where $0 \le t \le 1$.

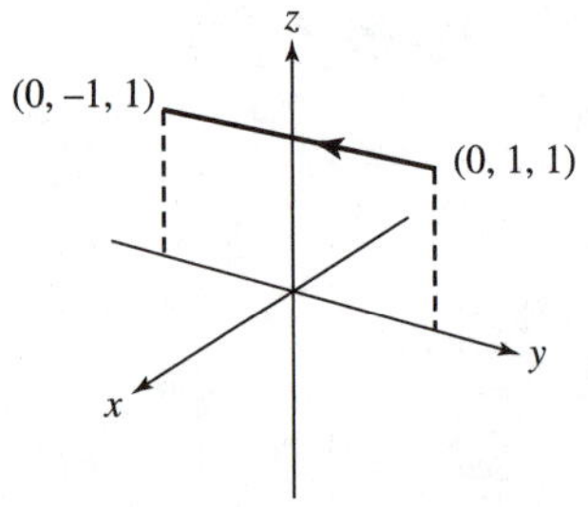

15. $3(x - 0) + (-2)(y - 2) + (-1)(z + 1) = 0$
$3x - 2y - z = -3$

17. $\overrightarrow{PQ} = \mathbf{i} - \mathbf{j} + 3\mathbf{k}$, $\overrightarrow{PS} = -\mathbf{i} - 3\mathbf{j} + 2\mathbf{k}$
$\overrightarrow{PQ} \times \overrightarrow{PS} = \begin{vmatrix} \mathbf{i} & \mathbf{j} & \mathbf{k} \\ 1 & -1 & 3 \\ -1 & -3 & 2 \end{vmatrix}$
$= 7\mathbf{i} - 5\mathbf{j} - 4\mathbf{k}$ is normal to the plane.
$7(x - 2) + (-5)(y - 0) + (-4)(z - 2) = 0$
$7x - 5y - 4z = 6$

19. $\mathbf{n} = \mathbf{i} + 3\mathbf{j} + 4\mathbf{k}$, $P(2, 4, 5)$
$(1)(x - 2) + (3)(y - 4) + (4)(z - 5) = 0$
$x + 3y + 4z = 34$

21. $\begin{cases} x = 2t + 1 = s + 2 \\ y = 3t + 2 = 2s + 4 \end{cases}$
$\begin{cases} 2t - s = 1 \\ 3t - 2s = 2 \end{cases}$
$\begin{cases} 4t - 2s = 2 \\ 3t - 2s = 2 \end{cases}$
Subtracting the equations yields $t = 0$ and $s = -1$.
Then $z = 4t + 3 = -4s - 1$,
$4(0) + 3 = (-4)(-1) - 1$ is satisfied.
The lines intersect when $t = 0$ and $s = -1$.
The point of intersection is $x = 1$, $y = 2$, and $z = 3$ or $P(1, 2, 3)$. A vector normal to the plane determined by these lines is
$\mathbf{n}_1 \times \mathbf{n}_2 = \begin{vmatrix} \mathbf{i} & \mathbf{j} & \mathbf{k} \\ 2 & 3 & 4 \\ 1 & 2 & -4 \end{vmatrix} = -20\mathbf{i} + 12\mathbf{j} + \mathbf{k}$,
where $\mathbf{n}_1$ and $\mathbf{n}_2$ are directions of the lines.
The plane containing the lines is represented by
$(-20)(x - 1) + (12)(y - 2) + (1)(z - 3) = 0$
or $-20x + 12y + z = 7$.

23. The cross product of $\mathbf{i} + \mathbf{j} - \mathbf{k}$ and $-4\mathbf{i} + 2\mathbf{j} - 2\mathbf{k}$ has the same direction as the normal to the plane.
$\mathbf{n} = \begin{vmatrix} \mathbf{i} & \mathbf{j} & \mathbf{k} \\ 1 & 1 & -1 \\ -4 & 2 & -2 \end{vmatrix} = 6\mathbf{j} + 6\mathbf{k}$
Select a point on either line, such as $P(-1, 2, 1)$. Since the lines are given to intersect, the desired plane is
$0(x + 1) + 6(y - 2) + 6(z - 1) = 0$
$6y + 6z = 18$
$y + z = 3$.

25. $\mathbf{n}_1 \times \mathbf{n}_2 = \begin{vmatrix} \mathbf{i} & \mathbf{j} & \mathbf{k} \\ 2 & 1 & -1 \\ 1 & 2 & 1 \end{vmatrix} = 3\mathbf{i} - 3\mathbf{j} + 3\mathbf{k}$
is a vector in the direction of the line of intersection of the planes.
$3(x - 2) + (-3)(y - 1) + 3(z + 1) = 0$
$3x - 3y + 3z = 0$
$x - y + z = 0$ is the desired plane containing $P_0(2, 1, -1)$.

27. a. Let d represent the distance from S to the line. (This distance is also the length of the component of $\overrightarrow{PQ}$ normal to the line.)
Since $\sin\theta = \dfrac{\text{opposite}}{\text{hypotenuse}} = \dfrac{d}{|\overrightarrow{PS}|}$, $d = |\overrightarrow{PS}| \sin\theta$.

b. Note that $|\overrightarrow{PS} \times \mathbf{v}| = |\overrightarrow{PS}||\mathbf{v}| \sin\theta = |\mathbf{v}|\, d$.
Therefore, $d = \dfrac{|\overrightarrow{PS} \times \mathbf{v}|}{|\mathbf{v}|}$.

29. $S(2, 1, 3)$, $P(2, 1, 3)$ and $\mathbf{v} = 2\mathbf{i} + 6\mathbf{j}$
$\overrightarrow{PS} \times \mathbf{v} = \begin{vmatrix} \mathbf{i} & \mathbf{j} & \mathbf{k} \\ 0 & 0 & 0 \\ 2 & 6 & 0 \end{vmatrix} = \mathbf{0}$
$d = \dfrac{|\overrightarrow{PS} \times \mathbf{v}|}{|\mathbf{v}|} = \dfrac{0}{\sqrt{40}} = 0$ is the distance from S to the line, i.e., the point S lies on the line.

31. a. Assuming $A \ne 0$, one point on the plane is $P\left(\frac{D}{A}, 0, 0\right)$.

b. Let $S = (x_0, y_0, z_0)$.
Then $\overrightarrow{PS} = \left(x_0 - \frac{D}{A}\right)\mathbf{i} + y_0\mathbf{j} + z_0\mathbf{k}$.

c. Since the vector $\mathbf{n} = A\mathbf{i} + B\mathbf{j} + C\mathbf{k}$ is normal to the plane, the distance d is the length of the projection of $\overrightarrow{PS}$ onto $\mathbf{n}$.

Therefore, $d = \left|\left(\overrightarrow{PS} \cdot \frac{\mathbf{n}}{|\mathbf{n}|}\right)\frac{\mathbf{n}}{|\mathbf{n}|}\right| = \left|\overrightarrow{PS} \cdot \frac{\mathbf{n}}{|\mathbf{n}|}\right|.$

Note: To arrive at a general formula, we may continue as follows:

$$d = \left|\overrightarrow{PS} \cdot \frac{\mathbf{n}}{|\mathbf{n}|}\right|$$

$$d = \left|\left[\left(x_0 - \frac{D}{A}\right)\mathbf{i} + y_0\mathbf{j} + z_0\mathbf{k}\right] \cdot \frac{(A\mathbf{i} + B\mathbf{j} + C\mathbf{k})}{\sqrt{A^2 + B^2 + C^2}}\right|$$

$$d = \frac{Ax_0 + By_0 + Cz_0 - D}{\sqrt{A^2 + B^2 + C^2}}$$

This formula is valid provided A, B, and C are not all 0.

33. $S(0, 1, 1)$, $4y + 3z = -12$ and $P(0, -3, 0)$ is on the plane. $\overrightarrow{PS} = 4\mathbf{j} + \mathbf{k}$ and $\mathbf{n} = 4\mathbf{j} + 3\mathbf{k}$

$$d = \left|\overrightarrow{PS} \cdot \frac{\mathbf{n}}{|\mathbf{n}|}\right| = \left|\frac{16 + 3}{\sqrt{16 + 9}}\right| = \frac{19}{5}$$

35. The point $P(1, 0, 0)$ is on the first plane and $S(10, 0, 0)$ is a point on the second plane.
$\overrightarrow{PS} = 9\mathbf{i}$, and $\mathbf{n} = \mathbf{i} + 2\mathbf{j} + 6\mathbf{k}$ is normal to the first plane.
The distance from S to the first plane is

$d = \left|\overrightarrow{PS} \cdot \frac{\mathbf{n}}{|\mathbf{n}|}\right| = \left|\frac{9}{\sqrt{1 + 4 + 36}}\right| = \frac{9}{\sqrt{41}}$, which is also the distance between the planes.

37. a. If $\mathbf{n}_1$ and $\mathbf{n}_2$ are chosen so that the angle between them is acute, then by definition, θ is the angle between $\mathbf{n}_1$ and $\mathbf{n}_2$.

Therefore, $\theta = \cos^{-1}\left(\frac{|\mathbf{n}_1 \cdot \mathbf{n}_2|}{|\mathbf{n}_1||\mathbf{n}_2|}\right)$, and $\mathbf{n}_1 \cdot \mathbf{n}_2 > 0$.

If the angle between $\mathbf{n}_1$ and $\mathbf{n}_2$ is obtuse, then the angle between $-\mathbf{n}_1$ and $\mathbf{n}_2$ is acute, so

$\theta = \cos^{-1}\left(\frac{|-\mathbf{n}_1 \cdot \mathbf{n}_2|}{|-\mathbf{n}_1||\mathbf{n}_2|}\right) = \cos^{-1}\left(\frac{|\mathbf{n}_1 \cdot \mathbf{n}_2|}{|\mathbf{n}_1||\mathbf{n}_2|}\right)$. If $\mathbf{n}_1$ and $\mathbf{n}_2$ are perpendicular, $\theta = \frac{\pi}{2} = \cos^{-1} 0 = \cos^{-1}\left(\frac{|\mathbf{n}_1 \cdot \mathbf{n}_2|}{|\mathbf{n}_1||\mathbf{n}_2|}\right)$.

So, in general, $\theta = \cos^{-1}\left(\frac{|\mathbf{n}_1 \cdot \mathbf{n}_2|}{|\mathbf{n}_1||\mathbf{n}_2|}\right)$.

b. The vectors $\mathbf{n}_1 = 3\mathbf{i} - 6\mathbf{j} - 2\mathbf{k}$ and $\mathbf{n}_2 = 2\mathbf{i} + \mathbf{j} - 2\mathbf{k}$ are normals to the planes. The angle between them is

$$\theta = \cos^{-1}\left(\frac{|\mathbf{n}_1 \cdot \mathbf{n}_2|}{|\mathbf{n}_1||\mathbf{n}_2|}\right)$$
$$= \cos^{-1}\left(\frac{|6 - 6 + 4|}{\sqrt{49}\sqrt{9}}\right)$$
$$= \cos^{-1}\left(\frac{4}{21}\right) \approx 1.38 \text{ rad}$$

39. $\mathbf{n}_1 = 2\mathbf{i} + 2\mathbf{j} - \mathbf{k}$ and $\mathbf{n}_2 = \mathbf{i} + 2\mathbf{j} + \mathbf{k}$

$$\theta = \cos^{-1}\left(\frac{|\mathbf{n}_1 \cdot \mathbf{n}_2|}{|\mathbf{n}_1||\mathbf{n}_2|}\right) = \cos^{-1}\left(\frac{|2 + 4 - 1|}{\sqrt{9}\sqrt{6}}\right) = \cos^{-1}\left(\frac{5}{3\sqrt{6}}\right)$$
$$\approx 0.82 \text{ rad}$$

41.
$$2x - y + 3z = 6$$
$$2(1 - t) - (3t) + 3(1 + t) = 6$$
$$-2t + 5 = 6$$
$$t = -\frac{1}{2}$$

$x = \frac{3}{2}$, $y = -\frac{3}{2}$ and $z = \frac{1}{2}$

$\left(\frac{3}{2}, -\frac{3}{2}, \frac{1}{2}\right)$ is the point.

43.
$$x + y + z = 2$$
$$(1 + 2t) + (1 + 5t) + (3t) = 2$$
$$10t + 2 = 2$$
$$t = 0$$

$x = 1$, $y = 1$, and $z = 0$

$(1, 1, 0)$ is the point.

45. $\mathbf{n}_1 = \mathbf{i} + \mathbf{j} + \mathbf{k}$ and $\mathbf{n}_2 = \mathbf{i} + \mathbf{j}$

$$\mathbf{n}_1 \times \mathbf{n}_2 = \begin{vmatrix} \mathbf{i} & \mathbf{j} & \mathbf{k} \\ 1 & 1 & 1 \\ 1 & 1 & 0 \end{vmatrix} = -\mathbf{i} + \mathbf{j},$$

the direction of the desired line.
$(1, 1, -1)$ is on both planes.
The desired line is $x = 1 - t$, $y = 1 + t$, $z = -1$.

47. $\mathbf{n}_1 = \mathbf{i} - 2\mathbf{j} + 4\mathbf{k}$ and $\mathbf{n}_2 = \mathbf{i} + \mathbf{j} - 2\mathbf{k}$

$$\mathbf{n}_1 \times \mathbf{n}_2 = \begin{vmatrix} \mathbf{i} & \mathbf{j} & \mathbf{k} \\ 1 & -2 & 4 \\ 1 & 1 & -2 \end{vmatrix} = 6\mathbf{j} + 3\mathbf{k},$$

the direction of the desired line.
$(4, 3, 1)$ is on both planes.
The desired line is $x = 4$, $y = 3 + 6t$, $z = 1 + 3t$.

49. L_1 and L_2: $x = 3 + 2t = 1 + 4s$ and $y = -1 + 4t = 1 + 2s$

$$\begin{cases} 2t - 4s = -2 \\ 4t - 2s = 2 \end{cases}$$
$$\begin{cases} 2t - 4s = -2 \\ 2t - s = 1 \end{cases}$$

$-3s = -3$
$s = 1$ and $t = 1$
On L_1, $z = 1$ and on L_2, $z = 1$
L_1 and L_2 intersect at $(5, 3, 1)$.

L_2 and L_3: The direction of L_2 is

$\frac{1}{6}(4\mathbf{i} + 2\mathbf{j} + 4\mathbf{k}) = \frac{1}{3}(2\mathbf{i} + \mathbf{j} + 2\mathbf{k})$ which is the same as the direction $\frac{1}{3}(2\mathbf{i} + \mathbf{j} + 2\mathbf{k})$ of L_3.

Hence, L_2 and L_3 are parallel.

L_1 and L_3: $x = 3 + 2t = 3 + 2r$ and $y = -1 + 4t = 2 + r$

$$\begin{cases} 2t - 2r = 0 \\ 4t - r = 3 \end{cases}$$
$$\begin{cases} t - r = 0 \\ 4t - r = 3 \end{cases}$$

$3t = 3$
$t = 1$ and $r = 1$
On L_1, $z = 1$ while on L_3, $z = 0$.
L_1 and L_2 do not intersect. The direction of L_1 is $\frac{1}{\sqrt{21}}(2\mathbf{i} + 4\mathbf{j} - \mathbf{k})$ while the direction of L_3 is $\frac{1}{3}(2\mathbf{i} + \mathbf{j} + 2\mathbf{k})$ and neither is a multiple of the other. Hence L_1 and L_3 are skew.

51. $x = 2 + 2t, y = -4 - t, z = 7 + 3t$

$x = -2 - s, y = -2 + \frac{1}{2}s, z = 1 - \frac{3}{2}s$

53. yz-plane: $x = 0$, hence $t = -\frac{1}{2}$, so $y = -\frac{1}{2}$ and $z = -\frac{3}{2}$.
$\left(0, -\frac{1}{2}, -\frac{3}{2}\right)$
xz-plane: $y = 0$, hence $t = -1$, so $x = -1$ and $z = -3$.
$(-1, 0, -3)$
xy-plane: $z = 0$, hence $t = 0$, so $x = 1$ and $y = -1$.
$(1, -1, 0)$

55. With substitution of the line into the plane we have

$$2(1 - 2t) + (2 + 5t) - (-3t) = 8$$
$$2 - 4t + 2 + 5t + 3t = 8$$
$$4t + 4 = 8$$
$$t = 1$$

The point $(-1, 7, -3)$ is contained in both the line and plane, so they are not parallel.

57. There are many possible answers. One is found as follows:

eliminate t to get $t = x - 1 = 2 - y = \frac{z - 3}{2}$

$x - 1 = 2 - y$ and $2 - y = \frac{z - 3}{2}$

$x + y = 3$ and $2y + z = 7$ are two such planes.

59. The points $(a, 0, 0)$, $(0, b, 0)$, and $(0, 0, c)$ are the x-, y-, and z-intercepts of the plane. Since a, b, and c are all nonzero, the plane must intersect all three coordinate axes and cannot pass through the origin. Thus, $\frac{x}{a} + \frac{y}{b} + \frac{z}{c} = 1$ describes all planes *except* those through the origin or parallel to a coordinate axis.

61. a. $\overrightarrow{EP} = c\overrightarrow{EP_1}$
$-x_0\mathbf{i} + y\mathbf{j} + z\mathbf{k} = c[(x_1 - x_0)\mathbf{i} + y_1\mathbf{j} + z_1\mathbf{k}]$
$-x_0 = c(x_1 - x_0)$, $y = cy_1$ and $z = cz_1$, where c is a positive real number.

b. At $x_1 = 0$: $c = 1$, so $y = y_1$ and $z = z_1$.

At $x_1 = x_0$: $x_0 = 0$, $y = 0$ and $z = 0$.

$\lim_{x_0 \to \infty} c = \lim_{x_0 \to \infty} \frac{-x_0}{x_1 - x_0} = \lim_{x_0 \to \infty} \frac{-1}{-1} = 1$,
so $c \to 1$ and $y \to y_1$ and $z \to z_1$.

Section 11.5 Cylinders and Cylindrical Coordinates (pp. 607–614)

Exploration 1 Easy as Pi

1. $x = r \cos \theta = \pi \cos \pi = -\pi$
$y = r \sin \theta = \pi \sin \pi = 0$
$z = z = \pi$
The rectangular coordinates are $(-\pi, 0, \pi)$.

2. $r = \sqrt{x^2 + y^2} = \sqrt{\pi^2 + \pi^2} = \pi\sqrt{2}$
Since $\tan \theta = \frac{y}{x} = \frac{\pi}{\pi} = 1$, and $x > 0$, $\theta = \frac{\pi}{4}$.
$z = z = \pi$
The polar coordinates are $\left(\pi\sqrt{2}, \frac{\pi}{4}, \pi\right)$.

3. The graph of $r = \pi$ is a circular cylinder of radius π centered about the z-axis.

4. The graph of the equation $\theta = \pi$ is the xz-plane.

5. The intersection of the two graphs in parts 3 and 4 consists of two lines parallel to the z-axis.

6. The graph of all points satisfying the simultaneous equations $r = \pi$ and $\theta = \pi$ consists of a single line parallel to the z-axis.

7. They are different because, for example, the point whose polar coordinates can be written as either $(\pi, 0, 0)$ or $(-\pi, \pi, 0)$ is on each of the graphs in parts 3 and 4, but its coordinates cannot be given in a way that simultaneously satisfies $r = \pi$ and $\theta = \pi$.

Quick Review 11.5

1. a. $x^2 + 2x + y^2 = 0$
$x^2 + 2x + 1 + y^2 = 1$
$(x + 1)^2 + y^2 = 1$
Circle, center $= (-1, 0)$, radius $= 1$

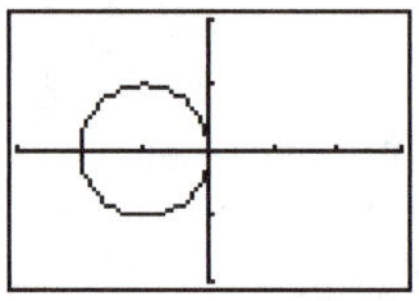

b. $x^2 + y^2 - 4y = 0$
$x^2 + y^2 - 4y + 4 = 4$
$x^2 + (y - 2)^2 = 4$
Circle, center $= (0, 2)$, radius $= 2$

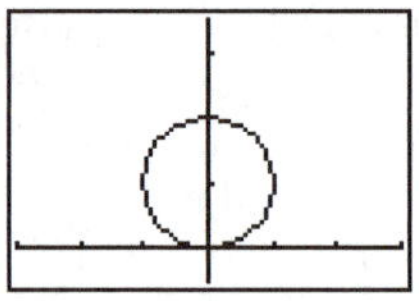

2. a. $x^2 + 2y^2 = 8$
$\frac{x^2}{8} + \frac{y^2}{4} = 1$
Ellipse, center $= (0, 0)$, vertices at $(\pm 2\sqrt{2}, 0)$, y-intercepts at $(0, \pm 2)$

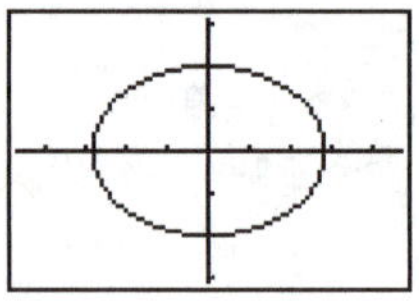

b. $4x^2 + y^2 = 36$
$\frac{x^2}{9} + \frac{y^2}{36} = 1$
Ellipse, center $= (0, 0)$, vertices at $(0, \pm 6)$, x-intercepts at $(\pm 3, 0)$

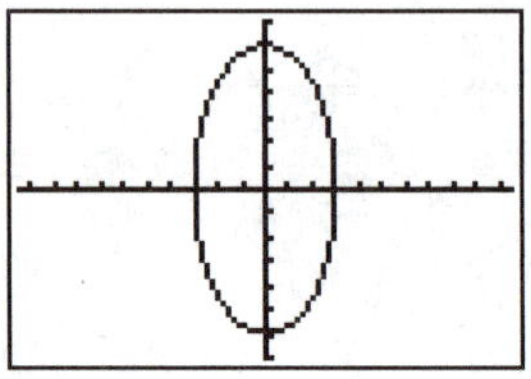

3. a. $xy = 4$

Hyperbola, center = (0, 0)

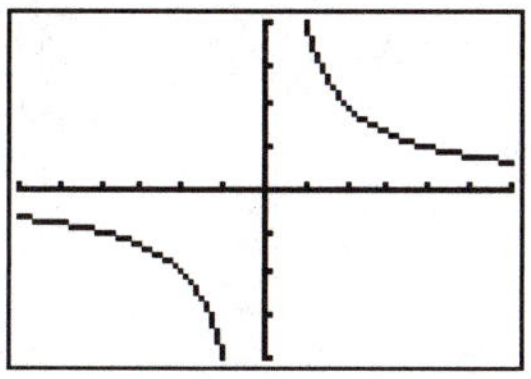

b. $x^2 - 4y^2 = 36$

$\frac{x^2}{36} - \frac{y^2}{9} = 1$

Hyperbola, center = (0, 0), vertices (± 6, 0)

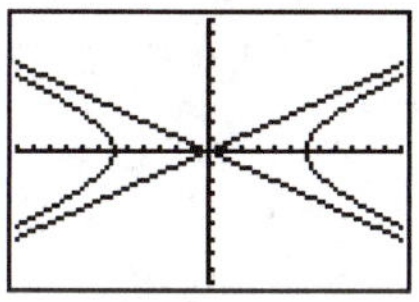

4. a. $y = x^2 - 1$

Parabola, vertex (0, -1), x-intercepts (± 1, 0)

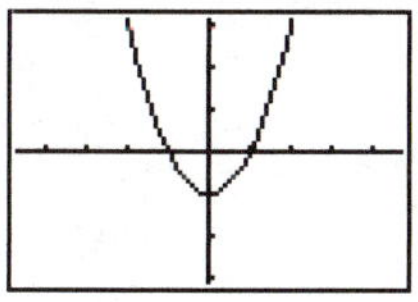

b. $x = 4 - y^2$

Parabola, vertex (4, 0), y-intercepts (0, ± 2)

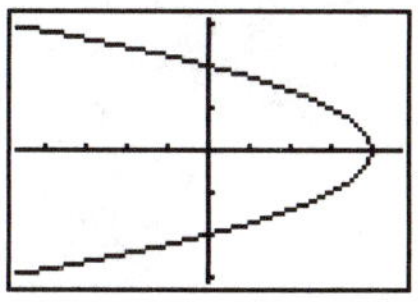

5. $x^2 + y^2 = 9$

$r^2 = 9$

$r = 3$

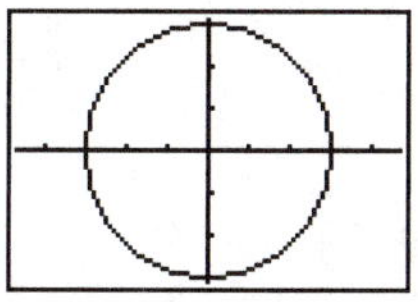

6. $(x - 1)^2 + y^2 = 1$

$(r\cos\theta - 1)^2 + (r\sin\theta)^2 = 1$

$r^2(\cos^2\theta + \sin^2\theta) - 2r\cos\theta + 1 = 1$

$r^2 - 2r\cos\theta = 0$

$r(r - 2\cos\theta) = 0$

$r = 0$ or $r = 2\cos\theta$

Note that $r = 0$ is redundant, so the polar equation can be given as $r = 2\cos\theta$.

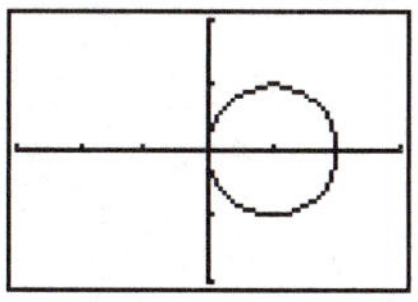

7. $x^2 + (y + 2)^2 = 4$

$(r\cos\theta)^2 + (r\sin\theta + 2)^2 = 4$

$r^2(\cos^2\theta + \sin^2\theta) + 4r\sin\theta + 4 = 4$

$r(r + 4\sin\theta) = 0$

$r = 0$ or $r = -4\sin\theta$

Note that $r = 0$ is redundant, so the polar equation can be given as $r = -4\sin\theta$.

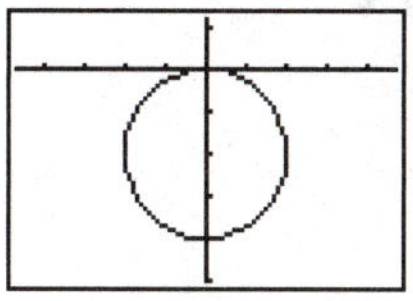

8. $(x + 3)^2 + y^2 = 9$

$(r\cos\theta + 3)^2 + (r\sin\theta)^2 = 9$

$r^2(\cos^2\theta + \sin^2\theta) + 6r\cos\theta + 9 = 9$

$r(r + 6\cos\theta) = 0$

$r = 0$ or $r = -6\cos\theta$

Note that $r = 0$ is redundant, so the polar equation can be given as $r = -6\cos\theta$.

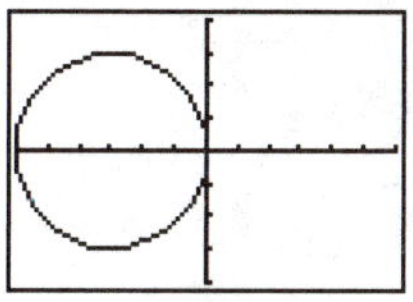

9. a. Cardioid

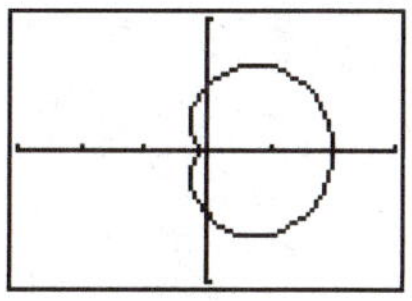

b. Cardioid

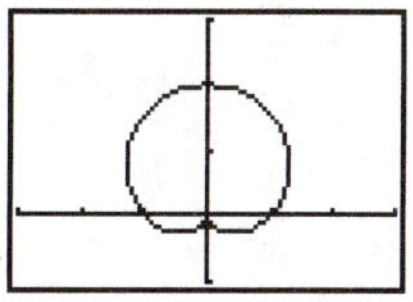

10. a. Horizontal line $y = 2$

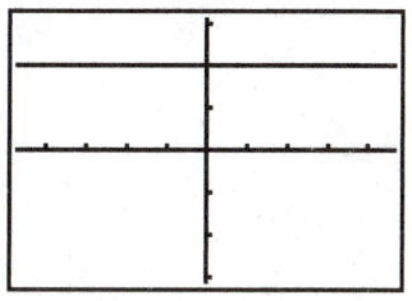

b. Vertical line $x = -1$

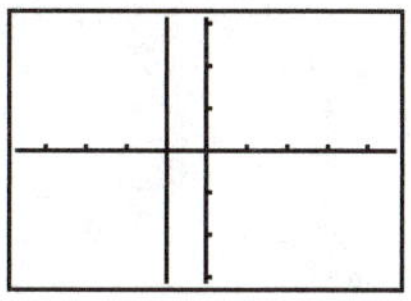

Section 11.5 Exercises

1. a, elliptical cylinder

3. $x^2 + y^2 = 4$

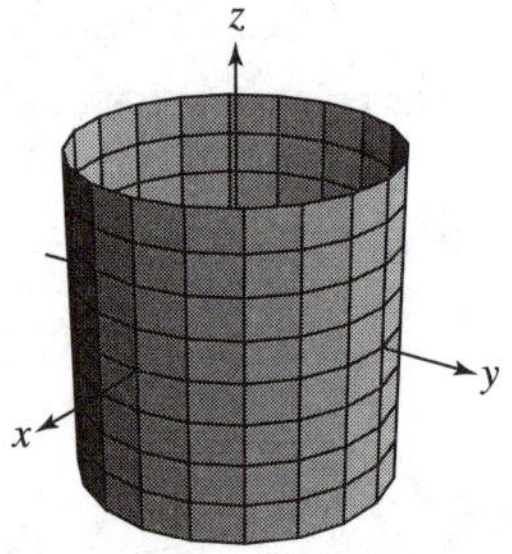

5. $z^2 - y^2 = 1$

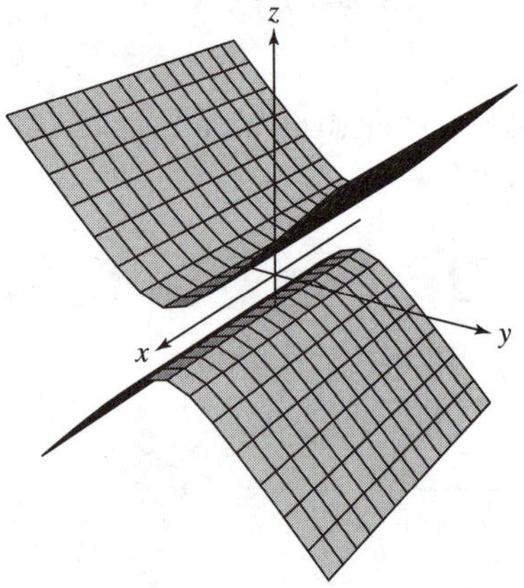

7. $x^2 + 4z^2 = 16$

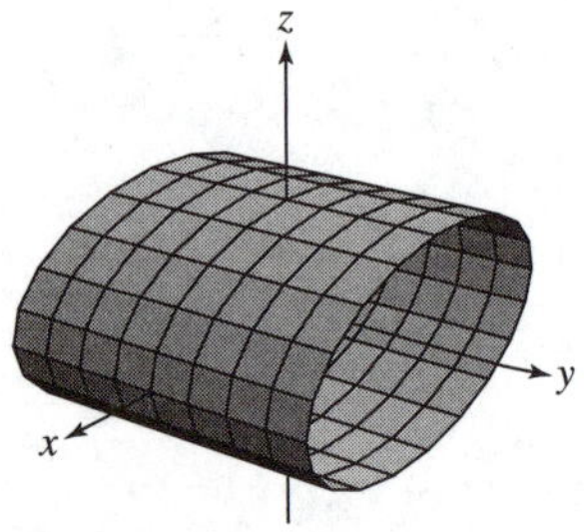

9. $16x^2 + 4y^2 = 1$

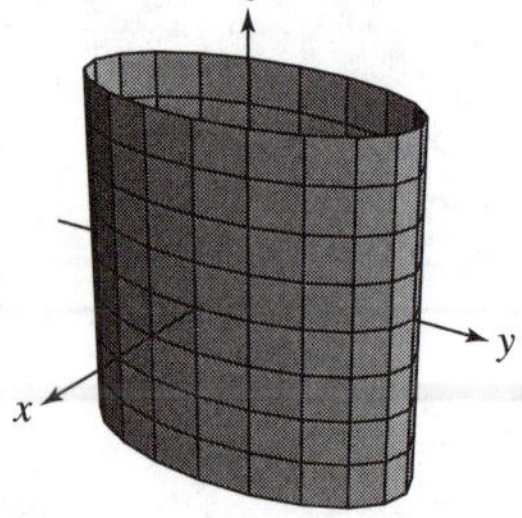

11. $z^2 + 4y^2 = 9$

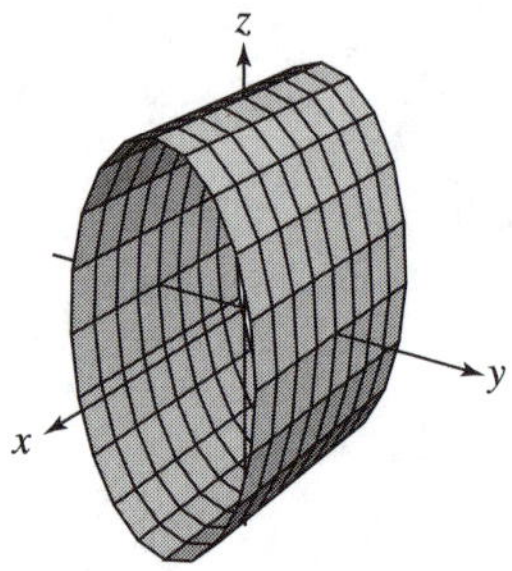

13. d

15. a

	Rectangular	Cylindrical
17.	(0, 0, 0)	(0, 0, 0)
19.	(0, 1, 0)	$\left(1, \frac{\pi}{2}, 0\right)$
21.	(1, 0, 0)	(1, 0, 0)
23.	(0, 1, 1)	$\left(1, \frac{\pi}{2}, 1\right)$

25. $r = 0$
rectangular, $x^2 + y^2 = 0$
the z-axis

27. $z = 0$
cylindrical, $z = 0$
the xy-plane

29. $x^2 + y^2 + z^2 = 4$
cylindrical, $r^2 + z^2 = 4$
a sphere of radius 2 centered at the origin

31. $r = \csc\theta$

rectangular, $r = \frac{r}{y}$

$y = 1$ since $r \neq 0$

the plane $y = 1$

33. $x^2 + y^2 + (z - 1)^2 = 1,\ z \leq 1$
cylindrical, $r^2 + (z - 1)^2 = 1$
$r^2 + z^2 - 2z + 1 = 1$
$r^2 + z^2 = 2z,\ z \leq 1$
the lower half (hemisphere) of the sphere of radius 1 centered at (0, 0, 1) (rectangular)

35. Right circular cylinder parallel to the z-axis generated by the circle $r = -2\sin\theta$ in the $r\theta$-plane

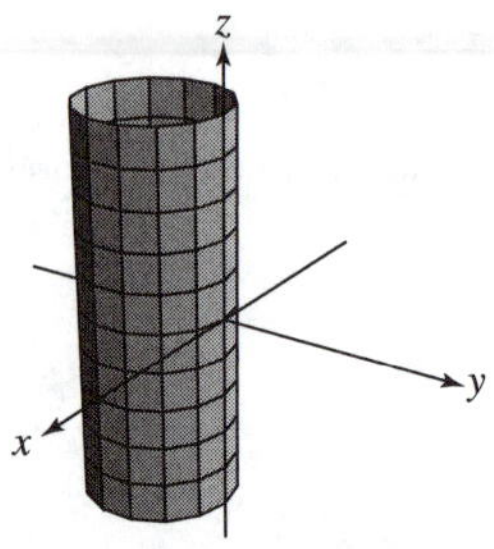

37. Cylinder of lines parallel to the z-axis generated by the cardioid $r = 1 - \cos\theta$ in the $r\theta$-plane

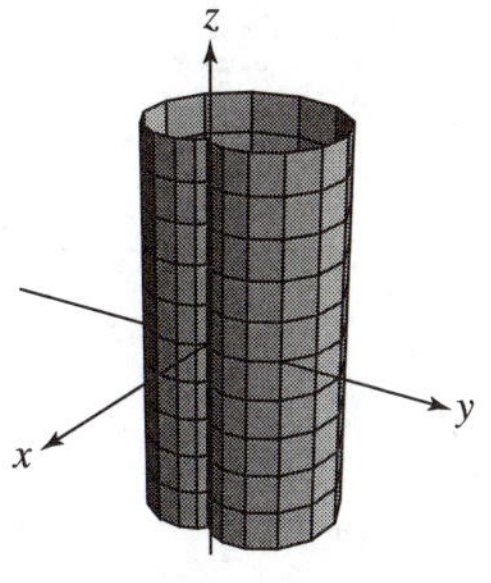

39. $r^2 + z^2 = 4r\cos\theta + 6r\sin\theta + 2z$

$x^2 + y^2 + z^2 = 4x + 6y + 2z$

$(x^2 - 4x + 4) + (y^2 - 6y + 9) + (z^2 - 2z + 1) = 14$

$(x - 2)^2 + (y - 3)^2 + (z - 1)^2 = 14$

The center is located at (2, 3, 1) in rectangular coordinates.

41. A plane perpendicular to the x-axis has the form $x = a$ in rectangular coordinates.

$r\cos\theta = a$

$$r = \frac{a}{\cos\theta}$$

$r = a\sec\theta$, in cylindrical coordinates

43. The equation $r = f(z)$ implies that the point $(r, \theta, z) = (f(z), \theta, z)$ will lie on the surface for all θ.

In particular $(f(z), \theta + \pi, z)$ lies on the surface whenever $(f(z), \theta, z)$ does.

The surface is symmetric with respect to the z-axis.

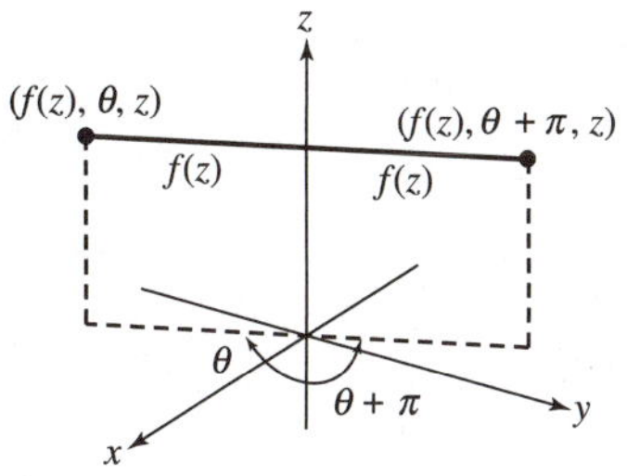

45. $z = y^2$

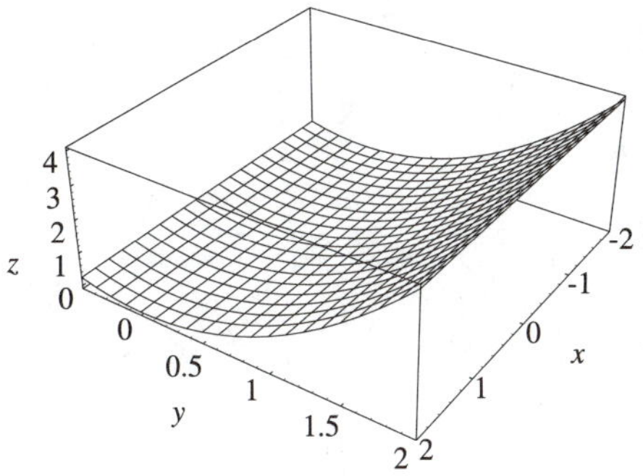

47.

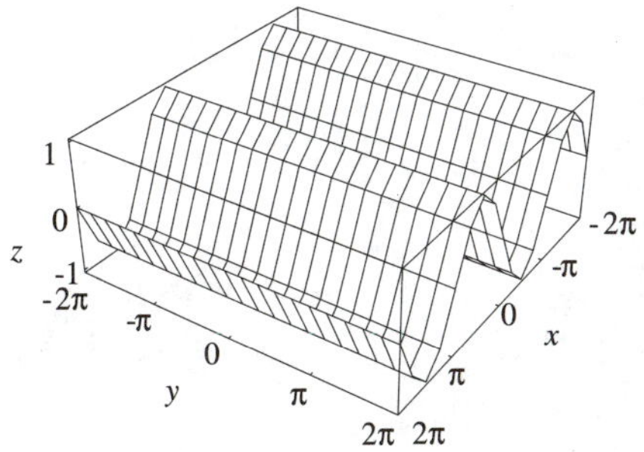

49. Example CAS commands:

Maple:

```
with(plots):
eq:= r^2+z^2=2*r*(cos(theta)
  +sin(theta))+2;
subs(r=sqrt(x^2+y^2),
  theta=arctan(y/x),eq);
simplify(%,trig);
eq2:=x^2+y^2+z^2-2*x-2*y=0;
implicitplot3d(eq2,x=0..3,y=0..3,
  z=-2..2);
are:=solve(eq,r);
simplify(are[1],trig);
r:=unapply(%,(theta,z));
cylinderplot(r,Pi/4..9*Pi/4,
  -2..2,grid = [100,100]);
```

Mathematica:

```
(We need the ParametricPlot3D package
for SphericalPlot3D)
<<Graphics`ParametricPlot3D`

(ContourPlot3D allows implicit
plotting in 3D)
<<Graphics`ContourPlot3D`

Clear[r,theta,x,y,z]
eqn=r^2+z^2== 2r
  (Cos[theta]+Sin[theta])+2
Solve[ eqn, r ]
r[theta_,z_] = r /. %[[2]] //
  Simplify
```

```
Note: the CylindricalPlot3D function only
      handles plotting z(r,theta), not
      r(theta,z), so we must use the more
      general ParametricPlot3D.
```

```
ParametricPlot3D[
  {r[theta,z] Cos[theta], r[theta,z]
  Sin[theta], z},
{theta,Pi/4,9Pi/4}, {z,-2,2} ]
Map[Expand, eqn]
% /. {r Cos[theta] -> x,
  r Sin[theta] -> y, r^2 -> x^2+y^2}
eqn2 = Map[ (# -2x -2y +2)&, % ]
ContourPlot3D[ eqn2[[1]],
  {x,-1,3}, {y,-1,3}, {z,-2,2},
  Contours -> {eqn2[[2]]} ]
```

Section 11.6 Quadric Surfaces
(pp. 614–625)

Exploration 1 Identifying Shapes of Familiar Objects

1. B
2. A
3. D
4. C

Quick Review 11.6

1. $\frac{x^2}{4} + \frac{y^2}{9} = 1$

$$y^2 = 9\left(1 - \frac{x^2}{4}\right)$$

$$y = \pm 3\sqrt{1 - \frac{x^2}{4}}$$

2. The equation of the ellipse can be written as $\left(\frac{x}{2}\right)^2 + \left(\frac{y}{3}\right)^2 = 1$, which corresponds to $\cos^2 t + \sin^2 t = 1$. Letting $\frac{x}{2} = \cos t$ and $\frac{y}{3} = \sin t$, the parametric equations are $x = 2 \cos t$, $y = 3 \sin t$, $0 \le t \le 2\pi$. (Other answers are possible.)

3. $\frac{x^2}{4} - \frac{y^2}{9} = 1$

$$y^2 = 9\left(\frac{x^2}{4} - 1\right)$$

$$y = \pm 3\sqrt{\frac{x^2}{4} - 1}$$

4. The equation of the hyperbola can be written as $\left(\frac{x}{2}\right)^2 - \left(\frac{y}{3}\right)^2 = 1$, which corresponds to $\sec^2 t - \tan^2 t = 1$. Letting $\frac{x}{2} = \sec t$ and $\frac{y}{3} = \tan t$, the parametric equations are $x = 2 \sec t$, $y = 3 \tan t$, $0 \le t \le 2\pi \left(t \ne \frac{\pi}{2}, t \ne \frac{3\pi}{2}\right)$. (Other answers are possible.)

5. x-intercepts: $(\pm 3, 0)$
   y-intercepts: $(0, \pm 2)$

6. x-intercepts: $(\pm 3, 0)$
   y-intercepts: None

7. **a.** Yes
   b. Yes
   c. Yes

8. **a.** Yes
   b. Yes
   c. Yes

9. **a.** No
   b. Yes
   c. No

10. **a.** Yes
    b. No
    c. No

Section 11.6 Exercises

1. b, ellipsoid
3. e, cone
5. c, paraboloid
7. i, hyperbolic paraboloid
9. f, cone
11. $9x^2 + y^2 + z^2 = 9$

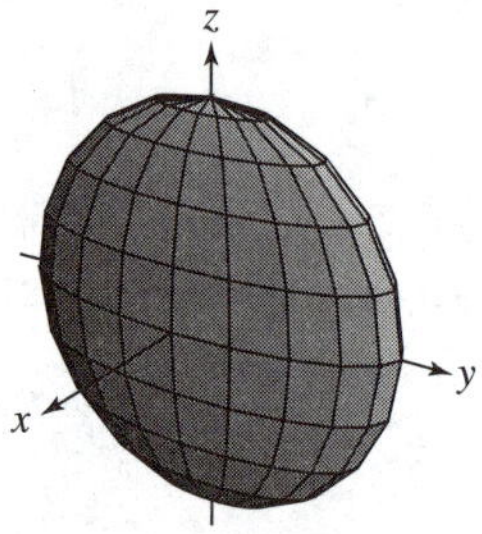

13. $4x^2 + 9y^2 + 4z^2 = 36$

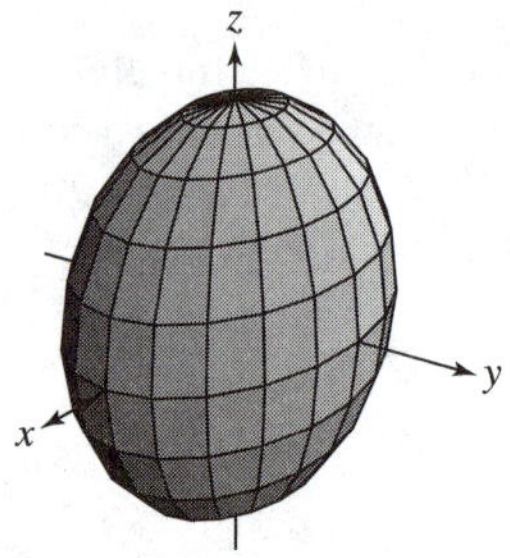

15. $x^2 + 4y^2 = z$

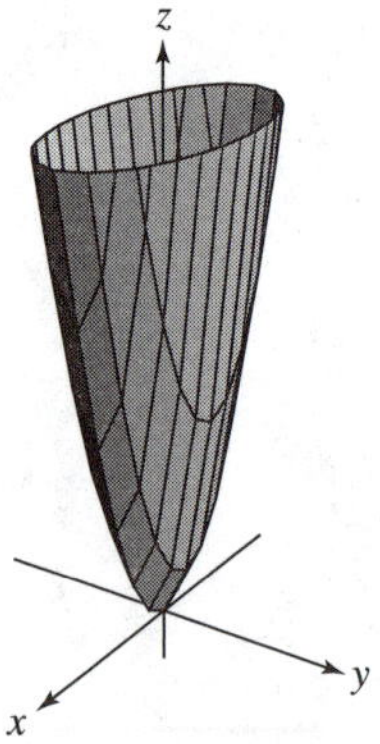

17. $z = 8 - x^2 - y^2$

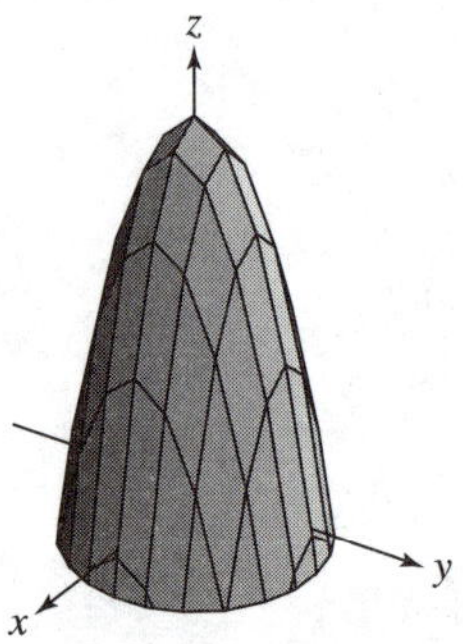

19. $x = 4 - 4y^2 - z^2$

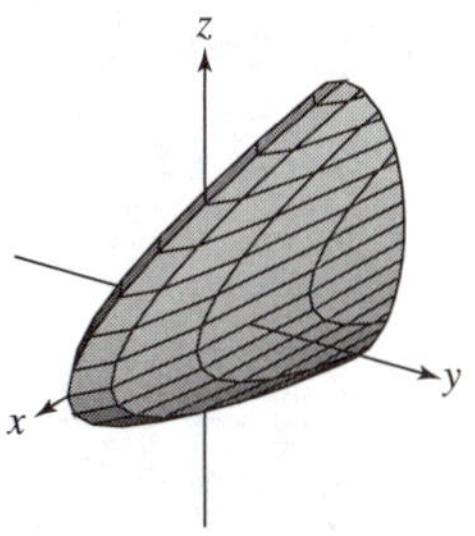

21. $x^2 + y^2 = z^2$

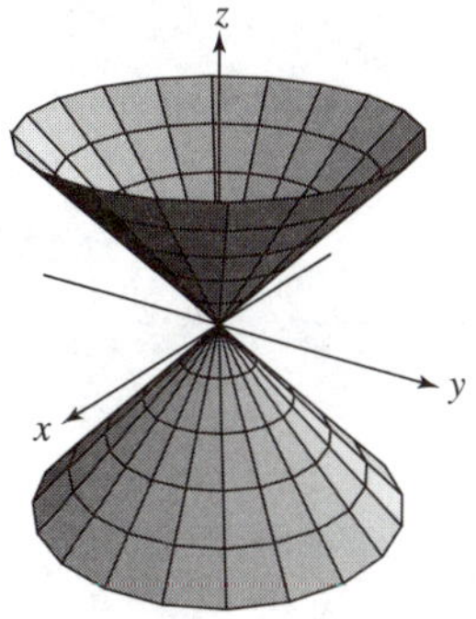

23. $4x^2 + 9z^2 = 9y^2$

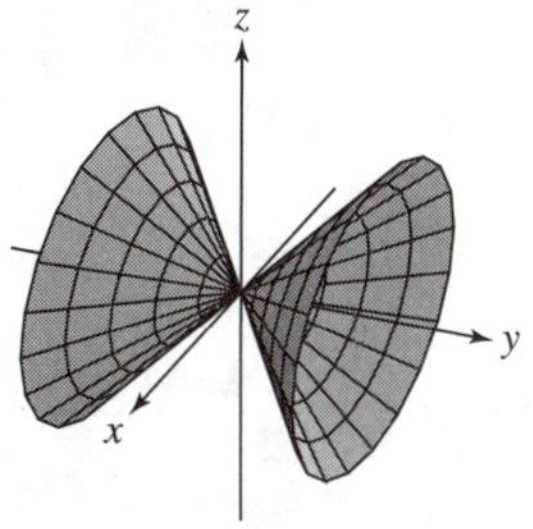

25. $x^2 + y^2 - z^2 = 1$

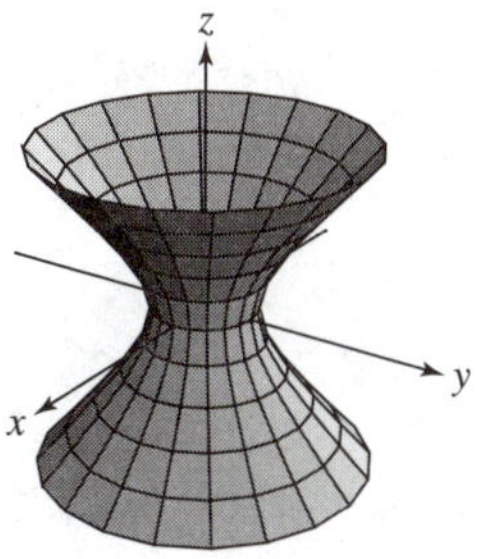

27. $\frac{y^2}{4} + \frac{z^2}{9} - \frac{x^2}{4} = 1$

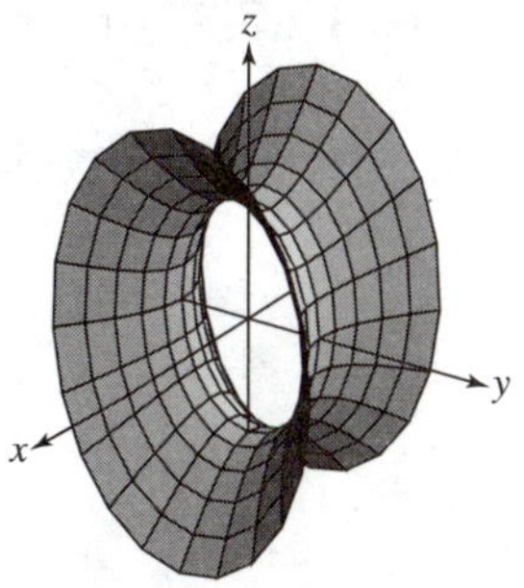

29. $z^2 - x^2 - y^2 = 1$

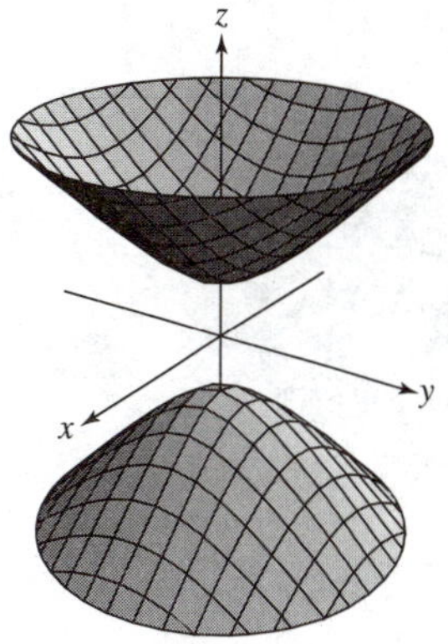

31. $x^2 - y^2 - \frac{z^2}{4} = 1$

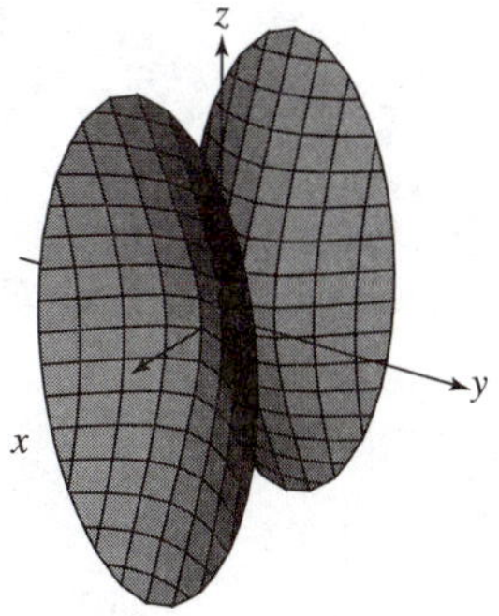

33. $y^2 - x^2 = z$

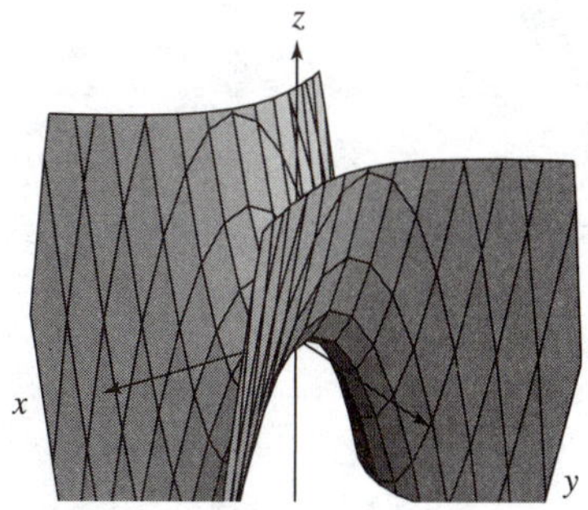

35. $x^2 + y^2 + z^2 = 4$

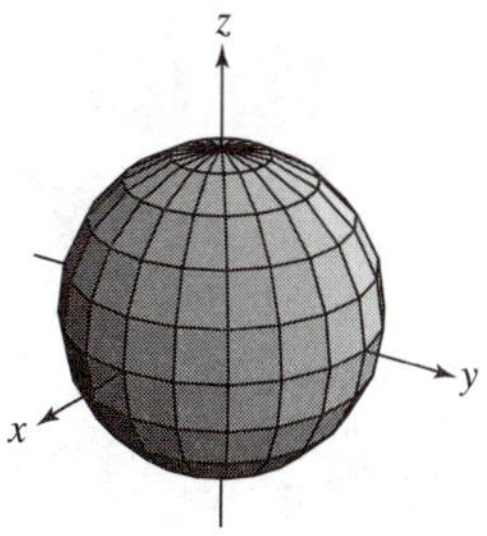

37. $z = 1 + y^2 - x^2$

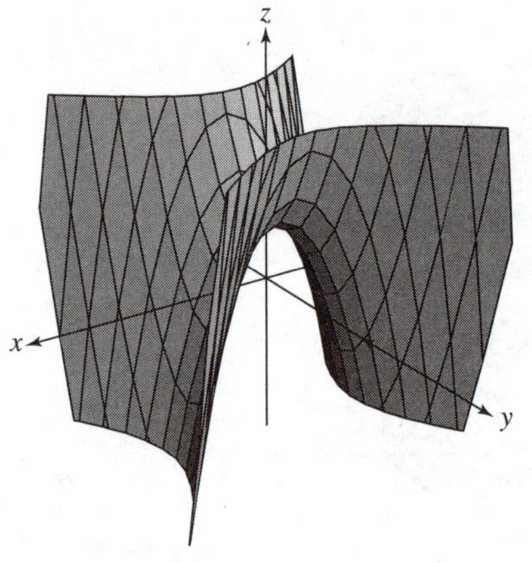

39. $y = -(x^2 + z^2)$

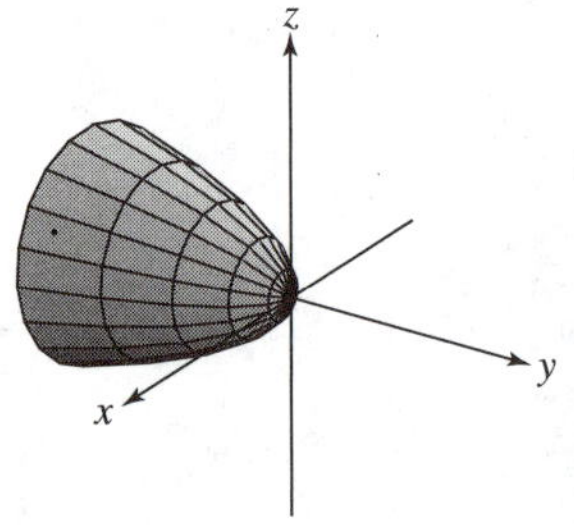

41. $x^2 + y^2 - z^2 = 4$

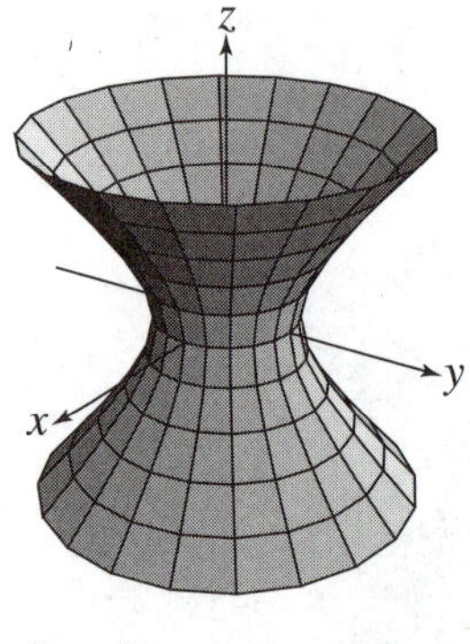

43. $x^2 + z^2 = 2$

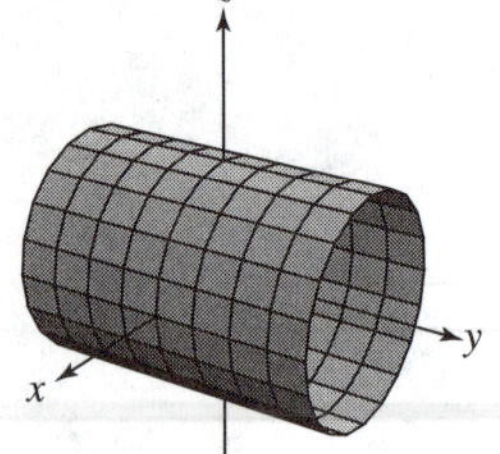

45. $16y^2 + 9z^2 = 4x^2$

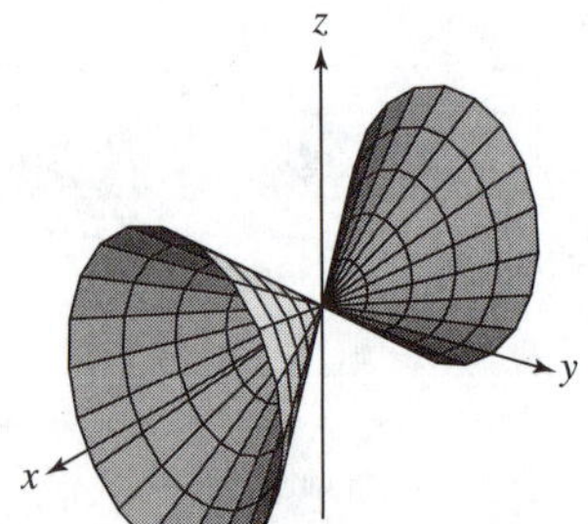

47. $z = -(x^2 + y^2)$

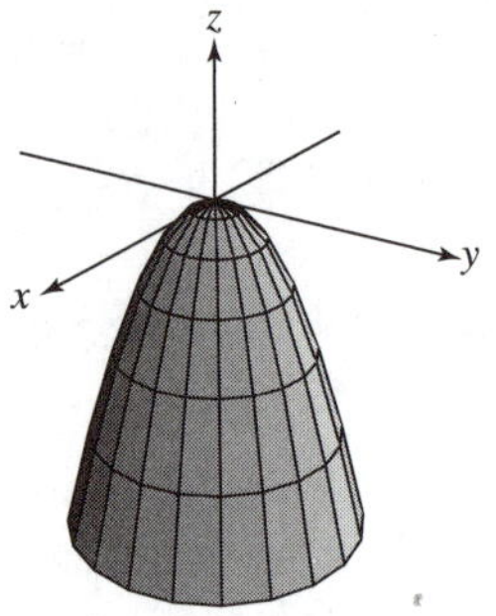

49. $36x^2 + 9y^2 + 4z^2 = 36$

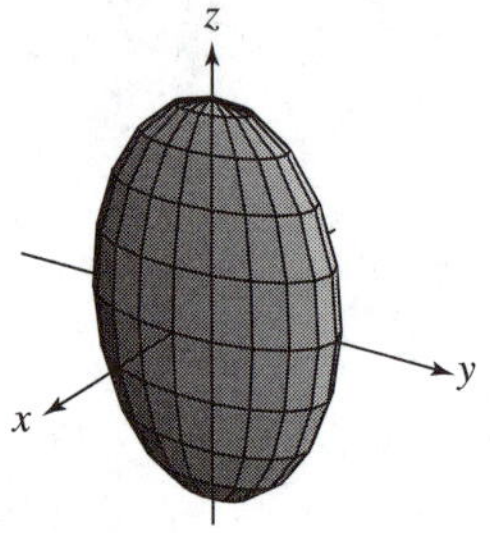

51. a. If $x^2 + \frac{y^2}{4} + \frac{z^2}{9} = 1$ and $z = c$, then $x^2 + \frac{y^2}{4} = \frac{9 - c^2}{9}$.

$$\frac{x^2}{\left(\frac{9 - c^2}{9}\right)} + \frac{y^2}{\left[\frac{4(9 - c^2)}{9}\right]} = 1$$

$$A = ab\pi = \pi\left(\frac{\sqrt{9 - c^2}}{3}\right)\left(\frac{2\sqrt{9 - c^2}}{3}\right) = \frac{2\pi(9 - c^2)}{9}$$

b. From part (a), each slice has the area $\frac{2\pi(9 - z^2)}{9}$, where $-3 \le z \le 3$.

$$\text{Thus } V = 2\int_0^3 \frac{2\pi}{9}(9 - z^2)\,dz = \frac{4\pi}{9}\int_0^3 (9 - z^2)\,dz$$
$$= \frac{4\pi}{9}\left[9z - \frac{z^3}{3}\right]_0^3 = \frac{4\pi}{9}(27 - 9) = 8\pi$$

c. $\frac{x^2}{a^2} + \frac{y^2}{b^2} + \frac{z^2}{c^2} = 1$

$$\frac{x^2}{\left[\frac{a^2(c^2 - z^2)}{c^2}\right]} + \frac{y^2}{\left[\frac{b^2(c^2 - z^2)}{c^2}\right]} = 1$$

$$A = \pi\left(\frac{a\sqrt{c^2 - z^2}}{c}\right)\left(\frac{b\sqrt{c^2 - z^2}}{c}\right) = \frac{ab\sqrt{c^2 - z^2}}{c^2}$$

$$V = 2\int_0^c \frac{\pi ab}{c^2}(c^2 - z^2)\,dz = \frac{2\pi ab}{c^2}\left[c^2 z - \frac{z^3}{3}\right]_0^c$$
$$= \frac{2\pi ab}{c^2}\left(\frac{2}{3}c^3\right) = \frac{4\pi abc}{3}.$$

Note that if $r = a = b = c$, then $V = \frac{4\pi r^3}{3}$, which is the volume of a sphere.

53. We calculate the volume by the slicing method, taking slices parallel to the xy-plane. For fixed z, $\frac{x^2}{a^2} + \frac{y^2}{b^2} = \frac{z}{c}$ gives the ellipse $\frac{x^2}{\left(\frac{za^2}{c}\right)} + \frac{y^2}{\left(\frac{zb^2}{c}\right)} = 1$. The area of this ellipse is $\pi\left(a\sqrt{\frac{z}{c}}\right)\left(b\sqrt{\frac{z}{c}}\right) = \frac{\pi abz}{c}$ (see Exercise 51a). Hence the volume is given by $V = \int_0^h \frac{\pi abz}{c}\,dz$ $= \left[\frac{\pi abz^2}{2c}\right]_0^h = \frac{\pi abh^2}{c}$. Now the area of the elliptic base when $z = h$ is $A = \frac{\pi abh}{c}$, as determined previously. Thus, $V = \frac{\pi abh^2}{c} = \frac{1}{2}\left(\frac{\pi abh}{c}\right)h = \frac{1}{2}$(base)(altitude), as claimed.

55. No, it is not mere coincidence. A plane parallel to one of the coordinate planes will set one of the variables x, y, or z equal to a constant in the general equation $Ax^2 + By^2 + Cz^2 + Dxy + Eyz + Fxz + Gx + Hy + Jz + K = 0$ for a quadric surface. The resulting equation then has the general form for a conic in that parallel plane. For example, setting $y = y_1$ results in the equation $Ax^2 + Cz^2 + D'x + E'z + Fxz + Gx + Jz + K' = 0$ where $D' = Dy_1$, $E' = Ey_1$, and $K' = K + By_1^2 + Hy_1$, which is the general form of a conic section in the plane $y = y_1$.

57. When $y = y_1$, $\frac{z}{c} = \frac{y_1^2}{b^2} - \frac{x^2}{a^2}$, a parabola in the plane $y = y_1$. Writing the parabola as $(x - 0)^2 = -\frac{a^2}{c}z + \frac{cy_1^2}{b^2}$ we see that the vertex is $\left(0, y, \frac{cy_1^2}{b^2}\right)$. Also, $4p = -\frac{a^2}{c}$ or $p = -\frac{a^2}{4c}$ and the focus is $\left(0, y_1, \frac{cy_1^2}{b^2} - \frac{a^2}{4c}\right)$.

59–63. Example CAS commands:

Maple:

```
with(plots):
eq1:=x^2/9-y^2/16-z^2/2=1;
implicitplot3d(eq1,x=-15..15,
  y=-9..9,z=-7..7, title =
  'Hyperboloid of Two Sheets');
```

Mathematica:

```
<<Graphics 'ContourPlot3D'
ContourPlot3D[x^2/9-y^2/16-z^2/2-1,
  {x,-9,9},{y,-12,12},{z,-5,5},
  PlotLabel -> "Elliptic Hyperboloid
  of Two Sheets" ]
```

Chapter 11 Review Exercises

(pp. 626–629)

1. length $= |2\mathbf{i} - 3\mathbf{j} + 6\mathbf{k}| = \sqrt{4 + 9 + 36} = 7$

The direction is $\frac{2}{7}\mathbf{i} - \frac{3}{7}\mathbf{j} + \frac{6}{7}\mathbf{k}$

$2\mathbf{i} - 3\mathbf{j} + 6\mathbf{k} = 7\left(\frac{2}{7}\mathbf{i} - \frac{3}{7}\mathbf{j} + \frac{6}{7}\mathbf{k}\right)$

2. length $= |\mathbf{i} + 2\mathbf{j} - \mathbf{k}| = \sqrt{1 + 4 + 1} = \sqrt{6}$

The direction is $\frac{1}{\sqrt{6}}\mathbf{i} + \frac{2}{\sqrt{6}}\mathbf{j} - \frac{1}{\sqrt{6}}\mathbf{k}$

$\mathbf{i} + 2\mathbf{j} - \mathbf{k} = \sqrt{6}\left(\frac{1}{\sqrt{6}}\mathbf{i} + \frac{2}{\sqrt{6}}\mathbf{j} - \frac{1}{\sqrt{6}}\mathbf{k}\right)$

3. $2\frac{\mathbf{A}}{|\mathbf{A}|} = 2 \cdot \frac{4\mathbf{i} - \mathbf{j} + 4\mathbf{k}}{\sqrt{4^2 + (-1)^2 + 4^2}} = 2 \cdot \frac{4\mathbf{i} - \mathbf{j} + 4\mathbf{k}}{\sqrt{33}}$

$= \frac{8}{\sqrt{33}}\mathbf{i} - \frac{2}{\sqrt{33}}\mathbf{j} + \frac{8}{\sqrt{33}}\mathbf{k}$

4. $\overrightarrow{OD} = \overrightarrow{OC} + \overrightarrow{OB}$

$\overrightarrow{OE} = \overrightarrow{OD} + \overrightarrow{OA}$

$= \overrightarrow{OC} + \overrightarrow{OB} + \overrightarrow{OA}$

5.

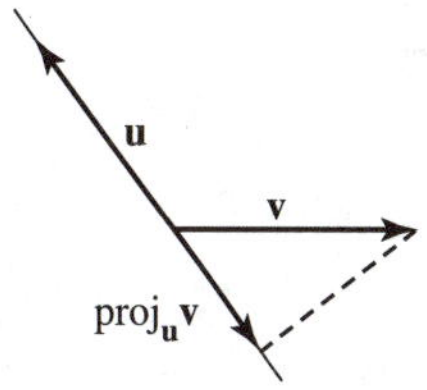

6. $\mathbf{a} = \text{proj}_{\mathbf{v}}\,\mathbf{u}$, $\mathbf{b} = \text{proj}_{\mathbf{u}}\,\mathbf{v}$, $\mathbf{c} = \mathbf{v} - \mathbf{b} = \mathbf{v} - \text{proj}_{\mathbf{u}}\,\mathbf{v}$

7. $|\mathbf{A}| = \sqrt{1 + 1} = \sqrt{2}$, $|\mathbf{B}| = \sqrt{4 + 1 + 4} = 3$,

$\mathbf{A} \cdot \mathbf{B} = (1)(2) + (1)(1) + (0)(-2) = 3$,

$\mathbf{B} \cdot \mathbf{A} = (2)(1) + (1)(1) + (-2)(0) = 3$

$\mathbf{A} \times \mathbf{B} = \begin{vmatrix} \mathbf{i} & \mathbf{j} & \mathbf{k} \\ 1 & 1 & 0 \\ 2 & 1 & -2 \end{vmatrix} = -2\mathbf{i} + 2\mathbf{j} - \mathbf{k}$,

$\mathbf{B} \times \mathbf{A} = -(\mathbf{A} \times \mathbf{B}) = 2\mathbf{i} - 2\mathbf{j} + \mathbf{k}$,

$|\mathbf{A} \times \mathbf{B}| = \sqrt{4 + 4 + 1} = 3$,

$\theta = \cos^{-1}\left(\frac{\mathbf{A} \cdot \mathbf{B}}{|\mathbf{A}||\mathbf{B}|}\right) = \cos^{-1}\left(\frac{1}{\sqrt{2}}\right) = \frac{\pi}{4}$, $|\mathbf{B}|\cos\theta = \frac{3}{\sqrt{2}}$,

$\text{proj}_{\mathbf{A}}\,\mathbf{B} = \left(\frac{\mathbf{A} \cdot \mathbf{B}}{\mathbf{A} \cdot \mathbf{A}}\right)\mathbf{A} = \frac{3}{2}(\mathbf{i} + \mathbf{j})$

8. $|\mathbf{A}| = \sqrt{1^2 + 1^2 + 2^2} = \sqrt{6}$,

$|\mathbf{B}| = \sqrt{(-1)^2 + (-1)^2} = \sqrt{2}$,

$\mathbf{A} \cdot \mathbf{B} = (1)(-1) + (1)(0) + (2)(-1) = -3$,

$\mathbf{B} \cdot \mathbf{A} = (-1)(1) + (0)(1) + (-1)(2) = -3$

$\mathbf{A} \times \mathbf{B} = \begin{vmatrix} \mathbf{i} & \mathbf{j} & \mathbf{k} \\ 1 & 1 & 2 \\ -1 & 0 & -1 \end{vmatrix} = -\mathbf{i} - \mathbf{j} + \mathbf{k}$,

$\mathbf{B} \times \mathbf{A} = -(\mathbf{A} \times \mathbf{B}) = \mathbf{i} + \mathbf{j} - \mathbf{k}$,

$|\mathbf{A} \times \mathbf{B}| = \sqrt{(-1)^2 + (-1)^2 + 1^2} = \sqrt{3}$,

$\theta = \cos^{-1}\left(\frac{\mathbf{A} \cdot \mathbf{B}}{|\mathbf{A}||\mathbf{B}|}\right) = \cos^{-1}\left(\frac{-3}{\sqrt{6}\sqrt{2}}\right) = \cos^{-1}\left(\frac{-3}{\sqrt{12}}\right)$

$= \cos^{-1}\left(-\frac{\sqrt{3}}{2}\right) = \frac{5\pi}{6}$,

$|\mathbf{B}|\cos\theta = \frac{-3}{\sqrt{6}} = -\sqrt{\frac{9}{6}} = -\sqrt{\frac{3}{2}}$,

$\text{proj}_{\mathbf{A}}\,\mathbf{B} = \left(\frac{\mathbf{A} \cdot \mathbf{B}}{\mathbf{A} \cdot \mathbf{A}}\right)\mathbf{A} = \frac{-3}{6}(\mathbf{i} + \mathbf{j} + 2\mathbf{k}) = -\frac{1}{2}(\mathbf{i} + \mathbf{j} + 2\mathbf{k})$

9. $\mathbf{B} = \left(\frac{\mathbf{A}\cdot\mathbf{B}}{\mathbf{A}\cdot\mathbf{A}}\right)\mathbf{A} + \left[\mathbf{B} - \left(\frac{\mathbf{A}\cdot\mathbf{B}}{\mathbf{A}\cdot\mathbf{A}}\right)\mathbf{A}\right]$

$= \frac{8}{6}(2\mathbf{i} + \mathbf{j} - \mathbf{k}) + \left[(\mathbf{i} + \mathbf{j} - 5\mathbf{k}) - \frac{4}{3}(2\mathbf{i} + \mathbf{j} - \mathbf{k})\right]$

$= \frac{4}{3}(2\mathbf{i} + \mathbf{j} - \mathbf{k}) + \frac{1}{3}(-5\mathbf{i} - \mathbf{j} - 11\mathbf{k}),$

10. $\mathbf{B} = \left(\frac{\mathbf{A}\cdot\mathbf{B}}{\mathbf{A}\cdot\mathbf{A}}\right)\mathbf{A} + \left[\mathbf{B} - \left(\frac{\mathbf{A}\cdot\mathbf{B}}{\mathbf{A}\cdot\mathbf{A}}\right)\mathbf{A}\right]$

$= \frac{-1}{5}(\mathbf{i} - 2\mathbf{j}) + \left[(\mathbf{i} + \mathbf{j} + \mathbf{k}) - \left(\frac{-1}{5}\right)(\mathbf{i} - 2\mathbf{j})\right]$

$= -\frac{1}{5}(\mathbf{i} - 2\mathbf{j}) + \left(\frac{6}{5}\mathbf{i} + \frac{3}{5}\mathbf{j} + \mathbf{k}\right),$

11. $\mathbf{A}\times\mathbf{B} = \begin{vmatrix} \mathbf{i} & \mathbf{j} & \mathbf{k} \\ 1 & 0 & 0 \\ 1 & 1 & 0 \end{vmatrix} = \mathbf{k}$

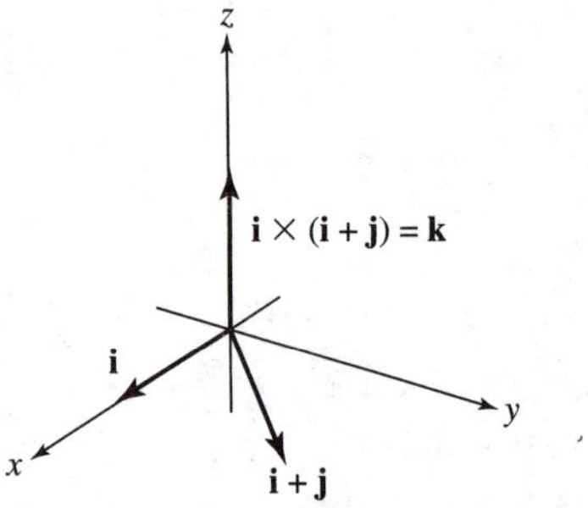

12. $\mathbf{A}\times\mathbf{B} = \begin{vmatrix} \mathbf{i} & \mathbf{j} & \mathbf{k} \\ 1 & -1 & 0 \\ 1 & 1 & 0 \end{vmatrix} = 2\mathbf{k}$

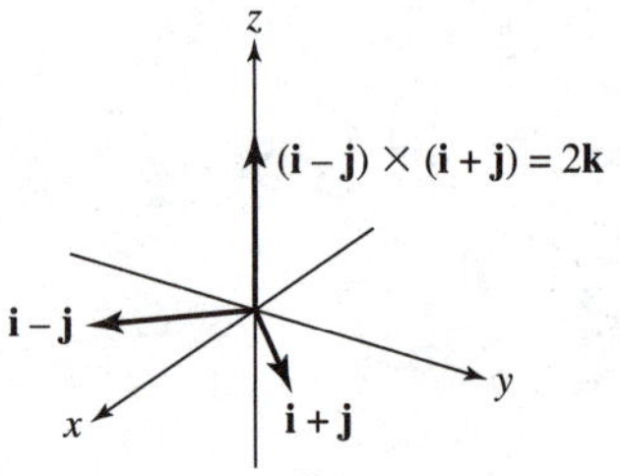

13. $y = \tan x$

$[y']_{\pi/4} = [\sec^2 x]_{\pi/4} = 2 = \frac{2}{1}$

$\mathbf{T} = \mathbf{i} + 2\mathbf{j}$

The unit tangents are $\pm\left(\frac{1}{\sqrt{5}}\mathbf{i} + \frac{2}{\sqrt{5}}\mathbf{j}\right)$ and the unit normals are $\pm\left(-\frac{2}{\sqrt{5}}\mathbf{i} + \frac{1}{\sqrt{5}}\mathbf{j}\right)$.

14. $x^2 + y^2 = 25$

$[y']_{(3,4)} = \left[-\frac{x}{y}\right]_{(3,4)} = \frac{-3}{4}$

$\mathbf{T} = 4\mathbf{i} - 3\mathbf{j}$

The unit tangents are $\pm\frac{1}{5}(4\mathbf{i} - 3\mathbf{j})$ and the unit normals are $\pm\frac{1}{5}(3\mathbf{i} + 4\mathbf{j})$.

15. Let $\mathbf{A} = a_1\mathbf{i} + a_2\mathbf{j} + a_3\mathbf{k}$ and $\mathbf{B} = b_1\mathbf{i} + b_2\mathbf{j} + b_3\mathbf{k}$. Then

$|\mathbf{A} + \mathbf{B}|^2 + |\mathbf{A} - \mathbf{B}|^2$

$= [(a_1 + b_1)^2 + (a_2 + b_2)^2 + (a_3 + b_3)^2]$

$+ [(a_1 - b_1)^2 + (a_2 - b_2)^2 + (a_3 - b_3)^2]$

$= (a_1^2 + 2a_1b_1 + b_1^2 + a_2^2 + 2a_2b_2 + b_2^2 + a_3^2 + 2a_3b_3 + b_3^2) + (a_1^2 - 2a_1b_1 + b_1^2 + a_2^2 - 2a_2b_2 + b_2^2 + a_3^2 - 2a_3b_3 + b_3^2)$

$= 2(a_1^2 + a_2^2 + a_3^2) + 2(b_1^2 + b_2^2 + b_3^2) = 2|\mathbf{A}|^2 + 2|\mathbf{B}|^2$

16. a. area $= \frac{1}{2}|\mathbf{v}|\,h = \frac{1}{2}|\mathbf{v}||\mathbf{u}|\sin(\angle BAC) = \frac{1}{2}|\mathbf{u}\times\mathbf{v}|$ since $0 < \angle BAC < \frac{\pi}{2}$

b. $h = |\mathbf{u}|\sin(\angle BAC) = \frac{|\mathbf{v}||\mathbf{u}|\sin(\angle BAC)}{|\mathbf{v}|} = \frac{|\mathbf{u}\times\mathbf{v}|}{|\mathbf{v}|}$

c. area $= \frac{1}{2}|(\mathbf{i} - \mathbf{j} + \mathbf{k}) \times (2\mathbf{i} + \mathbf{k})|$

$= \frac{1}{2}\begin{vmatrix} \mathbf{i} & \mathbf{j} & \mathbf{k} \\ 1 & -1 & 1 \\ 2 & 0 & 1 \end{vmatrix} = \frac{1}{2}|-\mathbf{i} + \mathbf{j} + 2\mathbf{k}|$

$= \frac{1}{2}\sqrt{(-1)^2 + 1^2 + 2^2} = \frac{1}{2}\sqrt{6} = \frac{\sqrt{6}}{2}$ and

$h = \frac{|(\mathbf{i} - \mathbf{j} + \mathbf{k}) \times (2\mathbf{i} + \mathbf{k})|}{|2\mathbf{i} + \mathbf{k}|} = \frac{|-\mathbf{i} + \mathbf{j} + 2\mathbf{k}|}{|2\mathbf{i} + \mathbf{k}|}$

$= \frac{\sqrt{6}}{\sqrt{2^2 + 1^2}} = \frac{\sqrt{6}}{\sqrt{5}} = \sqrt{\frac{6}{5}}$

17. Let $\mathbf{v} = v_1\mathbf{i} + v_2\mathbf{j} + v_3\mathbf{k}$ and $\mathbf{w} = w_1\mathbf{i} + w_2\mathbf{j} + w_3\mathbf{k}$.

Then $|\mathbf{v} - 2\mathbf{w}|^2$

$= |(v_1\mathbf{i} + v_2\mathbf{j} + v_3\mathbf{k}) - 2(w_1\mathbf{i} + w_2\mathbf{j} + w_3\mathbf{k})|^2$

$= |(v_1 - 2w_1)\mathbf{i} + (v_2 - 2w_2)\mathbf{j} + (v_3 - 2w_3)\mathbf{k}|^2$

$= \left(\sqrt{(v_1 - 2w_1)^2 + (v_2 - 2w_2)^2 + (v_3 - 2w_3)^2}\right)^2$

$= (v_1^2 + v_2^2 + v_3^2) - 4(v_1w_1 + v_2w_2 + v_3w_3) + 4(w_1^2 + w_2^2 + w_3^2)$

$= |\mathbf{v}|^2 - 4\mathbf{v}\cdot\mathbf{w} + 4|\mathbf{w}|^2$

$= |\mathbf{v}|^2 - 4|\mathbf{v}||\mathbf{w}|\cos\theta + 4|\mathbf{w}|^2$

$= 4 - 4(2)(3)\left(\cos\frac{\pi}{3}\right) + 36$

$= 40 - 24\left(\frac{1}{2}\right) = 40 - 12 = 28$

$|\mathbf{v} - 2\mathbf{w}| = \sqrt{28} = 2\sqrt{7}$

18. $\mathbf{u}$ and $\mathbf{v}$ are parallel when $\mathbf{u}\times\mathbf{v} = \mathbf{0}$

$\begin{vmatrix} \mathbf{i} & \mathbf{j} & \mathbf{k} \\ 2 & 4 & -5 \\ -4 & -8 & a \end{vmatrix} = \mathbf{0}$

$(4a - 40)\mathbf{i} + (20 - 2a)\mathbf{j} + (0)\mathbf{k} = \mathbf{0}$

$4a - 40 = 0$ and $20 - 2a = 0$

$a = 10$

19. a. area $= |\mathbf{A}\times\mathbf{B}| = \text{abs}\begin{vmatrix} \mathbf{i} & \mathbf{j} & \mathbf{k} \\ 1 & 1 & -1 \\ 2 & 1 & 1 \end{vmatrix}$

$= |2\mathbf{i} - 3\mathbf{j} - \mathbf{k}| = \sqrt{4 + 9 + 1} = \sqrt{14}$

b. volume $= \mathbf{A}\cdot(\mathbf{B}\times\mathbf{C}) = \begin{vmatrix} 1 & 1 & -1 \\ 2 & 1 & 1 \\ -1 & -2 & 3 \end{vmatrix}$

$$= 1(3 + 2) - 1(6 + 1) - 1(-4 + 1)$$
$$= 5 - 7 + 3 = 1$$

20. a. area $= |\mathbf{A} \times \mathbf{B}| = \text{abs} \begin{vmatrix} \mathbf{i} & \mathbf{j} & \mathbf{k} \\ 1 & 1 & 0 \\ 0 & 1 & 0 \end{vmatrix} = |\mathbf{k}| = 1$

b. volume $= \mathbf{A} \cdot (\mathbf{B} \times \mathbf{C}) = \begin{vmatrix} 1 & 1 & 0 \\ 0 & 1 & 0 \\ 1 & 1 & 1 \end{vmatrix}$

$$= 1(1 - 0) + 1(0 - 0) + 0 = 1$$

21. The desired vector is $\mathbf{n} \times \mathbf{v}$ or $\mathbf{v} \times \mathbf{n}$ since $\mathbf{n} \times \mathbf{v}$ is perpendicular to both $\mathbf{n}$ and $\mathbf{v}$ and, therefore, also parallel to the plane.

22. If $a = 0$ and $b \neq 0$, then the line $by = c$ and $\mathbf{i}$ are parallel.

If $a \neq 0$ and $b = 0$, then the line $ax = c$ and $\mathbf{j}$ are parallel.

If a and b are both $\neq 0$, then $ax + by = c$ contains the points $\left(\frac{c}{a}, 0\right)$ and $\left(0, \frac{c}{b}\right)$.

The vector $ab\left(\frac{c}{a}\mathbf{i} - \frac{c}{b}\mathbf{j}\right) = c(b\mathbf{i} - a\mathbf{j})$ and the line are parallel. Therefore, the vector $b\mathbf{i} - a\mathbf{j}$ is parallel to the line $ax + by = c$ in every case.

23. The line L passes through the point $P(0, 0, -1)$ and is parallel to $\mathbf{v} = -\mathbf{i} + \mathbf{j} + \mathbf{k}$. With $S(2, 2, 0)$, so $\overrightarrow{PS} = 2\mathbf{i} + 2\mathbf{j} + \mathbf{k}$ and

$$\overrightarrow{PS} \times \mathbf{v} = \begin{vmatrix} \mathbf{i} & \mathbf{j} & \mathbf{k} \\ 2 & 2 & 1 \\ -1 & 1 & 1 \end{vmatrix}$$
$$= (2 - 1)\mathbf{i} - (2 + 1)\mathbf{j} + (2 + 2)\mathbf{k}$$
$$= \mathbf{i} - 3\mathbf{j} + 4\mathbf{k},$$

we find the distance

$$d = \frac{|\overrightarrow{PS} \times \mathbf{v}|}{|\mathbf{v}|} = \frac{\sqrt{1 + 9 + 16}}{\sqrt{1 + 1 + 1}} = \frac{\sqrt{26}}{\sqrt{3}} = \frac{\sqrt{78}}{3}$$

24. The line L passes through the point $P(2, 2, 0)$ and is parallel to $\mathbf{v} = \mathbf{i} + \mathbf{j} + \mathbf{k}$. With $S = (0, 4, 1)$, so $\overrightarrow{PS} = -2\mathbf{i} + 2\mathbf{j} + \mathbf{k}$ and

$$\overrightarrow{PS} \times \mathbf{v} = \begin{vmatrix} \mathbf{i} & \mathbf{j} & \mathbf{k} \\ -2 & 2 & 1 \\ 1 & 1 & 1 \end{vmatrix}$$
$$= (2 - 1)\mathbf{i} - (-2 - 1)\mathbf{j} + (-2 - 2)\mathbf{k}$$
$$= \mathbf{i} + 3\mathbf{j} - 4\mathbf{k},$$

we find the distance

$$d = \frac{|\overrightarrow{PS} \times \mathbf{v}|}{|\mathbf{v}|} = \frac{\sqrt{1 + 9 + 16}}{\sqrt{1 + 1 + 1}} = \frac{\sqrt{26}}{\sqrt{3}} = \frac{\sqrt{78}}{3}$$

25. Parametric equations for the line are $x = 1 - 3t$, $y = 2$, $z = 3 + 7t$.

26. The line is parallel to $\overrightarrow{PQ} = 0\mathbf{i} + \mathbf{j} - \mathbf{k}$ and contains the point $P(1, 2, 0)$.
Parametric equations are $x = 1$, $y = 2 + t$, $z = -t$ for $0 \le t \le 1$.

27. The point $P(4, 0, 0)$ lies on the plane $x - y = 4$, and with $S(6, 0, -6)$, $\overrightarrow{PS} = (6 - 4)\mathbf{i} + 0\mathbf{j} + (-6 + 0)\mathbf{k} = 2\mathbf{i} - 6\mathbf{k}$ with $\mathbf{n} = \mathbf{i} - \mathbf{j}$

$$d = \frac{|\mathbf{n} \cdot \overrightarrow{PS}|}{|\mathbf{n}|} = \left|\frac{2 + 0 + 0}{\sqrt{1 + 1 + 0}}\right| = \frac{2}{\sqrt{2}} = \sqrt{2}$$

28. The point $P(0, 0, 2)$ lies on the plane $2x + 3y + z = 2$, and $S(3, 0, 10)$, $\overrightarrow{PS} = (3 - 0)\mathbf{i} + (0 - 0)\mathbf{j} + (10 - 2)\mathbf{k}$ $= 3\mathbf{i} + 8\mathbf{k}$ with $\mathbf{n} = 2\mathbf{i} + 3\mathbf{j} + \mathbf{k}$

$$d = \frac{|\mathbf{n} \cdot \overrightarrow{PS}|}{|\mathbf{n}|} = \left|\frac{6 + 0 + 8}{\sqrt{4 + 9 + 1}}\right| = \frac{14}{\sqrt{14}} = \sqrt{14}$$

29. $P(3, -2, 1)$ and $\mathbf{n} = 2\mathbf{i} + \mathbf{j} + \mathbf{k}$
$(2)(x - 3) + (1)(y - (-2)) + (1)(z - 1) = 0$
$2x + y + z = 5$

30. $P(-1, 6, 0)$ and $\mathbf{n} = \mathbf{i} - 2\mathbf{j} + 3\mathbf{k}$
$(1)(x - (-1)) + (-2)(y - 6) + (3)(z - 0) = 0$
$x - 2y + 3z = -13$

31. $P(1, -1, 2)$, $Q(2, 1, 3)$ and $R(-1, 2, -1)$
$\overrightarrow{PQ} = \mathbf{i} + 2\mathbf{j} + \mathbf{k}$, $\overrightarrow{PR} = -2\mathbf{i} + 3\mathbf{j} - 3\mathbf{k}$ and

$$\overrightarrow{PQ} \times \overrightarrow{PR} = \begin{vmatrix} \mathbf{i} & \mathbf{j} & \mathbf{k} \\ 1 & 2 & 1 \\ -2 & 3 & -3 \end{vmatrix} = -9\mathbf{i} + \mathbf{j} + 7\mathbf{k}$$

is normal to the plane.
$(-9)(x - 1) + (1)(y + 1) + (7)(z - 2) = 0$
$-9x + y + 7z = 4$

32. $P(1, 0, 0)$, $Q(0, 1, 0)$ and $R(0, 0, 1)$
$\overrightarrow{PQ} = -\mathbf{i} + \mathbf{j}$, $\overrightarrow{PR} = -\mathbf{i} + \mathbf{k}$ and

$$\overrightarrow{PQ} \times \overrightarrow{PR} = \begin{vmatrix} \mathbf{i} & \mathbf{j} & \mathbf{k} \\ -1 & 1 & 0 \\ -1 & 0 & 1 \end{vmatrix} = \mathbf{i} + \mathbf{j} + \mathbf{k}$$

is normal to the plane.
$(1)(x - 1) + (1)(y - 0) + (1)(z - 0) = 0$
$x + y + z = 1$

33. yz-plane: $\left(0, -\frac{1}{2}, -\frac{3}{2}\right)$, since $t = -\frac{1}{2}$, $y = -\frac{1}{2}$ and $z = -\frac{3}{2}$ when $x = 0$

xz-plane: $(-1, 0, -3)$, since $t = -1$, $x = -1$ and $z = -3$ when $y = 0$
xy-plane: $(1, -1, 0)$, since $t = 0$, $x = 1$ and $y = -1$ when $z = 0$

34. $x = 2t$, $y = -t$, $z = -t$ represents a line containing the origin and perpendicular to the plane $2x - y - z = 4$
This line intersects the plane $3x - 5y + 2z = 6$ when t is the solution of $3(2t) - 5(-t) + 2(-t) = 6$

$t = \frac{2}{3}$
$\left(\frac{4}{3}, -\frac{2}{3}, -\frac{2}{3}\right)$ is the point of intersection.

35. $\mathbf{n}_1 = \mathbf{i}$ and $\mathbf{n}_2 = \mathbf{i} + \mathbf{j} + \sqrt{2}\mathbf{k}$ are the normals.

The desired angle is $\cos^{-1}\left(\frac{\mathbf{n}_1 \cdot \mathbf{n}_2}{|\mathbf{n}_1||\mathbf{n}_2|}\right) = \cos^{-1}\left(\frac{1}{2}\right) = \frac{\pi}{3}$.

36. $\mathbf{n}_1 = \mathbf{i} + \mathbf{j}$ and $\mathbf{n}_2 = \mathbf{j} + \mathbf{k}$ are the normals.

The desired angle is $\cos^{-1}\left(\frac{\mathbf{n}_1 \cdot \mathbf{n}_2}{|\mathbf{n}_1||\mathbf{n}_2|}\right) = \cos^{-1}\left(\frac{1}{2}\right) = \frac{\pi}{3}$.

37. The direction of the line is $\mathbf{n}_1 \times \mathbf{n}_2 = \begin{vmatrix} \mathbf{i} & \mathbf{j} & \mathbf{k} \\ 1 & 2 & 1 \\ 1 & -1 & 2 \end{vmatrix}$

$$= 5\mathbf{i} - \mathbf{j} - 3\mathbf{k}.$$

Since the point $(-5, 3, 0)$ is on both planes, the desired line is $x = -5 + 5t$, $y = 3 - t$, $z = -3t$.

38. The direction of the line of intersection is

$$\mathbf{n}_1 \times \mathbf{n}_2 = \begin{vmatrix} \mathbf{i} & \mathbf{j} & \mathbf{k} \\ 1 & 2 & -2 \\ 5 & -2 & -1 \end{vmatrix}$$

$= -6\mathbf{i} - 9\mathbf{j} - 12\mathbf{k}$

$= -3(2\mathbf{i} + 3\mathbf{j} + 4\mathbf{k})$ which is the same as the direction of the given line.

39. a. The corresponding normals are $\mathbf{n}_1 = 3\mathbf{i} + 6\mathbf{k}$ and $\mathbf{n}_2 = 2\mathbf{i} + 2\mathbf{j} - \mathbf{k}$ and since $\mathbf{n}_1 \cdot \mathbf{n}_2 = (3)(2) + (0)(2) + (6)(-1) = 6 + 0 - 6 = 0$, the planes are orthogonal.

b. The line of intersection is parallel to

$$\mathbf{n}_1 \times \mathbf{n}_2 = \begin{vmatrix} \mathbf{i} & \mathbf{j} & \mathbf{k} \\ 3 & 0 & 6 \\ 2 & 2 & -1 \end{vmatrix} = -12\mathbf{i} + 15\mathbf{j} + 6\mathbf{k}$$

Now to find a point in the intersection,

$$\text{solve} \begin{cases} 3x + 6z = 1 \\ 2x + 2y - z = 3 \end{cases}$$

$$\begin{cases} 3x + 6z = 1 \\ 12x + 12y - 6z = 18 \end{cases}$$

$15x + 12y = 19$

$x = 0$ and $y = \dfrac{19}{12}$

$\left(0, \dfrac{19}{12}, \dfrac{1}{6}\right)$ is a point on the line we seek. Therefore, the line is $x = -12t$, $y = \dfrac{19}{12} + 15t$ and $z = \dfrac{1}{6} + 6t$.

40. A vector in the direction of the plane's normal is

$$\mathbf{n} = \mathbf{u} \times \mathbf{v} = \begin{vmatrix} \mathbf{i} & \mathbf{j} & \mathbf{k} \\ 2 & 3 & 1 \\ 1 & -1 & 2 \end{vmatrix} = 7\mathbf{i} - 3\mathbf{j} - 5\mathbf{k}$$

and $P(1, 2, 3)$ is on the plane.

$7(x - 1) - 3(y - 2) - 5(z - 3) = 0$

$7x - 3y - 5z = -14$

41. Yes; $\mathbf{v} \cdot \mathbf{n} = (2\mathbf{i} - 4\mathbf{j} + \mathbf{k}) \cdot (2\mathbf{i} + \mathbf{j} + 0\mathbf{k})$

$= 2 \cdot 2 - 4 \cdot 1 + 1 \cdot 0 = 0$

The vector is orthogonal to the plane's normal. $\mathbf{v}$ is parallel to the plane.

42. $\mathbf{n} \cdot \overrightarrow{PP_0} > 0$ represents the half-space of points lying on one side of the plane in the direction of $\mathbf{n}$.

43. A normal to the plane is

$$\mathbf{n} = \overrightarrow{AB} \times \overrightarrow{AC} = \begin{vmatrix} \mathbf{i} & \mathbf{j} & \mathbf{k} \\ 2 & 0 & -1 \\ 2 & -1 & 0 \end{vmatrix} = -\mathbf{i} - 2\mathbf{j} - 2\mathbf{k}$$

The distance is $d = \left|\dfrac{\overrightarrow{AP} \cdot \mathbf{n}}{\mathbf{n}}\right| = \left|\dfrac{(\mathbf{i} + 4\mathbf{j}) \cdot (-\mathbf{i} - 2\mathbf{j} - 2\mathbf{k})}{\sqrt{1 + 4 + 4}}\right|$

$= \left|\dfrac{-1 - 8 + 0}{3}\right| = 3$

44. $P(0, 0, 0)$ lies on the plane $2x + 3y + 5z = 0$, and $\overrightarrow{PS} = 2\mathbf{i} + 2\mathbf{j} + 3\mathbf{k}$ with $\mathbf{n} = 2\mathbf{i} + 3\mathbf{j} + 5\mathbf{k}$.

$$d = \frac{|\mathbf{n} \cdot \overrightarrow{PS}|}{|\mathbf{n}|} = \left|\frac{4 + 6 + 15}{\sqrt{4 + 9 + 25}}\right| = \frac{25}{\sqrt{38}}$$

45. $\mathbf{n} = 2\mathbf{i} - \mathbf{j} - \mathbf{k}$ is normal to the plane.

$$\mathbf{n} \times \mathbf{v} = \begin{vmatrix} \mathbf{i} & \mathbf{j} & \mathbf{k} \\ 2 & -1 & -1 \\ 1 & 1 & 1 \end{vmatrix} = 0\mathbf{i} - 3\mathbf{j} + 3\mathbf{k} = -3\mathbf{j} + 3\mathbf{k}$$

is orthogonal to $\mathbf{v} = \mathbf{i} + \mathbf{j} + \mathbf{k}$ and parallel to the plane.

46. The vector $\mathbf{B} \times \mathbf{C}$ is normal to the plane of $\mathbf{B}$ and $\mathbf{C}$. $\mathbf{A} \times (\mathbf{B} \times \mathbf{C})$ is orthogonal to $\mathbf{A}$ and parallel to the plane:

$$\mathbf{B} \times \mathbf{C} = \begin{vmatrix} \mathbf{i} & \mathbf{j} & \mathbf{k} \\ 1 & 2 & 1 \\ 1 & 1 & -2 \end{vmatrix} = -5\mathbf{i} + 3\mathbf{j} - \mathbf{k}$$

and $\mathbf{A} \times (\mathbf{B} \times \mathbf{C}) = \begin{vmatrix} \mathbf{i} & \mathbf{j} & \mathbf{k} \\ 2 & -1 & 1 \\ -5 & 3 & -1 \end{vmatrix} = -2\mathbf{i} - 3\mathbf{j} + \mathbf{k}$

$|\mathbf{A} \times (\mathbf{B} \times \mathbf{C})| = \sqrt{4 + 9 + 1} = \sqrt{14}$

and $\mathbf{u} = \dfrac{1}{\sqrt{14}}(-2\mathbf{i} - 3\mathbf{j} + \mathbf{k})$ is the desired unit vector.

47. A vector parallel to the line of intersection is

$$\mathbf{v} = \mathbf{n}_1 \times \mathbf{n}_2 = \begin{vmatrix} \mathbf{i} & \mathbf{j} & \mathbf{k} \\ 1 & 2 & 1 \\ 1 & -1 & 2 \end{vmatrix} = 5\mathbf{i} - \mathbf{j} - 3\mathbf{k}$$

$|\mathbf{v}| = \sqrt{25 + 1 + 9} = \sqrt{35}$

$2\left(\dfrac{\mathbf{v}}{|\mathbf{v}|}\right) = \dfrac{2}{\sqrt{35}}(5\mathbf{i} - \mathbf{j} - 3\mathbf{k})$ is the desired vector.

48. The line containing $(0, 0, 0)$ normal to the plane is represented by $x = 2t$, $y = -t$, and $z = -t$. This line intersects the plane $3x - 5y + 2z = 6$ when $3(2t) - 5(-t) + 2(-t) = 6$.

$t = \dfrac{2}{3}$

The point is $\left(\dfrac{4}{3}, -\dfrac{2}{3}, -\dfrac{2}{3}\right)$.

49. The line is represented by $x = 3 + 2t$, $y = 2 - t$, and $z = 1 + 2t$. It meets the plane $2x - y + 2z = -2$ when $2(3 + 2t) - (2 - t) + 2(1 + 2t) = -2$.

$9t + 6 = -2$

$t = -\dfrac{8}{9}$

The point is $\left(\dfrac{11}{9}, \dfrac{26}{9}, -\dfrac{7}{9}\right)$.

50. The direction of the line of intersection is

$$\mathbf{v} = \mathbf{n}_1 \times \mathbf{n}_2 = \begin{vmatrix} \mathbf{i} & \mathbf{j} & \mathbf{k} \\ 2 & 1 & -1 \\ 1 & 1 & 2 \end{vmatrix} = 3\mathbf{i} - 5\mathbf{j} + \mathbf{k}$$

$\theta = \cos^{-1}\left(\dfrac{\mathbf{v} \cdot \mathbf{i}}{|\mathbf{v}||\mathbf{i}|}\right) = \cos^{-1}\left(\dfrac{3}{\sqrt{35}}\right) \approx 59.5°$

51. The intersection occurs when $(3 + 2t) + 3(2t) - t = -4$.

$7t + 3 = -4$

$t = -1$

The point is $(1, -2, -1)$. The required line must be perpendicular to both the given line and to the normal of the plane, and hence is parallel to

$$\begin{vmatrix} \mathbf{i} & \mathbf{j} & \mathbf{k} \\ 2 & 2 & 1 \\ 1 & 3 & -1 \end{vmatrix} = -5\mathbf{i} + 3\mathbf{j} + 4\mathbf{k}.$$

The line is represented by $x = 1 - 5t$, $y = -2 + 3t$, and $z = -1 + 4t$.

52. If $P(a, b, c)$ is a point on the line of intersection, then P lies in both planes so $a - 2b + c + 3 = 0$ and $2a - b - c + 1 = 0$, hence $(a - 2b + c + 3) + k(2a - b - c + 1) = 0$ and the point is on the given plane.

53. The vector $\overrightarrow{AB} \times \overrightarrow{CD} = \begin{vmatrix} \mathbf{i} & \mathbf{j} & \mathbf{k} \\ 3 & -2 & 4 \\ \frac{26}{5} & 0 & -\frac{26}{5} \end{vmatrix}$

$= \frac{26}{5}(2\mathbf{i} + 7\mathbf{j} + 2\mathbf{k})$

is normal to the plane and $\mathbf{A}(-2, 0, -3)$ lies on the plane.
$2(x + 2) + 7(y - 0) + 2(z - (-3)) = 0$
$2x + 7y + 2z + 10 = 0$ is an equation of the plane.

54. Yes; the line's direction vector is $2\mathbf{i} + 3\mathbf{j} - 5\mathbf{k}$ which is parallel to the line and also parallel to the normal $-4\mathbf{i} - 6\mathbf{j} + 10\mathbf{k}$ to the plane.
The line is orthogonal to the plane.

55. The vector $\overrightarrow{PQ} \times \overrightarrow{PR} = \begin{vmatrix} \mathbf{i} & \mathbf{j} & \mathbf{k} \\ 2 & -1 & 3 \\ -3 & 0 & 1 \end{vmatrix}$

$= -\mathbf{i} - 11\mathbf{j} - 3\mathbf{k}$
is normal to the plane.

a. No, the plane is not orthogonal to $\overrightarrow{PQ} \times \overrightarrow{PR}$.

b. No, these equations represent a line, not a plane.

c. No, the plane $(x + 2) + 11(y - 1) - 3z = 0$ has normal $\mathbf{i} + 11\mathbf{j} - 3\mathbf{k}$ which is not parallel to $\overrightarrow{PQ} \times \overrightarrow{PR}$.

d. No, this vector equation is equivalent to the equations $3y + 3z = 3$, $3x - 2z = -6$, and $3x + 2y = -4$, or $x = -\frac{4}{3} - \frac{2}{3}t$, $y = t$, $z = 1 - t$, which represents a line, not a plane.

e. Yes, this is a plane containing the point $R(-2, 1, 0)$ with normal $\overrightarrow{PQ} \times \overrightarrow{PR}$.

56. a. $\overrightarrow{\mathbf{BA}} = \mathbf{i} - \mathbf{j} + 5\mathbf{k}$

The line through A and B is $x = 1 + t$, $y = -t$, $z = -1 + 5t$.

The line through C and D must be parallel and is L_1: $x = 1 + t$, $y = 2 - t$, $z = 3 + 5t$.

$\overrightarrow{\mathbf{BC}} = 2\mathbf{j} + 4\mathbf{k}$ The line through B and C is $x = 1$, $y = 2 + 2s$, $z = 3 + 4s$.

The line through A and D must be parallel and is L_2: $x = 2$, $y = -1 + 2s$, $z = 4 + 4s$. The lines L_1 and L_2 intersect at $D(2, 1, 8)$ where $t = 1$ and $s = 1$.

b. $\cos\theta = \dfrac{(2\mathbf{j} + 4\mathbf{k}) \cdot (\mathbf{i} - \mathbf{j} + 5\mathbf{k})}{\sqrt{20}\sqrt{27}} = \dfrac{18}{(2\sqrt{5})(3\sqrt{3})} = \dfrac{3}{\sqrt{15}}$

c. $\left(\dfrac{\overrightarrow{BA} \cdot \overrightarrow{BC}}{\overrightarrow{BC} \cdot \overrightarrow{BC}}\right)\overrightarrow{BC} = \dfrac{18}{20}\overrightarrow{BC} = \dfrac{9}{10}(2\mathbf{j} + 4\mathbf{k}) = \dfrac{9}{5}(\mathbf{j} + 2\mathbf{k})$

d. area $= |(2\mathbf{j} + 4\mathbf{k}) \times (\mathbf{i} - \mathbf{j} + 5\mathbf{k})| = |14\mathbf{i} + 4\mathbf{j} - 2\mathbf{k}|$
$= 6\sqrt{6}$

e. From part (d), $\mathbf{n} = 14\mathbf{i} + 4\mathbf{j} - 2\mathbf{k}$ is normal to the plane. Using the point $B(1, 0, -1)$ we get
$14(x - 1) + 4(y - 0) - 2(z + 1) = 0$
$14x + 4y - 2z = 16$ or $7x + 2y - z = 8$

f. From part (d), $\mathbf{n} = 14\mathbf{i} + 4\mathbf{j} - 2\mathbf{k}$.
The area of the projection on the yz-plane is $|\mathbf{n} \cdot \mathbf{i}| = 14$. The area of the projection on the xz-plane is $|\mathbf{n} \cdot \mathbf{j}| = 4$, and the area of the projection on the xy-plane is $|\mathbf{n} \cdot \mathbf{k}| = 2$.

57. $\overrightarrow{AB} = -2\mathbf{i} + \mathbf{j} + \mathbf{k}$, $\overrightarrow{CD} = \mathbf{i} + 4\mathbf{j} - \mathbf{k}$, and $\overrightarrow{AC} = 2\mathbf{i} + \mathbf{j}$

$\mathbf{n} = \begin{vmatrix} \mathbf{i} & \mathbf{j} & \mathbf{k} \\ -2 & 1 & 1 \\ 1 & 4 & -1 \end{vmatrix} = -5\mathbf{i} - \mathbf{j} - 9\mathbf{k}$

The distance is $d = \left|\dfrac{(2\mathbf{i} + \mathbf{j}) \cdot (-5\mathbf{i} - \mathbf{j} - 9\mathbf{k})}{\sqrt{25 + 1 + 81}}\right| = \dfrac{11}{\sqrt{107}}$.

58. $\overrightarrow{AB} = -2\mathbf{i} + 4\mathbf{j} - \mathbf{k}$, $\overrightarrow{CD} = \mathbf{i} - \mathbf{j} + 2\mathbf{k}$, and $\overrightarrow{AC} = -3\mathbf{i} + 3\mathbf{j}$

$\mathbf{n} = \begin{vmatrix} \mathbf{i} & \mathbf{j} & \mathbf{k} \\ -2 & 4 & -1 \\ 1 & -1 & 2 \end{vmatrix} = 7\mathbf{i} + 3\mathbf{j} - 2\mathbf{k}$

The distance is $d = \left|\dfrac{(-3\mathbf{i} + 3\mathbf{j}) \cdot (7\mathbf{i} + 3\mathbf{j} - 2\mathbf{k})}{\sqrt{49 + 9 + 4}}\right| = \dfrac{12}{\sqrt{62}}$

59. The y-axis in the xy-plane
The yz-plane in three-dimensional space

60. The line $x + y = 1$ in the xy-plane
The plane $x + y = 1$ in three-dimensional space

61. $2x^2 + 2y^2 + 2z^2 - 12x + 4y - 8z + 10 = 0$
$x^2 + y^2 + z^2 - 6x + 2y - 4z + 5 = 0$
$(x^2 - 6x) + (y^2 + 2y) + (z^2 - 4z) = -5$
$(x^2 - 6x + 9) + (y^2 + 2y + 1) + (z^2 - 4z + 4)$
$= -5 + 9 + 1 + 4$
$(x - 3)^2 + (y + 1)^2 + (z - 2)^2 = 9$
Center: $(3, -1, 2)$; radius: 3

62. Work $= \mathbf{F} \cdot \overrightarrow{PQ} = |\mathbf{F}||\overrightarrow{PQ}|\cos\theta = |160||250|\cos\frac{\pi}{6}$
$= (40{,}000)\left(\frac{\sqrt{3}}{2}\right) \approx 34{,}641$ J

63. Torque $= |\overrightarrow{PQ} \times \mathbf{F}|$

15 ft-lb $= |\overrightarrow{PQ}||\mathbf{F}|\sin\frac{\pi}{2} = \frac{3}{4}$ ft $\cdot |\mathbf{F}|$

$|\mathbf{F}| = 20$ lb

64. Extend $\overrightarrow{CD}$ to $\overrightarrow{CG}$ so that $\overrightarrow{CD} = \overrightarrow{DG}$. Then $\overrightarrow{CG} = t\,\overrightarrow{CF} = \overrightarrow{CB} + \overrightarrow{BG}$ and $t\,\overrightarrow{CF} = 3\,\overrightarrow{CE} + \overrightarrow{CA}$, since $ACBG$ is a parallelogram. Since F is on line segment AE, $\overrightarrow{AF} = s\overrightarrow{AE} = s(\overrightarrow{CE} - \overrightarrow{CA})$.
$\overrightarrow{CF} = \overrightarrow{CA} + \overrightarrow{AF} = \overrightarrow{CA} + s(\overrightarrow{CE} - \overrightarrow{CA})$
$= (1 - s)\overrightarrow{CA} + s\overrightarrow{CE}$
Thus, $t\overrightarrow{CF} = t(1 - s)\overrightarrow{CA} + st\overrightarrow{CE}$ and equating the expressions for $t\overrightarrow{CF}$, we get
$3\overrightarrow{CE} + \overrightarrow{CA} = t(1 - s)\overrightarrow{CA} + st\overrightarrow{CE}$
$(3 - st)\overrightarrow{CE} = (t - st - 1)\overrightarrow{CA}$
Since $\overrightarrow{CE}$ and $\overrightarrow{CA}$ have different directions, the only way that scalar multiples can be equal is if the scalars are both 0.
$3 - st = 0$
$3 = st$

$t - st - 1 = 0$
$t - 3 - 1 = 0$
$t = 4$

Then $\overrightarrow{CG} = 4\overrightarrow{CF}$ and since $\overrightarrow{CG} = 2\overrightarrow{CD}$, $2\overrightarrow{CD} = 4\overrightarrow{CF}$ or $\overrightarrow{CD} = 2\overrightarrow{CF}$ so F is the midpoint of line segment CD.

65. a. If $P(x, y, z)$ is a point in the plane determined by the three points $P_1(x_1, y_1, z_1)$, $P_2(x_2, y_2, z_2)$, and $P_3(x_3, y_3, z_3)$, then the vectors $\overrightarrow{PP_1}$, $\overrightarrow{PP_2}$ and $\overrightarrow{PP_3}$ all lie in the plane. Thus $\overrightarrow{PP_1} \cdot (\overrightarrow{PP_2} \times \overrightarrow{PP_3}) = 0$. Thus,

$$\begin{vmatrix} x_1 - x & y_1 - y & z_1 - z \\ x_2 - x & y_2 - y & z_2 - z \\ x_3 - x & y_3 - y & z_3 - z \end{vmatrix} = 0$$

by the determinant formula for the triple scalar product in Section 11.3.

b. Subtract row 1 from rows 2, 3, and 4, and evaluate the resulting determinant (which has the same value as the given determinant) by cofactor expansion about column 4. This expansion is exactly the determinant in part **a.** so we have all points $P(x, y, z)$ in the plane determined by $P_1(x_1, y_1, z_1)$, $P_2(x_2, y_2, z_2)$, and $P_3(x_3, y_3, z_3)$.

66. If $Q(x, y)$ is a point on the line $ax + by = c$, then $\overrightarrow{QP_1} = (x_1 - x)\mathbf{i} + (y_1 - y)\mathbf{j}$, and $\mathbf{n} = a\mathbf{i} + b\mathbf{j}$ is normal to the line. The distance is

$$|\text{proj}_{\mathbf{n}} \overrightarrow{P_1Q}| = \left| \frac{[(x_1 - x)\mathbf{i} + (y_1 - y)\mathbf{j}] \cdot (a\mathbf{i} + b\mathbf{j})}{\sqrt{a^2 + b^2}} \right|$$

$$= \frac{|a(x_1 - x) + b(y_1 - y)|}{\sqrt{a^2 + b^2}}$$

$$= \frac{|ax_1 + by_1 - c|}{\sqrt{a^2 + b^2}}, \quad \text{since } c = ax + by.$$

	Rectangular	Cylindrical
67.	$(1, 0, 0)$	$(1, 0, 0)$
68.	$(0, 1, 0)$	$\left(1, \frac{\pi}{2}, 0\right)$
69.	$(-1, 0, -1)$	$(1, \pi, -1)$
70.	$(0, -1, 1)$	$\left(1, \frac{3\pi}{2}, 1\right)$

71. $z = 2$
cylindrical: $z = 2$
a plane parallel to the xy-plane

72. $x^2 + y^2 + (z - 3)^2 = 9$
cylindrical: $r^2 + (z - 3)^2 = 9$
$r^2 + z^2 - 6z + 9 = 9$
$r^2 + z^2 = 6z$
a sphere of radius 3 centered at $(0, 0, 3)$ (rectangular)

73. $r = 7 \sin \theta$
rectangular: $r = 7 \sin \theta$

$r = 7\left(\frac{y}{r}\right)$

$r^2 = 7y$

$x^2 + y^2 - 7y = 0$

$x^2 + y^2 - 7y + \frac{49}{4} = \frac{49}{4}$

$x^2 + \left(y - \frac{7}{2}\right)^2 = \frac{49}{4}$

a circular cylinder parallel to the z-axis generated by the circle $x^2 + \left(y - \frac{7}{2}\right)^2 = \frac{49}{4}$.

74. $r = 4 \cos \theta$
rectangular: $r = 4 \cos \theta$

$r = 4\left(\frac{x}{r}\right)$

$r^2 = 4x$

$x^2 - 4x + y^2 = 0$

$x^2 - 4x + 4 + y^2 = 4$

$(x - 2)^2 + y^2 = 2^2$

a circular cylinder parallel to the z-axis generated by the circle $(x - 2)^2 + y^2 = 4$.

75. $x^2 + y^2 + z^2 = 4$

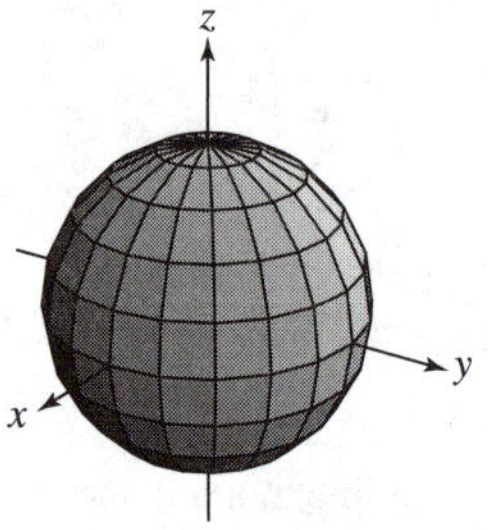

76. $x^2 + (y - 1)^2 + z^2 = 1$

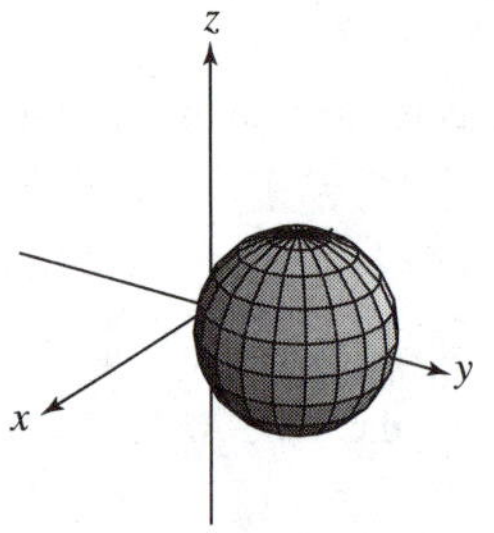

77. $x^2 + z^2 = y^2$

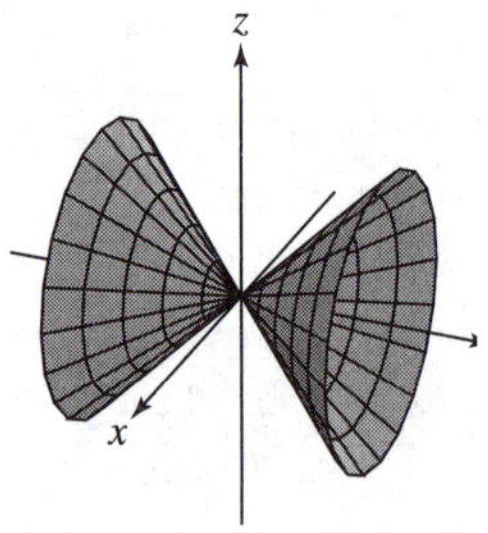

78. $4y^2 + z^2 - 4x^2 = 4$

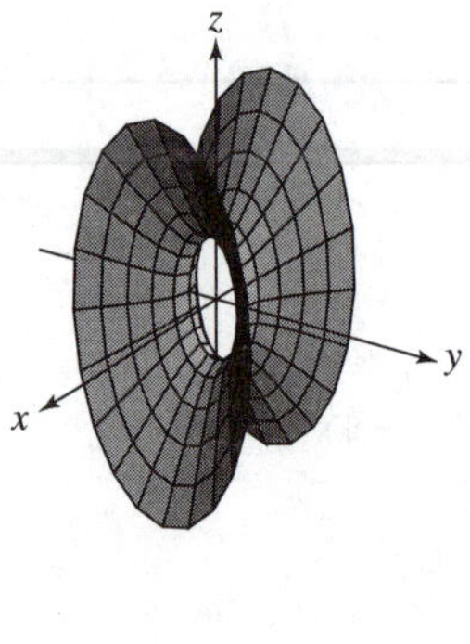

79. $y^2 - x^2 - z^2 = 1$

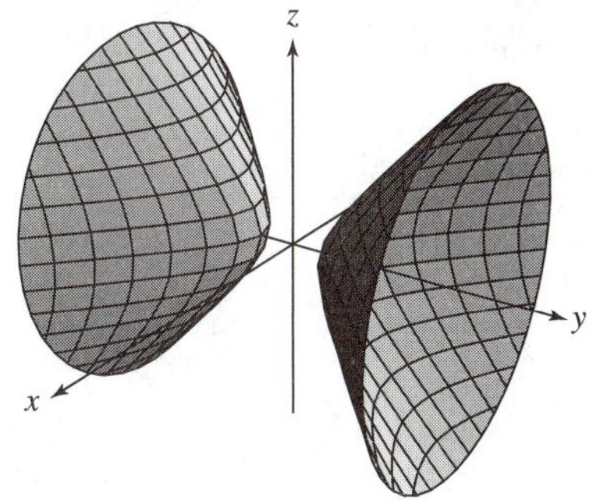

80. $z = x^2 + y^2 + 1$

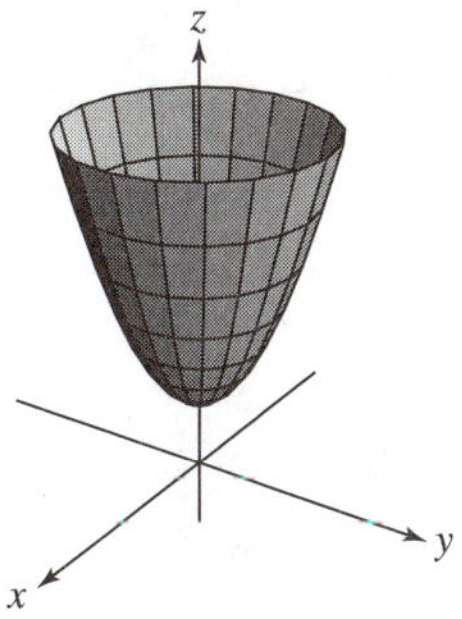

81. $z^2 - \frac{x^2}{4} - y^2 = 1$

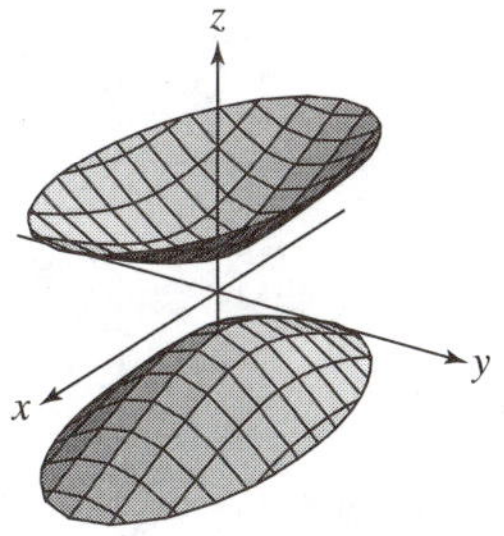

82. $9x^2 + 4y^2 + z^2 = 36$

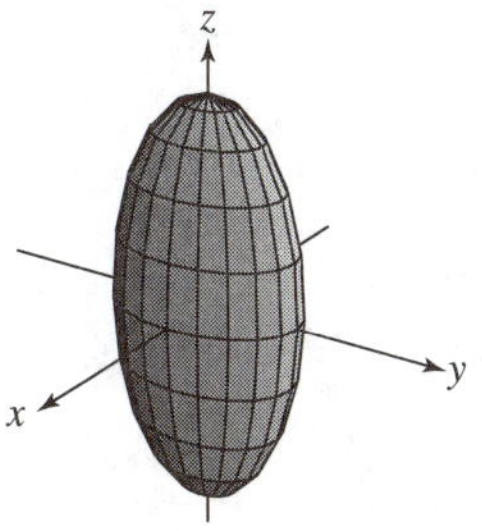

83. $4x^2 + 9z^2 = y^2$

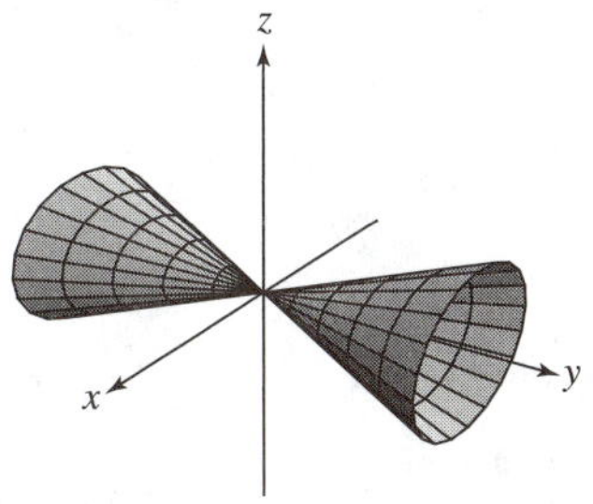

84. $z = 4x^2 + y^2 - 4$

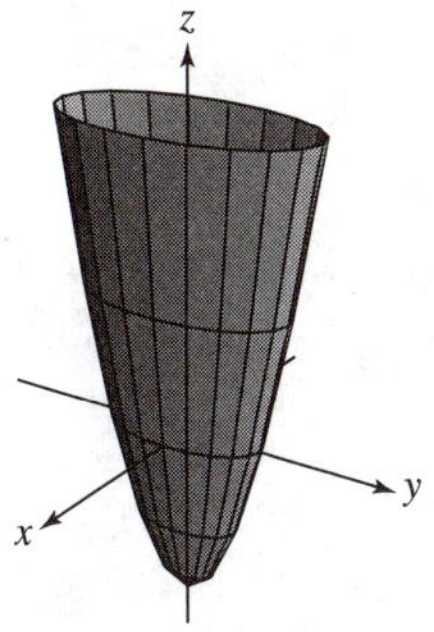

Chapter 12
Vector-Valued Functions and Motion in Space

Chapter Opener Solution

$\mathbf{r}(t) = (t - 30\cos t)\mathbf{i} + 2t\mathbf{j} + (-45\sin t)\mathbf{k}$

$\mathbf{v}(t) = (1 + 30\sin t)\mathbf{i} + 2\mathbf{j} - (45\cos t)\mathbf{k}$

$$\begin{aligned}|\mathbf{v}(t)| &= \sqrt{(1 + 30\sin t)^2 + 2^2 + (-45\cos t)^2}\\ &= \sqrt{1 + 60\sin t + 900\sin^2 t + 4 + 2025\cos^2 t}\\ &= \sqrt{1125\cos^2 t + 60\sin t + 905}\end{aligned}$$

$$\begin{aligned}\text{Length} &= \int_{\pi/2}^{5\pi/2} |\mathbf{v}(t)|\,dt\\ &= \int_{\pi/2}^{5\pi/2} \sqrt{1125\cos^2 t + 60\sin t + 905}\,dt\\ &\approx 238.37 \text{ ft (using NINT)}\end{aligned}$$

■ Section 12.1 Vector-Valued Functions and Space Curves (pp.631-644)

Exploration 1: How Vector-Valued Functions in Space Ought to Work

1. Since $\lim_{t\to 0}(t^2 + 3) = 3$, $\lim_{t\to 0}(\cos t) = 1$, and $\lim_{t\to 0}\frac{\sin t}{t} = 1$, the limit is $3\mathbf{i} + \mathbf{j} + \mathbf{k}$.

2. It is continuous at $t = 0$ because the components $x(t) = e^{3t+1}$, $y(t) = |t|$, and $z(t) = \tan^2 t$ are all continuous at $t = 0$. It is not differentiable there, because $y(t) = |t|$ is not differentiable at $t = 0$.

3. $$\begin{aligned}\frac{d\mathbf{r}}{dt} &= \left(\frac{d}{dt}t^3\right)\mathbf{i} + \left(\frac{d}{dt}\ln(t+3)\right)\mathbf{j} + \left(\frac{d}{dt}\tan t\right)\mathbf{k}\\ &= 3t^2\mathbf{i} + \frac{1}{t+3}\mathbf{j} + (\sec^2 t)\mathbf{k}\end{aligned}$$

4. $$\begin{aligned}&\int_0^4 [(t^2)\mathbf{i} + (\sin \pi t)\mathbf{j} + (\sqrt{t})\mathbf{k}]\,dt\\ &= \frac{t^3}{3}\Big]_0^4\mathbf{i} - \frac{\cos \pi t}{\pi}\Big]_0^4\mathbf{j} + \frac{2}{3}t^{3/2}\Big]_0^4\mathbf{k}\\ &= \frac{64}{3}\mathbf{i} - \left(\frac{1}{\pi} - \frac{1}{\pi}\right)\mathbf{j} + \frac{2}{3}(8)\mathbf{k}\\ &= \frac{64}{3}\mathbf{i} + \frac{16}{3}\mathbf{k}\end{aligned}$$

5. $\mathbf{v}(t) = \frac{d}{dt}\mathbf{r}(t) = 6t^2\mathbf{i} + (\cos t)\mathbf{j} + 2e^{2t}\mathbf{k}$

6. $\mathbf{a}(t) = \frac{d}{dt}\mathbf{v}(t) = 12t\mathbf{i} - (\sin t)\mathbf{j} + 4e^{2t}\mathbf{k}$

7. $\mathbf{v}(0) = 0\mathbf{i} + 1\mathbf{j} + 2\mathbf{k} = \mathbf{j} + 2\mathbf{k}$
   Speed $= \sqrt{0^2 + 1^2 + 2^2} = \sqrt{5}$
   Direction $= \frac{\mathbf{v}(0)}{|\mathbf{v}(0)|} = \frac{\mathbf{j} + 2\mathbf{k}}{\sqrt{5}} = \frac{1}{\sqrt{5}}\mathbf{j} + \frac{2}{\sqrt{5}}\mathbf{k}$

8. $\frac{d}{dt}\mathbf{u}\cdot\mathbf{v} = \frac{d\mathbf{u}}{dt}\cdot\mathbf{v} + \mathbf{u}\cdot\frac{d\mathbf{v}}{dt}$

9. $\frac{d\mathbf{r}}{ds} = \frac{d\mathbf{r}}{dt}\frac{dt}{ds}$

10. $\frac{d}{dt}(\mathbf{u}\times\mathbf{v}) = \frac{d\mathbf{u}}{dt}\times\mathbf{v} + \mathbf{u}\times\frac{d\mathbf{v}}{dt}$

Quick Review 12.1

1. Since int $x = -3$ for $-3 \le x < -2$, $\lim_{x\to -2^-}$ int $x = -3$.
2. Since int $x = -2$ for $-2 \le x < -1$, $\lim_{x\to -2^+}$ int $x = -2$.
3. The function $f(x) = \text{int } x$ is continuous at all noninteger values of x.
4. Every integer is a point of discontinuity of $f(x) = \text{int } x$.

5. $f(x) = \frac{\cot x}{x}$

   $f'(x) = \frac{(x)(-\csc^2 x) - (1)(\cot x)}{x^2} = \frac{-x\csc^2 x - \cot x}{x^2}$

   Slope $= f'\left(\frac{\pi}{4}\right)$

   $= \frac{-\frac{\pi}{4}(2) - 1}{\left(\frac{\pi}{4}\right)^2} = -\frac{8(\pi + 2)}{\pi^2}$

   $y - y_1 = m(x - x_1)$ where $x_1 = \frac{\pi}{4}$ and $y_1 = f\left(\frac{\pi}{4}\right) = \frac{4}{\pi}$

   $y - \frac{4}{\pi} = -\frac{8(\pi + 2)}{\pi^2}\left(x - \frac{\pi}{4}\right)$

6. Slope $= -\frac{1}{f'\left(\frac{\pi}{4}\right)} = \frac{\pi^2}{8(\pi + 2)}$

   $y - y_1 = m(x - x_1)$ where $x_1 = \frac{\pi}{4}$ and $y_1 = f\left(\frac{\pi}{4}\right) = \frac{4}{\pi}$

   $y - \frac{4}{\pi} = \frac{\pi^2}{8(\pi + 2)}\left(x - \frac{\pi}{4}\right)$

7. $y = \frac{2x + 1}{x + 3}$

   $\frac{dy}{dx} = \frac{(x + 3)(2) - (2x + 1)(1)}{(x + 3)^2} = \frac{5}{(x + 3)^2} = 5(x + 3)^{-2}$

   $\frac{d^2y}{dx^2} = -10(x + 3)^{-3} = -\frac{10}{(x + 3)^3}$

8. $\int \frac{dx}{1 + x} = \ln|x + 1| + C$

9. $\int_0^2 \frac{dx}{1 + x} = \ln|x + 1|\Big]_0^2 = \ln 3 - \ln 1 = \ln 3$

10. $\frac{dy}{dx} = \cos x + \sin x$

    $\int dy = \int (\cos x + \sin x)\, dx$

    $y = \sin x - \cos x + C$

    Find C: $1 = \sin\frac{\pi}{2} - \cos\frac{\pi}{2} + C$

    $1 = 1 + C$

    $C = 0$

    $y = \sin x - \cos x$

Section 12.1 Exercises

1. $x = t + 1$ and $y = t^2 - 1$

   $y = (x - 1)^2 - 1 = x^2 - 2x$

   $\mathbf{v} = \frac{d\mathbf{r}}{dt} = \mathbf{i} + 2t\mathbf{j}$

   $\mathbf{a} = \frac{d\mathbf{v}}{dt} = 2\mathbf{j}$

   $\mathbf{v} = \mathbf{i} + 2\mathbf{j}$ and $\mathbf{a} = 2\mathbf{j}$ at $t = 1$

3. $x = e^t$ and $y = \frac{2}{9}e^{2t}$

   $y = \frac{2}{9}x^2$

   $\mathbf{v} = \frac{d\mathbf{r}}{dt} = e^t\mathbf{i} + \frac{4}{9}e^{2t}\mathbf{j}$

   $\mathbf{a} = e^t\mathbf{i} + \frac{8}{9}e^{2t}\mathbf{j}$

   $\mathbf{v} = 3\mathbf{i} + 4\mathbf{j}$ and $\mathbf{a} = 3\mathbf{i} + 8\mathbf{j}$ at $t = \ln 3$

5. $\mathbf{v} = \frac{d\mathbf{r}}{dt} = (\cos t)\mathbf{i} - (\sin t)\mathbf{j}$

   and $\mathbf{a} = \frac{d\mathbf{v}}{dt} = -(\sin t)\mathbf{i} - (\cos t)\mathbf{j}$

   for $t = \frac{\pi}{4}$, $\mathbf{v}\left(\frac{\pi}{4}\right) = \frac{\sqrt{2}}{2}\mathbf{i} - \frac{\sqrt{2}}{2}\mathbf{j}$ and $\mathbf{a}\left(\frac{\pi}{4}\right) = -\frac{\sqrt{2}}{2}\mathbf{i} - \frac{\sqrt{2}}{2}\mathbf{j}$

   for $t = \frac{\pi}{2}$, $\mathbf{v}\left(\frac{\pi}{2}\right) = -\mathbf{j}$ and $\mathbf{a}\left(\frac{\pi}{2}\right) = -\mathbf{i}$

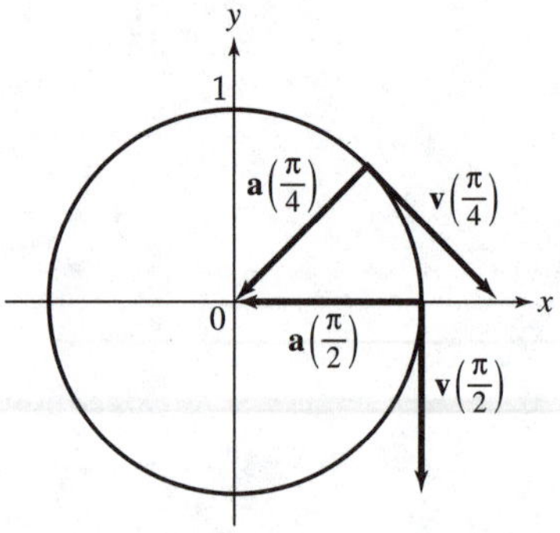

7. $\mathbf{v} = \frac{d\mathbf{r}}{dt} = (1 - \cos t)\mathbf{i} + (\sin t)\mathbf{j}$

   and $\mathbf{a} = \frac{d\mathbf{v}}{dt} = (\sin t)\mathbf{i} + (\cos t)\mathbf{j}$

   for $t = \pi$, $\mathbf{v}(\pi) = 2\mathbf{i}$ and $\mathbf{a}(\pi) = -\mathbf{j}$

   for $t = \frac{3\pi}{2}$, $\mathbf{v}\left(\frac{3\pi}{2}\right) = \mathbf{i} - \mathbf{j}$ and $\mathbf{a}\left(\frac{3\pi}{2}\right) = -\mathbf{i}$

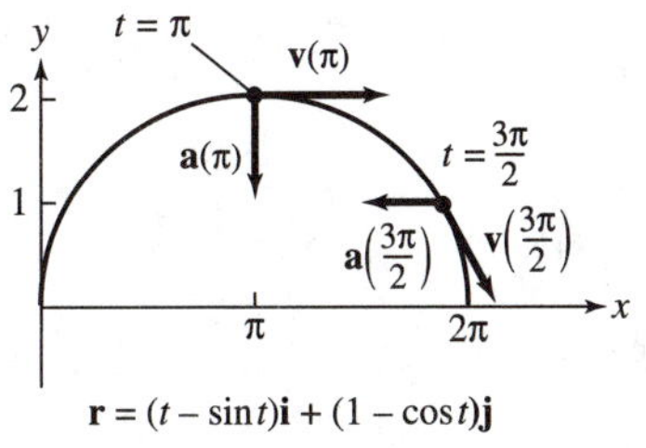

9\. a. $\mathbf{r} = (t+1)\mathbf{i} + (t^2-1)\mathbf{j} + 2t\mathbf{k}$

$\mathbf{v} = \frac{d\mathbf{r}}{dt} = \mathbf{i} + 2t\mathbf{j} + 2\mathbf{k}$

$\mathbf{a} = \frac{d^2\mathbf{r}}{dt^2} = 2\mathbf{j}$

b. Speed: $|\mathbf{v}(1)| = \sqrt{1^2 + (2(1))^2 + 2^2} = 3$

Direction: $\frac{\mathbf{v}(1)}{|\mathbf{v}(1)|} = \frac{\mathbf{i} + 2(1)\mathbf{j} + 2\mathbf{k}}{3} = \frac{1}{3}\mathbf{i} + \frac{2}{3}\mathbf{j} + \frac{2}{3}\mathbf{k}$

c. $\mathbf{v}(1) = 3\left(\frac{1}{3}\mathbf{i} + \frac{2}{3}\mathbf{j} + \frac{2}{3}\mathbf{k}\right)$

11\. a. $\mathbf{r} = (\sec t)\mathbf{i} + (\tan t)\mathbf{j} + \frac{4}{3}t\mathbf{k}$

$\mathbf{v} = \frac{d\mathbf{r}}{dt} = (\sec t \tan t)\mathbf{i} + (\sec^2 t)\mathbf{j} + \frac{4}{3}\mathbf{k}$

$\mathbf{a} = \frac{d^2\mathbf{r}}{dt^2} = (\sec t \tan^2 t + \sec^3 t)\mathbf{i} + (2\sec^2 t \tan t)\mathbf{j}$

b. Speed: $\left|\mathbf{v}\left(\frac{\pi}{6}\right)\right| = \sqrt{\left(\sec\frac{\pi}{6}\tan\frac{\pi}{6}\right)^2 + \left(\sec^2\frac{\pi}{6}\right)^2 + \left(\frac{4}{3}\right)^2}$
$= 2$

Direction: $\frac{\mathbf{v}\left(\frac{\pi}{6}\right)}{\left|\mathbf{v}\left(\frac{\pi}{6}\right)\right|} = \frac{\left(\sec\frac{\pi}{6}\tan\frac{\pi}{6}\right)\mathbf{i} + \left(\sec^2\frac{\pi}{6}\right)\mathbf{j} + \frac{4}{3}\mathbf{k}}{2}$
$= \frac{1}{3}\mathbf{i} + \frac{2}{3}\mathbf{j} + \frac{2}{3}\mathbf{k}$

c. $\mathbf{v}\left(\frac{\pi}{6}\right) = 2\left(\frac{1}{3}\mathbf{i} + \frac{2}{3}\mathbf{j} + \frac{2}{3}\mathbf{k}\right)$

13\. $\mathbf{v} = 3\mathbf{i} + \sqrt{3}\mathbf{j} + 2t\mathbf{k}$ and $\mathbf{a} = 2\mathbf{k}$
$\mathbf{v}(0) = 3\mathbf{i} + \sqrt{3}\mathbf{j}$ and $\mathbf{a}(0) = 2\mathbf{k}$
$|\mathbf{v}(0)| = \sqrt{3^2 + (\sqrt{3})^2 + 0^2} = \sqrt{12}$
and $|\mathbf{a}(0)| = \sqrt{2^2} = 2$
$\mathbf{v}(0) \cdot \mathbf{a}(0) = 0$
$\cos\theta = 0$
$\theta = \frac{\pi}{2}$

15\. $\mathbf{v} = \left(\frac{2t}{t^2+1}\right)\mathbf{i} + \left(\frac{1}{t^2+1}\right)\mathbf{j} + t(t^2+1)^{-1/2}\mathbf{k}$
and $\mathbf{a} = \left[\frac{-2t^2+2}{(t^2+1)^2}\right]\mathbf{i} - \left[\frac{2t}{(t^2+1)^2}\right]\mathbf{j} + \left[\frac{1}{(t^2+1)^{3/2}}\right]\mathbf{k}$

$\mathbf{v}(0) = \mathbf{j}$ and $\mathbf{a}(0) = 2\mathbf{i} + \mathbf{k}$
$|\mathbf{v}(0)| = 1$ and $|\mathbf{a}(0)| = \sqrt{2^2 + 1^2} = \sqrt{5}$
$\mathbf{v}(0) \cdot \mathbf{a}(0) = 0$
$\cos\theta = 0$
$\theta = \frac{\pi}{2}$

17\. $\mathbf{v} = (1 - \cos t)\mathbf{i} + (\sin t)\mathbf{j}$ and $\mathbf{a} = (\sin t)\mathbf{i} + (\cos t)\mathbf{j}$
$\mathbf{v} \cdot \mathbf{a} = (\sin t)(1 - \cos t) + (\sin t)(\cos t) = \sin t.$
Thus, $\mathbf{v} \cdot \mathbf{a} = 0$ when $\sin t = 0$
$t = 0, \pi$, or 2π

19\. $\int_0^1 [t^3\mathbf{i} + 7\mathbf{j} + (t+1)\mathbf{k}]\,dt = \left[\frac{t^4}{4}\right]_0^1\mathbf{i} + [7t]_0^1\mathbf{j} + \left[\frac{t^2}{2} + t\right]_0^1\mathbf{k}$
$= \frac{1}{4}\mathbf{i} + 7\mathbf{j} + \frac{3}{2}\mathbf{k}$

21\. $\int_1^4 \left(\frac{1}{t}\mathbf{i} + \frac{1}{5-t}\mathbf{j} + \frac{1}{2t}\mathbf{k}\right)dt$
$= [\ln t]_1^4\,\mathbf{i} + [-\ln(5-t)]_1^4\,\mathbf{j} + \left[\frac{1}{2}\ln t\right]_1^4\mathbf{k}$
$= (\ln 4)\mathbf{i} + (\ln 4)\mathbf{j} + (\ln 2)\mathbf{k}$

23\. $\mathbf{r} = \int \left[\left(\frac{3}{2}(t+1)^{1/2}\right)\mathbf{i} + e^{-t}\mathbf{j} + \left(\frac{1}{t+1}\right)\mathbf{k}\right]dt$
$= (t+1)^{3/2}\mathbf{i} - e^{-t}\mathbf{j} + \ln(t+1)\mathbf{k} + \mathbf{C}$
$\mathbf{r}(0) = (0+1)^{3/2}\mathbf{i} - e^{-0}\mathbf{j} + \ln(0+1)\mathbf{k} + \mathbf{C} = \mathbf{k}$
$\mathbf{C} = -\mathbf{i} + \mathbf{j} + \mathbf{k}$
$\mathbf{r} = [(t+1)^{3/2} - 1]\mathbf{i} + (1 - e^{-t})\mathbf{j} + [1 + \ln(t+1)]\mathbf{k}$

25\. $\frac{d\mathbf{r}}{dt} = \int (-32\mathbf{k})\,dt = -32t\mathbf{k} + \mathbf{C}_1$

$\frac{d\mathbf{r}}{dt}(0) = 8\mathbf{i} + 8\mathbf{j}$

$-32(0)\mathbf{k} + \mathbf{C}_1 = 8\mathbf{i} + 8\mathbf{j}$

$\mathbf{C}_1 = 8\mathbf{i} + 8\mathbf{j}$

$\frac{d\mathbf{r}}{dt} = 8\mathbf{i} + 8\mathbf{j} - 32t\mathbf{k}$

$\mathbf{r} = \int (8\mathbf{i} + 8\mathbf{j} - 32t\mathbf{k})\,dt = 8t\mathbf{i} + 8t\mathbf{j} - 16t^2\mathbf{k} + \mathbf{C}_2$

$\mathbf{r}(0) = 100\mathbf{k}$

$8(0)\mathbf{i} + 8(0)\mathbf{j} - 16(0)^2\mathbf{k} + \mathbf{C}_2 = 100\mathbf{k}$

$\mathbf{C}_2 = 100\mathbf{k}$

$\mathbf{r} = 8t\mathbf{i} + 8t\mathbf{j} + (100 - 16t^2)\mathbf{k}$

27\. $\mathbf{r}(t) = (\sin t)\mathbf{i} + (t^2 - \cos t)\mathbf{j} + e^t\mathbf{k}$
$\mathbf{v}(t) = (\cos t)\mathbf{i} + (2t + \sin t)\mathbf{j} + e^t\mathbf{k}$
$t_0 = 0$
$\mathbf{v}(0) = \mathbf{i} + \mathbf{k}$ and $\mathbf{r}(0) = -\mathbf{j} + \mathbf{k}$, so $P_0 = (0, -1, 1)$
$x = 0 + t = t$, $y = -1$, and $z = 1 + t$ are parametric equations of the tangent line.

29\. $\mathbf{r}(t) = (a\sin t)\mathbf{i} + (a\cos t)\mathbf{j} + bt\mathbf{k}$
$\mathbf{v}(t) = (a\cos t)\mathbf{i} - (a\sin t)\mathbf{j} + b\mathbf{k}$
$t_0 = 2\pi$
$\mathbf{v}(0) = a\mathbf{i} + b\mathbf{k}$ and $\mathbf{r}(0) = a\mathbf{j} + 2b\pi\mathbf{k}$,
so $P_0 = (0, a, 2b\pi)$
$x = 0 + at = at$, $y = a$, and $z = 2\pi b + bt$ are parametric equations of the tangent line.

31\. a. $\mathbf{v}(t) = -(\sin t)\mathbf{i} + (\cos t)\mathbf{j}$
$\mathbf{a}(t) = -(\cos t)\mathbf{i} - (\sin t)\mathbf{j}$
(i) $|\mathbf{v}(t)| = \sqrt{(-\sin t)^2 + (\cos t)^2} = 1$
constant speed
(ii) $\mathbf{v} \cdot \mathbf{a} = (\sin t)(\cos t) - (\cos t)(\sin t) = 0$
yes, orthogonal
(iii) counterclockwise movement
(iv) yes, $\mathbf{r}(0) = \mathbf{i} + 0\mathbf{j}$

b. $\mathbf{v}(t) = -(2\sin 2t)\mathbf{i} + (2\cos 2t)\mathbf{j}$

$\mathbf{a}(t) = -(4\cos 2t)\mathbf{i} - (4\sin 2t)\mathbf{j}$

(i) $|\mathbf{v}(t)| = \sqrt{4\sin^2 2t + 4\cos^2 2t} = 2$

constant speed

(ii) $\mathbf{v}\cdot\mathbf{a} = 8\sin 2t\cos 2t - 8\cos 2t\sin 2t = 0$

yes, orthogonal

(iii) counterclockwise movement

(iv) yes, $\mathbf{r}(0) = \mathbf{i} + 0\mathbf{j}$

c. $\mathbf{v}(t) = -\sin\left(t - \frac{\pi}{2}\right)\mathbf{i} + \cos\left(t - \frac{\pi}{2}\right)\mathbf{j}$

$\mathbf{a}(t) = -\cos\left(t - \frac{\pi}{2}\right)\mathbf{i} - \sin\left(t - \frac{\pi}{2}\right)\mathbf{j}$

(i) $|\mathbf{v}(t)| = \sqrt{\sin^2\left(t - \frac{\pi}{2}\right) + \cos^2\left(t - \frac{\pi}{2}\right)} = 1$

constant speed

(ii) $\mathbf{v}\cdot\mathbf{a} = \sin\left(t - \frac{\pi}{2}\right)\cos\left(t - \frac{\pi}{2}\right) - \cos\left(t - \frac{\pi}{2}\right)\sin\left(t - \frac{\pi}{2}\right) = 0$

yes, orthogonal

(iii) counterclockwise movement

(iv) no, $\mathbf{r}(0) = 0\mathbf{i} - \mathbf{j}$ instead of $\mathbf{i} + 0\mathbf{j}$, so the particle starts at $(0, -1)$.

d. $\mathbf{v}(t) = -(\sin t)\mathbf{i} - (\cos t)\mathbf{j}$

$\mathbf{a}(t) = -(\cos t)\mathbf{i} + (\sin t)\mathbf{j}$

(i) $|\mathbf{v}(t)| = \sqrt{(-\sin t)^2 + (-\cos t)^2} = 1$

constant speed

(ii) $\mathbf{v}\cdot\mathbf{a} = (\sin t)(\cos t) - (\cos t)(\sin t) = 0$

yes, orthogonal

(iii) clockwise movement

(iv) yes, $\mathbf{r}(0) = \mathbf{i} - 0\mathbf{j}$

e. $\mathbf{v}(t) = -(2t\sin t^2)\mathbf{i} + (2t\cos t^2)\mathbf{j}$

$\mathbf{a}(t) = -(4t^2\cos t^2 + 2\sin t^2)\mathbf{i} + (2\cos t^2 - 4t^2\sin t^2)\mathbf{j}$

(i) $|\mathbf{v}(t)| = \sqrt{2(t\sin t^2)^2 + (t\cos t^2)^2}$
$= 2\sqrt{t^2}$
$= 2|t|$

variable speed

(ii) $\mathbf{v}\cdot\mathbf{a} = 8t^3\cos t^2\sin t^2 + 4t\sin^2 t^2 + 4t\cos^2 t^2 - 8t^3\cos t^2\sin t^2$
$= 4t \neq 0$ in general

not orthogonal in general

(iii) counterclockwise movement

(iv) yes, $\mathbf{r}(0) = \mathbf{i} + 0\mathbf{j}$

33. $\frac{d\mathbf{v}}{dt} = \mathbf{a} = 3\mathbf{i} - \mathbf{j} + \mathbf{k}$

$\mathbf{v}(t) = 3t\mathbf{i} - t\mathbf{j} + t\mathbf{k} + \mathbf{C}_1$

The particle travels in the direction of the vector $(4-1)\mathbf{i} + (1-2)\mathbf{j} + (4-3)\mathbf{k} = 3\mathbf{i} - \mathbf{j} + \mathbf{k}$ (since it travels in a straight line), and at time $t = 0$ it has speed 2

$$\mathbf{v}(0) = \frac{2}{\sqrt{9+1+1}}(3\mathbf{i} - \mathbf{j} + \mathbf{k}) = \mathbf{C}_1$$

$$\frac{d\mathbf{r}}{dt} = \mathbf{v}(t) = \left(3t + \frac{6}{\sqrt{11}}\right)\mathbf{i} - \left(t + \frac{2}{\sqrt{11}}\right)\mathbf{j} + \left(t + \frac{2}{\sqrt{11}}\right)\mathbf{k}$$

$$\mathbf{r}(t) = \left(\frac{3}{2}t^2 + \frac{6}{\sqrt{11}}t\right)\mathbf{i} - \left(\frac{1}{2}t^2 + \frac{2}{\sqrt{11}}t\right)\mathbf{j} + \left(\frac{1}{2}t^2 + \frac{2}{\sqrt{11}}t\right)\mathbf{k} + \mathbf{C}_2$$

$$\mathbf{r}(0) = \mathbf{i} + 2\mathbf{j} + 3\mathbf{k} = \mathbf{C}_2$$

$$\mathbf{r}(t) = \left(\frac{3}{2}t^2 + \frac{6}{\sqrt{11}}t + 1\right)\mathbf{i} - \left(\frac{1}{2}t^2 + \frac{2}{\sqrt{11}}t - 2\right)\mathbf{j} + \left(\frac{1}{2}t^2 + \frac{2}{\sqrt{11}}t + 3\right)\mathbf{k}$$
$$= \left(\frac{1}{2}t^2 + \frac{2}{\sqrt{11}}t\right)(3\mathbf{i} - \mathbf{j} + \mathbf{k}) + (\mathbf{i} + 2\mathbf{j} + 3\mathbf{k})$$

35. The velocity vector is tangent to the graph of $y^2 = 2x$ at the point $(2, 2)$, has length 5, and a positive $\mathbf{i}$ component. Now, $y^2 = 2x$

$$2y\frac{dy}{dx} = 2$$

$$\left.\frac{dy}{dx}\right|_{(2,2)} = \frac{2}{2\cdot 2} = \frac{1}{2}$$

The tangent vector lies in the direction of the vector $\mathbf{i} + \frac{1}{2}\mathbf{j}$.

The velocity vector is $\mathbf{v} = \frac{5}{\sqrt{1 + \frac{1}{4}}}\left(\mathbf{i} + \frac{1}{2}\mathbf{j}\right)$
$= \frac{5}{\left(\frac{\sqrt{5}}{2}\right)}\left(\mathbf{i} + \frac{1}{2}\mathbf{j}\right)$
$= 2\sqrt{5}\mathbf{i} + \sqrt{5}\mathbf{j}$

37. $\mathbf{v} = (-3\sin t)\mathbf{j} + (2\cos t)\mathbf{k}$
and $\mathbf{a} = (-3\cos t)\mathbf{j} - (2\sin t)\mathbf{k}$
$|\mathbf{v}|^2 = 9\sin^2 t + 4\cos^2 t$

$\frac{d}{dt}(|\mathbf{v}|^2) = 18\sin t\cos t - 8\cos t\sin t = 10\sin t\cos t$

$\frac{d}{dt}(|\mathbf{v}|^2) = 0$ when $10\sin t\cos t = 0$

$\sin t = 0$ or $\cos t = 0$

$t = 0, \pi$ or $t = \frac{\pi}{2}, \frac{3\pi}{2}$.

When $t = 0$ or π, $|\mathbf{v}|^2 = 4$

$|\mathbf{v}| = \sqrt{4} = 2$

When $t = \frac{\pi}{2}$ or $\frac{3\pi}{2}$, $|\mathbf{v}| = \sqrt{9} = 3$.

Therefore max $|\mathbf{v}|$ is 3 when $t = \frac{\pi}{2}, \frac{3\pi}{2}$, and min $|\mathbf{v}| = 2$ when $t = 0, \pi$. Next, $|\mathbf{a}|^2 = 9\cos^2 t + 4\sin^2 t$

$\frac{d}{dt}(|\mathbf{a}|^2) = -18\cos t\sin t + 8\sin t\cos t = -10\sin t\cos t$

$\frac{d}{dt}(|\mathbf{a}|^2) = 0$ when $-10\sin t\cos t = 0$

$\sin t = 0$ or $\cos t = 0$

$t = 0, \pi$ or $t = \frac{\pi}{2}, \frac{3\pi}{2}$.

When $t = 0$ or π, $|\mathbf{a}|^2 = 9$

$|\mathbf{a}| = 3$

When $t = \frac{\pi}{2}$ or $\frac{3\pi}{2}$, $|\mathbf{a}|^2 = 4$

$|\mathbf{a}| = 2$.

Therefore, max $|\mathbf{a}| = 3$ when $t = 0, \pi$, and min $|\mathbf{a}| = 2$ when $t = \frac{\pi}{2}, \frac{3\pi}{2}$.

39. $\mathbf{u} = \mathbf{C} = a\mathbf{i} + b\mathbf{j} + c\mathbf{k}$ with a, b, c real constants

$$\frac{d\mathbf{u}}{dt} = \frac{da}{dt}\mathbf{i} + \frac{db}{dt}\mathbf{j} + \frac{dc}{dt}\mathbf{k} = 0\mathbf{i} + 0\mathbf{j} + 0\mathbf{k} = \mathbf{0}$$

41. Let $\mathbf{u} = f_1(t)\mathbf{i} + f_2(t)\mathbf{j} + f_3(t)\mathbf{k}$ and $\mathbf{v} = g_1(t)\mathbf{i} + g_2(t)\mathbf{j} + g_3(t)\mathbf{k}$.

Then $\mathbf{u} + \mathbf{v} = [f_1(t) + g_1(t)]\mathbf{i} + [f_2(t) + g_2(t)]\mathbf{j} + [f_3(t) + g_3(t)]\mathbf{k}$

$$\begin{aligned}\frac{d}{dt}(\mathbf{u} + \mathbf{v}) &= [f_1'(t) + g_1'(t)]\mathbf{i} + [f_2'(t) + g_2'(t)]\mathbf{j} + [f_3'(t) + g_3'(t)]\mathbf{k} \\ &= [f_1'(t)\mathbf{i} + f_2'(t)\mathbf{j} + f_3'(t)\mathbf{k}] + [g_1'(t)\mathbf{i} + g_2'(t)\mathbf{j} + g_3'(t)\mathbf{k}] \\ &= \frac{d\mathbf{u}}{dt} + \frac{d\mathbf{v}}{dt}\end{aligned}$$

$$\mathbf{u} - \mathbf{v} = [f_1(t) - g_1(t)]\mathbf{i} + [f_2(t) - g_2(t)]\mathbf{j} + [f_3(t) - g_3(t)]\mathbf{k}$$

$$\begin{aligned}\frac{d}{dt}(\mathbf{u} - \mathbf{v}) &= [f_1'(t) - g_1'(t)]\mathbf{i} + [f_2'(t) - g_2'(t)]\mathbf{j} + [f_3'(t) - g_3'(t)]\mathbf{k} \\ &= [f_1'(t)\mathbf{i} + f_2'(t)\mathbf{j} + f_3'(t)\mathbf{k}] - [g_1'(t)\mathbf{i} + g_2'(t)\mathbf{j} + g_3'(t)\mathbf{k}] \\ &= \frac{d\mathbf{u}}{dt} - \frac{d\mathbf{v}}{dt}\end{aligned}$$

43. $$\lim_{t\to t_0} [\mathbf{r}_1(t) \times \mathbf{r}_2(t)] = \lim_{t\to t_0} \begin{vmatrix} \mathbf{i} & \mathbf{j} & \mathbf{k} \\ f_1(t) & g_1(t) & h_1(t) \\ f_2(t) & g_2(t) & h_2(t) \end{vmatrix}$$

$$= \begin{vmatrix} \mathbf{i} & \mathbf{j} & \mathbf{k} \\ \lim_{t\to t_0} f_1(t) & \lim_{t\to t_0} g_1(t) & \lim_{t\to t_0} h_1(t) \\ \lim_{t\to t_0} f_2(t) & \lim_{t\to t_0} g_2(t) & \lim_{t\to t_0} h_2(t) \end{vmatrix}$$

$$= \lim_{t\to t_0} \mathbf{r}_1(t) \times \lim_{t\to t_0} \mathbf{r}_2(t) = \mathbf{A} \times \mathbf{B}$$

45. a. Let $\mathbf{r}(t) = f(t)\mathbf{i} + g(t)\mathbf{j} + h(t)\mathbf{k}$.

$$\begin{aligned}\int_a^b k\mathbf{r}(t)\,dt &= \int_a^b [kf(t)\mathbf{i} + kg(t)\mathbf{j} + kh(t)\mathbf{k}]\,dt \\ &= \int_a^b [kf(t)]\,dt\,\mathbf{i} + \int_a^b [kg(t)]\,dt\,\mathbf{j} + \int_a^b [kh(t)]\,dt\,\mathbf{k} \\ &= k\left(\int_a^b f(t)\,dt\,\mathbf{i} + \int_a^b g(t)\,dt\,\mathbf{j} + \int_a^b h(t)\,dt\,\mathbf{k}\right) \\ &= k\int_a^b \mathbf{r}(t)\,dt\end{aligned}$$

b. Let $\mathbf{r}_1(t) = f_1(t)\mathbf{i} + g_1(t)\mathbf{j} + h_2(t)\mathbf{k}$ and $\mathbf{r}_2(t) = f_2(t)\mathbf{i} + g_2(t)\mathbf{j} + h_2(t)\mathbf{k}$.

$$\begin{aligned}\int_a^b [\mathbf{r}_1(t) \pm \mathbf{r}_2(t)]\,dt &= \int_a^b ([f_1(t)\mathbf{i} + g_1(t)\mathbf{j} + h_1(t)\mathbf{k}] \pm [f_2(t)\mathbf{i} + g_2(t)\mathbf{j} + h_2(t)\mathbf{k}])\,dt \\ &= \int_a^b ([f_1(t) \pm f_2(t)]\mathbf{i} + [g_1(t) \pm g_2(t)]\mathbf{j} + [h_1(t) \pm h_2(t)]\mathbf{k})\,dt \\ &= \int_a^b [f_1(t) \pm f_2(t)]\,dt\,\mathbf{i} + \int_a^b [g_1(t) \pm g_2(t)]\,dt\,\mathbf{j} + \int_a^b [h_1(t) \pm h_2(t)]\,dt\,\mathbf{k} \\ &= \left[\int_a^b f_1(t)\,dt\,\mathbf{i} \pm \int_a^b f_2(t)\,dt\,\mathbf{i}\right] + \left[\int_a^b g_1(t)\,dt\,\mathbf{j} \pm \int_a^b g_2(t)\,dt\,\mathbf{j}\right] + \left[\int_a^b h_1(t)\,dt\,\mathbf{k} \pm \int_a^b h_2(t)\,dt\,\mathbf{k}\right] \\ &= \int_a^b \mathbf{r}_1(t)\,dt \pm \int_a^b \mathbf{r}_2(t)\,dt\end{aligned}$$

c. Let $\mathbf{C} = c_1\mathbf{i} + c_2\mathbf{j} + c_3\mathbf{k}$.

Then $$\begin{aligned}\int_a^b \mathbf{C}\cdot\mathbf{r}(t)\,dt &= \int_a^b [c_1f(t) + c_2g(t) + c_3h(t)]\,dt \\ &= c_1\int_a^b f(t)\,dt + c_2\int_a^b g(t)\,dt + c_3\int_a^b h(t)\,dt \\ &= \mathbf{C}\cdot\int_a^b \mathbf{r}(t)\,dt\end{aligned}$$

$$\begin{aligned}\int_a^b \mathbf{C} \times \mathbf{r}(t)\,dt &= \int_a^b [c_2h(t) - c_3g(t)]\mathbf{i} + [c_3f(t) - c_1h(t)]\mathbf{j} + [c_1g(t) - c_2f(t)]\mathbf{k}\,dt \\ &= \left[c_2\int_a^b h(t)\,dt - c_3\int_a^b g(t)\,dt\right]\mathbf{i} + \left[c_3\int_a^b f(t)\,dt - c_1\int_a^b h(t)\,dt\right]\mathbf{j} + \left[c_1\int_a^b g(t)\,dt - c_2\int_a^b f(t)\,dt\right]\mathbf{k} \\ &= \mathbf{C} \times \int_a^b \mathbf{r}(t)\,dt\end{aligned}$$

47. a. If $\mathbf{R}_1(t)$ and $\mathbf{R}_2(t)$ have identical derivatives on I, then

$$\frac{d\mathbf{R}_1}{dt} = \frac{df_1}{dt}\mathbf{i} + \frac{dg_1}{dt}\mathbf{j} + \frac{dh_1}{dt}\mathbf{k} = \frac{df_2}{dt}\mathbf{i} + \frac{dg_2}{dt}\mathbf{j} + \frac{dh_2}{dt}\mathbf{k} = \frac{d\mathbf{R}_2}{dt}$$

Thus, $\frac{df_1}{dt} = \frac{df_2}{dt}, \frac{dg_1}{dt} = \frac{dg_2}{dt}, \frac{dh_1}{dt} = \frac{dh_2}{dt}$, or

$f_1(t) = f_2(t) + c_1$, $g_1(t) = g_2(t) + c_2$,

$h_1(t) = h_2(t) + c_3$

Hence, $f_1(t)\mathbf{i} + g_1(t)\mathbf{j} + h_1(t)\mathbf{k}$

$= [f_2(t) + c_1]\mathbf{i} + [g_2(t) + c_2]\mathbf{j} + [h_2(t) + c_3]\mathbf{k}$, so

$\mathbf{R}_1(t) = \mathbf{R}_2(t) + \mathbf{C}$, where $\mathbf{C} = c_1\mathbf{i} + c_2\mathbf{j} + c_3\mathbf{k}$.

b. Let $\mathbf{R}(t)$ be an antiderivative of $\mathbf{r}(t)$ on I. Then $\mathbf{R}'(t) = \mathbf{r}(t)$. If $\mathbf{U}(t)$ is an antiderivative of $\mathbf{r}(t)$ on I, then $\mathbf{U}'(t) = \mathbf{r}(t)$. Thus $\mathbf{U}'(t) = \mathbf{R}'(t)$ on I and $\mathbf{U}(t) = \mathbf{R}(t) + \mathbf{C}$.

49. For Exercise 5:

Show that the components $x(t) = \sin t$ and $y(t) = \cos t$ satisfy $x^2 + y^2 = 1$.

$\sin^2 t + \cos^2 t = 1$ True

For Exercise 6:

Show that the components $x(t) = 4\cos\frac{t}{2}$ and $y(t) = 4\sin\frac{t}{2}$ satisfy $x^2 + y^2 = 16$.

$$\left(4\cos\frac{t}{2}\right)^2 + \left(4\sin\frac{t}{2}\right)^2 \stackrel{?}{=} 16$$

$$16\left(\cos^2\frac{t}{2} + \sin^2\frac{t}{2}\right) = 16 \quad \text{True}$$

For Exercise 7:

The components $x(t) = t - \sin t$ and $y(t) = 1 - \cos t$ are the same as the expressions defining the parametric curve.

For Exercise 8:

Show that the components $x(t) = t$ and $y(t) = t^2 + 1$ satisfy $y = x^2 + 1$.

$t^2 + 1 = t^2 + 1$ True

51. $$\frac{d}{dt}\left[\mathbf{r}\cdot\left(\frac{d\mathbf{r}}{dt}\times\frac{d^2\mathbf{r}}{dt^2}\right)\right]$$

$$= \frac{d\mathbf{r}}{dt}\cdot\left(\frac{d\mathbf{r}}{dt}\times\frac{d^2\mathbf{r}}{dt^2}\right) + \mathbf{r}\cdot\left(\frac{d^2\mathbf{r}}{dt^2}\times\frac{d^2\mathbf{r}}{dt^2}\right) + \mathbf{r}\cdot\left(\frac{d\mathbf{r}}{dt}\times\frac{d^3\mathbf{r}}{dt^3}\right)$$

$$= \mathbf{r}\cdot\left(\frac{d\mathbf{r}}{dt}\times\frac{d^3\mathbf{r}}{dt^3}\right), \text{ since } \mathbf{A}\cdot(\mathbf{A}\times\mathbf{B}) = 0$$

and $\mathbf{A}\cdot(\mathbf{B}\times\mathbf{B}) = 0$ for any vectors $\mathbf{A}$ and $\mathbf{B}$

53–55. Sample CAS commands:

Maple:

```
with(plots):
x:= t -> sin(t)-t*cos(t);
y:= t -> cos(t)+t*sin(t);
z:= t -> t^2
s1:= spacecurve([x(t),y(t),z(t)],
t=0..6*Pi, numpoints = 120,
axes=NORMAL):
dx:= t -> D(x)(t);
dy:= t -> D(y)(t);
dz:= t -> D(z)(t);
t0:= 3*Pi/2:
s2:=spacecurve([x(t0)+t*dx(t0),y(t0)
+t*dy(t0),z(t0)+t*dz(t0),t=-2..2]):
display([s1,s2],title = 'Space Curve
and Tangent Line at t0=3 Pi/2');
```

Mathematica:

```
Clear[x,y,z,t]
r[t_] = {x[t],y[t],z[t]}
x[t_] = Sin[t]-t Cos[t]
y[t_] = Cos[t]+t Sin[t]
z[t_] = t^2
{a,b} = {0, 6 Pi};
t0 = 3/2 Pi;
p1 = ParametricPlot3D
  [{x[t],y[t],z[t]}, {t,a,b}]
v[t_] = r'[t]
v0 = v[t0]
line[t_] = r[t0] + t v0
p2 = ParametricPlot3D[ Evaluate
  [line[t] ], {t,-2,2} ]
Show[ p1, p2 ]
```

57. Example CAS commands:

Maple:

```
with(plots):
x:= t -> cos(a*t):
y:= t -> sin(a*t):
z:= t -> b*t: a:=2: b:=1:
s1:= spacecurve([x(t),y(t),z(t)],
  t=0..4*Pi, numpoints = 400,
  axes=NORMAL):
dx:= t -> D(x)(t);
dy:= t -> D(y)(t);
dz:= t -> D(z)(t);
t0:= 3*Pi/2:
s2:=spacecurve([x(t0)+t*dx(t0),y(t0)
  +t*dy(t0),z(t0)+t*dz(t0),t=-2..2]):
display([s1,s2],title = 'Helix With
a = 2 and b = 1');
```

Mathematica:

```
Clear[a,b]
x[t_] = Cos[a t]
y[t_] = Sin[a t]
z[t_] = b t
t0 = 3/2 Pi;
v[t_] = r'[t]
v0 = v[t0]
line[t_] = r[t0] + t v0
b = 1
a = 2
p1 = ParametricPlot3D
  [{x[t],y[t],z[t]}, {t,0,4Pi} ]
p2 = ParametricPlot3D
  [Evaluate[ line[t] ], {t,-2,2} ]
Show[ p1, p2 ]
```

■ Section 12.2 Arc Length and the Unit Tangent Vector T (pp. 645–649)

Quick Review 12.2

1. $x^2 + y^2 = 9$

$$\frac{d}{dx}(x^2 + y^2) = \frac{d}{dx}(9)$$

$$2x + 2y\frac{dy}{dx} = 0$$

$$\frac{dy}{dx} = -\frac{x}{y}$$

At $(3, 0)$, $\frac{dy}{dx} = -\frac{3}{0}$ (undefined).

The tangent line is the vertical line $x = 3$.

2. From the solution to Exercise 1, $\frac{dy}{dx} = -\frac{x}{y}$.

At $(0, 3)$, $\frac{dy}{dx} = -\frac{0}{3} = 0$.

The tangent line is the horizontal line $y = 3$.

3. From the solution to Exercise 1, $\frac{dy}{dx} = -\frac{x}{y}$.

At $\left(-\frac{3}{\sqrt{2}}, \frac{3}{\sqrt{2}}\right)$, $\frac{dy}{dx} = -\frac{-3/\sqrt{2}}{3/\sqrt{2}} = 1$.

$$y - y_1 = m(x - x_1)$$

$$y - \frac{3}{\sqrt{2}} = 1\left(x + \frac{3}{\sqrt{2}}\right)$$

$$y = x + 3\sqrt{2}$$

4. From the solution to Exercise 1, $\frac{dy}{dx} = -\frac{x}{y}$.

At $\left(\frac{3}{\sqrt{2}}, \frac{3}{\sqrt{2}}\right)$, $\frac{dy}{dx} = -\frac{3/\sqrt{2}}{3/\sqrt{2}} = -1$.

$y - y_1 = m(x - x_1)$

$y - \frac{3}{\sqrt{2}} = -1\left(x - \frac{3}{\sqrt{2}}\right)$

$y = -x + 3\sqrt{2}$

5. a. $\frac{dy}{dx} = \frac{dy/dt}{dx/dt} = \frac{3\cos t}{-2\sin t}\bigg|_{t=\pi/3} = -\frac{\sqrt{3}}{2}$

A tangent vector is $\mathbf{v} = 2\mathbf{i} - \sqrt{3}\mathbf{j}$

and $|\mathbf{v}| = \sqrt{2^2 + (-\sqrt{3})^2} = \sqrt{7}$,

so $\frac{\mathbf{v}}{|\mathbf{v}|} = \frac{2}{\sqrt{7}}\mathbf{i} - \frac{\sqrt{3}}{\sqrt{7}}\mathbf{j} = \frac{2\sqrt{7}}{7}\mathbf{i} - \frac{\sqrt{21}}{7}\mathbf{j}$.

The desired unit vectors are $\pm\left(\frac{2\sqrt{7}}{7}\mathbf{i} - \frac{\sqrt{21}}{7}\mathbf{j}\right)$.

b. Interchange the components of $\frac{\mathbf{v}}{|\mathbf{v}|}$ and change the sign of one of them. The desired unit vectors are

$\pm\left(\frac{\sqrt{21}}{7}\mathbf{i} + \frac{2\sqrt{7}}{7}\mathbf{j}\right)$.

6. Length $= \int_0^{\pi/2} \sqrt{1 + \left(\frac{dy}{dx}\right)^2}\,dx$

$= \int_0^{\pi/2} \sqrt{1 + \cos^2 x}\,dx$

≈ 1.91 (using NINT)

7. $\left(\frac{dx}{dt}\right)^2 + \left(\frac{dy}{dt}\right)^2 = (-3\cos^2 t \sin t)^2 + (3\sin^2 t \cos t)^2$

$= 9\cos^4 t \sin^2 t + 9\sin^4 t \cos^2 t$

$= 9\cos^2 t \sin^2 t(\cos^2 t + \sin^2 t)$

$= 9\cos^2 t \sin^2 t$

Length $= \int_0^{\pi} \sqrt{\left(\frac{dx}{dt}\right)^2 + \left(\frac{dy}{dt}\right)^2}$

$= \int_0^{\pi} \sqrt{9\cos^2 t \sin^2 t}\,dt$

$= \int_0^{\pi} |3\cos t \sin t|\,dt$

$= 6\int_0^{\pi/2} \cos t \sin t\,dt$

$= 3\sin^2 t\Big]_0^{\pi/2}$

$= 3 - 0 = 3$

8. $\frac{dy}{dx} = \frac{d}{dx}\left(\int_2^x t\sin t\,dt\right) = x\sin x$

9. Let $u = x^2$.

$\frac{dy}{dx} = \frac{dy}{du}\frac{du}{dx} = \frac{d}{du}\left(\int_0^u \cos(t^2)\,dt\right)\cdot\frac{d}{dx}x^2$

$= (\cos u^2)(2x)$

$= 2x\cos x^4$

10. Let $u = \cos x$ and $v = \sin x$.

$\frac{dy}{dx} = \frac{d}{dx}\left(\int_{\cos x}^{\sin x} \ln(1 + t^2)\,dt\right)$

$= \frac{d}{dx}\left(\int_0^{\sin x} \ln(1 + t^2)\,dt - \int_0^{\cos x} \ln(1 + t^2)\,dt\right)$

$= \frac{d}{dv}\left(\int_0^v \ln(1 + t^2)\,dt\right)\frac{dv}{dx} - \frac{d}{du}\left(\int_0^u \ln(1 + t^2)\,dt\right)\frac{du}{dx}$

$= \ln(1 + v^2)\cdot\cos x - \ln(1 + u^2)\cdot(-\sin x)$

$= (\cos x)\ln(1 + \sin^2 x) + (\sin x)\ln(1 + \cos^2 x)$

Section 12.2 Exercises

1. a. $\mathbf{r} = (2\cos t)\mathbf{i} + (2\sin t)\mathbf{j} + \sqrt{5}t\mathbf{k}$

$\mathbf{v} = (-2\sin t)\mathbf{i} + (2\cos t)\mathbf{j} + \sqrt{5}\mathbf{k}$

$|\mathbf{v}| = \sqrt{(-2\sin t)^2 + (2\cos t)^2 + (\sqrt{5})^2}$

$= \sqrt{4\sin^2 t + 4\cos^2 t + 5} = 3$

$\mathbf{T} = \frac{\mathbf{v}}{|\mathbf{v}|} = \left(-\frac{2}{3}\sin t\right)\mathbf{i} + \left(\frac{2}{3}\cos t\right)\mathbf{j} + \frac{\sqrt{5}}{3}\mathbf{k}$

b. Length $= \int_0^{\pi} |\mathbf{v}|\,dt = \int_0^{\pi} 3\,dt = [3t]_0^{\pi} = 3\pi$

3. a. $\mathbf{r} = t\mathbf{i} + \frac{2}{3}t^{3/2}\mathbf{k}$

$\mathbf{v} = \mathbf{i} + t^{1/2}\mathbf{k}$

$|\mathbf{v}| = \sqrt{1^2 + (t^{1/2})^2} = \sqrt{1 + t}$

$\mathbf{T} = \frac{\mathbf{v}}{|\mathbf{v}|} = \frac{1}{\sqrt{1+t}}\mathbf{i} + \frac{t}{\sqrt{1+t}}\mathbf{k}$

b. Length $= \int_0^8 \sqrt{1+t}\,dt = \left[\frac{2}{3}(1+t)^{3/2}\right]_0^8 = \frac{52}{3}$

5. a. $\mathbf{r} = (\cos^3 t)\mathbf{j} + (\sin^3 t)\mathbf{k}$

$\mathbf{v} = (-3\cos^2 t \sin t)\mathbf{j} + (3\sin^2 t \cos t)\mathbf{k}$

$|\mathbf{v}| = \sqrt{(-3\cos^2 t \sin t)^2 + (3\sin^2 t \cos t)^2}$

$= \sqrt{(9\cos^2 t \sin^2 t)(\cos^2 t + \sin t)}$

$= 3|\cos t \sin t|$

$\mathbf{T} = \frac{\mathbf{v}}{|\mathbf{v}|} = \frac{-3\cos^2 t \sin t}{3|\cos t \sin t|}\mathbf{j} + \frac{3\sin^2 t \cos t}{3|\cos t \sin t|}\mathbf{k}$

$= (-\cos t)\mathbf{j} + (\sin t)\mathbf{k}$, if $0 \le t \le \frac{\pi}{2}$

b. Length $= \int_0^{\pi/2} 3|\cos t \sin t|\,dt = \int_0^{\pi/2} 3\cos t \sin t\,dt$

$= \int_0^{\pi/2} \frac{3}{2}\sin 2t\,dt = \left[-\frac{3}{4}\cos 2t\right]_0^{\pi/2} = \frac{3}{2}$

7. a. $\mathbf{r} = (t\cos t)\mathbf{i} + (t\sin t)\mathbf{j} + \frac{2\sqrt{2}}{3}t^{3/2}\mathbf{k}$

$\mathbf{v} = (\cos t - t\sin t)\mathbf{i} + (\sin t + t\cos t)\mathbf{j} + (\sqrt{2}t^{1/2})\mathbf{k}$

$|\mathbf{v}| = \sqrt{(\cos t - t\sin t)^2 + (\sin t + t\cos t)^2 + (\sqrt{2t})^2}$

$= \sqrt{1 + t^2 + 2t} = \sqrt{(t+1)^2} = |t + 1| = t + 1$,

if $t \ge 0$

$\mathbf{T} = \frac{\mathbf{v}}{|\mathbf{v}|}$

$= \left(\frac{\cos t - t\sin t}{t+1}\right)\mathbf{i} + \left(\frac{\sin t + t\cos t}{t+1}\right)\mathbf{j} + \left(\frac{\sqrt{2t}}{t+1}\right)\mathbf{k}$

b. Length $= \int_0^{\pi} (t + 1)\,dt = \left[\frac{t^2}{2} + t\right]_0^{\pi} = \frac{\pi^2}{2} + \pi$

9. Let $P(t_0)$ denote the point.
Then $\mathbf{v} = (5\cos t)\mathbf{i} - (5\sin t)\mathbf{j} + 12\mathbf{k}$

$$\text{and } 26\pi = \int_0^{t_0} \sqrt{25\cos^2 t + 25\sin^2 t + 144}\, dt$$
$$= \int_0^{t_0} 13\, dt = 13t_0$$

$t_0 = 2\pi$, and the point is
$P(2\pi) = (5\sin 2\pi, 5\cos 2\pi, 24\pi) = (0, 5, 24\pi)$

11. a. $\mathbf{r} = (4\cos t)\mathbf{i} + (4\sin t)\mathbf{j} + 3t\mathbf{k}$
$\mathbf{v} = (-4\sin t)\mathbf{i} + (4\cos t)\mathbf{j} + 3\mathbf{k}$
$|\mathbf{v}| = \sqrt{(-4\sin t)^2 + (4\cos t)^2 + 3^2} = \sqrt{25} = 5$
$s(t) = \int_0^t 5\, d\tau = 5t$

b. Length $= s\left(\frac{\pi}{2}\right) = \frac{5\pi}{2}$

13. a. $\mathbf{r} = (e^t\cos t)\mathbf{i} + (e^t\sin t)\mathbf{j} + e^t\mathbf{k}$
$\mathbf{v} = (e^t\cos t - e^t\sin t)\mathbf{i} + (e^t\sin t + e^t\cos t)\mathbf{j} + e^t\mathbf{k}$
$|\mathbf{v}|$
$= \sqrt{(e^t\cos t - e^t\sin t)^2 + (e^t\sin t + e^t\cos t)^2 + (e^t)^2}$
$= \sqrt{3e^{2t}} = \sqrt{3}e^t$
$s(t) = \int_0^t \sqrt{3}e^\tau\, d\tau = \sqrt{3}e^t - \sqrt{3}$

b. Length $= s(0) - s(-\ln 4) = 0 - (\sqrt{3}e^{-\ln 4} - \sqrt{3})$
$= \frac{3\sqrt{3}}{4}$

15. $\mathbf{r} = (\sqrt{2}t)\mathbf{i} + (\sqrt{2}t)\mathbf{j} + (1 - t^2)\mathbf{k}$
$\mathbf{v} = \sqrt{2}\mathbf{i} + \sqrt{2}\mathbf{j} - 2t\mathbf{k}$
$|\mathbf{v}| = \sqrt{(\sqrt{2})^2 + (\sqrt{2})^2 + (-2t)^2} = \sqrt{4 + 4t^2}$
$= 2\sqrt{1 + t^2}$

$$\text{Length} = \int_0^1 2\sqrt{1 + t^2}\, dt$$
$$= \left[2\left(\frac{t}{2}\sqrt{1 + t^2} + \frac{1}{2}\ln(t + \sqrt{1 + t^2})\right)\right]_0^1$$
$$= \sqrt{2} + \ln(1 + \sqrt{2}) \approx 2.30$$

17. $\mathbf{r} = (\cos t)\mathbf{i} + (\sin t)\mathbf{j} + (1 - \cos t)\mathbf{k}, 0 \le t \le 2\pi$
$x = \cos t, y = \sin t, z = 1 - \cos t$
$x^2 + y^2 = \cos^2 t + \sin^2 t = 1$, a right circular cylinder with the z-axis as the axis and radius = 1.
Therefore $P(\cos t, \sin t, 1 - \cos t)$ lies on the cylinder $x^2 + y^2 = 1$
At $t = 0$, $\mathbf{r}(0) = \mathbf{i}$, thus
$P(1, 0, 0)$ is on the curve.
At $t = \frac{\pi}{2}$, $\mathbf{r}\left(\frac{\pi}{2}\right) = \mathbf{j} + \mathbf{k}$, thus
$Q(0, 1, 1)$ is also on the curve.
At $t = \pi$, $\mathbf{r}(\pi) = -\mathbf{i} + 2\mathbf{k}$, thus
$R(-1, 0, 2)$ is on the curve.
Then $\overrightarrow{PQ} = -\mathbf{i} + \mathbf{j} + \mathbf{k}$ and $\overrightarrow{PR} = -2\mathbf{i} + 2\mathbf{k}$

$$\overrightarrow{PQ} \times \overrightarrow{PR} = \begin{bmatrix} \mathbf{i} & \mathbf{j} & \mathbf{k} \\ -1 & 1 & 1 \\ -2 & 0 & 2 \end{bmatrix} = 2\mathbf{i} + 2\mathbf{k} \text{ is a vector normal to}$$

the plane of P, Q, and R. Then the plane containing P, Q, and R has an equation $2x + 2z = 2(1) + 2(0)$ or $x + z = 1$. Any point on the curve will satisfy this equation since $x + z = \cos t + (1 - \cos t) = 1$. Therefore, any point on the curve lies on the intersection of the cylinder $x^2 + y^2 = 1$ and the plane $x + z = 1$.
The curve is an ellipse.

19. $\mathbf{a} = (-\cos t)\mathbf{i} - (\sin t)\mathbf{j} + (\cos t)\mathbf{k}$
$\mathbf{n} = \mathbf{i} + \mathbf{k}$ is normal to the plane $x + z = 1$
$\mathbf{n} \cdot \mathbf{a} = -\cos t + \cos t = 0$
a is orthogonal to **n**
a is parallel to the plane

$$\mathbf{a}(0) = -\mathbf{i} + \mathbf{k}, \mathbf{a}\left(\frac{\pi}{2}\right) = -\mathbf{j}, \mathbf{a}(\pi) = \mathbf{i} - \mathbf{k}, \mathbf{a}\left(\frac{3\pi}{2}\right) = \mathbf{j}$$

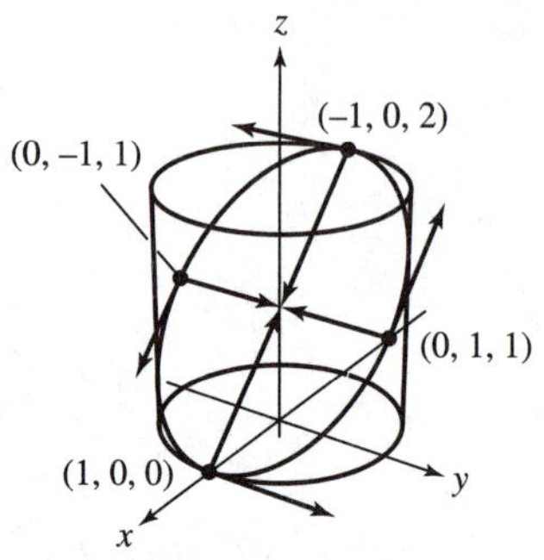

21. Graph parametrically. For example, for Exercise 3, use
$x = t, y = \frac{2}{3}t^{3/2}, 0 \le t \le 8$.

23. a. $x = r\cos\theta$
$dx = \cos\theta\, dr - r\sin\theta\, d\theta$
$y = r\sin\theta$
$dy = \sin\theta\, dr + r\cos\theta\, d\theta$ thus
$dx^2 = \cos^2\theta\, dr^2 - 2r\sin\theta\cos\theta\, dr\, d\theta + r^2\sin^2\theta\, d\theta^2$
and
$dy^2 = \sin^2\theta\, dr^2 + 2r\sin\theta\cos\theta\, dr\, d\theta + r^2\cos^2\theta\, d\theta^2$
$dx^2 + dy^2 + dz^2 = dr^2 + r^2\, d\theta^2 + dz^2$

b.

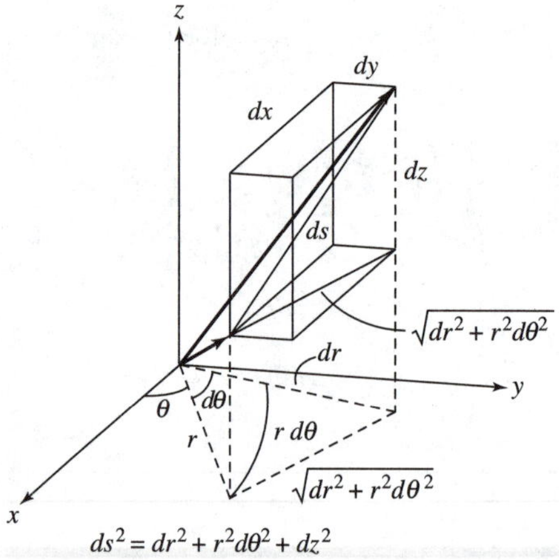

$ds^2 = dr^2 + r^2 d\theta^2 + dz^2$

c. $r = e^\theta$
$dr = e^\theta\, d\theta$

$$L = \int_0^{\ln 8} \sqrt{dr^2 + r^2\, d\theta^2 + dz^2}$$
$$= \int_0^{\ln 8} \sqrt{e^{2\theta} + e^{2\theta} + e^{2\theta}}\, d\theta$$
$$= \int_0^{\ln 8} \sqrt{3}e^\theta\, d\theta = [\sqrt{3}e^\theta]_0^{\ln 8}$$
$$= 8\sqrt{3} - \sqrt{3} = 7\sqrt{3}$$

■ Section 12.3 Curvature, Torsion, and the TNB Frame (pp. 649–661)

Quick Review 12.3

1. $\mathbf{u}\cdot\mathbf{v} = \left(\frac{1}{\sqrt{2}}\mathbf{i} + \frac{1}{\sqrt{2}}\mathbf{j}\right)\cdot\left(\frac{1}{\sqrt{2}}\mathbf{i} - \frac{1}{\sqrt{2}}\mathbf{j}\right) = \frac{1}{2} - \frac{1}{2} = 0$

2. $2\sqrt{2}\,\mathbf{u} + 3\sqrt{2}\,\mathbf{v}$

$= 2\sqrt{2}\left(\frac{1}{\sqrt{2}}\mathbf{i} + \frac{1}{\sqrt{2}}\mathbf{j}\right) + 3\sqrt{2}\left(\frac{1}{\sqrt{2}}\mathbf{i} - \frac{1}{\sqrt{2}}\mathbf{j}\right)$

$= (2\mathbf{i} + 2\mathbf{j}) + (3\mathbf{i} - 3\mathbf{j})$

$= 5\mathbf{i} - \mathbf{j}$

$= \mathbf{w}$

3. $\mathbf{w}\cdot\mathbf{u} = (5\mathbf{i} - \mathbf{j})\cdot\left(\frac{1}{\sqrt{2}}\mathbf{i} + \frac{1}{\sqrt{2}}\mathbf{j}\right) = \frac{5-1}{\sqrt{2}} = \frac{4}{\sqrt{2}} = 2\sqrt{2}$

4. $\mathbf{w}\cdot\mathbf{v} = (5\mathbf{i} - \mathbf{j})\cdot\left(\frac{1}{\sqrt{2}}\mathbf{i} - \frac{1}{\sqrt{2}}\mathbf{j}\right) = \frac{5+1}{\sqrt{2}} = \frac{6}{\sqrt{2}} = 3\sqrt{2}$

5. $\mathbf{w} = a\mathbf{u} + b\mathbf{v}$

$5\mathbf{i} - \mathbf{j} = a\left(\frac{1}{\sqrt{2}}\mathbf{i} + \frac{1}{\sqrt{2}}\mathbf{j}\right) + b\left(\frac{1}{\sqrt{2}}\mathbf{i} - \frac{1}{\sqrt{2}}\mathbf{j}\right)$

$5\mathbf{i} - \mathbf{j} = \frac{a+b}{\sqrt{2}}\mathbf{i} + \frac{a-b}{\sqrt{2}}\mathbf{j}$

$a + b = 5\sqrt{2}$

$a - b = -\sqrt{2}$

$2a = 4\sqrt{2}$

$a = 2\sqrt{2}$

Since $a + b = 5\sqrt{2}$, we have $b = 5\sqrt{2} - a = 5\sqrt{2} - 2\sqrt{2} = 3\sqrt{2}$.
Therefore, $a = 2\sqrt{2} = \mathbf{w}\cdot\mathbf{u}$ and $b = 3\sqrt{2} = \mathbf{w}\cdot\mathbf{v}$.

6. $\mathbf{v} = \frac{d}{dt}\mathbf{r}(t) = (\cos t)\mathbf{i} + 1\mathbf{j} = (\cos t)\mathbf{i} + \mathbf{j}$

7. $\mathbf{a} = \frac{d}{dt}\mathbf{v}(t) = (-\sin t)\mathbf{i} + 0\mathbf{j} = -(\sin t)\mathbf{i}$

8. $\mathbf{v}\cdot\mathbf{a} = 0$

$[(\cos t)\mathbf{i} + \mathbf{j}]\cdot[-(\sin t)\mathbf{i}] = 0$ when

$-\cos t \sin t = 0$

Note that $\mathbf{v}$ is never zero; since we require $\mathbf{a} \neq 0$, we exclude the values where $\sin t = 0$. The desired values occur where $\cos t = 0$, or at $t = (2k+1)\frac{\pi}{2}$.
At these values, $\mathbf{v} = 0\mathbf{i} + \mathbf{j} = \mathbf{j}$ and

$\mathbf{a} = -\sin\left((2k+1)\frac{\pi}{2}\right)\mathbf{i} = \begin{cases} -\mathbf{i}, & \text{if } k \text{ is even} \\ \mathbf{i}, & \text{if } k \text{ is odd} \end{cases}$.

9. $\mathbf{u}\times\mathbf{v} = \begin{vmatrix} \mathbf{i} & \mathbf{j} & \mathbf{k} \\ 2 & -3 & -1 \\ -2 & 7 & -3 \end{vmatrix} = 16\mathbf{i} + 8\mathbf{j} + 8\mathbf{k}$

10. $\mathbf{u}\times\mathbf{v} = \begin{vmatrix} \mathbf{i} & \mathbf{j} & \mathbf{k} \\ -1 & 0 & 1 \\ 0 & 3 & -4 \end{vmatrix} = -3\mathbf{i} - 4\mathbf{j} - 3\mathbf{k}$

Section 12.3 Exercises

1. $\mathbf{r} = t\mathbf{i} + \ln(\cos t)\mathbf{j}$

$\mathbf{v} = \mathbf{i} + \left(\frac{-\sin t}{\cos t}\right)\mathbf{j} = \mathbf{i} - (\tan t)\mathbf{j}$

$|\mathbf{v}| = \sqrt{1^2 + (-\tan t)^2} = \sqrt{\sec^2 t} = |\sec t| = \sec t,$

since $-\frac{\pi}{2} < t < \frac{\pi}{2}$

$\mathbf{T} = \frac{\mathbf{v}}{|\mathbf{v}|} = \left(\frac{1}{\sec t}\right)\mathbf{i} - \left(\frac{\tan t}{\sec t}\right)\mathbf{j} = (\cos t)\mathbf{i} - (\sin t)\mathbf{j}$

$\frac{d\mathbf{T}}{dt} = (-\sin t)\mathbf{i} - (\cos t)\mathbf{j}$

$\left|\frac{d\mathbf{T}}{dt}\right| = \sqrt{(-\sin t)^2 + (-\cos t)^2} = 1$

$\mathbf{N} = \frac{\left(\frac{d\mathbf{T}}{dt}\right)}{\left|\frac{d\mathbf{T}}{dt}\right|} = (-\sin t)\mathbf{i} - (\cos t)\mathbf{j}$

$\mathbf{a} = (-\sec^2 t)\mathbf{j}$

$\mathbf{v}\times\mathbf{a} = \begin{vmatrix} \mathbf{i} & \mathbf{j} & \mathbf{k} \\ 1 & -\tan t & 0 \\ 0 & -\sec^2 t & 0 \end{vmatrix} = (-\sec^2 t)\mathbf{k}$

$|\mathbf{v}\times\mathbf{a}| = \sqrt{(-\sec^2 t)^2} = \sec^2 t$

$\kappa = \frac{|\mathbf{v}\times\mathbf{a}|}{|\mathbf{v}|^3} = \frac{\sec^2 t}{\sec^3 t} = \cos t$

3. $\mathbf{r} = (2t + 3)\mathbf{i} + (5 - t^2)\mathbf{j}$

$\mathbf{v} = 2\mathbf{i} - 2t\mathbf{j}$

$|\mathbf{v}| = \sqrt{2^2 + (-2t)^2} = 2\sqrt{1 + t^2}$

$\mathbf{T} = \frac{\mathbf{v}}{|\mathbf{v}|} = \frac{2t}{2\sqrt{1+t^2}}\mathbf{i} + \frac{-2t}{2\sqrt{1+t^2}}\mathbf{j}$

$= \frac{1}{\sqrt{1+t^2}}\mathbf{i} - \frac{t}{\sqrt{1+t^2}}\mathbf{j}$

$\frac{d\mathbf{T}}{dt} = \frac{-t}{(\sqrt{1+t^2})^3}\mathbf{i} - \frac{1}{(\sqrt{1+t^2})^3}\mathbf{j}$

$\left|\frac{d\mathbf{T}}{dt}\right| = \sqrt{\left(\frac{-t}{(\sqrt{1+t^2})^3}\right)^2 + \left(-\frac{1}{(\sqrt{1+t^2})^3}\right)^2}$

$= \sqrt{\frac{1}{(1+t^2)^2}} = \frac{1}{1+t^2}$

$\mathbf{N} = \frac{\left(\frac{d\mathbf{T}}{dt}\right)}{\left|\frac{d\mathbf{T}}{dt}\right|} = \frac{-t}{\sqrt{1+t^2}}\mathbf{i} - \frac{1}{\sqrt{1+t^2}}\mathbf{j}$

$\mathbf{a} = -2\mathbf{j}$

$\mathbf{v}\times\mathbf{a} = \begin{vmatrix} \mathbf{i} & \mathbf{j} & \mathbf{k} \\ 2 & -2t & 0 \\ 0 & -2 & 0 \end{vmatrix} = -4\mathbf{k}$

$|\mathbf{v}\times\mathbf{a}| = \sqrt{(-4)^2} = 4$

$\kappa = \frac{|\mathbf{v}\times\mathbf{a}|}{|\mathbf{v}|^3} = \frac{4}{(2\sqrt{1+t^2})^3} = \frac{1}{2}(1 + t^2)^{-3/2}$

5. $\mathbf{r} = (2t + 3)\mathbf{i} + (t^2 - 1)\mathbf{j}$

$\mathbf{v} = 2\mathbf{i} + 2t\mathbf{j}$

$|\mathbf{v}| = \sqrt{2^2 + (2t)^2} = 2\sqrt{1 + t^2}$

$a_{\mathrm{T}} = \frac{d}{dt}|\mathbf{v}| = 2\left(\frac{1}{2}\right)(1 + t^2)^{-1/2}(2t) = \frac{2t}{\sqrt{1+t^2}}$

$\mathbf{a} = 2\mathbf{j}$

$|\mathbf{a}| = 2$

$a_{\mathrm{N}} = \sqrt{|\mathbf{a}|^2 - a_{\mathrm{T}}^2} = \sqrt{2^2 - \left(\frac{2t}{\sqrt{1+t^2}}\right)^2} = \frac{2}{\sqrt{1+t^2}}$

$\mathbf{a} = \frac{2t}{\sqrt{1+t^2}}\mathbf{T} + \frac{2}{\sqrt{1+t^2}}\mathbf{N}$

7. a. $\mathbf{r} = x\mathbf{i} + f(x)\mathbf{j}$

$\mathbf{v} = \mathbf{i} + f'(x)\mathbf{j}$

$\mathbf{a} = f''(x)\mathbf{j}$

$$\mathbf{v}\times\mathbf{a}=\begin{vmatrix}\mathbf{i} & \mathbf{j} & \mathbf{k}\\ 1 & f'(x) & 0\\ 0 & f''(x) & 0\end{vmatrix}=f''(x)\mathbf{k}$$

$|\mathbf{v}\times\mathbf{a}|=\sqrt{(f''(x))^2}=|f''(x)|$

and $|\mathbf{v}|=\sqrt{1^2+[f'(x)]^2}=\sqrt{1+[f'(x)]^2}$

$\kappa=\dfrac{|\mathbf{v}\times\mathbf{a}|}{|\mathbf{v}|^3}=\dfrac{|f''(x)|}{[1+(f'(x))^2]^{3/2}}$

b. $y=\ln(\cos x)$

$\dfrac{dy}{dx}=\left(\dfrac{1}{\cos x}\right)(-\sin x)=-\tan x$

$\dfrac{d^2y}{dx^2}=-\sec^2 x$

$\kappa=\dfrac{|-\sec^2 x|}{[1+(-\tan x)^2]^{3/2}}=\dfrac{\sec^2 x}{[\sec^3 x]}=\dfrac{1}{\sec x}=\cos x,$

since $-\dfrac{\pi}{2}<x<\dfrac{\pi}{2}$

c. If $x=x_0$ is a point of inflection, then $f''(x_0)=0$ (since f is twice differentiable), so $\kappa=0$.

9. a. $\mathbf{r}(t)=f(t)\mathbf{i}+g(t)\mathbf{j}$

$\mathbf{v}=f'(t)\mathbf{i}+g'(t)\mathbf{j}$ is tangent to the curve at the point $(f(t),g(t))$.

$\mathbf{n}\cdot\mathbf{v}=[-g'(t)\mathbf{i}+f'(t)\mathbf{j}]\cdot[f'(t)\mathbf{i}+g'(t)\mathbf{j}]$

$=-g'(t)f'(t)+f'(t)g'(t)=0$

$-\mathbf{n}\cdot\mathbf{v}=-(\mathbf{n}\cdot\mathbf{v})=0$

Thus, $\mathbf{n}$ and $-\mathbf{n}$ are both normal to the curve at the point $(f(t),g(t))$.

b. $\mathbf{r}(t)=t\mathbf{i}+e^{2t}\mathbf{j}$

$\mathbf{v}=\mathbf{i}+2e^{2t}\mathbf{j}$

$\mathbf{n}=-2e^{2t}\mathbf{i}+\mathbf{j}$ points toward the concave side of the curve

$\mathbf{N}=\dfrac{\mathbf{n}}{|\mathbf{n}|}$ and $|\mathbf{n}|=\sqrt{4e^{4t}+1}$

$\mathbf{N}=\dfrac{-2e^{2t}}{\sqrt{1+4e^{4t}}}\mathbf{i}+\dfrac{1}{\sqrt{1+4e^{4t}}}\mathbf{j}$

c. $\mathbf{r}(t)=\sqrt{4-t^2}\,\mathbf{i}+t\mathbf{j}$

$\mathbf{v}=\dfrac{-t}{\sqrt{4-t^2}}\mathbf{i}+\mathbf{j}$

$\mathbf{n}=-\mathbf{i}-\dfrac{t}{\sqrt{4-t^2}}\mathbf{j}$ points toward the concave side of the curve

$\mathbf{N}=\dfrac{\mathbf{n}}{|\mathbf{n}|}$ and $|\mathbf{n}|=\sqrt{1+\dfrac{t^2}{4-t^2}}=\dfrac{2}{\sqrt{4-t^2}}$

$\mathbf{N}=-\dfrac{1}{2}(\sqrt{4-t^2}\,\mathbf{i}+t\mathbf{j})$

11. $\mathbf{r}=(3\sin t)\mathbf{i}+(3\cos t)\mathbf{j}+4t\mathbf{k}$

$\mathbf{v}=(3\cos t)\mathbf{i}+(-3\sin t)\mathbf{j}+4\mathbf{k}$

$|\mathbf{v}|=\sqrt{(3\cos t)^2+(-3\sin t)^2+4^2}=\sqrt{25}=5$

$\mathbf{T}=\dfrac{\mathbf{v}}{|\mathbf{v}|}=\left(\dfrac{3}{5}\cos t\right)\mathbf{i}-\left(\dfrac{3}{5}\sin t\right)\mathbf{j}+\dfrac{4}{5}\mathbf{k}$

$\dfrac{d\mathbf{T}}{dt}=\left(-\dfrac{3}{5}\sin t\right)\mathbf{i}-\left(\dfrac{3}{5}\cos t\right)\mathbf{j}$

$\left|\dfrac{d\mathbf{T}}{dt}\right|=\sqrt{\left(-\dfrac{3}{5}\sin t\right)^2+\left(-\dfrac{3}{5}\cos t\right)^2}=\dfrac{3}{5}$

$\mathbf{N}=\dfrac{\left(\frac{d\mathbf{T}}{dt}\right)}{\left|\frac{d\mathbf{T}}{dt}\right|}=(-\sin t)\mathbf{i}-(\cos t)\mathbf{j}$

$$\mathbf{B}=\mathbf{T}\times\mathbf{N}=\begin{vmatrix}\mathbf{i} & \mathbf{j} & \mathbf{k}\\ \frac{3}{5}\cos t & -\frac{3}{5}\sin t & \frac{4}{5}\\ -\sin t & -\cos t & 0\end{vmatrix}$$

$=\left(\dfrac{4}{5}\cos t\right)\mathbf{i}-\left(\dfrac{4}{5}\sin t\right)\mathbf{j}-\dfrac{3}{5}\mathbf{k}$

$\mathbf{a}=(-3\sin t)\mathbf{i}+(-3\cos t)\mathbf{j}$

$$\mathbf{v}\times\mathbf{a}=\begin{vmatrix}\mathbf{i} & \mathbf{j} & \mathbf{k}\\ 3\cos t & -3\sin t & 4\\ -3\sin t & -3\cos t & 0\end{vmatrix}$$

$=(12\cos t)\mathbf{i}-(12\sin t)\mathbf{j}-9\mathbf{k}$

$|\mathbf{v}\times\mathbf{a}|=\sqrt{(12\cos t)^2+(-12\sin t)^2+(-9)^2}$

$=\sqrt{225}=15$

$\kappa=\dfrac{|\mathbf{v}\times\mathbf{a}|}{|\mathbf{v}|^3}=\dfrac{15}{5^3}=\dfrac{3}{25}$

$\dfrac{d\mathbf{a}}{dt}=(-3\cos t)\mathbf{i}+(3\sin t)\mathbf{j}$

$$\tau=\frac{\begin{vmatrix}3\cos t & -3\sin t & 4\\ -3\sin t & -3\cos t & 0\\ -3\cos t & 3\sin t & 0\end{vmatrix}}{|\mathbf{v}\times\mathbf{a}|^2}$$

$=\dfrac{-36\sin^2 t-36\cos^2 t}{15^2}=-\dfrac{4}{25}$

13. $\mathbf{r}=(e^t\cos t)\mathbf{i}+(e^t\sin t)\mathbf{j}+2\mathbf{k}$

$\mathbf{v}=(e^t\cos t-e^t\sin t)\mathbf{i}+(e^t\sin t+e^t\cos t)\mathbf{j}$

$|\mathbf{v}|=\sqrt{(e^t\cos t-e^t\sin t)^2+(e^t\sin t+e^t\cos t)^2}$

$=\sqrt{2e^{2t}}=e^t\sqrt{2}$

$\mathbf{T}=\dfrac{\mathbf{v}}{|\mathbf{v}|}=\left(\dfrac{\cos t-\sin t}{\sqrt{2}}\right)\mathbf{i}+\left(\dfrac{\sin t+\cos t}{\sqrt{2}}\right)\mathbf{j}$

$\dfrac{d\mathbf{T}}{dt}=\left(\dfrac{-\sin t-\cos t}{\sqrt{2}}\right)\mathbf{i}+\left(\dfrac{\cos t-\sin t}{\sqrt{2}}\right)\mathbf{j}$

$\left|\dfrac{d\mathbf{T}}{dt}\right|=\sqrt{\left(\dfrac{-\sin t-\cos t}{\sqrt{2}}\right)^2+\left(\dfrac{\cos t-\sin t}{\sqrt{2}}\right)^2}=1$

$\mathbf{N}=\dfrac{\left(\frac{d\mathbf{T}}{dt}\right)}{\left|\frac{d\mathbf{T}}{dt}\right|}=\left(\dfrac{-\cos t-\sin t}{\sqrt{2}}\right)\mathbf{i}+\left(\dfrac{-\sin t+\cos t}{\sqrt{2}}\right)\mathbf{j}$

$$\mathbf{B}=\mathbf{T}\times\mathbf{N}=\begin{vmatrix}\mathbf{i} & \mathbf{j} & \mathbf{k}\\ \frac{\cos t-\sin t}{\sqrt{2}} & \frac{\sin t+\cos t}{\sqrt{2}} & 0\\ \frac{-\cos t-\sin t}{\sqrt{2}} & \frac{-\sin t+\cos t}{\sqrt{2}} & 0\end{vmatrix}$$

$=\left[\dfrac{1}{2}(\cos t-\sin t)(-\sin t+\cos t)\right.$

$\left.-\dfrac{1}{2}(-\cos t-\sin t)(\sin t+\cos t)\right]\mathbf{k}$

$=\left[\dfrac{1}{2}(\cos^2 t-2\cos t\sin t+\sin^2 t)\right.$

$\left.+\dfrac{1}{2}(\cos^2 t+2\sin t\cos t+\sin^2 t)\right]\mathbf{k}=\mathbf{k}$

$\mathbf{a}=(-2e^t\sin t)\mathbf{i}+(2e^t\cos t)\mathbf{j}$

$$\mathbf{v}\times\mathbf{a} = \begin{vmatrix} \mathbf{i} & \mathbf{j} & \mathbf{k} \\ e^t\cos t - e^t\sin t & e^t\sin t + e^t\cos t & 0 \\ -2e^t\sin t & 2e^t\cos t & 0 \end{vmatrix} = 2e^{2t}\mathbf{k}$$

$$|\mathbf{v}\times\mathbf{a}| = \sqrt{(2e^{2t})^2} = 2e^{2t}$$

$$\kappa = \frac{|\mathbf{v}\times\mathbf{a}|}{|\mathbf{v}|^3} = \frac{2e^{2t}}{(e^t\sqrt{2})^3} = \frac{1}{e^t\sqrt{2}}$$

$$\frac{d\mathbf{a}}{dt} = (-2e^t\sin t - 2e^t\cos t)\mathbf{i} + (2e^t\cos t - 2e^t\sin t)\mathbf{j}$$

$$\tau = \frac{\begin{vmatrix} e^t\cos t - e^t\sin t & e^t\sin t + e^t\cos t & 0 \\ -2e^t\sin t & 2e^t\cos t & 0 \\ -2e^t\sin t - 2e^t\cos t & 2e^t\cos t - 2e^t\sin t & 0 \end{vmatrix}}{|\mathbf{v}\times\mathbf{a}|^2} = 0$$

15. $\mathbf{r} = \left(\frac{t^3}{3}\right)\mathbf{i} + \left(\frac{t^2}{2}\right)\mathbf{j},\ t > 0$

$$\mathbf{v} = t^2\mathbf{i} + t\mathbf{j}$$

$$|\mathbf{v}| = \sqrt{t^4 + t^2} = t\sqrt{t^2+1}, \text{ since } t > 0$$

$$\mathbf{T} = \frac{\mathbf{v}}{|\mathbf{v}|} = \frac{t}{\sqrt{t^2+1}}\mathbf{i} + \frac{1}{\sqrt{t^2+1}}\mathbf{j}$$

$$\frac{d\mathbf{T}}{dt} = \frac{1}{(t^2+1)^{3/2}}\mathbf{i} - \frac{t}{(t^2+1)^{3/2}}\mathbf{j}$$

$$\left|\frac{d\mathbf{T}}{dt}\right| = \sqrt{\left(\frac{1}{(t^2+1)^{3/2}}\right)^2 + \left(\frac{-t}{(t^2+1)^{3/2}}\right)^2} = \sqrt{\frac{1+t^2}{(t^2+1)^3}} = \frac{1}{t^2+1}$$

$$\mathbf{N} = \frac{\left(\frac{d\mathbf{T}}{dt}\right)}{\left|\frac{d\mathbf{T}}{dt}\right|} = \frac{1}{\sqrt{t^2+1}}\mathbf{i} - \frac{t}{\sqrt{t^2+1}}\mathbf{j}$$

$$\mathbf{B} = \mathbf{T}\times\mathbf{N} = \begin{vmatrix} \mathbf{i} & \mathbf{j} & \mathbf{k} \\ \frac{t}{\sqrt{t^2+1}} & \frac{1}{\sqrt{t^2+1}} & 0 \\ \frac{1}{\sqrt{t^2+1}} & \frac{-t}{\sqrt{t^2+1}} & 0 \end{vmatrix} = -\mathbf{k}$$

$$\mathbf{a} = 2t\mathbf{i} + \mathbf{j}$$

$$\mathbf{v}\times\mathbf{a} = \begin{vmatrix} \mathbf{i} & \mathbf{j} & \mathbf{k} \\ t^2 & t & 0 \\ 2t & 1 & 0 \end{vmatrix} = -t^2\mathbf{k}$$

$$|\mathbf{v}\times\mathbf{a}| = \sqrt{(-t^2)^2} = t^2$$

$$\kappa = \frac{|\mathbf{v}\times\mathbf{a}|}{|\mathbf{v}|^3} = \frac{t^2}{(t\sqrt{t^2+1})^3} = \frac{1}{t(t^2+1)^{3/2}}$$

$$\frac{d\mathbf{a}}{dt} = 2\mathbf{i}$$

$$\tau = \frac{\begin{vmatrix} t^2 & t & 0 \\ 2t & 1 & 0 \\ 2 & 0 & 0 \end{vmatrix}}{|\mathbf{v}\times\mathbf{a}|^2} = 0$$

17. $\mathbf{r} = (a\cos t)\mathbf{i} + (a\sin t)\mathbf{j} + bt\mathbf{k}$

$$\mathbf{v} = (-a\sin t)\mathbf{i} + (a\cos t)\mathbf{j} + b\mathbf{k}$$

$$|\mathbf{v}| = \sqrt{(-a\sin t)^2 + (a\cos t)^2 + b^2} = \sqrt{a^2+b^2}$$

$$a_{\mathrm{T}} = \frac{d}{dt}|\mathbf{v}| = 0$$

$$\mathbf{a} = (-a\cos t)\mathbf{i} + (-a\sin t)\mathbf{j}$$

$$|\mathbf{a}| = \sqrt{(-a\cos t)^2 + (-a\sin t)^2} = \sqrt{a^2} = |a|$$

$$a_{\mathrm{N}} = \sqrt{|\mathbf{a}|^2 - a_{\mathrm{T}}^2} = \sqrt{|\mathbf{a}|^2 - 0^2} = |\mathbf{a}| = |a|$$

$$\mathbf{a} = (0)\mathbf{T} + |a|\mathbf{N} = |a|\mathbf{N}$$

19. $\mathbf{r} = (t+1)\mathbf{i} + 2t\mathbf{j} + t^2\mathbf{k}$

$$\mathbf{v} = \mathbf{i} + 2\mathbf{j} + 2t\mathbf{k}$$

$$|\mathbf{v}| = \sqrt{1^2 + 2^2 + (2t)^2} = \sqrt{5+4t^2}$$

$$a_{\mathrm{T}} = \frac{d}{dt}|\mathbf{v}| = \frac{1}{2}(5+4t^2)^{-1/2}(8t) = \frac{4t}{\sqrt{5+4t^2}}$$

$$a_{\mathrm{T}}(1) = \frac{4}{\sqrt{9}} = \frac{4}{3}$$

$$\mathbf{a} = 2\mathbf{k}$$

$$\mathbf{a}(1) = 2\mathbf{k}$$

$$|\mathbf{a}(1)| = 2$$

$$a_{\mathrm{N}}(1) = \sqrt{|\mathbf{a}|^2 - a_{\mathrm{T}}^2} = \sqrt{2^2 - \left(\frac{4}{3}\right)^2} = \sqrt{\frac{20}{9}} = \frac{2\sqrt{5}}{3}$$

$$\mathbf{a}(1) = \frac{4}{3}\mathbf{T} + \frac{2\sqrt{5}}{3}\mathbf{N}$$

21. $\mathbf{r} = t^2\mathbf{i} + \left(t + \frac{1}{3}t^3\right)\mathbf{j} + \left(t - \frac{1}{3}t^3\right)\mathbf{k}$

$$\mathbf{v} = 2t\mathbf{i} + (1+t^2)\mathbf{j} + (1-t^2)\mathbf{k}$$

$$|\mathbf{v}| = \sqrt{(2t)^2 + (1+t^2)^2 + (1-t^2)^2} = \sqrt{2(t^4+2t^2+1)} = \sqrt{2}(1+t^2)$$

$$a_{\mathrm{T}} = 2t\sqrt{2}$$

$$a_{\mathrm{T}}(0) = 0$$

$$\mathbf{a} = 2\mathbf{i} + 2t\mathbf{j} - 2t\mathbf{k}$$

$$\mathbf{a}(0) = 2\mathbf{i}$$

$$|\mathbf{a}(0)| = 2$$

$$a_{\mathrm{N}}(0) = \sqrt{|\mathbf{a}|^2 - a_{\mathrm{T}}^2} = \sqrt{2^2 - 0^2} = 2$$

$$\mathbf{a}(0) = (0)\mathbf{T} + 2\mathbf{N} = 2\mathbf{N}$$

23. $\mathbf{r} = (\cos t)\mathbf{i} + (\sin t)\mathbf{j} - \mathbf{k}$

$$\mathbf{v} = (-\sin t)\mathbf{i} + (\cos t)\mathbf{j}$$

$$|\mathbf{v}| = \sqrt{(-\sin t)^2 + (\cos t)^2} = 1$$

$$\mathbf{T} = \frac{\mathbf{v}}{|\mathbf{v}|} = (-\sin t)\mathbf{i} + (\cos t)\mathbf{j}$$

$$\frac{d\mathbf{T}}{dt} = (-\cos t)\mathbf{i} - (\sin t)\mathbf{j}$$

$$\left|\frac{d\mathbf{T}}{dt}\right| = \sqrt{(-\cos t)^2 + (-\sin t)^2} = 1$$

$$\mathbf{N} = \frac{\left(\frac{d\mathbf{T}}{dt}\right)}{\left|\frac{d\mathbf{T}}{dt}\right|} = (-\cos t)\mathbf{i} - (\sin t)\mathbf{j}$$

Thus, $\mathbf{T}\left(\frac{\pi}{4}\right) = -\frac{\sqrt{2}}{2}\mathbf{i} + \frac{\sqrt{2}}{2}\mathbf{j}$ and $\mathbf{N}\left(\frac{\pi}{4}\right) = -\frac{\sqrt{2}}{2}\mathbf{i} - \frac{\sqrt{2}}{2}\mathbf{j}$.

$$\mathbf{B} = \mathbf{T}\times\mathbf{N} = \begin{vmatrix} \mathbf{i} & \mathbf{j} & \mathbf{k} \\ -\sin t & \cos t & 0 \\ -\cos t & -\sin t & 0 \end{vmatrix} = \mathbf{k}$$

$\mathbf{B}\left(\frac{\pi}{4}\right) = \mathbf{k}$, the normal to the osculating plane

$$\mathbf{r}\left(\frac{\pi}{4}\right) = \frac{\sqrt{2}}{2}\mathbf{i} + \frac{\sqrt{2}}{2}\mathbf{j} - \mathbf{k}$$

$P = \left(\frac{\sqrt{2}}{2}, \frac{\sqrt{2}}{2}, -1\right)$ lies on the osculating plane

$$0\left(x - \frac{\sqrt{2}}{2}\right) + 0\left(y - \frac{\sqrt{2}}{2}\right) + (z - (-1)) = 0$$

The osculating plane is $z = -1$.

$\mathbf{T}$ is normal to the normal plane.

23. (continued)

$$\left(-\frac{\sqrt{2}}{2}\right)\left(x-\frac{\sqrt{2}}{2}\right)+\left(\frac{\sqrt{2}}{2}\right)\left(y-\frac{\sqrt{2}}{2}\right)+0(z-(-1))=0$$

$$-\frac{\sqrt{2}}{2}x+\frac{\sqrt{2}}{2}y=0$$

The normal plane is $-x+y=0$.

N is normal to the rectifying plane.

$$\left(-\frac{\sqrt{2}}{2}\right)\left(x-\frac{\sqrt{2}}{2}\right)+\left(-\frac{\sqrt{2}}{2}\right)\left(y-\frac{\sqrt{2}}{2}\right)+0(z-(-1))=0$$

$$-\frac{\sqrt{2}}{2}x-\frac{\sqrt{2}}{2}y=-1$$

The rectifying plane is $x+y=\sqrt{2}$.

25. Yes. If the car is moving along a curved path, then $\kappa \neq 0$ and $a_N = \kappa|\mathbf{v}|^2 \neq 0$.

$\mathbf{a} = a_T\mathbf{T} + a_N\mathbf{N} \neq 0$.

27. $\mathbf{a} \perp \mathbf{v} = \mathbf{a} \perp \mathbf{T}$

$a_T = 0$

$\frac{d}{dt}|\mathbf{v}| = 0$

The speed $|\mathbf{v}|$ is constant.

29. $\mathbf{a} = a_T\mathbf{T} + a_N\mathbf{N}$, where $a_T = \frac{d}{dt}|\mathbf{v}| = \frac{d}{dt}(\text{constant}) = 0$ and $a_N = \kappa|\mathbf{v}|^2$

$\mathbf{F} = m\mathbf{a} = m\kappa|\mathbf{v}|^2\mathbf{N}$

$|\mathbf{F}| = m\kappa|\mathbf{v}|^2 = (m|\mathbf{v}|^2)\kappa$, a constant multiple of the curvature κ of the trajectory.

31. $y = ax^2$

$y' = 2ax$

$y'' = 2a$

From Exercise 7(a), $\kappa(x) = \frac{|2a|}{(1+4a^2x^2)^{3/2}}$

$= |2a|(1+4a^2x^2)^{-3/2}$

$\kappa'(x) = -\frac{3}{2}|2a|(1+4a^2x^2)^{-5/2}(8a^2x)$

Thus, $\kappa'(x) = 0$ only when $x = 0$.

Now, $\kappa'(x) > 0$ for $x < 0$ and $\kappa'(x) < 0$ for $x > 0$ so that $\kappa(x)$ has an absolute maximum at $x = 0$ which is the vertex of the parabola. Since $x = 0$ is the only critical point for $\kappa(x)$, the curvature has no minimum value.

33. $\kappa = \frac{a}{a^2+b^2}$

$\frac{d\kappa}{da} = \frac{-a^2+b^2}{(a^2+b^2)^2}$

$\frac{d\kappa}{da} = 0$ when $a^2+b^2=0$, which occurs when $a = \pm b$.

$a = b$ since $a, b > 0$.

Now, $\frac{d\kappa}{da} > 0$ if $a < b$ and $\frac{d\kappa}{da} < 0$ if $a > b$.

κ is at a maximum for $a = b$ and $\kappa(b) = \frac{b}{b^2+b^2} = \frac{1}{2b}$ is the maximum value.

35. $\mathbf{r} = (x_0 + At)\mathbf{i} + (y_0 + Bt)\mathbf{j} + (z_0 + Ct)\mathbf{k}$

$\mathbf{v} = A\mathbf{i} + B\mathbf{j} + C\mathbf{k}$

$\mathbf{a} = \mathbf{0}$

$\mathbf{v} \times \mathbf{a} = \mathbf{0}$

$\kappa = 0$

Since the curve is a plane curve, $\tau = 0$.

37. a. From Exercise 34, $\kappa = \frac{1}{t}$ and $|\mathbf{v}| = t$.

$$K = \int_a^b \left(\frac{1}{t}\right)(t)\,dt = b - a$$

b. $y = x^2$

$x = t$ and $y = t^2$, $-\infty < t < \infty$

$\mathbf{r}(t) = t\mathbf{i} + t^2\mathbf{j}$

$\mathbf{v} = \mathbf{i} + 2t\mathbf{j}$

$|\mathbf{v}| = \sqrt{1+4t^2}$

Also, $\mathbf{a} = 2\mathbf{j}$.

$$\mathbf{v}\times\mathbf{a} = \begin{vmatrix} \mathbf{i} & \mathbf{j} & \mathbf{k} \\ 1 & 2t & 0 \\ 0 & 2 & 0 \end{vmatrix} = 2\mathbf{k}$$

$|\mathbf{v}\times\mathbf{a}| = 2$

$$\kappa = \frac{|\mathbf{v}\times\mathbf{a}|}{|\mathbf{v}|^3} = \frac{2}{(\sqrt{1+4t^2})^3}$$

$$\text{Then } K = \int_{-\infty}^{\infty} \frac{2}{(\sqrt{1+4t^2})^3}(\sqrt{1+4t^2})\,dt$$

$$= \int_{-\infty}^{\infty} \frac{2}{1+4t^2}\,dt$$

$$= \lim_{a\to-\infty}\int_a^0 \frac{2}{1+4t^2}\,dt + \lim_{b\to\infty}\int_0^b \frac{2}{1+4t^2}\,dt$$

$$= \lim_{a\to-\infty}[\tan^{-1}2t]_a^0 + \lim_{b\to\infty}[\tan^{-1}2t]_0^b$$

$$= \lim_{a\to-\infty}(-\tan^{-1}2a) + \lim_{b\to\infty}(\tan^{-1}2b)$$

$$= \frac{\pi}{2} + \frac{\pi}{2} = \pi$$

39. If a plane curve is sufficiently differentiable, the torsion is zero as the following argument shows:

$\mathbf{r} = f(t)\mathbf{i} + g(t)\mathbf{j}$

$\mathbf{v} = f'(t)\mathbf{i} + g'(t)\mathbf{j}$

$\mathbf{a} = f''(t)\mathbf{i} + g''(t)\mathbf{j}$

$\frac{d\mathbf{a}}{dt} = f'''(t)\mathbf{i} + g'''(t)\mathbf{j}$

$$\tau = \frac{\begin{vmatrix} f'(t) & g'(t) & 0 \\ f''(t) & g''(t) & 0 \\ f'''(t) & g'''(t) & 0 \end{vmatrix}}{|\mathbf{v}\times\mathbf{a}|^2} = 0$$

41. $y = x^2$

$f'(x) = 2x$ and $f''(x) = 2$

$$\kappa = \frac{|2|}{(1+(2x)^2)^{3/2}} = \frac{2}{(1+4x^2)^{3/2}}$$

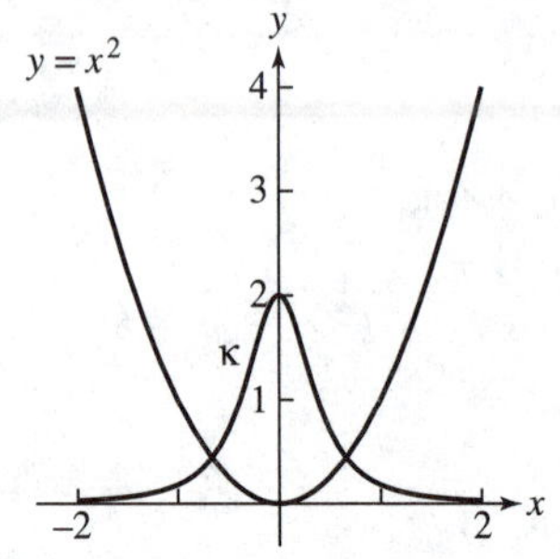

43. $y = \sin x$

$f'(x) = \cos x$ and $f''(x) = -\sin x$

$$\kappa = \frac{|-\sin x|}{(1 + \cos^2 x)^{3/2}} = \frac{|\sin x|}{(1 + \cos^2 x)^{3/2}}$$

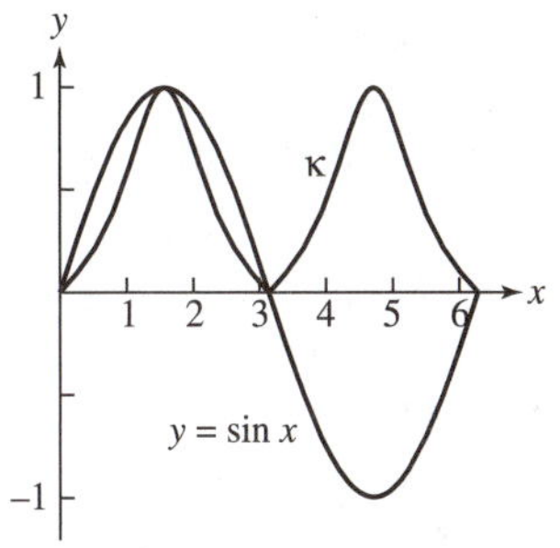

45. $\mathbf{r}(t) = f(t)\mathbf{i} + g(t)\mathbf{j} + h(t)\mathbf{k}$
$\mathbf{v} = f'(t)\mathbf{i} + g'(t)\mathbf{j} + h'(t)\mathbf{k}$
Since $\mathbf{v} \cdot \mathbf{k} = 0$, $h'(t) = 0$ and $h(t) = C$.
Thus, $\mathbf{r}(t) = f(t)\mathbf{i} + g(t)\mathbf{j} + C\mathbf{k}$
and $\mathbf{r}(a) = f(a)\mathbf{i} + g(a)\mathbf{j} + C\mathbf{k} = \mathbf{0}$
$f(a) = 0$, $g(a) = 0$ and $C = 0$, so $h(t) = 0$.

47–50. Example CAS commands:

Maple:

```
with(plots):
x:= t -> t^3 - 2*t^2 - t;
y:= t -> 3*t/sqrt(1+t^2);
dx:= t -> D(x)(t);
dy:= t -> D(y)(t);
ds:= t -> sqrt((dx^2)(t)+(dy^2)(t));
d2x:= t -> D(dx)(t);
d2y:= t -> D(dy)(t);
kap:= t ->
  abs(dx(t)*d2y(t)-dy(t)*d2x(t))
  /((ds)(t))^3;
a:= t -> x0-(1/kap(t))*(dy(t)/ds(t));
b:= t -> x0+(1/kap(t))*(dx(t)/ds(t));
s1:= plot([x(t),y(t),t = -2..5],
  -15..5, -10..4,
  scaling=CONSTRAINED):
display(s1);
t0:=1: x0:= x(t0): y0:= y(t0):
circle:= ((x-a(t0))^2+(y-b(t0))^2
  = (1/kap(t0))^2);
s2:=implicitplot(circle,
  x=-15..6,y=-10..5,
  scaling=CONSTRAINED):
s3:=plot([a(t0),b(t0),x0,y0]):
display({s1,s2,s3});
```

Mathematica:

```
Clear[x,y,t]
r[t_] = {x[t],y[t]}
x[t_] = t^3 - 2 t^2 - t
y[t_] = 3 t / Sqrt[1+t^2]
{a,b} = {-2,5};
t0 = 1;
p1 = ParametricPlot[ {x[t],y[t]},
  {t,a,b}, AspectRatio -> Automatic ]
v0 = r'[t0]
s0 = Sqrt[ v0 . v0]
k0 = Abs[ x' [t0] y''[t0] - y' [t0]
  x''[t0] ]/s0^3
N[%]
n0 = {-y'[t0], x' [t0] } / s0
r0 = r[t0]
c0 = r0 + 1/k0 n0

Note: Plot the circle parametrically
rather than implicitly.

circ = ParametricPlot[ Evaluate[c0 +
  1/k0 {Cos[t],Sin[t]}], {t,0,2Pi},
  AspectRatio -> Automatic ]
line = Graphics[{Line[{c0,r0}]}]
Show[ p1, circ, line ]
```

■ Section 12.4 Planetary Motion and Satellites (pp. 661–670)

Quick Review 12.4

1. $4x^2 + 9y^2 = 36$

$$\frac{x^2}{9} + \frac{y^2}{4} = 1$$

$$e = \frac{c}{a} = \frac{\sqrt{a^2 - b^2}}{a} = \frac{\sqrt{9 - 4}}{3} = \frac{\sqrt{5}}{3}$$

2. $12x^2 - 3y^2 = 24$

$$\frac{x^2}{2} - \frac{y^2}{8} = 1$$

$$e = \frac{c}{a} = \frac{\sqrt{a^2 + b^2}}{a} = \frac{\sqrt{2 + 8}}{\sqrt{2}} = \frac{\sqrt{10}}{\sqrt{2}} = \sqrt{5}$$

3. Since $y^2 + 4y - x + 6 = 0$ represents a parabola, the eccentricity is $e = 1$.

4. Since $e = 1$, the conic is a parabola with focus (0, 0) and directrix $x = -2$.
For (x, y) on the parabola, note that $x \geq -1$, and the distance from (x, y) to the directrix is
$PD = x + 2 = r\cos\theta + 2$.

$$PF = e \cdot PD$$
$$r = 1 \cdot (r\cos\theta + 2)$$
$$r - r\cos\theta = 2$$
$$r = \frac{2}{1 - \cos\theta}$$

5. Since $e = \frac{5}{4}$, the conic is a hyperbola with focus at (0, 0) and directrix $x = 4$. For (x, y) on the hyperbola, the distance from (x, y) to the directrix is

$PD = |x - 4| = |r\cos\theta - 4|$.

$PF = e \cdot PD$

$$r = \frac{5}{4}|r\cos\theta - 4|$$

$4r = 5r\cos\theta - 20$ or $4r = -5r\cos\theta + 20$

$$r = \frac{-20}{4 - 5\cos\theta} \quad \text{or} \quad r = \frac{20}{4 + 5\cos\theta}$$

Note: Either equation, taken by itself, represents the entire hyperbola, so either equation is an acceptable answer for this Exercise.

6. Since $e = \frac{2}{3}$, the conic is an ellipse with focus at (0, 0) and directrix $x = -5$. For (x, y) on the ellipse, the distance from (x, y) to the directrix is $PD = x + 5 = r\cos\theta + 5$.

$$PF = e \cdot PD$$
$$r = \frac{2}{3}(r\cos\theta + 5)$$
$$3r = 2r\cos\theta + 10$$
$$3r - 2r\cos\theta = 10$$
$$r = \frac{10}{3 - 2\cos\theta}$$

7. Major: $(0, \pm 8)$; minor: $(\pm 6, 0)$

8. Major: $(\pm 7, 0)$; minor: $(0, \pm 4)$

9. $$\frac{dy}{dx} = 2(x + y^2x)$$
$$\frac{dy}{dx} = 2x(1 + y^2)$$
$$\int \frac{dy}{1 + y^2} = \int 2x\,dx$$
$$\tan^{-1} y = x^2 + C$$
$$y = \tan(x^2 + C)$$

10. $$(y + 1)\frac{dy}{dx} = y(x - 1)$$
$$\int \frac{y+1}{y}\,dy = \int (x - 1)\,dx$$
$$\int \left(1 + \frac{1}{y}\right) dy = \int (x - 1)\,dx$$
$$y + \ln(|y|) = \frac{x^2}{2} - x + C$$

Section 12.4 Exercises

1. $$\frac{T^2}{a^3} = \frac{4\pi^2}{GM}$$
$$T^2 = \frac{4\pi^2}{GM}a^3$$
$$T^2 = \frac{4\pi^2}{(6.6720 \times 10^{-11}\ \text{Nm}^2\ \text{kg}^{-2})(5.975 \times 10^{24}\ \text{kg})} \cdot (6{,}808{,}000\ \text{m})^3 \approx 3.125 \times 10^7\ \text{sec}^2$$
$$T \approx \sqrt{3125 \times 10^4\ \text{sec}^2} \approx 55.90 \times 10^2\ \text{sec} \approx 93.2\ \text{min}$$

3. 92.25 min = 5535 sec and $\frac{T^2}{a^3} = \frac{4\pi^2}{GM}$
$$a^3 = \frac{GM}{4\pi^2}T^2$$
$$a^3 = \frac{(6.6720 \times 10^{-11}\ \text{Nm}^2\text{kg}^{-2})(5.975 \times 10^{24}\ \text{kg})}{4\pi^2}(5535\ \text{sec})^2$$
$$\approx 3.094 \times 10^{20}\ \text{m}^3$$
$$a \approx \sqrt[3]{3.094 \times 10^{20}\ \text{m}^3} = 6.763 \times 10^6\ \text{m} \approx 6763\ \text{km}$$

The mean distance from center of the Earth
$$= \frac{12{,}757\ \text{km} + 183\ \text{km} + 589\ \text{km}}{2} = 6764.5\ \text{km}.$$

5. From Exercise 4, use $a = 21{,}900$ km.
$2a$ = diameter of Mars + perigee height + apogee height
$= D + 1499\ \text{km} + 35{,}800\ \text{km}$
$2(21{,}900)\ \text{km} = D + 37{,}299\ \text{km}$
$D = 6501\ \text{km}$

7. **a.** Period of the satellite = rotational period of the earth
period of the satellite = 1436.1 min = 86,166 sec
$$a^3 = \frac{GMT^2}{4\pi^2}$$
$$= \frac{(6.6720 \times 10^{-11}\ \text{Nm}^2\ \text{kg}^{-2})(5.975 \times 10^{24}\ \text{kg})(86{,}166\ \text{sec})^2}{4\pi^2}$$
$$\approx 7.4973 \times 10^{22}\ \text{m}^3$$
$$a \approx \sqrt[3]{74.973 \times 10^{21}\ \text{m}^3} \approx 4.2167 \times 10^7\ \text{m}$$
$$\approx 42{,}167\ \text{km}$$

b. The equatorial radius of the earth is approximately 6379 km. The height of the orbit is
42,167 − 6379 = 35,788 km.

c. Symcom 3, GOES 4, and Intelsat 5

9. Period of the Moon = 2.36055×10^6 sec
$$a^3 = \frac{GMT^2}{4\pi^2}$$
$$= \frac{(6.6720 \times 10^{-11}\ \text{Nm}^2\text{kg}^{-2})(5.975 \times 10^{24}\ \text{kg})(2.36055 \times 10^6\ \text{sec})^2}{4\pi^2}$$
$$\approx 5.627 \times 10^{25}\ \text{m}^3$$
$$a \approx \sqrt[3]{5.627 \times 10^{25}\ \text{m}^3} \approx 3.832 \times 10^8\ \text{m} \approx 383{,}200\ \text{km}$$
The moon is about 383,200 km from the center of the earth, or about 376,821 km from the surface.

11. Solar System:
$$\frac{T^2}{a^3} = \frac{4\pi^2}{(6.6720 \times 10^{-11}\ \text{Nm}^2\ \text{kg}^{-2})(1.99 \times 10^{30}\ \text{kg})}$$
$$\approx 2.97 \times 10^{-19}\ \text{sec}^2/\text{m}^3;$$
Earth: $$\frac{T^2}{a^3} = \frac{4\pi^2}{(6.6720 \times 10^{-11}\ \text{Nm}^2\ \text{kg}^{-2})(5.975 \times 10^{24}\ \text{kg})}$$
$$\approx 9.903 \times 10^{-14}\ \text{sec}^2/\text{m}^3;$$
Moon: $$\frac{T^2}{a^3} = \frac{4\pi^2}{(6.6720 \times 10^{-11}\ \text{Nm}^2\ \text{kg}^{-2})(7.354 \times 10^{22}\ \text{kg})}$$
$$\approx 8.046 \times 10^{-12}\ \text{sec}^2/\text{m}^3$$

13. Since the eccentricity of a circle is 0, $r = \frac{GM}{v^2}$.
$$v^2 = \frac{GM}{r}$$
$v = \sqrt{\frac{GM}{r}}$ which is constant since G, M, and r (the radius of orbit) are constant.

15. $$T = \left(\frac{2\pi a^2}{r_0 v_0}\right)\sqrt{1 - e^2}$$
$$T^2 = \left(\frac{4\pi^2 a^4}{r_0^2 v_0^2}\right)(1 - e^2)$$
$$= \left(\frac{4\pi^2 a^4}{r_0^2 v_0^2}\right)\left[1 - \left(\frac{r_0 v_0^2}{GM} - 1\right)^2\right] \text{ (from Equation 32)}$$
$$= \left(\frac{4\pi^2 a^4}{r_0^2 v_0^2}\right)\left[-\frac{r_0^2 v_0^4}{G^2M^2} + 2\left(\frac{r_0 v_0^2}{GM}\right)\right]$$
$$= \left(\frac{4\pi^2 a^4}{r_0^2 v_0^2}\right)\left[\frac{2GMr_0 v_0^2 - r_0^2 v_0^4}{G^2M^2}\right]$$
$$= \frac{(4\pi^2 a^4)(2GM - r_0 v_0^2)}{r_0 G^2 M^2}$$

$$= (4\pi^2 a^4)\left(\frac{2GM - r_0 v_0^2}{2r_0 GM}\right)\left(\frac{2}{GM}\right)$$

$$= (4\pi^2 a^4)\left(\frac{1}{2a}\right)\left(\frac{2}{GM}\right) \text{ (from Equation 35)}$$

$$T^2 = \frac{4\pi^2 a^3}{GM}$$

$$\frac{T^2}{a^3} = \frac{4\pi^2}{GM}$$

17. Setting $\theta = \pi t$ and $r = 3 - 4\cos\theta$, we see that $x - 2 = r\cos\theta$ and $y = r\sin\theta$.
The graph of the path of planet B is the limacon $r = 3 - 4\cos\theta$ shown below. The planet A is located at $x = -2$.

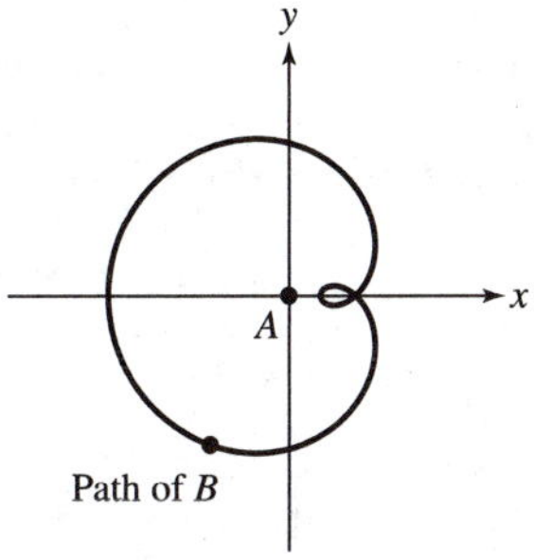

Chapter 12 Review Exercises

(pp. 671–673)

1. $\mathbf{r}(t) = (4\cos t)\mathbf{i} + (\sqrt{2}\sin t)\mathbf{j}$

$x = \cos t$ and $y = \sqrt{2}\sin t$

$$\frac{x^2}{16} + \frac{y^2}{2} = 1$$

$\mathbf{v} = (-4\sin t)\mathbf{i} + (\sqrt{2}\cos t)\mathbf{j}$

and $\mathbf{a} = (-4\cos t)\mathbf{i} - (\sqrt{2}\sin t)\mathbf{j}$

$\mathbf{r}(0) = 4\mathbf{i}$, $\mathbf{v}(0) = \sqrt{2}\mathbf{j}$, $\mathbf{a}(0) = -4\mathbf{i}$

$\mathbf{r}\left(\frac{\pi}{4}\right) = 2\sqrt{2}\mathbf{i} + \mathbf{j}$, $\mathbf{v}\left(\frac{\pi}{4}\right) = -2\sqrt{2}\mathbf{i} + \mathbf{j}$,

$\mathbf{a}\left(\frac{\pi}{4}\right) = -2\sqrt{2}\mathbf{i} - \mathbf{j}$

$|\mathbf{v}| = \sqrt{16\sin^2 t + 2\cos^2 t}$

$$a_T = \frac{d}{dt}|\mathbf{v}| = \frac{14\sin t\cos t}{\sqrt{16\sin^2 t + 2\cos^2 t}}$$

At $t = 0$: $a_T = 0$, $a_N = \sqrt{|\mathbf{a}|^2 - 0} = 4$, $\mathbf{a} = 4\mathbf{N}$

$$\kappa = \frac{a_N}{|\mathbf{v}|^2} = \frac{4}{2} = 2$$

At $t = \frac{\pi}{4}$: $a_T = \frac{7}{\sqrt{8+1}} = \frac{7}{3}$, $a_N = \sqrt{9 - \frac{49}{9}} = \frac{4\sqrt{2}}{3}$,

$\mathbf{a} = \frac{7}{3}\mathbf{T} + \frac{4\sqrt{2}}{3}\mathbf{N}$, $\kappa = \frac{a_N}{|\mathbf{v}|^2} = \frac{4\sqrt{2}}{27}$

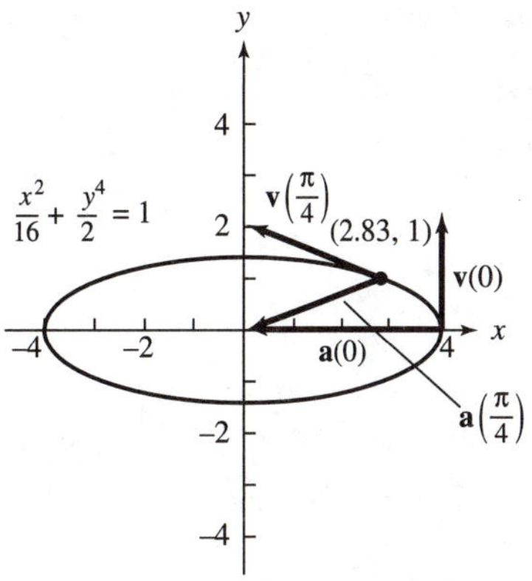

2. $\mathbf{r}(t) = (\sqrt{3}\sec t)\mathbf{i} + (\sqrt{3}\tan t)\mathbf{j}$

$x = \sqrt{3}\sec t$ and $y = \sqrt{3}\tan t$

$$\frac{x^2}{3} - \frac{y^2}{3} = \sec^2 t - \tan^2 t = 1$$

$x^2 - y^2 = 3$

$\mathbf{v} = (\sqrt{3}\sec t\tan t)\mathbf{i} + (\sqrt{3}\sec^2 t)\mathbf{j}$ and

$\mathbf{a} = (\sqrt{3}\sec t\tan^2 t + \sqrt{3}\sec^3 t)\mathbf{i} - (2\sqrt{3}\sec^2 t\tan t)\mathbf{j}$

$\mathbf{r}(0) = \sqrt{3}\mathbf{i}$, $\mathbf{v}(0) = \sqrt{3}\mathbf{j}$, $\mathbf{a}(0) = \sqrt{3}\mathbf{i}$

$|\mathbf{v}| = \sqrt{3\sec^2 t\tan^2 t + 3\sec^4 t}$

$$a_T = \frac{d}{dt}|\mathbf{v}| = \frac{6\sec^2 t\tan^3 t + 18\sec^4 t\tan t}{2\sqrt{3\sec^2 t\tan^2 t + 3\sec^4 t}}$$

At $t = 0$: $a_T = 0$, $a_N = \sqrt{|\mathbf{a}|^2 - 0} = \sqrt{3}$, $\mathbf{a} = \sqrt{3}\mathbf{N}$,

$$\kappa = \frac{a_N}{|\mathbf{v}|^2} = \frac{\sqrt{3}}{3} = \frac{1}{\sqrt{3}}$$

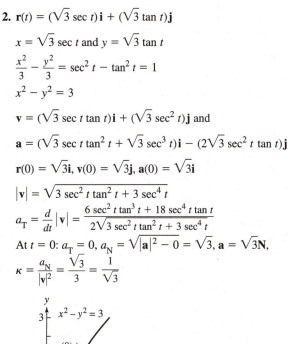

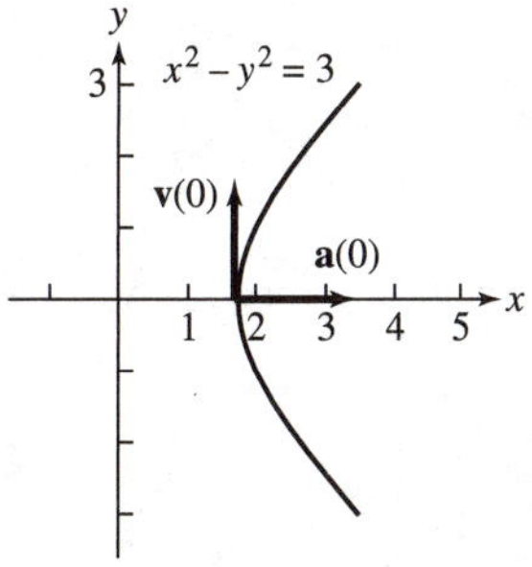

3. $\mathbf{r} = \frac{1}{\sqrt{1+t^2}}\mathbf{i} + \frac{t}{\sqrt{1+t^2}}\mathbf{j}$

$\mathbf{v} = -t(1+t^2)^{-3/2}\mathbf{i} + (1+t^2)^{-3/2}\mathbf{j}$

$$|\mathbf{v}| = \sqrt{[-t(1+t^2)^{-3/2}]^2 + [(1+t^2)^{-3/2}]^2} = \frac{1}{1+t^2}$$

We want to maximize $|\mathbf{v}|$: $\frac{d|\mathbf{v}|}{dt} = \frac{-2t}{(1+t^2)^2}$ and $\frac{d|\mathbf{v}|}{dt} = 0$

$$\frac{-2t}{(1+t^2)^2} = 0$$

$t = 0$

For $t < 0$, $\frac{-2t}{(1+t^2)^2} > 0$.

For $t > 0$, $\frac{-2t}{(1+t^2)^2} < 0$.

$|\mathbf{v}|_{max}$ occurs when $t = 0$

$|\mathbf{v}|_{max} = 1$

4. $\mathbf{r} = (e^t \cos t)\mathbf{i} + (e^t \sin t)\mathbf{j}$

$\mathbf{v} = (e^t \cos t - e^t \sin t)\mathbf{i} + (e^t \sin t + e^t \cos t)\mathbf{j}$

$\mathbf{a} = (e^t \cos t - e^t \sin t - e^t \sin t - e^t \cos t)\mathbf{i} + (e^t \sin t + e^t \cos t + e^t \cos t - e^t \sin t)\mathbf{j}$

$= (-2e^t \sin t)\mathbf{i} + (2e^t \cos t)\mathbf{j}$

Let θ be the angle between $\mathbf{r}$ and $\mathbf{a}$.

Then $\theta = \cos^{-1}\left(\dfrac{\mathbf{r}\cdot\mathbf{a}}{|\mathbf{r}||\mathbf{a}|}\right)$

$$= \cos^{-1}\left(\frac{-2e^{2t}\sin t\cos t + 2e^{2t}\sin t\cos t}{\sqrt{(e^t\cos t)^2 + (e^t \sin t)^2}\sqrt{(-2e^t\sin t)^2 + (2e^t\cos t)^2}}\right)$$

$= \cos^{-1}\left(\dfrac{0}{2e^{2t}}\right) = \cos^{-1} 0 = \dfrac{\pi}{2}$ for all t

5. $\mathbf{v} = 3\mathbf{i} + 4\mathbf{j}$ and $\mathbf{a} = 5\mathbf{i} + 15\mathbf{j}$

$$\mathbf{v}\times\mathbf{a} = \begin{vmatrix} \mathbf{i} & \mathbf{j} & \mathbf{k} \\ 3 & 4 & 0 \\ 5 & 15 & 0 \end{vmatrix} = 25\mathbf{k}$$

$|\mathbf{v}\times\mathbf{a}| = 25$

$|\mathbf{v}| = \sqrt{3^2 + 4^2} = 5$

$\kappa = \dfrac{|\mathbf{v}\times\mathbf{a}|}{|\mathbf{v}|^3} = \dfrac{25}{5^3} = \dfrac{1}{5}$

6. $\kappa = \dfrac{|y''|}{[1 + (y')^2]^{3/2}} = e^x(1 + e^{2x})^{-3/2}$

$\dfrac{d\kappa}{dx} = e^x(1 + e^{2x})^{-3/2} + e^x\left[-\dfrac{3}{2}(1 + e^{2x})^{-5/2}(2e^{2x})\right]$

$= e^x(1 + e^{2x})^{-3/2} - 3e^{3x}(1 + e^{2x})^{-5/2}$

$= e^x(1 + e^{2x})^{-5/2}[(1 + e^{2x}) - 3e^{2x}]$

$= e^x(1 + e^{2x})^{-5/2}(1 - 2e^{2x})$

$\dfrac{d\kappa}{dx} = 0$ when $(1 - 2e^{2x}) = 0$.

$e^{2x} = \dfrac{1}{2}$

$2x = -\ln 2$

$x = -\dfrac{1}{2}\ln 2 = -\ln\sqrt{2}$

$y = \dfrac{1}{\sqrt{2}}$

Therefore κ is at a maximum at the point $\left(-\ln\sqrt{2}, \dfrac{1}{\sqrt{2}}\right)$.

7. $\mathbf{r} = x\mathbf{i} + y\mathbf{j}$

$\mathbf{v} = \dfrac{dx}{dt}\mathbf{i} + \dfrac{dy}{dt}\mathbf{j}$ and $\mathbf{v}\cdot\mathbf{i} = y$, hence

$\dfrac{dx}{dt} = y$

Since the particle moves around the unit circle,

$x^2 + y^2 = 1,\ 2x\dfrac{dx}{dt} + 2y\dfrac{dy}{dt} = 0$

$\dfrac{dy}{dt} = -\dfrac{x}{y}\dfrac{dx}{dt}$

$\dfrac{dy}{dt} = -\dfrac{x}{y}(y) = -x.$

Since $\dfrac{dx}{dt} = y$ and $\dfrac{dy}{dt} = -x$, we have $\mathbf{v} = y\mathbf{i} - x\mathbf{j}$

At $(1, 0)$, $\mathbf{v} = -\mathbf{j}$ and the motion is clockwise.

8. $9y = x^3$

$9\dfrac{dy}{dt} = 3x^2\dfrac{dx}{dt}$

$\dfrac{dy}{dt} = \dfrac{1}{3}x^2\dfrac{dx}{dt}$

If $\mathbf{r} = x\mathbf{i} + y\mathbf{j}$, where x and y are differentiable functions of t, then $\mathbf{v} = \dfrac{dx}{dt}\mathbf{i} + \dfrac{dy}{dt}\mathbf{j}$. Hence $\mathbf{v}\cdot\mathbf{i} = 4$ means that $\dfrac{dx}{dt} = 4$ and $\mathbf{v}\cdot\mathbf{j} = \dfrac{dy}{dt} = \dfrac{1}{3}x^2\dfrac{dx}{dt} = \dfrac{1}{3}(3)^2(4) = 12$ at $(3, 3)$. Also, $\mathbf{a} = \dfrac{d^2x}{dt^2}\mathbf{i} + \dfrac{d^2y}{dt^2}\mathbf{j}$ and $\dfrac{d^2y}{dt^2} = \left(\dfrac{2}{3}x\right)\left(\dfrac{dx}{dt}\right)^2 + \left(\dfrac{1}{3}x^2\right)\dfrac{d^2x}{dt^2}$.

Hence $\mathbf{a}\cdot\mathbf{i} = -2$ means that

$\dfrac{d^2x}{dt^2} = -2$ and $\mathbf{a}\cdot\mathbf{j} = \dfrac{d^2y}{dt^2} = \dfrac{2}{3}(3)(4)^2 + \dfrac{1}{3}(3)^2(-2) = 26$ at the point $(x, y) = (3, 3)$.

9. $\dfrac{d\mathbf{r}}{dt} = -t\mathbf{i} - t\mathbf{j} - t\mathbf{k}$

$\mathbf{r} = -\dfrac{t^2}{2}\mathbf{i} - \dfrac{t^2}{2}\mathbf{j} - \dfrac{t^2}{2}\mathbf{k} + \mathbf{C}$

At $t = 0$, $\mathbf{r} = \mathbf{r}(0) = \mathbf{C} = \mathbf{i} + 2\mathbf{j} + 3\mathbf{k}$.

$\mathbf{r} = -\dfrac{t^2}{2}\mathbf{i} - \dfrac{t^2}{2}\mathbf{j} - \dfrac{t^2}{2}\mathbf{k} + (\mathbf{i} + 2\mathbf{j} + 3\mathbf{k})$

$= \left(-\dfrac{t^2}{2} + 1\right)\mathbf{i} + \left(-\dfrac{t^2}{2} + 2\right)\mathbf{j} + \left(-\dfrac{t^2}{2} + 3\right)\mathbf{k}$

10. $\dfrac{d\mathbf{r}}{dt} = (180t)\mathbf{i} + (180t - 16t^2)\mathbf{j}$

$\mathbf{r} = 90t^2\mathbf{i} + \left(90t^2 - \dfrac{16}{3}t^3\right)\mathbf{j} + \mathbf{C}$

At $t = 0$, $\mathbf{r} = \mathbf{r}(0) = \mathbf{C} = 100\mathbf{j}$.

$\mathbf{r} = 90t^2\mathbf{i} + \left(90t^2 - \dfrac{16}{3}t^3\right)\mathbf{j} + 100\mathbf{j}$

$= 90t^2\mathbf{i} + \left(90t^2 - \dfrac{16}{3}t^3 + 100\right)\mathbf{j}$

11. $\dfrac{d\mathbf{r}}{dt}$ orthogonal to $\mathbf{r}$

$0 = \dfrac{d\mathbf{r}}{dt}\cdot\mathbf{r} = \dfrac{1}{2}\dfrac{d\mathbf{r}}{dt}\cdot\mathbf{r} + \dfrac{1}{2}\mathbf{r}\cdot\dfrac{d\mathbf{r}}{dt} = \dfrac{1}{2}\dfrac{d}{dt}(\mathbf{r}\cdot\mathbf{r})$

$\mathbf{r}\cdot\mathbf{r} = K$, a constant. If $\mathbf{r} = x\mathbf{i} + y\mathbf{j}$, where x and y are differentiable functions of t, then $\mathbf{r}\cdot\mathbf{r} = x^2 + y^2$

$x^2 + y^2 = K$, which is the equation of a circle centered at the origin.

12. $\ddot{s} = \dfrac{d}{dt}\sqrt{\dot{x}^2 + \dot{y}^2} = \dfrac{\dot{x}\ddot{x} + \dot{y}\ddot{y}}{\sqrt{\dot{x}^2 + \dot{y}^2}}$

$\ddot{x}^2 + \ddot{y}^2 - \ddot{s}^2 = \ddot{x}^2 + \ddot{y}^2 - \dfrac{(\dot{x}\ddot{x} + \dot{y}\ddot{y})^2}{\dot{x}^2 + \dot{y}^2}$

$= \dfrac{(\ddot{x}^2 + \ddot{y}^2)(\dot{x}^2 + \dot{y}^2) - (\dot{x}^2\ddot{x}^2 + 2\dot{x}\ddot{x}\dot{y}\ddot{y} + \dot{y}^2\ddot{y}^2)}{\dot{x}^2 + \dot{y}^2}$

$= \dfrac{\dot{x}^2\ddot{y}^2 + \dot{y}^2\ddot{x}^2 + 2\dot{x}\ddot{x}\dot{y}\ddot{y}}{\dot{x}^2 + \dot{y}^2} = \dfrac{(\dot{x}\ddot{y} - \dot{y}\ddot{x})^2}{\dot{x}^2 + \dot{y}^2}$

$\sqrt{\ddot{x}^2 + \ddot{y}^2 - \ddot{s}^2} = \dfrac{|\dot{x}\ddot{y} - \dot{y}\ddot{x}|}{\sqrt{\dot{x}^2 - \dot{y}^2}}$

$\dfrac{\dot{x}^2 + \dot{y}^2}{\sqrt{\ddot{x}^2 + \ddot{y}^2 - \ddot{s}^2}} = \dfrac{(\dot{x}^2 + \dot{y}^2)^{3/2}}{|\dot{x}\ddot{y} - \dot{y}\ddot{x}|} = \dfrac{1}{\kappa} = \rho$

13. $\mathbf{r} = (2\cos t)\mathbf{i} + (2\sin t)\mathbf{j} + t^2\mathbf{k}$

$\mathbf{v} = (-2\sin t)\mathbf{i} + (2\cos t)\mathbf{j} + 2t\mathbf{k}$

$|\mathbf{v}| = \sqrt{(-2\sin t)^2 + (2\cos t)^2 + (2t)^2} = 2\sqrt{1 + t^2}$

Length $= \displaystyle\int_0^{\pi/4} 2\sqrt{1 + t^2}\,dt$

$$= \left[t\sqrt{1+t^2} + \ln\left|t+\sqrt{1+t^2}\right|\right]_0^{\pi/4}$$

$$= \frac{\pi}{4}\sqrt{1+\frac{\pi^2}{16}} + \ln\left(\frac{\pi}{4} + \sqrt{1+\frac{\pi^2}{16}}\right) \approx 1.72$$

14. $\mathbf{r} = (3\cos t)\mathbf{i} + (3\sin t)\mathbf{j} + 2t^{3/2}\mathbf{k}$

$\mathbf{v} = (-3\sin t)\mathbf{i} + (3\cos t)\mathbf{j} + 3t^{1/2}\mathbf{k}$

$|\mathbf{v}| = \sqrt{(-3\sin t)^2 + (3\cos t)^2 + (3t^{1/2})^2}$

$= 3\sqrt{1+t}$

$\text{Length} = \int_0^3 3\sqrt{1+t}\,dt = \left[2(1+t)^{3/2}\right]_0^3 = 14$

15. $\mathbf{r} = \frac{4}{9}(1+t)^{3/2}\mathbf{i} + \frac{4}{9}(1-t)^{3/2}\mathbf{j} + \frac{1}{3}t\mathbf{k}$

$\mathbf{v} = \frac{2}{3}(1+t)^{1/2}\mathbf{i} - \frac{2}{3}(1-t)^{1/2}\mathbf{j} + \frac{1}{3}\mathbf{k}$

$|\mathbf{v}| = \sqrt{\left[\frac{2}{3}(1+t)^{1/2}\right]^2 + \left[-\frac{2}{3}(1-t)^{1/2}\right]^2 + \left(\frac{1}{3}\right)^2} = 1$

$\mathbf{T} = \frac{\mathbf{v}}{|\mathbf{v}|} = \frac{2}{3}(1+t)^{1/2}\mathbf{i} - \frac{2}{3}(1-t)^{1/2}\mathbf{j} + \frac{1}{3}\mathbf{k}$

$\mathbf{T}(0) = \frac{2}{3}\mathbf{i} - \frac{2}{3}\mathbf{j} + \frac{1}{3}\mathbf{k}$

$\frac{d\mathbf{T}}{dt} = \frac{1}{3}(1+t)^{-1/2}\mathbf{i} + \frac{1}{3}(1-t)^{-1/2}\mathbf{j}$

$\frac{d\mathbf{T}}{dt}(0) = \frac{1}{3}\mathbf{i} + \frac{1}{3}\mathbf{j}$

$\left|\frac{d\mathbf{T}}{dt}(0)\right| = \frac{\sqrt{2}}{3}$

$\mathbf{N}(0) = \frac{\frac{d\mathbf{T}}{dt}(0)}{\left|\frac{d\mathbf{T}}{dt}(0)\right|} = \frac{1}{\sqrt{2}}\mathbf{i} + \frac{1}{\sqrt{2}}\mathbf{j}$

$$\mathbf{B}(0) = \mathbf{T}(0) \times \mathbf{N}(0) = \begin{vmatrix} \mathbf{i} & \mathbf{j} & \mathbf{k} \\ \frac{2}{3} & -\frac{2}{3} & \frac{1}{3} \\ \frac{1}{\sqrt{2}} & \frac{1}{\sqrt{2}} & 0 \end{vmatrix}$$

$$= -\frac{1}{3\sqrt{2}}\mathbf{i} + \frac{1}{3\sqrt{2}}\mathbf{j} + \frac{4}{3\sqrt{2}}\mathbf{k}$$

$\mathbf{a} = \frac{1}{3}(1+t)^{-1/2}\mathbf{i} + \frac{1}{3}(1-t)^{-1/2}\mathbf{j}$

$\mathbf{a}(0) = \frac{1}{3}\mathbf{i} + \frac{1}{3}\mathbf{j}$ and $\mathbf{v}(0) = \frac{2}{3}\mathbf{i} - \frac{2}{3}\mathbf{j} + \frac{1}{3}\mathbf{k}$

$$\mathbf{v}(0) \times \mathbf{a}(0) = \begin{vmatrix} \mathbf{i} & \mathbf{j} & \mathbf{k} \\ \frac{2}{3} & -\frac{2}{3} & \frac{1}{3} \\ \frac{1}{3} & \frac{1}{3} & 0 \end{vmatrix}$$

$$= -\frac{1}{9}\mathbf{i} + \frac{1}{9}\mathbf{j} + \frac{4}{9}\mathbf{k}$$

$|\mathbf{v}\times\mathbf{a}| = \frac{\sqrt{2}}{3}$

$\kappa(0) = \frac{|\mathbf{v}\times\mathbf{a}|}{|\mathbf{v}|^3} = \frac{\left(\frac{\sqrt{2}}{3}\right)}{1^3} = \frac{\sqrt{2}}{3}$

$\dot{\mathbf{a}} = -\frac{1}{6}(1+t)^{-3/2}\mathbf{i} + \frac{1}{6}(1-t)^{-3/2}\mathbf{j}$

$\dot{\mathbf{a}}(0) = -\frac{1}{6}\mathbf{i} + \frac{1}{6}\mathbf{j}$

$$\tau(0) = \frac{\begin{vmatrix} \frac{2}{3} & -\frac{2}{3} & \frac{1}{3} \\ \frac{1}{3} & \frac{1}{3} & 0 \\ -\frac{1}{6} & \frac{1}{6} & 0 \end{vmatrix}}{|\mathbf{v}\times\mathbf{a}|^2} = \frac{\left(\frac{1}{3}\right)\left(\frac{2}{18}\right)}{\left(\frac{\sqrt{2}}{3}\right)^2} = \frac{1}{6}$$

When $t = 0$, $\left(\frac{4}{9}, \frac{4}{9}, 0\right)$ is the point on the curve.

16. $\mathbf{r} = (e^t\sin 2t)\mathbf{i} + (e^t\cos 2t)\mathbf{j} + 2e^t\mathbf{k}$

$\mathbf{v} = (e^t\sin 2t + 2e^t\cos 2t)\mathbf{i}$
$\quad + (e^t\cos 2t - 2e^t\sin 2t)\mathbf{j} + 2e^t\mathbf{k}$

$|\mathbf{v}|$

$= \sqrt{(e^t\sin 2t + 2e^t\cos 2t)^2 + (e^t\cos 2t - 2e^t\sin 2t)^2 + (2e^t)^2}$

$= 3e^t$

$\mathbf{T} = \frac{\mathbf{v}}{|\mathbf{v}|} = \left(\frac{1}{3}\sin 2t + \frac{2}{3}\cos 2t\right)\mathbf{i} + \left(\frac{1}{3}\cos 2t - \frac{2}{3}\sin 2t\right)\mathbf{j}$
$\quad + \frac{2}{3}\mathbf{k}$

$\mathbf{T}(0) = \frac{2}{3}\mathbf{i} + \frac{1}{3}\mathbf{j} + \frac{2}{3}\mathbf{k}$

$\frac{d\mathbf{T}}{dt} = \left(\frac{2}{3}\cos 2t - \frac{4}{3}\sin 2t\right)\mathbf{i} + \left(-\frac{2}{3}\sin 2t - \frac{4}{3}\cos 2t\right)\mathbf{j}$

$\frac{d\mathbf{T}}{dt}(0) = \frac{2}{3}\mathbf{i} - \frac{4}{3}\mathbf{j}$

$\left|\frac{d\mathbf{T}}{dt}(0)\right| = \frac{2\sqrt{5}}{3}$

$\mathbf{N}(0) = \frac{\frac{d\mathbf{T}}{dt}(0)}{\left|\frac{d\mathbf{T}}{dt}(0)\right|} = \frac{\left(\frac{2}{3}\mathbf{i} - \frac{4}{3}\mathbf{j}\right)}{\left(\frac{2\sqrt{5}}{3}\right)} = \frac{1}{\sqrt{5}}\mathbf{i} - \frac{2}{\sqrt{5}}\mathbf{j}$

$$\mathbf{B}(0) = \mathbf{T}(0) \times \mathbf{N}(0) = \begin{vmatrix} \mathbf{i} & \mathbf{j} & \mathbf{k} \\ \frac{2}{3} & \frac{1}{3} & \frac{2}{3} \\ \frac{1}{\sqrt{5}} & -\frac{2}{\sqrt{5}} & 0 \end{vmatrix}$$

$$= \frac{4}{3\sqrt{5}}\mathbf{i} + \frac{2}{3\sqrt{5}}\mathbf{j} - \frac{5}{3\sqrt{5}}\mathbf{k}$$

$\mathbf{a} = (4e^t\cos 2t - 3e^t\sin 2t)\mathbf{i} + (-3e^t\cos 2t - 4e^t\sin 2t)\mathbf{j}$
$\quad + 2e^t\mathbf{k}$

$\mathbf{a}(0) = 4\mathbf{i} - 3\mathbf{j} + 2\mathbf{k}$ and $\mathbf{v}(0) = 2\mathbf{i} + \mathbf{j} + 2\mathbf{k}$

$$\mathbf{v}(0) \times \mathbf{a}(0) = \begin{vmatrix} \mathbf{i} & \mathbf{j} & \mathbf{k} \\ 2 & 1 & 2 \\ 4 & -3 & 2 \end{vmatrix} = 8\mathbf{i} + 4\mathbf{j} - 10\mathbf{k}$$

$|\mathbf{v}\times\mathbf{a}| = \sqrt{64+16+100} = 6\sqrt{5}$ and $|\mathbf{v}(0)| = 3$

$\kappa(0) = \frac{|\mathbf{v}\times\mathbf{a}|}{|\mathbf{v}|^3} = \frac{6\sqrt{5}}{3^3} = \frac{2\sqrt{5}}{9}$

$\dot{\mathbf{a}} = (4e^t\cos 2t - 8e^t\sin 2t - 3e^t\sin 2t - 6e^t\cos 2t)\mathbf{i}$
$\quad + (-3e^t\cos 2t + 6e^t\sin 2t - 4e^t\sin 2t - 8e^t\cos 2t)\mathbf{j}$
$\quad + 2e^t\mathbf{k}$

$= (-2e^t\cos 2t - 11e^t\sin 2t)\mathbf{i}$
$\quad + (-11e^t\cos 2t + 2e^t\sin 2t)\mathbf{j} + 2e^t\mathbf{k}$

$\dot{\mathbf{a}}(0) = -2\mathbf{i} - 11\mathbf{j} + 2\mathbf{k}$

$$\tau(0) = \frac{\begin{vmatrix} 2 & 1 & 2 \\ 4 & -3 & 2 \\ -2 & -11 & 2 \end{vmatrix}}{|\mathbf{v} \times \mathbf{a}|^2} = \frac{-80}{180} = -\frac{4}{9}$$

When $t = 0$, $(0, 1, 2)$ is on the curve.

17. $\mathbf{r} = t\mathbf{i} + \frac{1}{2}e^{2t}\mathbf{j}$

$\mathbf{v} = \mathbf{i} + e^{2t}\mathbf{j}$

$|\mathbf{v}| = \sqrt{1 + e^{4t}}$

$$\mathbf{T} = \frac{\mathbf{v}}{|\mathbf{v}|} = \frac{1}{\sqrt{1+e^{4t}}}\mathbf{i} + \frac{e^{2t}}{\sqrt{1+e^{4t}}}\mathbf{j}$$

$$\mathbf{T}(\ln 2) = \frac{1}{\sqrt{17}}\mathbf{i} + \frac{4}{\sqrt{17}}\mathbf{j}$$

$$\frac{d\mathbf{T}}{dt} = \frac{-2e^{4t}}{(1+e^{4t})^{3/2}}\mathbf{i} + \frac{2e^{2t}}{(1+e^{4t})^{3/2}}\mathbf{j}$$

$$\frac{d\mathbf{T}}{dt}(\ln 2) = \frac{-32}{17\sqrt{17}}\mathbf{i} + \frac{8}{17\sqrt{17}}\mathbf{j}$$

$$\left|\frac{d\mathbf{T}}{dt}(\ln 2)\right| = \sqrt{\left(\frac{-32}{17\sqrt{17}}\right)^2 + \left(\frac{8}{17\sqrt{17}}\right)^2} = \frac{8}{17}$$

$$\mathbf{N}(\ln 2) = \frac{\frac{d\mathbf{T}}{dt}(\ln 2)}{\left|\frac{d\mathbf{T}}{dt}(\ln 2)\right|} = -\frac{4}{\sqrt{17}}\mathbf{i} + \frac{1}{\sqrt{17}}\mathbf{j}$$

$$\mathbf{B}(\ln 2) = \mathbf{T}(\ln 2) \times \mathbf{N}(\ln 2) = \begin{vmatrix} \mathbf{i} & \mathbf{j} & \mathbf{k} \\ \frac{1}{\sqrt{17}} & \frac{4}{\sqrt{17}} & 0 \\ -\frac{4}{\sqrt{17}} & \frac{1}{\sqrt{17}} & 0 \end{vmatrix} = \mathbf{k}$$

$\mathbf{a} = 2e^{2t}\mathbf{j}$

$\mathbf{a}(\ln 2) = 8\mathbf{j}$ and $\mathbf{v}(\ln 2) = \mathbf{i} + 4\mathbf{j}$

$$\mathbf{v}(\ln 2) \times \mathbf{a}(\ln 2) = \begin{vmatrix} \mathbf{i} & \mathbf{j} & \mathbf{k} \\ 1 & 4 & 0 \\ 0 & 8 & 0 \end{vmatrix} = 8\mathbf{k}$$

$|\mathbf{v} \times \mathbf{a}| = 8$ and $|\mathbf{v}(\ln 2)| = \sqrt{17}$

$$\kappa(\ln 2) = \frac{|\mathbf{v} \times \mathbf{a}|}{|\mathbf{v}|^3} = \frac{8}{17\sqrt{17}}$$

$\dot{\mathbf{a}} = 4e^{2t}\mathbf{j}$

$\dot{\mathbf{a}}(\ln 2) = 16\mathbf{j}$

$$\tau(\ln 2) = \frac{\begin{vmatrix} 1 & 4 & 0 \\ 0 & 8 & 0 \\ 0 & 16 & 0 \end{vmatrix}}{|\mathbf{v} \cdot \mathbf{a}|^2} = 0$$

When $t = \ln 2$, $(\ln 2, 2, 0)$ is on the curve.

18. $\mathbf{r} = \frac{3}{2}(e^{2t} + e^{-2t})\mathbf{i} + \frac{3}{2}(e^{2t} - e^{-2t})\mathbf{j} + 6t\mathbf{k}$

$\mathbf{v} = 3(e^{2t} - e^{-2t})\mathbf{i} + 3(e^{2t} + e^{-2t})\mathbf{j} + 6\mathbf{k}$

$$\begin{aligned} |\mathbf{v}| &= \sqrt{9(e^{4t} - 2 + e^{-4t}) + 9(e^{4t} + 2 + e^{-4t}) + 36} \\ &= \sqrt{18(e^{4t} + e^{-4t} + 2)} \\ &= \sqrt{18(e^{2t} + e^{-2t})^2} \\ &= 3\sqrt{2}\,(e^{2t} + e^{-2t}) \end{aligned}$$

$$\mathbf{T} = \frac{\mathbf{v}}{|\mathbf{v}|} = \frac{1}{\sqrt{2}}\left(\frac{e^{2t} - e^{-2t}}{e^{2t} + e^{-2t}}\right)\mathbf{i} + \frac{1}{\sqrt{2}}\mathbf{j} + \frac{\sqrt{2}}{e^{2t} + e^{-2t}}\mathbf{k}$$

$$\begin{aligned} \mathbf{T}(\ln 2) &= \frac{1}{\sqrt{2}}\left(\frac{4 - \frac{1}{4}}{4 + \frac{1}{4}}\right)\mathbf{i} + \frac{1}{\sqrt{2}}\mathbf{j} + \frac{\sqrt{2}}{4 + \frac{1}{4}}\mathbf{k} \\ &= \frac{15}{17\sqrt{2}}\mathbf{i} + \frac{1}{\sqrt{2}}\mathbf{j} + \frac{4\sqrt{2}}{17}\mathbf{k} \end{aligned}$$

$$\begin{aligned} \frac{d\mathbf{T}}{dt} &= \frac{1}{\sqrt{2}}\left(\frac{(e^{2t} + e^{-2t})(2e^{2t} + 2e^{-2t}) - (e^{2t} - e^{-2t})(2e^{2t} - 2e^{-2t})}{(e^{2t} + e^{-2t})^2}\right)\mathbf{i} \\ &\quad + \left(\frac{-\sqrt{2}(2e^{2t} - 2e^{-2t})}{(e^{2t} + e^{-2t})^2}\right)\mathbf{k} \\ &= \frac{4\sqrt{2}}{(e^{2t} + e^{-2t})^2}\mathbf{i} - \frac{2\sqrt{2}(e^{2t} - e^{-2t})}{(e^{2t} + e^{-2t})^2}\mathbf{k} \end{aligned}$$

$$\begin{aligned} \frac{d\mathbf{T}}{dt}(\ln 2) &= \frac{4\sqrt{2}}{\left(4 + \frac{1}{4}\right)^2}\mathbf{i} - \frac{2\sqrt{2}\left(4 - \frac{1}{4}\right)}{\left(4 + \frac{1}{4}\right)^2}\mathbf{k} \\ &= \frac{64\sqrt{2}}{289}\mathbf{i} - \frac{120\sqrt{2}}{289}\mathbf{k} \end{aligned}$$

$$\left|\frac{d\mathbf{T}}{dt}(\ln 2)\right| = \sqrt{\left(\frac{64\sqrt{2}}{289}\right)^2 + \left(-\frac{120\sqrt{2}}{289}\right)^2} = \frac{8\sqrt{2}}{17}$$

$$\mathbf{N}(\ln 2) = \frac{\frac{d\mathbf{T}}{dt}(\ln 2)}{\left|\frac{d\mathbf{T}}{dt}(\ln 2)\right|} = \frac{8}{17}\mathbf{i} - \frac{15}{17}\mathbf{k}$$

$$\begin{aligned} \mathbf{B}(\ln 2) &= \mathbf{T}(\ln 2) \times \mathbf{N}(\ln 2) = \begin{vmatrix} \mathbf{i} & \mathbf{j} & \mathbf{k} \\ \frac{15}{17\sqrt{2}} & \frac{1}{\sqrt{2}} & \frac{4\sqrt{2}}{17} \\ \frac{8}{17} & 0 & -\frac{15}{17} \end{vmatrix} \\ &= -\frac{15}{17\sqrt{2}}\mathbf{i} + \frac{1}{\sqrt{2}}\mathbf{j} - \frac{4\sqrt{2}}{17}\mathbf{k} \end{aligned}$$

$\mathbf{a} = 6(e^{2t} + e^{-2t})\mathbf{i} + 6(e^{2t} - e^{-2t})\mathbf{j}$

$$\mathbf{a}(\ln 2) = 6\left(4 + \frac{1}{4}\right)\mathbf{i} + 6\left(4 - \frac{1}{4}\right)\mathbf{j} = \frac{51}{2}\mathbf{i} + \frac{45}{2}\mathbf{j}$$

$$\begin{aligned} \mathbf{v}(\ln 2) &= 3\left(4 - \frac{1}{4}\right)\mathbf{i} + 3\left(4 + \frac{1}{4}\right)\mathbf{j} + 6\mathbf{k} \\ &= \frac{45}{4}\mathbf{i} + \frac{51}{4}\mathbf{j} + 6\mathbf{k} \end{aligned}$$

$$\begin{aligned} \mathbf{v}(\ln 2) \times \mathbf{a}(\ln 2) &= \begin{vmatrix} \mathbf{i} & \mathbf{j} & \mathbf{k} \\ \frac{45}{4} & \frac{51}{4} & 6 \\ \frac{51}{2} & \frac{45}{2} & 0 \end{vmatrix} \\ &= -135\mathbf{i} + 153\mathbf{j} - 72\mathbf{k} \end{aligned}$$

$|\mathbf{v} \times \mathbf{a}| = 153\sqrt{2}$ and $|\mathbf{v}(\ln 2)| = \frac{51}{4}\sqrt{2}$

$$\kappa(\ln 2) = \frac{|\mathbf{v} \times \mathbf{a}|}{|\mathbf{v}|^3} = \frac{153\sqrt{2}}{\left(\frac{51}{4}\sqrt{2}\right)^3} = \frac{32}{867}$$

$\dot{\mathbf{a}} = 12(e^{2t} - e^{-2t})\mathbf{i} + 12(e^{2t} + e^{-2t})\mathbf{j}$

$\dot{\mathbf{a}}(\ln 2) = 45\mathbf{i} + 51\mathbf{j}$

$$\tau(\ln 2) = \frac{\begin{vmatrix} \frac{45}{4} & \frac{51}{4} & 6 \\ \frac{51}{2} & \frac{45}{2} & 0 \\ 45 & 51 & 0 \end{vmatrix}}{|\mathbf{v} \times \mathbf{a}|^2} = \frac{32}{867}$$

When $t = \ln 2$, $\left(\frac{51}{8}, \frac{45}{8}, 6 \ln 2\right)$ is on the curve.

19. $\mathbf{r} = (2 + 3t + 3t^2)\mathbf{i} + (4t + 4t^2)\mathbf{j} - (6 \cos t)\mathbf{k}$

$\mathbf{v} = (3 + 6t)\mathbf{i} + (4 + 8t)\mathbf{j} + (6 \sin t)\mathbf{k}$

$|\mathbf{v}| = \sqrt{(3 + 6t)^2 + (4 + 8t)^2 + (6 \sin t)^2}$

$= \sqrt{25 + 100t + 100t^2 + 36 \sin^2 t}$

$\frac{d|\mathbf{v}|}{dt} = \frac{100 + 200t + 72 \sin t \cos t}{2\sqrt{25 + 100t + 100t^2 + 36 \sin^2 t}}$

$a_T(0) = \frac{d|\mathbf{v}|}{dt}(0) = 10$

$\mathbf{a} = 6\mathbf{i} + 8\mathbf{j} + (6 \cos t)\mathbf{k}$

$|\mathbf{a}| = \sqrt{6^2 + 8^2 + (6 \cos t)^2} = \sqrt{100 + 36 \cos^2 t}$

$|\mathbf{a}(0)| = \sqrt{136}$

$a_N = \sqrt{|\mathbf{a}|^2 - a_T^2} = \sqrt{136 - 10^2} = \sqrt{36} = 6$

$\mathbf{a}(0) = 10\mathbf{T} + 6\mathbf{N}$

20. $\mathbf{r} = (2 + t)\mathbf{i} + (t + 2t^2)\mathbf{j} + (1 + t^2)\mathbf{k}$

$\mathbf{v} = \mathbf{i} + (1 + 4t)\mathbf{j} + 2t\mathbf{k}$

$|\mathbf{v}| = \sqrt{1^2 + (1 + 4t)^2 + (2t)^2} = \sqrt{2 + 8t + 20t^2}$

$\frac{d|\mathbf{v}|}{dt} = \frac{1}{2}(2 + 8t + 20t^2)^{-1/2}(8 + 40t)$

$a_T = \frac{d|\mathbf{v}|}{dt}(0) = 2\sqrt{2}$

$\mathbf{a} = 4\mathbf{j} + 2\mathbf{k}$

$|\mathbf{a}| = \sqrt{4^2 + 2^2} = \sqrt{20}$

$a_N = \sqrt{|\mathbf{a}|^2 - a_T^2} = \sqrt{20 - (2\sqrt{2})^2} = \sqrt{12} = 2\sqrt{3}$

$\mathbf{a}(0) = 2\sqrt{2}\mathbf{T} + 2\sqrt{3}\mathbf{N}$

21. $\mathbf{r}(t) = \left[\int_0^t \cos\left(\frac{1}{2}\pi\theta^2\right) d\theta\right]\mathbf{i} + \left[\int_0^t \sin\left(\frac{1}{2}\pi\theta^2\right) d\theta\right]\mathbf{j}$

$\mathbf{v}(t) = \cos\left(\frac{\pi t^2}{2}\right)\mathbf{i} + \sin\left(\frac{\pi t^2}{2}\right)\mathbf{j}$

$|\mathbf{v}| = 1$

$\mathbf{a}(t) = -\pi t \sin\left(\frac{\pi t^2}{2}\right)\mathbf{i} + \pi t \cos\left(\frac{\pi t^2}{2}\right)\mathbf{j}$

$$\mathbf{v} \times \mathbf{a} = \begin{vmatrix} \mathbf{i} & \mathbf{j} & \mathbf{k} \\ \cos\left(\frac{\pi t^2}{2}\right) & \sin\left(\frac{\pi t^2}{2}\right) & 0 \\ -\pi t \sin\left(\frac{\pi t^2}{2}\right) & \pi t \cos\left(\frac{\pi t^2}{2}\right) & 0 \end{vmatrix} = \pi t\mathbf{k}$$

$\kappa = \frac{|\mathbf{v} \times \mathbf{a}|}{|\mathbf{v}|^3} = \pi t$

$|\mathbf{v}(t)| = \frac{ds}{dt} = 1$, so $s = t + C$.

Since $\mathbf{r}(0) = \mathbf{0}$, $s(0) = 0$, thus, $C = 0$.

$\kappa = \pi s$

22. $s = a\theta$

$\theta = \frac{s}{a}$

$\phi = \frac{s}{a} + \frac{\pi}{2}$

$\frac{d\phi}{ds} = \frac{1}{a}$

$\kappa = \left|\frac{1}{a}\right| = \frac{1}{a}$ since $a > 0$

23. $\mathbf{r}(\sin t)\mathbf{i} + (\sqrt{2} \cos t)\mathbf{j} + (\sin t)\mathbf{k}$

$\mathbf{v} = (\cos t)\mathbf{i} - (\sqrt{2} \sin t)\mathbf{j} + (\cos t)\mathbf{k}$

$|\mathbf{v}| = \sqrt{(\cos t)^2 + (-\sqrt{2} \sin t)^2 + (\cos t)^2} = \sqrt{2}$

$\mathbf{T} = \frac{\mathbf{v}}{|\mathbf{v}|} = \left(\frac{1}{\sqrt{2}} \cos t\right)\mathbf{i} - (\sin t)\mathbf{j} + \left(\frac{1}{\sqrt{2}} \cos t\right)\mathbf{k}$

$\frac{d\mathbf{T}}{dt} = \left(-\frac{1}{\sqrt{2}} \sin t\right)\mathbf{i} - (\cos t)\mathbf{j} - \left(\frac{1}{\sqrt{2}} \sin t\right)\mathbf{k}$

$\left|\frac{d\mathbf{T}}{dt}\right| = \sqrt{\left(-\frac{1}{\sqrt{2}} \sin t\right)^2 + (-\cos t)^2 + \left(-\frac{1}{\sqrt{2}} \sin t\right)^2} = 1$

$\mathbf{N} = \frac{\left(\frac{d\mathbf{T}}{dt}\right)}{\left|\frac{d\mathbf{T}}{dt}\right|} = \left(-\frac{1}{\sqrt{2}} \sin t\right)\mathbf{i} - (\cos t)\mathbf{j} - \left(\frac{1}{\sqrt{2}} \sin t\right)\mathbf{k}$

$$\mathbf{B} = \mathbf{T} \times \mathbf{N} = \begin{vmatrix} \mathbf{i} & \mathbf{j} & \mathbf{k} \\ \frac{1}{\sqrt{2}} \cos t & -\sin t & \frac{1}{\sqrt{2}} \cos t \\ -\frac{1}{\sqrt{2}} \sin t & -\cos t & -\frac{1}{\sqrt{2}} \sin t \end{vmatrix}$$

$= \frac{1}{\sqrt{2}}\mathbf{i} - \frac{1}{\sqrt{2}}\mathbf{k}$

$\mathbf{a} = (-\sin t)\mathbf{i} - (\sqrt{2} \cos t)\mathbf{j} - (\sin t)\mathbf{k}$

$$\mathbf{v} \times \mathbf{a} = \begin{vmatrix} \mathbf{i} & \mathbf{j} & \mathbf{k} \\ \cos t & -\sqrt{2} \sin t & \cos t \\ -\sin t & -\sqrt{2} \cos t & -\sin t \end{vmatrix} = \sqrt{2}\mathbf{i} - \sqrt{2}\mathbf{k}$$

$|\mathbf{v} \times \mathbf{a}| = \sqrt{4} = 2$

$\kappa = \frac{|\mathbf{v} \times \mathbf{a}|}{|\mathbf{v}|^3} = \frac{2}{(\sqrt{2})^3} = \frac{1}{\sqrt{2}}$

$\dot{\mathbf{a}} = (-\cos t)\mathbf{i} + (\sqrt{2} \sin t)\mathbf{j} - (\cos t)\mathbf{k}$

$$\tau = \frac{\begin{vmatrix} \cos t & -\sqrt{2} \sin t & \cos t \\ -\sin t & -\sqrt{2} \cos t & -\sin t \\ -\cos t & \sqrt{2} \sin t & -\cos t \end{vmatrix}}{|\mathbf{v} \times \mathbf{a}|^2}$$

$= \frac{(\cos t)(\sqrt{2}) - (\sqrt{2} \sin t)(0) + (\cos t)(-\sqrt{2})}{4} = 0$

24. $\mathbf{r} = \mathbf{i} + (5\cos t)\mathbf{j} + (3\sin t)\mathbf{k}$

$\mathbf{v} = (-5\sin t)\mathbf{j} + (3\cos t)\mathbf{k}$

$\mathbf{a} = (-5\cos t)\mathbf{j} - (3\sin t)\mathbf{k}$

$\mathbf{v}\cdot\mathbf{a} = 25\sin t\cos t - 9\sin t\cos t = 16\sin t\cos t$

$\mathbf{v}\cdot\mathbf{a} = 0$ when $16\sin t\cos t = 0$, or $\sin t = 0$ or $\cos t = 0$.

This occurs when $t = 0, \frac{\pi}{2}$ or π.

25. $\mathbf{r} = 2\mathbf{i} + \left(4\sin\frac{t}{2}\right)\mathbf{j} + \left(3 - \frac{t}{\pi}\right)\mathbf{k}$

$0 = \mathbf{r}\cdot(\mathbf{i} - \mathbf{j}) = 2(1) + \left(4\sin\frac{t}{2}\right)(-1)$

$0 = 2 - 4\sin\frac{t}{2}$

$\sin\frac{t}{2} = \frac{1}{2}$

$\frac{t}{2} = \frac{\pi}{6}$

The first time is when $t = \frac{\pi}{3}$.

26. $\mathbf{r}(t) = t\mathbf{i} + t^2\mathbf{j} + t^3\mathbf{k}$

$\mathbf{v} = \mathbf{i} + 2t\mathbf{j} + 3t^2\mathbf{k}$

$|\mathbf{v}| = \sqrt{1 + 4t^2 + 9t^4}$

$|\mathbf{v}(1)| = \sqrt{14}$

$\mathbf{T}(1) = \frac{1}{\sqrt{14}}\mathbf{i} + \frac{2}{\sqrt{14}}\mathbf{j} + \frac{3}{\sqrt{14}}\mathbf{k}$, which is normal to the normal plane.

$\frac{1}{\sqrt{14}}(x - 1) + \frac{2}{\sqrt{14}}(y - 1) + \frac{3}{\sqrt{14}}(z - 1) = 0$

or $x + 2y + 3z = 6$ is an equation of the normal plane.

Next we calculate $\mathbf{N}(1)$, which is normal to the rectifying plane.

$\mathbf{a} = 2\mathbf{j} + 6t\mathbf{k}$

$\mathbf{a}(1) = 2\mathbf{j} + 6\mathbf{k}$

$$\mathbf{v}(1)\times\mathbf{a}(1) = \begin{vmatrix}\mathbf{i} & \mathbf{j} & \mathbf{k}\\ 1 & 2 & 3\\ 0 & 2 & 6\end{vmatrix} = 6\mathbf{i} - 6\mathbf{j} + 2\mathbf{k}$$

$|\mathbf{v}(1)\times\mathbf{a}(1)| = \sqrt{76}$

$\kappa(1) = \frac{\sqrt{76}}{(\sqrt{14})^3} = \frac{\sqrt{19}}{7\sqrt{14}}$

$\frac{ds}{dt} = |\mathbf{v}(t)|$, so

$$\left.\frac{d^2s}{dt^2}\right|_{t=1} = \left.\frac{1}{2}(1 + 4t^2 + 9t^4)^{-1/2}(8t + 36t^3)\right|_{t=1} = \frac{22}{\sqrt{14}}.$$

Since $\mathbf{a} = \frac{d^2s}{dt^2}\mathbf{T} + \kappa\left(\frac{ds}{dt}\right)^2\mathbf{N}$, at $t = 1$ we have

$$2\mathbf{j} + 6\mathbf{k} = \frac{22}{\sqrt{14}}\left(\frac{\mathbf{i} + 2\mathbf{j} + 3\mathbf{k}}{\sqrt{14}}\right) + \frac{\sqrt{19}}{7\sqrt{14}}(\sqrt{14})^2\mathbf{N}$$

$$\mathbf{N} = \frac{\sqrt{14}}{2\sqrt{19}}\left(-\frac{11}{7}\mathbf{i} - \frac{8}{7}\mathbf{j} + \frac{9}{7}\mathbf{k}\right)$$

$-\frac{11}{7}(x - 1) - \frac{8}{7}(y - 1) + \frac{9}{7}(z - 1) = 0$

or $11x + 8y - 9z = 10$ is an equation of the rectifying plane. Finally,

$$\mathbf{B}(1) = \mathbf{T}(1)\times\mathbf{N}(1) = \left(\frac{\sqrt{14}}{2\sqrt{19}}\right)\left(\frac{1}{\sqrt{14}}\right)\left(\frac{1}{7}\right)\begin{vmatrix}\mathbf{i} & \mathbf{j} & \mathbf{k}\\ 1 & 2 & 3\\ -11 & -8 & 9\end{vmatrix}$$

$$= \frac{1}{14\sqrt{19}}(42\mathbf{i} - 42\mathbf{j} + 14\mathbf{k})$$

$$= \frac{1}{\sqrt{19}}(3\mathbf{i} - 3\mathbf{j} + \mathbf{k})$$

$3(x - 1) - 3(y - 1) + (z - 1) = 0$ or $3x - 3y + z = 1$ is an equation of the osculating plane.

27. $\mathbf{r} = e^t\mathbf{i} + (\sin t)\mathbf{j} + \ln(1 - t)\mathbf{k}$

$\mathbf{v} = e^t\mathbf{i} + (\cos t)\mathbf{j} - \left(\frac{1}{1 - t}\right)\mathbf{k}$

$\mathbf{v}(0) = \mathbf{i} + \mathbf{j} - \mathbf{k}$ is a vector tangent to the curve when $t = 0$.

$\mathbf{r}(0) = \mathbf{i}$

The point $(1, 0, 0)$ is on the line.

$x = 1 + t$, $y = t$, and $z = -t$ are parametric equations of the line.

28. $\mathbf{r} = (\sqrt{2}\cos t)\mathbf{i} + (\sqrt{2}\sin t)\mathbf{j} + t\mathbf{k}$

$\mathbf{v} = (-\sqrt{2}\sin t)\mathbf{i} + (\sqrt{2}\cos t)\mathbf{j} + \mathbf{k}$

$$\mathbf{v}\left(\frac{\pi}{4}\right) = \left(-\sqrt{2}\sin\frac{\pi}{4}\right)\mathbf{i} + \left(\sqrt{2}\cos\frac{\pi}{4}\right)\mathbf{j} + \mathbf{k} = -\mathbf{i} + \mathbf{j} + \mathbf{k}$$

is a vector tangent to the helix when $t = \frac{\pi}{4}$.

$$\mathbf{r}\left(\frac{\pi}{4}\right) = \left(\sqrt{2}\cos\frac{\pi}{4}\right)\mathbf{i} + \left(\sqrt{2}\sin\frac{\pi}{4}\right)\mathbf{j} + \frac{\pi}{4}\mathbf{k} = \mathbf{i} + \mathbf{j} + \frac{\pi}{4}\mathbf{k}$$

The point $\left(1, 1, \frac{\pi}{4}\right)$ is on the line.

$x = 1 - t$, $y = 1 + t$, and $z = \frac{\pi}{4} + t$ are parametric equations of the line.

29. a. $\triangle SOT \approx \triangle TOD$

$\frac{DO}{OT} = \frac{OT}{SO}$

$\frac{y_0}{6380} = \frac{6380}{6380 + 437}$

$y_0 = \frac{6380^2}{6817}$

$y_0 \approx 5971$ km

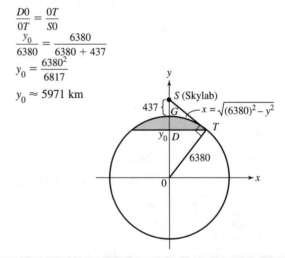

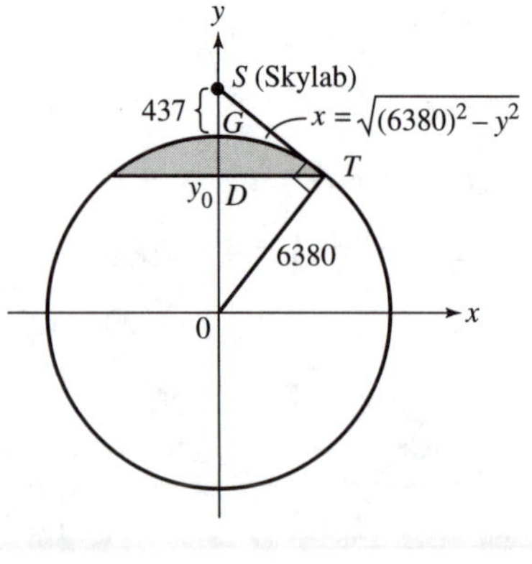

b. $$VA = \int_{5971}^{6380} 2\pi x\sqrt{1 + \left(\frac{dx}{dy}\right)^2}\,dy$$

$$= 2\pi\int_{5971}^{6380}\sqrt{6380^2 - y^2}\left(\frac{6380}{\sqrt{6380^2 - y^2}}\right)dy$$

$$= 2\pi\int_{5971}^{6380} 6380\,dy = 2\pi[6380y]_{5971}^{6380}$$

$\approx 16{,}395{,}469\text{ km}^2 \approx 1.640\times 10^7\text{ km}^2$

c. percentage visible $\approx \frac{16{,}395{,}469\text{ km}^2}{4\pi(6380\text{ km})^2} \approx 3.21\%$

30. $r = \dfrac{(1+e)r_0}{1 + e\cos\theta}$

$\dfrac{dr}{d\theta} = \dfrac{(1+e)r_0(e\sin\theta)}{(1+e\cos\theta)^2}$

$\dfrac{dr}{d\theta} = 0$ when $\dfrac{(1+e)r_0(e\sin\theta)}{(1+e\cos\theta)^2} = 0$,

or $(1+e)r_0(e\sin\theta) = 0$, so $\sin\theta = 0$.

Thus, $\theta = 0$ or π.

Note that $\dfrac{dr}{d\theta} > 0$ when $\sin\theta > 0$ and $\dfrac{dr}{d\theta} < 0$ when $\sin\theta < 0$. Since $\sin\theta < 0$ on $-\pi < \theta < 0$ and $\sin\theta > 0$ on $0 < \theta < \pi$, r is a minimum when $\theta = 0$ and $r(0) = \dfrac{(1+e)r_0}{1+e\cos 0} = r_0$.

31. a. $f(x) = x - 1 - \dfrac{1}{2}\sin x$

$f(0) = -1$ and $f(2) = 2 - 1 - \dfrac{1}{2}\sin 2 \geq \dfrac{1}{2}$ since $|\sin 2| \leq 1$.

Since f is continuous on $[0, 2]$, the Intermediate Value Theorem implies there is a root between 0 and 2.

b. Root ≈ 1.498701134

32. a. $\mathbf{v} = \dot{x}\mathbf{i} + \dot{y}\mathbf{j}$ and $\mathbf{v} = \dot{r}\,\mathbf{u}_r + r\,\dot{\theta}\,\mathbf{u}_\theta$

$= (\dot{r})[(\cos\theta)\mathbf{i} + (\sin\theta)\mathbf{j}] + (r\dot{\theta})[(-\sin\theta)\mathbf{i} + (\cos\theta)\mathbf{j}]$

$\mathbf{v}\cdot\mathbf{i} = \dot{x}$ and $\mathbf{v}\cdot\mathbf{i} = \dot{r}\cos\theta - r\,\dot{\theta}\sin\theta$

$\dot{x} = \dot{r}\cos\theta - r\,\dot{\theta}\sin\theta$

$\mathbf{v}\cdot\mathbf{j} = \dot{y}$ and $\mathbf{v}\cdot\mathbf{j} = \dot{r}\sin\theta + r\,\dot{\theta}\cos\theta$

$\dot{y} = \dot{r}\sin\theta + r\,\dot{\theta}\cos\theta$

b. $\mathbf{u}_r = (\cos\theta)\mathbf{i} + (\sin\theta)\mathbf{j}$

$\mathbf{v}\cdot\mathbf{u}_r = \dot{x}\cos\theta + \dot{y}\sin\theta$

$= (\dot{r}\cos\theta - r\,\dot{\theta}\sin\theta)(\cos\theta)$
$+ (\dot{r}\sin\theta + r\,\dot{\theta}\cos\theta)(\sin\theta)$ by part (a).

$\mathbf{v}\cdot\mathbf{u}_r = \dot{r}$

Therefore, $\dot{r} = \dot{x}\cos\theta + \dot{y}\sin\theta$.

$\mathbf{u}_\theta = -(\sin\theta)\mathbf{i} + (\cos\theta)\mathbf{j}$

$\mathbf{v}\cdot\mathbf{u}_\theta = -\dot{x}\sin\theta + \dot{y}\cos\theta$

$= (\dot{r}\cos\theta - r\,\dot{\theta}\sin\theta)(-\sin\theta)$
$+ (\dot{r}\sin\theta + r\,\dot{\theta}\cos\theta)(\cos\theta)$ by part (a)

$\mathbf{v}\cdot\mathbf{u}_\theta = r\,\dot{\theta}$

Therefore, $r\,\dot{\theta} = -\dot{x}\sin\theta + \dot{y}\cos\theta$.

33. $r = f(\theta)$

$\dot{r} = f'(\theta)\dot{\theta}$

$\ddot{r} = f''(\theta)(\dot{\theta})^2 + f'(\theta)(\ddot{\theta})$

$\mathbf{v} = \dot{r}\mathbf{u}_r + r\dot{\theta}\mathbf{u}_\theta$

$= (\cos\theta\,\dot{r} - r\sin\theta\,\dot{\theta})\,\mathbf{i} + (\sin\theta\,\dot{r} + r\cos\theta\,\dot{\theta})\,\mathbf{j}$

$|\mathbf{v}| = [(\dot{r})^2 + r^2(\dot{\theta})^2]^{1/2} = [(f')^2 + f^2]^{1/2}(\dot{\theta})$

$|\mathbf{v}\times\mathbf{a}| = |\dot{x}\ddot{y} - \dot{y}\ddot{x}|$, where $x = r\cos\theta$ and $y = r\sin\theta$.

Then $\dot{x} = (-r\sin\theta)\dot{\theta} + (\cos\theta)\dot{r}$

$\ddot{x} = (-2\sin\theta)\,\dot{\theta}\,\dot{r} - (r\cos\theta)(\dot{\theta})^2 - (r\sin\theta)\ddot{\theta} + (\cos\theta)\ddot{r}$

$\dot{y} = (r\cos\theta)\dot{\theta} + (\sin\theta)\dot{r}$

$\ddot{y} = (2\cos\theta)\dot{\theta}\,\dot{r} - (r\sin\theta)(\dot{\theta})^2 + (r\cos\theta)\ddot{\theta} + (\sin\theta)\ddot{r}$

Then, after *much* algebra,

$|\mathbf{v}\times\mathbf{a}| = r^2(\dot{\theta})^3 + r\,\ddot{\theta}\,\dot{r} - r\,\dot{\theta}\,\ddot{r} + 2\,\dot{\theta}\,\dot{r}^2$

$$\kappa = \frac{r^2(\dot{\theta})^3 + r\,\ddot{\theta}\,\dot{r} - r\,\dot{\theta}\,\ddot{r} + 2r\,\dot{\theta}\,\dot{r}^2}{[(f')^2 + f^2]^{3/2}(\dot{\theta})^3}$$

$$= \frac{r^2 - r\left[\ddot{r}\left(\frac{dt}{d\theta}\right)^2 - \dot{r}\,\ddot{\theta}\left(\frac{dt}{d\theta}\right)^3\right] + 2\left(\dot{r}\,\frac{dt}{d\theta}\right)^2}{[(f')^2 + f^2]^{3/2}}$$

$$= \frac{r^2 - r(f'') + 2(f')^2}{[(f')^2 + f^2]^{3/2}}$$

$$= \frac{f^2 - ff'' + 2(f')^2}{[(f')^2 + f^2]^{3/2}}$$

34. a. Let $r = 2 - t$ and $\theta = 3t$

$\dfrac{dr}{dt} = -1$ and $\dfrac{d\theta}{dt} = 3$

$\dfrac{d^2r}{dt^2} = \dfrac{d^2\theta}{dt^2} = 0$

The halfway point is (1, 3).

At $t = 1$:

$\mathbf{v} = \dfrac{dr}{dt}\mathbf{u}_r + r\dfrac{d\theta}{dt}\mathbf{u}_\theta$

$\mathbf{v}(1) = -\mathbf{u}_r + 3\mathbf{u}_\theta$

$\mathbf{a} = \left[\dfrac{d^2r}{dt^2} - r\left(\dfrac{d\theta}{dt}\right)^2\right]\mathbf{u}_r + \left[r\dfrac{d^2\theta}{dt^2} + 2\dfrac{dr}{dt}\dfrac{d\theta}{dt}\right]\mathbf{u}_\theta$

$\mathbf{a}(1) = -9\mathbf{u}_r - 6\mathbf{u}_\theta$

b. It takes the beetle 2 min to crawl to the origin. The rod has revolved 6 radians.

$$L = \int_0^6 \sqrt{[f(\theta)]^2 + [f'(\theta)]^2}\,d\theta$$

$$= \int_0^6 \sqrt{\left(2 - \frac{\theta}{3}\right)^2 + \left(-\frac{1}{3}\right)^2}\,d\theta$$

$$= \int_0^6 \sqrt{4 - \frac{4\theta}{3} + \frac{\theta^2}{9} + \frac{1}{9}}\,d\theta$$

$$= \int_0^6 \sqrt{\frac{37 - 12\theta + \theta^2}{9}}\,d\theta$$

$$= \frac{1}{3}\int_0^6 \sqrt{(\theta - 6)^2 + 1}\,d\theta$$

$$= \frac{1}{3}\left[\frac{(\theta - 6)}{2}\sqrt{(\theta - 6)^2 + 1} + \frac{1}{2}\ln\left|\theta - 6 + \sqrt{(\theta - 6)^2 + 1}\right|\right]_0^6$$

$$= \sqrt{37} - \frac{1}{6}\ln(\sqrt{37} - 6) \approx 6.5 \text{ in.}$$

Chapter 13
Multivariable Functions and Partial Derivatives

Chapter Opener

$f_x(x, y) = -4x^3 - 28xy^2 + 64x, f_y(x, y) = -4y^3 - 28x^2y + 256y; f_x(x, y) = 0, f_y(x, y) = 0 \Rightarrow$ (x, y) is one of the following: $(-4, 0)$, $(0, 0)$ Use a CAS to find that $(x, y) = (4, 0), (0, -8), (0, 8), (-3, -1), (-3, 1), (3, -1)$, or $(3, 1)$. Since for these (x, y) pairs, the maximum value of $z = f(x, y)$ is $f(0, -8) = f(0, 8) = 4096$, the highest elevation is 4096.

■ Section 13.1 Functions of Several Variables (pp. 675–685)

Exploration 1: Finding Domains and Ranges

Function	Domain	Range
$w=\sqrt{y-x^2}$	$y \geq x^2$	$[0, \infty)$
$w = \frac{1}{xy}$	$xy \neq 0$	$(-\infty, 0) \cup (0, \infty)$
$w = x \ln y$	Half-plane $y > 0$	$(-\infty, \infty)$
$w = \sqrt{1 - (x^2 + y^2)}$	Disk $x^2 + y^2 \leq 1$	$[0, 1]$
$w=\sqrt{x^2 + y^2 + z^2}$	xyz-space	$[0, \infty)$
$w = \frac{1}{x^2 + y^2 + z^2}$	$x^2 + y^2 + z^2 \neq 0$	$(0, \infty)$
$w = \sin xyz$	xyz-space	$[-1, 1]$
$w = \sqrt{4 - x^2 - y^2 - z^2}$	sphere $x^2 + y^2 + z^2 \leq 4$	$[0, 2]$

Exploration 2: Visualizing the Complex Zeros of a Function

1. Surface g touches or crosses the xy-plane when $g(x, y) = |f(x + yi)| = 0$, which implies $f(x + yi) = 0$.

2. $g(x, y) = |(x + yi)^3 + 1| = |x^3 - 3xy^2 + 1 + (3x^2y - y^3)i| = \sqrt{(x^3 - 3xy^2 + 1)^2 + (3x^2y - y^3)^2}$. Expanding the radicand gives the result.

3. $x^3 + 1 = (x + 1)(x^2 - x + 1)$
$= (x + 1)\left[x - \frac{1 + \sqrt{3}i}{2}\right]\left[x - \frac{1 - \sqrt{3}i}{2}\right] \Rightarrow$
zeros of f: $-1, \frac{1 + \sqrt{3}i}{2}, \frac{1 - \sqrt{3}i}{2}$

4-5. The estimated zeros should approach -1, $0.5 + 0.866i$, $0.5 - 0.866i$.

Quick Review 13.1

1. $V = \frac{1}{3}\pi r^2 h$

2. $S = 2\pi rh + 2\pi r^2$

3. $t = \frac{y}{5280x}$

4. $h = \sqrt{a^2 - \frac{b^2}{4}}$

5. $A = \frac{1}{2}ab \sin 20°$

6. $V = \frac{1}{2}\pi ab^2$

7. $V = \frac{x^2y}{4\pi}$ or $V = \frac{xy^2}{4\pi}$

8. $V = \frac{4\pi r^2h - \pi h^3}{4} = \frac{\pi h}{4}(4r^2 - h^2)$

9. $A = \frac{1}{2}a^2 \sin \theta$

10. $v = 2\pi mk$ ft/min

Section 13.1 Exercises

1. Domain: all points in the xy-plane
Range: all real numbers
unbounded

3. Domain: $(x, y) \neq (0, 0)$
Range: all real numbers
unbounded

5. Domain: all points in the xy-plane
Range: all real numbers
unbounded

7. Domain: all (x, y) satisfying $x^2 + y^2 < 16$
Range: $z \geq \frac{1}{4}$
bounded

9. a.

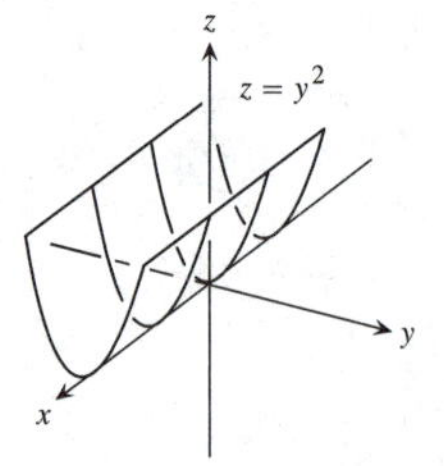

b.

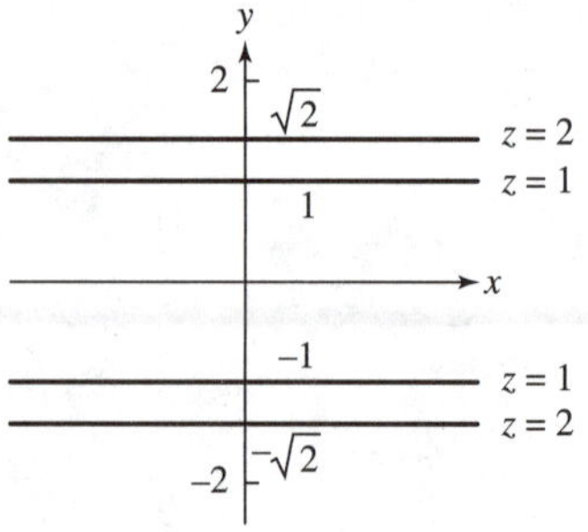

11. a.

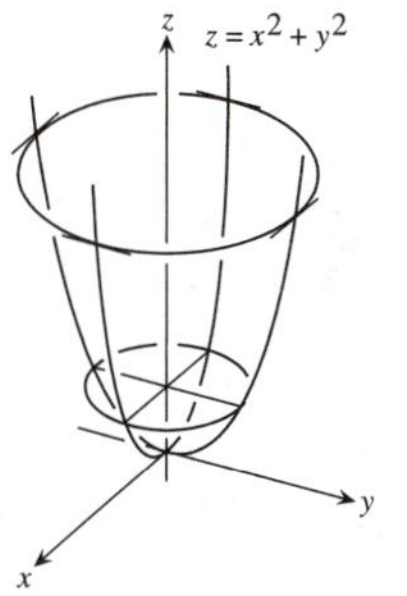

b.

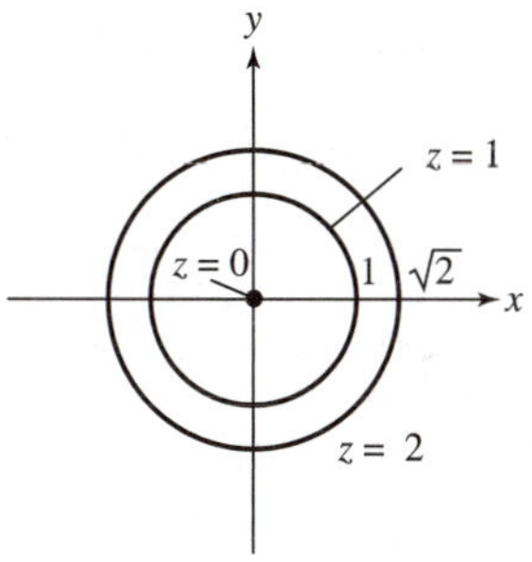

13. a.

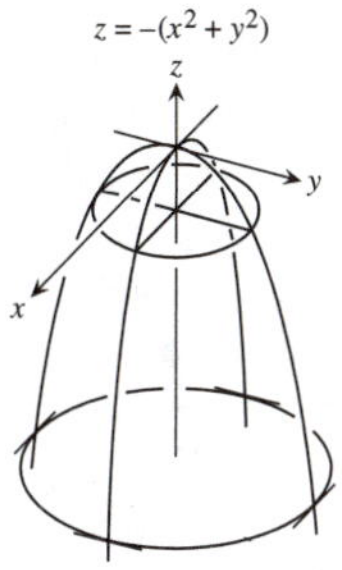

b.

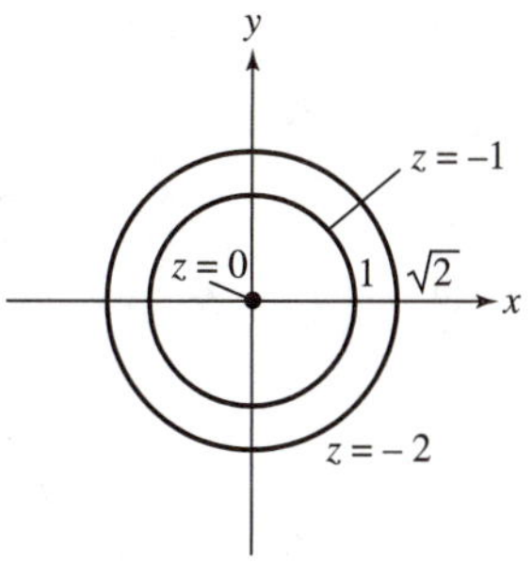

15. a.

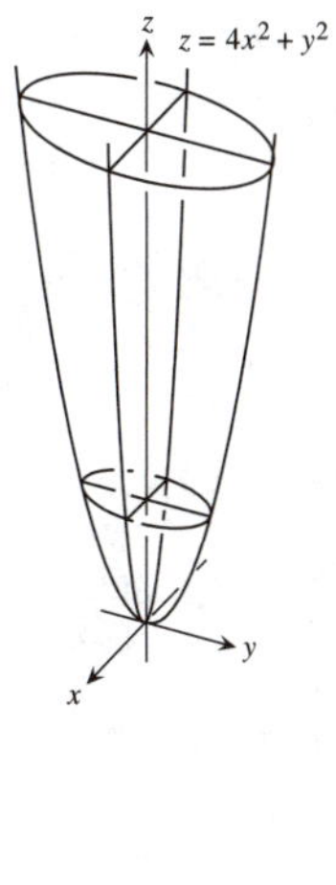

b.

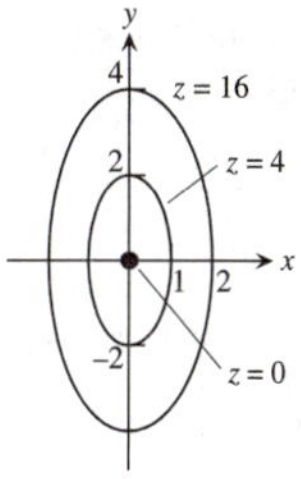

17. f

19. a

21. d

23.

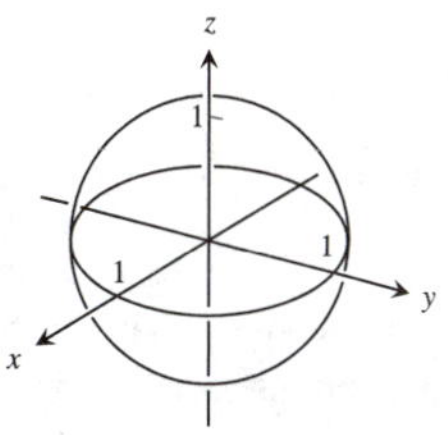

$f(x, y, z) = x^2 + y^2 + z^2 = 1$

25.

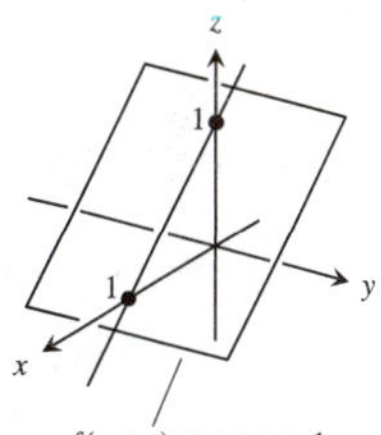

$f(x, y, z) = x + z = 1$

27.

$f(x, y, z) = x^2 + y^2 = 1$

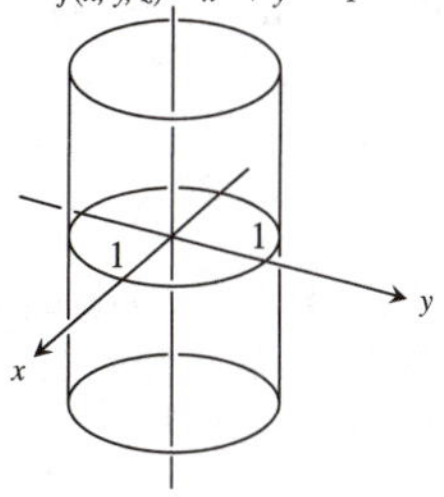

29.

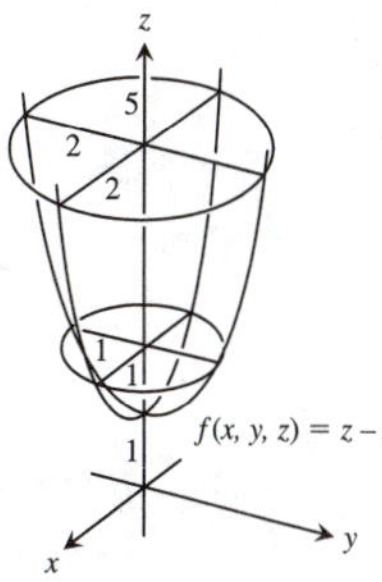

$f(x, y, z) = z - x^2 - y^2 = 1$

31. $f(x, y) = 16 - x^2 - y^2$ and $(2\sqrt{2}, \sqrt{2})$
$\Rightarrow z = 16 - (2\sqrt{2})^2 - (\sqrt{2})^2 = 6 \Rightarrow 6 = 16 - x^2 - y^2$
$\Rightarrow x^2 + y^2 = 10$

33. $f(x, y) = \int_x^y \frac{1}{1+t^2}\,dt$ at $(-\sqrt{2}, \sqrt{2}) \Rightarrow z = \tan^{-1}y$ $- \tan^{-1}x$; at $(-\sqrt{2}, \sqrt{2}) \Rightarrow z = \tan^{-1}\sqrt{2} - \tan^{-1}(-\sqrt{2})$ $= 2\tan^{-1}\sqrt{2} \Rightarrow \tan^{-1}y - \tan^{-1}x = 2\tan^{-1}\sqrt{2}$

35. $f(x, y, z) = \sqrt{x-y} - \ln z$ at $(3, -1, 1) \Rightarrow w = \sqrt{x-y}$ $- \ln z$; at $(3, -1, 1) \Rightarrow w = \sqrt{3-(-1)} - \ln 1 = 2$ $\Rightarrow \sqrt{x-y} - \ln z = 2$

37. $g(x, y, z) = \sum_{n=0}^{\infty} \frac{(x+y)^n}{n!z^n}$ at $(\ln 2, \ln 4, 3)$

$\Rightarrow w = \sum_{n=0}^{\infty} \frac{(x+y)^n}{n!z^n} = e^{(x+y)/z}$;

$\Rightarrow w = e^{(\ln 2 + \ln 4)/3} = e^{(\ln 8)/3} = e^{\ln 2} = 2$

$\Rightarrow 2 = e^{(x+y)/z} \Rightarrow \frac{x+y}{z} = \ln 2$

39. The graph of a real-valued function of four independent variables is a set in five-coordinate space. The graph of a real-valued function of n independent variables is a set in n-plus-1-coordinate space.

41. $f(x, y, z) = xyz$ and $x = 20 - t$, $y = t$, $z = 20$ $\Rightarrow w = (20-t)(t)(20)$ along the line $\Rightarrow w = 400t - 20t^2$

$\Rightarrow \frac{dw}{dt} = 400 - 40t$; $\frac{dw}{dt} = 0 \Rightarrow 400 - 40t = 0 \Rightarrow t = 10$

and $\frac{d^2w}{d^2t} = -40$ for all $t \Rightarrow$ yes, maximum at $t = 10$

$\Rightarrow x = 20 - 10 = 10$, $y = 10$, $z = 20 \Rightarrow$ maximum of f along the line is $f(10, 10, 20) = (10)(10)(20) = 2000$

43. $w = 4\left(\frac{Th}{d}\right)^{1/2} = 4\left[\frac{(290\text{ K})(16.8\text{ km})}{5\text{ K/km}}\right]^{1/2} \approx 124.86$ km

$\Rightarrow$ must be $\frac{1}{2}(124.86) \approx 63$ km south of Nantucket

44–47. Example CAS commands:

Maple:

```
with(plots):
f:= (x,y)->x*sin(y/2)+y*sin(2*x):
plot3d(f(x,y),x=0..3*Pi,y=0..3*Pi,
  axes=FRAMED,title = `x sin y/2 +
  y sin 2x`);
contourplot(f(x,y),x=0..5*Pi,
  y=0..5*Pi);
eq:=f(x,y)=f(3*Pi,3*Pi);
implicitplot(eq,x=0..3*Pi,
  y=0..10*Pi);
```

Mathematica:

```
Clear[x,y]
<<Graphics`ImplicitPlot`
SetOptions[Plot3D, PlotPoints -> 25];
SetOptions[ContourPlot, PlotPoints ->
  25, ContourShading -> False];
f[x_,y_] = x Sin[y/2] + y Sin[2x]
{xa,xb} = {0, 5Pi};
{ya,yb} = {0, 5Pi};
{x0,y0} = {3Pi, 3Pi};
Plot3D[f[x,y], {x,xa,xb}, {y,ya,yb}]
ContourPlot[f[x,y], {x,xa,xb},
  {y,ya,yb}]
ImplicitPlot[f[x,y]==f[x0,y0],
  {x,xa,xb}, {y,ya,yb}]
```

48–51. Example CAS commands:

Maple:

```
with(plots):
eq:=ln(x^2+y^2+z^2)=0.25;
implicitplot3d(eq,x=-1..1,y=-1..1,
  z=-1..1,axes=BOXED,scaling=
  CONSTRAINED);
```

Mathematica:

```
<<Graphics `ContourPlot 3D`
ContourPlot3D[4 Log[x^2+y^2+z^2],
  {x,-1.1,1.2}, {{y,-1.1,1.2},
  {z,-1.1,1.2}, Contours->{1.}]
```

52–55. Example CAS commands:

Maple:

```
with(plots):
x:=(u,v)->u*cos(v);  y:=(u,v)-
  >u*sin(v);  z:=(u,v)->u;
plot3d([x(u,v),y(u,v),z(u,v)],u=0..2,
  v=0..2*Pi,axes=FRAMED);
contourplot([x(u,v),y(u,v),z(u,v)],
u=0..2,v=0..2*Pi);
```

Mathematica:

Note: While in Maple it is trivial to get contours from any 3D surface, in Mathematica it is not obvious for parametric surfaces. In these examples, z only depends on one parameter, so we can solve for that parameter in terms of z, and substitute to get x and y in terms of z and the other parameter, then parametrically plot level curves for several equally spaced values of z (using "Table").

```
ParametricPlot3D[{u Cos[v],
  u Sin[v],u}, {u,0,2}, {v,0,2Pi}]
ParametricPlot[Evaluate[Table[
  {z Cos[v],z Sin[v]}, {z,0,2,1/3}]],
  {v,0,2Pi}, AspectRatio ->
  Automatic]
```

57. $x^4 + 1 = (x^2 + \sqrt{2}x + 1)(x^2 - \sqrt{2}x + 1)$

Using the quadratic formula twice we get four roots:

$\frac{\sqrt{2}}{2} + \frac{\sqrt{2}}{2}i \approx 0.71 + 0.71i$

$\frac{\sqrt{2}}{2} - \frac{\sqrt{2}}{2}i \approx 0.71 - 0.71i$

$-\frac{\sqrt{2}}{2} + \frac{\sqrt{2}}{2}i \approx -0.71 + 0.71i$

$-\frac{\sqrt{2}}{2} - \frac{\sqrt{2}}{2}i \approx -0.71 - 0.71i$

Section 13.2 Limits and Continuity (pp. 686–693)

Exploration 1: Exploring Pathways to the Origin

a. For $y = x^2$, $f(x, y) = \frac{2x^2(x^2)}{x^4 + (x^2)^2} = 1$ and $\lim_{(x,y)\to(0,0)} f(x, y) = 1$.

b. For $y = -x^2$, $f(x, y) = \frac{2x^2(-x^2)}{x^4 + (-x^2)^2} = -1$

and $\lim_{(x,y)\to(0,0)} f(x, y) = -1$

c. $f(x, y)$ has different limits along different paths of approach to $(0, 0)$.

d. For $y = kx^2$, $f(x, y) = \frac{2x^2(kx^2)}{x^4 + (kx^2)^2} = \frac{2k}{1 + k^2}$. Since this is a constant for a given k, a curve $y = kx^2$ is a curve along which $f(x, y)$ is constant, which fits the definition of a level curve.

e. A limit cannot exist where several level curves, representing distinct values of the function, converge.

Quick Review 13.2

1. $\lim_{x \to 2} \frac{3x^2 - 7x + 5}{x^3 + x^2 - 9} = \frac{3(4) - 7(2) + 5}{8 + 4 - 9} = \frac{3}{3} = 1$

2. $\lim_{x \to 0} \sqrt{3 + \sec x} = \sqrt{3 + 1} = 2$

3. $\lim_{x \to -4} \frac{x + 5}{\sqrt{x^2 + 9}} = \frac{1}{\sqrt{25}} = \frac{1}{5}$

4. $\lim_{x \to 3} \frac{2x^2 - 7x + 3}{x^2 - 9} = \lim_{x \to 3} \frac{(2x - 1)(x - 3)}{(x + 3)(x - 3)} = \lim_{x \to 3} \frac{2x - 1}{x + 3}$
$= \frac{5}{6}$

5. $\lim_{x \to \pi} \frac{\sin x}{x - \pi} = \lim_{y \to 0} \frac{\sin(y + \pi)}{y} = \lim_{y \to 0} \frac{-\sin y}{y} = -1$

6. All $x \neq \pm 2$

7. All $x > 0$

8. All $x \neq -\frac{3}{2}$

9. $\lim_{x \to 0} \frac{\sin 2x}{\sin x} = \lim_{x \to 0} \frac{2 \sin x \cos x}{\sin x} = 2$. So $f(x)$ is not continuous at $x = 0$. f is continuous at all $x \neq 0$.

10. All real numbers

Section 13.2 Exercises

1. $\lim_{(x, y) \to (0, 0)} \frac{3x^2 - y^2 + 5}{x^2 + y^2 + 2} = \frac{3(0)^2 - 0^2 + 5}{0^2 + 0^2 + 2} = \frac{5}{2}$

3. $\lim_{(x, y) \to (0, \ln 2)} e^{x - y} = e^{0 - \ln 2} = e^{\ln\left(\frac{1}{2}\right)} = \frac{1}{2}$

5. $\lim_{(x, y, z) \to (1, 3, 4)} \sqrt{x^2 + y^2 + z^2 - 1} = \sqrt{1 + 9 + 16 - 1}$
$= 5$

7. $\lim_{(x, y) \to \left(0, \frac{\pi}{4}\right)} \sec x \tan y = (\sec 0)\left(\tan \frac{\pi}{4}\right) = (1)(1) = 1$

9. $\lim_{(x, y) \to (1, 1)} \cos(\sqrt[3]{|xy|} - 1) = \cos(\sqrt[3]{(1)(1)} - 1)$
$= \cos 0 = 1$

11. $\lim_{(x, y) \to (0, 0)} \frac{e^y \sin x}{x} = \lim_{(x, y) \to (0, 0)} (e^y)\left(\frac{\sin x}{x}\right)$
$= e^0 \cdot \lim_{x \to 0} \left(\frac{\sin x}{x}\right) = 1 \cdot 1 = 1$

13. $\lim_{\substack{(x, y) \to (1, 1) \\ x \neq y}} \frac{x^2 - 2xy + y^2}{x - y} = \lim_{\substack{(x, y) \to (1, 1) \\ x \neq y}} \frac{(x - y)^2}{x - y}$
$= \lim_{(x, y) \to (1, 1)} (x - y) = (1 - 1) = 0$

15. $\lim_{\substack{(x, y) \to (1, 1) \\ x \neq 1}} \frac{xy - y - 2x + 2}{x - 1} = \lim_{\substack{(x, y) \to (1, 1) \\ x \neq 1}} \frac{(x - 1)(y - 2)}{x - 1}$
$= \lim_{(x, y) \to (1, 1)} (y - 2) = (1 - 2) = -1$

17. $\lim_{\substack{(x, y) \to (0, 0) \\ x \neq y}} \frac{x - y + 2\sqrt{x} - 2\sqrt{y}}{\sqrt{x} - \sqrt{y}}$
$= \lim_{\substack{(x, y) \to (0, 0) \\ x \neq y}} \frac{(\sqrt{x} - \sqrt{y})(\sqrt{x} + \sqrt{y} + 2)}{\sqrt{x} - \sqrt{y}}$
$= \lim_{(x, y) \to (0, 0)} (\sqrt{x} + \sqrt{y} + 2) = (\sqrt{0} + \sqrt{0} + 2) = 2$

Note: (x, y) must approach $(0, 0)$ through the first quadrant only with $x \neq y$.

19. Let $z = 2x - y$. Then
$\lim_{\substack{(x, y) \to (2, 0) \\ 2x - y \neq 4}} \frac{\sqrt{2x - y} - 2}{2x - y - 4} = \lim_{z \to 4} \frac{\sqrt{z} - 2}{z - 4}$
$= \lim_{z \to 4} \frac{\sqrt{z} - 2}{(\sqrt{z} - 2)(\sqrt{z} + 2)} = \lim_{z \to 4} \frac{1}{\sqrt{z} + 2} = \frac{1}{4}$

21. $\lim_{P \to (2, 3, -6)} \sqrt{x^2 + y^2 + z^2} = \sqrt{4 + 9 + 36} = 7$

23. $\lim_{P \to (3, 3, 0)} (\sin^2 x + \cos^2 y + \sec^2 z)$
$= (\sin^2 3 + \cos^2 3) + \sec^2 0 = 1 + 1^2 = 2$

25. $\lim_{P \to \left(-\frac{1}{4}, \frac{\pi}{2}, 2\right)} \tan^{-1}(xyz) = \tan^{-1}\left(-\frac{1}{4} \cdot \frac{\pi}{2} \cdot 2\right)$
$= \tan^{-1}\left(-\frac{\pi}{4}\right)$

27. a. All (x, y)

b. All (x, y) except $(0, 0)$

29. a. All (x, y) except where $x = 0$ or $y = 0$

b. All (x, y)

31. a. All (x, y, z)

b. All (x, y, z) except the interior of the cylinder $x^2 + y^2 = 1$

33. a. All (x, y, z) where $z \neq 0$

b. All (x, y, z) where $x^2 + y^2 + z^2 \neq 1$

35. $\lim_{\substack{(x, y) \to (0, 0) \\ \text{along } y = x \\ x > 0}} -\frac{x}{\sqrt{x^2 + y^2}} = \lim_{x \to 0^+} -\frac{x}{\sqrt{x^2 + x^2}}$
$= \lim_{x \to 0^+} -\frac{x}{\sqrt{2}|x|} = \lim_{x \to 0} -\frac{x}{\sqrt{2}x} = \lim_{x \to 0} -\frac{1}{\sqrt{2}} = -\frac{1}{\sqrt{2}}$;
$\lim_{\substack{(x, y) \to (0, 0) \\ \text{along } y = x \\ x < 0}} -\frac{x}{\sqrt{x^2 + y^2}} = \lim_{x \to 0^-} -\frac{x}{\sqrt{2}|x|} = \lim_{x \to 0} -\frac{x}{\sqrt{2}(-x)}$
$= \lim_{x \to 0} \frac{1}{\sqrt{2}} = \frac{1}{\sqrt{2}}$

37. $\lim_{\substack{(x, y) \to (0, 0) \\ \text{along } y = kx^2}} \frac{x^4 - y^2}{x^4 + y^2} = \lim_{x \to 0} \frac{x^4 - (kx^2)^2}{x^4 + (kx^2)^2} = \lim_{x \to 0} \frac{x^4 - k^2x^4}{x^4 + k^2x^4}$
$= \frac{1 - k^2}{1 + k^2} \Rightarrow$ different limits for different values of k

39. $\lim_{\substack{(x, y) \to (0, 0) \\ \text{along } y = kx \\ k \neq -1}} \frac{x - y}{x + y} = \lim_{x \to 0} \frac{x - kx}{x + kx} = \frac{1 - k}{1 + k}$

$\Rightarrow$ different limits for different values of k, $k \neq -1$

41. $\lim_{\substack{(x, y) \to (0, 0) \\ \text{along } y = kx^2 \\ k \neq 0}} \frac{x^2 + y}{y} = \lim_{x \to 0} \frac{x^2 + kx^2}{kx^2} = \frac{1 + k}{k}$

$\Rightarrow$ different limits for different values of k, $k \neq 0$

43. No, the limit depends only on the values $f(x, y)$ has when $(x, y) \neq (x_0, y_0)$.

45. a. $f(x, y)|_{y = mx} = \frac{2m}{1 + m^2} = \frac{2 \tan \theta}{1 + \tan^2 \theta} = \sin 2\theta$. The value of $f(x, y) = \sin 2\theta$ varies with θ, which is the line's angle of inclination.

b. Since $f(x, y)|_{y = mx} = \sin 2\theta$ and since $-1 \leq \sin 2\theta \leq 1$ for every θ, $\lim_{(x, y) \to (0, 0)} f(x, y)$ varies from -1 to 1 along $y = mx$.

47. $\lim_{(x, y) \to (0, 0)} \left(1 - \frac{x^2y^2}{3}\right) = 1$ and $\lim_{(x, y) \to (0, 0)} 1 = 1$

$\Rightarrow \lim_{(x, y) \to (0, 0)} \frac{\tan^{-1} xy}{xy} = 1$, by the Sandwich Theorem.

49. The limit is 0 since $\left|\sin\left(\frac{1}{x}\right)\right| \leq 1 \Rightarrow -1 \leq \sin\left(\frac{1}{x}\right) \leq 1$

$\Rightarrow -y \leq y \sin\left(\frac{1}{x}\right) \leq y$ for $y \geq 0$, and $-y \geq y \sin\left(\frac{1}{x}\right) \geq y$

for $y \leq 0$. Thus as $(x, y) \to (0, 0)$, both $-y$ and y approach

$0 \Rightarrow y \sin\left(\frac{1}{x}\right) \to 0$, by the Sandwich Theorem.

■ Section 13.3 Partial Derivatives
(pp. 693–702)

Exploration 1: An Intersection of Two Hiking Paths

1-5. The correct conclusion is that just as the slope of one path is independent of the slope of the other path, so one partial derivative cannot be used to predict the other.

Quick Review 13.3

1. $\frac{dy}{dx} = 2kx + 5, \frac{d^2y}{dx^2} = 2k$

2. $\frac{dy}{dx} = k \cos kx, \frac{d^2y}{dx^2} = -k^2 \sin kx$

3. $\frac{dy}{dx} = 20ke^{xt}, \frac{d^2y}{dx^2} = 20k^2e^{kx}$

4. $\frac{dy}{dx} = (x + k + 1)e^x, \frac{d^2y}{dx^2} = (x + k + 2)e^x$

5. $\frac{dy}{dx} = \frac{\sqrt{k}}{1 + kx^2}, \frac{d^2y}{dx^2} = -\frac{2k^{3/2}x}{(1 + kx^2)^2}$

6. $\frac{dy}{dx} = \frac{1}{x - k} + 3, \frac{d^2y}{dx^2} = -\frac{1}{(x - k)^2}$

7. $\frac{dy}{dx} = k^x \ln k, \frac{d^2y}{dx^2} = k^x(\ln k)^2$

8. $\frac{dy}{dx} = -ke^k \sin(kx), \frac{d^2y}{dx^2} = -k^2e^k \cos(kx)$

9. $\frac{dy}{dx} = -\frac{k}{(x - k)^2}, \frac{d^2y}{dx^2} = \frac{2k}{(x - k)^3}$

10. $\frac{dy}{dx} = \frac{k}{(x - k)^2}, \frac{d^2y}{dx^2} = \frac{-2k}{(x - k)^3}$

Section 13.3 Exercises

1. $\frac{\partial x}{\partial f} = 2, \frac{\partial f}{\partial y} = 0$

3. $\frac{\partial f}{\partial x} = y - 1, \frac{\partial f}{\partial y} = x$

5. $\frac{\partial f}{\partial x} = 2x(y + 2), \frac{\partial f}{\partial y} = x^2 - 1$

7. $\frac{\partial f}{\partial x} = -\frac{1}{(x + y)^2} \cdot \frac{\partial}{\partial x}(x + y) = -\frac{1}{(x + y)^2},$

$\frac{\partial f}{\partial y} = -\frac{1}{(x + y)^2} \cdot \frac{\partial}{\partial y}(x + y) = -\frac{1}{(x + y)^2}$

9. $\frac{\partial f}{\partial x} = \frac{x}{\sqrt{x^2 + y^2}}, \frac{\partial f}{\partial y} = \frac{y}{\sqrt{x^2 + y^2}}$

11. $\frac{\partial f}{\partial x} = \frac{(xy - 1)(1) - (x + y)(y)}{(xy - 1)^2} = \frac{-y^2 - 1}{(xy - 1)^2},$

$\frac{\partial f}{\partial y} = \frac{(xy - 1)(1) - (x + y)(x)}{(xy - 1)^2} = \frac{-x^2 - 1}{(xy - 1)^2}$

13. $\frac{\partial f}{\partial x} = e^x \frac{\partial}{\partial x} \sin(y + 1) + \sin(y + 1) \cdot \frac{\partial}{\partial x} e^x = e^x(0)$

$+ \sin(y + 1) \cdot e^x = e^x \sin(y + 1), \frac{\partial f}{\partial y} = e^x \frac{\partial}{\partial y} \sin(y + 1)$

$+ \sin(y + 1) \cdot \frac{\partial}{\partial y} e^x = e^x \cos(y + 1) + \sin(y + 1) \cdot 0$

$= e^x \cos(y + 1)$

15. $\frac{\partial f}{\partial x} = e^{(x + y + 1)} \cdot \frac{\partial}{\partial x}(x + y + 1) = e^{(x + y + 1)},$

$\frac{\partial f}{\partial y} = e^{(x + y + 1)} \cdot \frac{\partial}{\partial y}(x + y + 1) = e^{(x + y + 1)}$

17. $f(x, y) = \frac{\ln x}{\ln y} \Rightarrow \frac{\partial f}{\partial x} = \frac{1}{x \ln y}$ and $\frac{\partial f}{\partial y} = \frac{-\ln x}{y(\ln y)^2}$

19. $f_x = y + z, f_y = x + z, f_z = y + x$

21. $f_x = 0, f_y = 2y, f_z = 4z$

23. $f_x = \frac{1}{x + 2y + 3z}, f_y = \frac{2}{x + 2y + 3z}, f_z = \frac{3}{x + 2y + 3z}$

25. $\frac{\partial f}{\partial t} = -2\pi \sin(2\pi t - \alpha), \frac{\partial f}{\partial \alpha} = \sin(2\pi t - \alpha)$

27. $\frac{\partial h}{\partial \rho} = \sin \phi \cos \theta, \frac{\partial h}{\partial \phi} = \rho \cos \phi \cos \theta,$

$\frac{\partial h}{\partial \theta} = -\rho \sin \phi \sin \theta$

29. $W_P = V, W_V = P + \frac{\delta v^2}{2g}, W_\delta = \frac{Vv^2}{2g}, W_v = \frac{2V\delta v}{2g} = \frac{V\delta v}{g},$

$W_g = -\frac{V\delta v^2}{2g^2}$

31. $\frac{\partial f}{\partial x} = 1 + y, \frac{\partial f}{\partial y} = 1 + x, \frac{\partial^2 f}{\partial x^2} = 0, \frac{\partial^2 f}{\partial y^2} = 0,$
$\frac{\partial^2 f}{\partial y \partial x} = \frac{\partial^2 f}{\partial x \partial y} = 1$

33. $\frac{\partial f}{\partial x} = 2xy + y\cos x, \frac{\partial f}{\partial y} = x^2 - \sin y + \sin x,$
$\frac{\partial^2 f}{\partial x^2} = 2y - y\sin x, \frac{\partial^2 f}{\partial y^2} = -\cos y, \frac{\partial^2 f}{\partial y \partial x} = \frac{\partial^2 f}{\partial x \partial y}$
$= 2x + \cos x$

35. $\frac{\partial f}{\partial x} = \frac{1}{x+y}, \frac{\partial f}{\partial y} = \frac{1}{x+y}, \frac{\partial^2 f}{\partial x^2} = \frac{-1}{(x+y)^2}, \frac{\partial^2 f}{\partial y^2} = \frac{-1}{(x+y)^2},$
$\frac{\partial^2 f}{\partial y \partial x} = \frac{\partial^2 f}{\partial x \partial y} = \frac{-1}{(x+y)^2}$

37. $\frac{\partial w}{\partial x} = \frac{2}{2x+3y}, \frac{\partial w}{\partial y} = \frac{3}{2x+3y}, \frac{\partial^2 w}{\partial y \partial x} = \frac{-6}{(2x+3y)^2},$
$\frac{\partial^2 w}{\partial x \partial y} = \frac{-6}{(2x+3y)^2}$

39. $\frac{\partial w}{\partial x} = y^2 + 2xy^3 + 3x^2y^4, \frac{\partial w}{\partial y} = 2xy + 3x^2y^2 + 4x^3y^3,$
$\frac{\partial^2 w}{\partial y \partial x} = 2y + 6xy^2 + 12x^2y^3$, and $\frac{\partial^2 w}{\partial x \partial y} = 2y + 6xy^2 + 12x^2y^3$

41. **a.** x first **b.** y first
c. x first **d.** x first
e. y first **f.** y first

43. Differentiate implicitly with respect to x, holding y constant and treating z as a function of x. $\frac{\partial}{\partial x}(x^3 z) + \frac{\partial}{\partial x}(z^3 x)$
$- \frac{\partial}{\partial x}(2yz) = 0;\ x^3\frac{\partial z}{\partial x} + z(3x^2) + z^3(1) + x\left(3z^2\frac{\partial z}{\partial x}\right) - 2y\frac{\partial z}{\partial x}$
$= 0;\ \frac{\partial z}{\partial x}(x^3 + 3xz^2 - 2y) = -3x^2z - z^3;\ \frac{\partial z}{\partial x}$
$= \frac{-3x^2z - z^3}{x^3 + 3xz^2 - 2y};\ \left.\frac{\partial z}{\partial x}\right|_{(1,1,1)} = \frac{-3-1}{1+3-2} = \frac{-4}{2} = -2$

45. $f_z(x_0, y_0, z_0) = \lim_{h \to 0} \frac{f(x_0, y_0, z_0 + h) - f(x_0, y_0, z_0)}{h};$
$f_z(1, 2, 3) = \lim_{h \to 0} \frac{f(1, 2, 3+h) - f(1, 2, 3)}{h}$
$= \lim_{h \to 0} \frac{2(3+h)^2 - 2(9)}{h} = \lim_{h \to 0} \frac{12h + 2h^2}{h} = \lim_{h \to 0} (12 + 2h)$
$= 12$

47. **a.** $\frac{\partial f}{\partial x} = 2x, \frac{\partial f}{\partial y} = 2y, \frac{\partial f}{\partial z} = -4z \Rightarrow \frac{\partial^2 f}{\partial x^2} = 2, \frac{\partial^2 f}{\partial y^2} = 2,$
$\frac{\partial^2 f}{\partial z^2} = -4 \Rightarrow \frac{\partial^2 f}{\partial x^2} + \frac{\partial^2 f}{\partial y^2} + \frac{\partial^2 f}{\partial z^2} = 2 + 2 + (-4) = 0$

b. $\frac{\partial f}{\partial x} = -6xz, \frac{\partial f}{\partial y} = -6yz, \frac{\partial f}{\partial z} = 6z^2 - 3(x^2 + y^2),$
$\frac{\partial^2 f}{\partial x^2} = -6z, \frac{\partial^2 f}{\partial y^2} = -6z, \frac{\partial^2 f}{\partial z^2} = 12z \Rightarrow$
$\frac{\partial^2 f}{\partial x^2} + \frac{\partial^2 f}{\partial y^2} + \frac{\partial^2 f}{\partial z^2} = -6z - 6z + 12z = 0$

c. $\frac{\partial f}{\partial x} = -2e^{-2y}\sin 2x, \frac{\partial f}{\partial y} = -e^{-2y}\cos 2x, \frac{\partial^2 f}{\partial x^2}$
$= -4e^{-2y}\cos 2x, \frac{\partial^2 f}{\partial y^2} = 4e^{-2y}\cos 2x \Rightarrow \frac{\partial^2 f}{\partial x^2} + \frac{\partial^2 f}{\partial y^2}$
$= -4e^{-2y}\cos 2x + 4e^{-2y}\cos 2x = 0$

d. $\frac{\partial f}{\partial x} = \frac{x}{x^2 + y^2}, \frac{\partial f}{\partial y} = \frac{y}{x^2 + y^2}, \frac{\partial^2 f}{\partial x^2} = \frac{y^2 - x^2}{(x^2 + y^2)^2}, \frac{\partial^2 f}{\partial y^2}$
$= \frac{x^2 - y^2}{(x^2 + y^2)^2} \Rightarrow \frac{\partial^2 f}{\partial x^2} + \frac{\partial^2 f}{\partial y^2} = \frac{y^2 - x^2}{(x^2 + y^2)^2} + \frac{x^2 - y^2}{(x^2 + y^2)^2}$
$= 0$

e. $\frac{\partial f}{\partial x} = -\frac{1}{2}(x^2 + y^2 + z^2)^{-3/2}(2x)$
$= -x(x^2 + y^2 + z^2)^{-3/2},$
$\frac{\partial f}{\partial y} = -\frac{1}{2}(x^2 + y^2 + z^2)^{-3/2}(2y)$
$= -y(x^2 + y^2 + z^2)^{-3/2},$
$\frac{\partial f}{\partial z} = -\frac{1}{2}(x^2 + y^2 + z^2)^{-3/2}(2z)$
$= -z(x^2 + y^2 + z^2)^{-3/2};$
$\frac{\partial^2 f}{\partial x^2} = -(x^2 + y^2 + z^2)^{-3/2} + 3x^2(x^2 + y^2 + z^2)^{-5/2},$
$\frac{\partial^2 f}{\partial y^2} = -(x^2 + y^2 + z^2)^{-3/2} + 3y^2(x^2 + y^2 + z^2)^{-5/2},$
$\frac{\partial^2 f}{\partial z^2} = -(x^2 + y^2 + z^2)^{-3/2} + 3z^2(x^2 + y^2 + z^2)^{-5/2}$
$\Rightarrow \frac{\partial^2 f}{\partial x^2} + \frac{\partial^2 f}{\partial y^2} + \frac{\partial^2 f}{\partial z^2} = [-(x^2 + y^2 + z^2)^{-3/2}$
$+ 3x^2(x^2 + y^2 + z^2)^{-5/2}] + [-(x^2 + y^2 + z^2)^{-3/2}$
$+ 3y^2(x^2 + y^2 + z^2)^{-5/2}] + [-(x^2 + y^2 + z^2)^{-3/2}$
$+ 3z^2(x^2 + y^2 + z^2)^{-5/2}] = -3(x^2 + y^2 + z^2)^{-3/2}$
$+ (3x^2 + 3y^2 + 3z^2)(x^2 + y^2 + z^2)^{-5/2} = 0$

f. $\frac{\partial f}{\partial x} = 3e^{3x+4y}\cos 5z, \frac{\partial f}{\partial y} = 4e^{3x+4y}\cos 5z,$
$\frac{\partial f}{\partial z} = -5e^{3x+4y}\sin 5z; \frac{\partial^2 f}{\partial x^2} = 9e^{3x+4y}\cos 5z,$
$\frac{\partial^2 f}{\partial y^2} = 16e^{3x+4y}\cos 5z, \frac{\partial^2 f}{\partial z^2} = -25e^{3x+4y}\cos 5z$
$\Rightarrow \frac{\partial^2 f}{\partial x^2} + \frac{\partial^2 f}{\partial y^2} + \frac{\partial^2 f}{\partial z^2} = 9e^{3x+4y}\cos 5z$
$+ 16e^{3x+4y}\cos 5z - 25e^{3x+4y}\cos 5z = 0$

■ Section 13.4 Differentiability, Linearization, and Differentials (pp. 703–714)

Exploration 1: Sensitivity to Change

1. $V(r, h) = \pi r^2 h$

2. $dV = \frac{\partial V}{\partial r} \cdot dr + \frac{\partial V}{\partial h} \cdot dh = 2\pi rh \cdot dr + \pi r^2 \cdot dh$
$dV|_{r=5,\,h=25} = 2\pi(5)(25)dr + \pi(5)^2 dh$
$= 250\pi dr + 25\pi dh$

3. Since a 1-unit change in r produces a 250π-unit change in V, and a 1-unit change in h produces a 25π-unit change in V, you should be more concerned with variations in radii.

4. $dV|_{r=25,\,h=5} = 2\pi(25)(5)dr + \pi(25)^2 dh$
$= 250\pi dr + 625\pi dh$. Now you should be more concerned with variations in height.

5. Find r, h such that $2\pi rh = \pi r^2$: $2h = r$, and $\frac{h}{r} = \frac{1}{2}$.

Quick Review 13.4

1. $y = 1 - x$
2. 0.8
3. 0.01058
4. 0.013
5. 1.3%
6. Graph the error $y = |x - \sin(2x) + 1 - (1 - x)|$, along with the line $y = 0.02$ in $[-0.3, 0.3]$ by $[-0.01, 0.03]$:

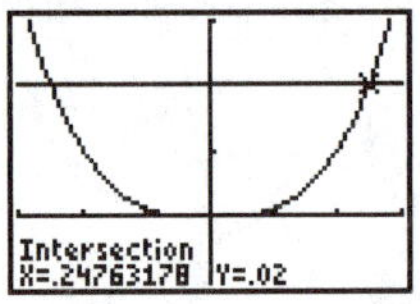

Interval: $\approx (-0.2476, 0.2476)$

7. $df = (1 - 2\cos(2x))dx$
8. $-\frac{1}{2}$
9. The graph is quite straight near $x = \pi$, so the linearization at $x = \pi$ should be quite close to the graph of f.
10. The graph is turning around near $x = \frac{1}{2}$, so a straight-line approximation will be less accurate.

Section 13.4 Exercises

1. a. $f(0, 0) = 1, f_x(x, y) = 2x \Rightarrow f_x(0, 0) = 0, f_y(x, y) = 2y \Rightarrow f_y(0, 0) = 0 \Rightarrow L(x, y) = 1 + 0(x - 0) + 0(y - 0) = 1$

   b. $f(1, 1) = 3, f_x(1, 1) = 2, f_y(1, 1) = 2 \Rightarrow L(x, y) = 3 + 2(x - 1) + 2(y - 1) = 2x + 2y - 1$

3. a. $f(0, 0) = 5, f_x(x, y) = 3$ for all (x, y), $f_y(x, y) = -4$ for all $(x, y) \Rightarrow L(x, y) = 5 + 3(x - 0) - 4(y - 0) = 3x - 4y + 5$

   b. $f(1, 1) = 4, f_x(1, 1) = 3, f_y(1, 1) = -4 \Rightarrow L(x, y) = 4 + 3(x - 1) - 4(y - 1) = 3x - 4y + 5$

5. a. $f(0, 0) = 1, f_x(x, y) = e^x \cos y \Rightarrow f_x(0, 0) = 1, f_y(x, y) = -e^x \sin y \Rightarrow f_y(0, 0) = 0 \Rightarrow L(x, y) = 1 + 1(x - 0) + 0(y - 0) = x + 1$

   b. $f\left(0, \frac{\pi}{2}\right) = 0, f_x\left(0, \frac{\pi}{2}\right) = 0, f_y\left(0, \frac{\pi}{2}\right) = -1 \Rightarrow L(x, y) = 0 + 0(x - 0) - 1\left(y - \frac{\pi}{2}\right) = -y + \frac{\pi}{2}$

7. $f(2, 1) = 3, f_x(x, y) = 2x - 3y \Rightarrow f_x(2, 1) = 1, f_y(x, y) = -3x \Rightarrow f_y(2, 1) = -6 \Rightarrow L(x, y) = 3 + 1(x - 2) - 6(y - 1) = 7 + x - 6y; f_{xx}(x, y) = 2, f_{yy}(x, y) = 0,$
$f_{xy}(x, y) = -3 \Rightarrow M = 3$; thus $|E(x, y)| \le \left(\frac{1}{2}\right)(3)(|x - 2| + |y - 1|)^2 \le \left(\frac{3}{2}\right)(0.1 + 0.1)^2 = 0.06$

9. $f(0, 0) = 1, f_x(x, y) = e^x \cos y \Rightarrow f_x(0, 0) = 1, f_y(x, y) = -e^x \sin y \Rightarrow f_y(0, 0) = 0 \Rightarrow L(x, y) = 1 + 1(x - 0) + 0(y - 0) = 1 + x; f_{xx}(x, y) = e^x \cos y, f_{yy}(x, y) = -e^x \cos y, f_{xy}(x, y) = -e^x \sin y; |x| \le 0.1 \Rightarrow -0.1 \le x \le 0.1$ and $|y| \le 0.1 \Rightarrow -0.1 \le y \le 0.1$; thus the max of $|f_{xx}(x, y)|$ on R is $e^{0.1} \cos(0.1) \le 1.11$, the max of $|f_{yy}(x, y)|$ on R is $e^{0.1} \cos(0.1) \le 1.11$, and the max of $|f_{xy}(x, y)|$ on R is $e^{0.1} \sin(0.1) \le 0.11 \Rightarrow M = 1.11$; thus $|E(x, y)| \le \left(\frac{1}{2}\right)(1.11)(|x| + |y|)^2 \le (0.555)(0.1 + 0.1)^2 = 0.0222$

11. $A = xy \Rightarrow dA = x\,dy + y\,dx$; if $x > y$ then a 1-unit change in y gives a greater change in dA than a 1-unit change in x. Thus, pay more attention to y which is the smaller of the two dimensions.

13. $T_x(x, y) = e^y + e^{-y}$ and $T_y(x, y) = x(e^y - e^{-y}) \Rightarrow dT = T_x(x, y)dx + T_y(x, y)dy = (e^y + e^{-y})dx + x(e^y - e^{-y})dy \Rightarrow dT|_{(2, \ln 2)} = 2.5\,dx + 3.0\,dy$. If $|dx| \le 0.1$ and $|dy| \le 0.02$, then the maximum possible error in the computed value of T is $(2.5)(0.1) + (3.0)(0.02) = 0.31$ in magnitude.

15. $V_r = 2\pi rh$ and $V_h = \pi r^2 \Rightarrow dV = V_r\,dr + V_h\,dh \Rightarrow dV = 2\pi rh\,dr + \pi r^2 dh \Rightarrow dV|_{(5, 12)} = 120\pi\,dr + 25\pi\,dh$; $|dr| \le 0.1$ cm and $|dh| \le 0.1$ cm $\Rightarrow dV \le (120\pi)(0.1) + (25\pi)(0.1) = 14.5\pi$ cm^3; $V(5, 12) = 300\pi$ cm^3 $\Rightarrow$ maximum percentage error is $\pm \frac{14.5\pi}{300\pi} \times 100 = \pm 4.83\%$

17. $df = f_x(x, y)dx + f_y(x, y)dy = 3x^2y^4dx + 4x^3y^3dy \Rightarrow df|_{(1, 1)} = 3\,dx + 4\,dy$; for a square, $dx = dy \Rightarrow df = 7\,dx$ so that $|df| \le 0.1 \Rightarrow 7|dx| \le 0.1 \Rightarrow |dx| \le \frac{0.1}{7} \approx 0.014 \Rightarrow$ for the square, $|x - 1| \le 0.014$ and $|y - 1| \le 0.014$

19. From Exercise 20, $dR = \left(\frac{R}{R_1}\right)^2 dR_1 + \left(\frac{R}{R_2}\right)^2 dR_2$ so that R_1 changing from 20 to 20.1 ohms $\Rightarrow dR_1 = 0.1$ ohm and R_2 changing from 25 to 24.9 ohms $\Rightarrow dR_2 = -0.1$ ohms; $\frac{1}{R} = \frac{1}{R_1} + \frac{1}{R_2} \Rightarrow R = \frac{100}{9}$ ohms $\Rightarrow dR|_{(20, 25)} = \frac{\left(\frac{100}{9}\right)^2(0.1)}{(20)^2} + \frac{\left(\frac{100}{9}\right)^2(-0.1)}{(25)^2} \approx 0.011$ ohms $\Rightarrow$ percentage change is $\left.\frac{dR}{R}\right|_{(20, 25)} \times 100 = \frac{0.011}{\left(\frac{100}{9}\right)} \times 100 \approx 0.1\%$

21. a. $f(1, 1, 1) = 3, f_x(1, 1, 1) = y + z|_{(1, 1, 1)} = 2, f_y(1, 1, 1) = x + z|_{(1, 1, 1)} = 2, f_z(1, 1, 1) = y + x|_{(1, 1, 1)} = 2 \Rightarrow L(x, y, z) = 3 + 2(x - 1) + 2(y - 1) + 2(z - 1) = 2x + 2y + 2z - 3$

   b. $f(1, 0, 0) = 0, f_x(1, 0, 0) = 0, f_y(1, 0, 0) = 1, f_z(1, 0, 0) = 1 \Rightarrow L(x, y, z) = 0 + 0(x - 1) + (y - 0) + (z - 0) = y + z$

   c. $f(0, 0, 0) = 0, f_x(0, 0, 0) = 0, f_y(0, 0, 0) = 0, f_z(0, 0, 0) = 0 \Rightarrow L(x, y, z) = 0$

23. a. $f(1, 0, 0) = 1, f_x(1, 0, 0) = \left.\frac{x}{\sqrt{x^2 + y^2 + z^2}}\right|_{(1, 0, 0)} = 1,$
$f_y(1, 0, 0) = \left.\frac{y}{\sqrt{x^2 + y^2 + z^2}}\right|_{(1, 0, 0)} = 0,$
$f_z(1, 0, 0) = \left.\frac{z}{\sqrt{x^2 + y^2 + z^2}}\right|_{(1, 0, 0)} = 0$
$\Rightarrow L(x, y, z) = 1 + 1(x - 1) + 0(y - 0) + 0(z - 0) = x$

b. $f(1, 1, 0) = \sqrt{2}, f_x(1, 1, 0) = \frac{1}{\sqrt{2}}, f_y(1, 1, 0) = \frac{1}{\sqrt{2}},$
$f_z(1, 1, 0) = 0 \Rightarrow L(x, y, z) = \sqrt{2} + \frac{1}{\sqrt{2}}(x - 1)$
$+ \frac{1}{\sqrt{2}}(y - 1) + 0(z - 0) = \frac{1}{\sqrt{2}}x + \frac{1}{\sqrt{2}}y$

c. $f(1, 2, 2) = 3, f_x(1, 2, 2) = \frac{1}{3}, f_y(1, 2, 2) = \frac{2}{3}, f_z(1, 2, 2)$
$= \frac{2}{3} \Rightarrow L(x, y, z) = 3 + \frac{1}{3}(x - 1) + \frac{2}{3}(y - 2)$
$+ \frac{2}{3}(z - 2) = \frac{1}{3}x + \frac{2}{3}y + \frac{2}{3}z$

25. a. $f(0, 0, 0) = 2, f_x(0, 0, 0) = e^x|_{(0, 0, 0)} = 1,$
$f_y(0, 0, 0) = -\sin(y + z)|_{(0, 0, 0)} = 0,$
$f_z(0, 0, 0) = -\sin(y + z)|_{(0, 0, 0)} = 0 \Rightarrow L(x, y, z) =$
$2 + 1(x - 0) + 0(y - 0) + 0(z - 0) = 2 + x$

b. $f\left(0, \frac{\pi}{2}, 0\right) = 1, f_x\left(0, \frac{\pi}{2}, 0\right) = 1, f_y\left(0, \frac{\pi}{2}, 0\right) = -1,$
$f_z\left(0, \frac{\pi}{2}, 0\right) = -1 \Rightarrow L(x, y, z) = 1 + 1(x - 0)$
$- 1\left(y - \frac{\pi}{2}\right) - 1(z - 0) = x - y - z + \frac{\pi}{2} + 1$

c. $f\left(0, \frac{\pi}{4}, \frac{\pi}{4}\right) = 1, f_x\left(0, \frac{\pi}{4}, \frac{\pi}{4}\right) = 1, f_y\left(0, \frac{\pi}{4}, \frac{\pi}{4}\right) = -1,$
$f_z\left(0, \frac{\pi}{4}, \frac{\pi}{4}\right) = -1 \Rightarrow L(x, y, z) = 1 + 1(x - 0)$
$- 1\left(y - \frac{\pi}{4}\right) - 1\left(z - \frac{\pi}{4}\right) = x - y - z + \frac{\pi}{2} + 1$

27. $f(x, y, z) = xz - 3yz + 2$ at $P_0(1, 1, 2) \Rightarrow f(1, 1, 2) = -2;$
$f_x = z, f_y = -3z, f_z = x - 3y \Rightarrow L(x, y, z) = -2$
$+ 2(x - 1) - 6(y - 1) - 2(z - 2) = 2x - 6y - 2z + 6;$
$f_{xx} = 0, f_{yy} = 0, f_{zz} = 0, f_{xy} = 0, f_{xz} = 1, f_{yz} = -3$
$\Rightarrow M = 3$; thus
$|E(x, y, z)| \le \left(\frac{1}{2}\right)(3)(0.01 + 0.01 + 0.02)^2 = 0.0024$

29. $f(x, y, z) = xy + 2yz - 3xz$ at $P_0(1, 1, 0) \Rightarrow f(1, 1, 0) = 1;$
$f_x = y - 3z, f_y = x + 2z, f_z = 2y - 3x \Rightarrow L(x, y, z) = 1$
$+ (x - 1) + (y - 1) - (z - 0) = x + y - z - 1; f_{xx} = 0,$
$f_{yy} = 0, f_{zz} = 0, f_{xy} = 1, f_{xz} = -3, f_{yz} = 2 \Rightarrow M = 3$; thus
$|E(x, y, z)| \le \left(\frac{1}{2}\right)(3)(0.01 + 0.01 + 0.01)^2 = 0.00135$

31. a. $dS = S_p\,dp + S_x\,dx + S_w\,dw + S_h\,dh$
$= C\left(\frac{x^4}{wh^3}dp + \frac{4px^3}{wh^3}dx - \frac{px^4}{w^2h^3}dw - \frac{3px^4}{wh^4}dh\right)$
$= C\left(\frac{px^4}{wh^3}\right)\left(\frac{1}{p}dp + \frac{4}{x}dx - \frac{1}{w}dw - \frac{3}{h}dh\right)$
$= S_0\left(\frac{1}{p_0}dp + \frac{4}{x_0}dx - \frac{1}{w_0}dw - \frac{3}{h_0}dh\right)$
$= S_0\left(\frac{1}{100}dp + dx - 5\,dw - 30\,dh\right)$, where p_0
$= 100$ N/m, $x_0 = 4$ m, $w_0 = 0.2$ m, $h_0 = 0.1$ m

b. More sensitive to a change in height

33. $u_x = e^y, u_y = xe^y + \sin z, u_z = y\cos z \Rightarrow du = e^y\,dx$
$+ (xe^y + \sin z)dy + (y\cos z)dz \Rightarrow du|_{\left(2, \ln 3, \frac{\pi}{2}\right)}$
$= 3\,dx + 7\,dy + 0\,dz = 3\,dx + 7\,dy \Rightarrow$ magnitude of the maximum possible error $\le 3(0.2) + 7(0.6) = 4.8$

35. Yes, since f_{xx}, f_{yy}, f_{xy}, and f_{yx} are all continuous on R, use the same reasoning as in Exercise 34 with $f_x(x, y)$
$= f_x(x_0, y_0) + f_{xx}(x_0, y_0)\Delta x + f_{xy}(x_0, y_0)\Delta y + \epsilon_1\Delta x + \epsilon_2\Delta y$
and $f_y(x, y) = f_y(x_0, y_0) + f_{yx}(x_0, y_0)\Delta x + f_{yy}(x_0, y_0)\Delta y$
$+\epsilon\,\Delta x + \epsilon_2\Delta y$. Then $\lim_{(x, y)\to(x_0, y_0)} f_x(x, y) = f_x(x_0, y_0)$ and
$\lim_{(x, y)\to(x_0, y_0)} f_y(x, y) = f_y(x_0, y_0)$.

37. a. $V = \pi r^2 h \Rightarrow dV = 2\pi rh\,dr + \pi r^2 dh \Rightarrow$ at $r = 1$ and $h = 5$ we have $dV = 10\pi\,dr + \pi\,dh \Rightarrow$ the volume is about 10 times more sensitive to a change in r.

b. $dV = 0 \Rightarrow 0 = 2\pi rh\,dr + \pi r^2 dh = 2h\,dr + r\,dh$
$= 10\,dr + dh \Rightarrow dr = -\frac{1}{10}dh$; choose $dh = 1.5 \Rightarrow dr$
$= -0.15 \Rightarrow h = 6.5$ in. and $r = 0.85$ in. is one solution for $\Delta V \approx dV = 0$.

39. $f(a, b, c, d) = \begin{vmatrix} a & b \\ c & d \end{vmatrix} = ad - bc \Rightarrow f_a = d, f_b = -c,$
$f_c = -b, f_d = a \Rightarrow df = d\,da - c\,db - b\,dc + a\,dd$; since $|a|$ is much greater than $|b|$, $|c|$, and $|d|$, the function f is most sensitive to a change in d.

Section 13.5 The Chain Rule
(pp. 714–723)

Exploration 1: Bypassing the Chain Rule

1. $w = xy = \cos t\sin t = \frac{1}{2}\sin 2t$

2. $\frac{d}{dt} = \left(\frac{1}{2}\sin 2t\right) = \frac{1}{2}(-\cos 2t)(2) = -\cos 2t$

3. For $w = x^2e^y, x = \cos t, y = \sin t$:
$\frac{dw}{dt} = \frac{\partial w}{\partial x}\frac{dx}{dt} + \frac{\partial w}{\partial y}\frac{dy}{dt} = 2xe^y(-\sin t) + x^2e^y(\cos t)$
$= -2\sin t\cos t(e^{\sin t}) + \cos^3 t(e^{\sin t})$
$= e^{\sin t}\cos t(\cos^2 t - 2\sin t)$; or, $w = \cos^2 t(e^{\sin t})$
$\Rightarrow \frac{dw}{dt} = \cos^2 t(e^{\sin t}\cos t) - 2\cos t\sin t\,(e^{\sin t})$
$= e^{\sin t}\cos t\,(\cos^2 t - 2\sin t)$

4. The two methods will always lead to the same solution.

Quick Review 13.5

1. $\frac{\partial f}{\partial x} = y; \frac{\partial f}{\partial y} = x + \cos y$

2. $\frac{\partial f}{\partial x} = 3x^2; \frac{\partial f}{\partial y} = -2y$

3. $\frac{\partial f}{\partial x} = \sec^2(x + y); \frac{\partial f}{\partial y} = \sec^2(x + y)$

4. $\frac{\partial f}{\partial x} = ye^{xy}; \frac{\partial f}{\partial y} = xe^{xy}$

5. $\frac{\partial f}{\partial x} = \sin y + y \sin x$; $\frac{\partial f}{\partial y} = x \cos y - \cos x$

6. 0

7. $\frac{\pi}{2}$

8. -2

9. -2

10. $\frac{2}{3}$

Section 13.5 Exercises

1. a. $\frac{\partial w}{\partial x} = 2x$, $\frac{\partial w}{\partial y} = 2y$, $\frac{dx}{dt} = -\sin t$, $\frac{dy}{dt} = \cos t \Rightarrow \frac{dw}{dt}$ $= -2x \sin t + 2y \cos t = -2 \cos t \sin t + 2 \sin t \cos t$ $= 0$; $w = x^2 + y^2 = \cos^2 t + \sin^2 t = 1 \Rightarrow \frac{dw}{dt} = 0$

   b. $\frac{dw}{dt}(\pi) = 0$

3. a. $\frac{\partial w}{\partial x} = \frac{1}{z}$, $\frac{\partial w}{\partial y} = \frac{1}{z}$, $\frac{\partial w}{\partial z} = \frac{-(x+y)}{z^2}$, $\frac{dx}{dt} = -2 \cos t \sin t$, $\frac{dy}{dt} = 2 \sin t \cos t$, $\frac{dz}{dt} = -\frac{1}{t^2} \Rightarrow \frac{dw}{dt} = -\frac{2}{z} \cos t \sin t$ $+ \frac{2}{z} \sin t \cos t + \frac{x+y}{z^2 t^2} = \frac{\cos^2 t + \sin^2 t}{\left(\frac{1}{t^2}\right)(t^2)} = 1$;

   $w = \frac{x}{z} + \frac{y}{z} = \frac{\cos^2 t}{\left(\frac{1}{t}\right)} + \frac{\sin^2 t}{\left(\frac{1}{t}\right)} = t \Rightarrow \frac{dw}{dt} = 1$

   b. $\frac{dw}{dt}(3) = 1$

5. a. $\frac{\partial w}{\partial x} = 2ye^x$, $\frac{\partial w}{\partial y} = 2e^x$, $\frac{\partial w}{\partial z} = -\frac{1}{z}$, $\frac{dx}{dt} = \frac{2t}{t^2+1}$, $\frac{dy}{dt} = \frac{1}{t^2+1}$, $\frac{dz}{dt} = e^t \Rightarrow \frac{dw}{dt} = \frac{4yte^x}{t^2+1} + \frac{2e^x}{t^2+1} - \frac{e^t}{z}$ $= \frac{(4t)(\tan^{-1} t)(t^2+1)}{t^2+1} + \frac{2(t^2+1)}{t^2+1} - \frac{e^t}{e^t} = 4t \tan^{-1} t$ $+ 1$; $w = 2ye^x - \ln z = (2 \tan^{-1} t)(t^2+1) - t$ $\Rightarrow \frac{dw}{dt} = \left(\frac{2}{t^2+1}\right)(t^2+1) + (2 \tan^{-1} t)(2t) - 1$ $= 4t \tan^{-1} t + 1$

   b. $\frac{dw}{dt}(1) = (4)(1)\left(\frac{\pi}{4}\right) + 1 = \pi + 1$

7. a. $\frac{\partial z}{\partial r} = \frac{\partial z}{\partial x}\frac{\partial x}{\partial r} + \frac{\partial z}{\partial y}\frac{\partial y}{\partial r} = (4e^x \ln y)\left(\frac{\cos \theta}{r \cos \theta}\right) + \left(\frac{4e^x}{y}\right)(\sin \theta)$ $= \frac{4e^x \ln y}{r} + \frac{4e^x \sin \theta}{y} = \frac{4(r \cos \theta) \ln (r \sin \theta)}{r}$ $+ \frac{4(r \cos \theta)(\sin \theta)}{r \sin \theta} = (4 \cos \theta) \ln(r \sin \theta) + 4 \cos \theta$;

   $\frac{\partial z}{\partial \theta} = \frac{\partial z}{\partial x}\frac{\partial x}{\partial \theta} + \frac{\partial z}{\partial y}\frac{\partial y}{\partial \theta} = (4e^x \ln y)\left(\frac{-r \sin \theta}{r \cos \theta}\right)$ $+ \left(\frac{4e^x}{y}\right)(r \cos \theta) = -(4e^x \ln y)(\tan \theta) + \frac{4e^x r \cos \theta}{y}$ $= [-4(r \cos \theta) \ln(r \sin \theta)](\tan \theta) + \frac{4(r \cos \theta)(r \cos \theta)}{r \sin \theta}$ $= (-4r \sin \theta) \ln (r \sin \theta) + \frac{4r \cos^2 \theta}{\sin \theta}$; $z = 4e^x \ln y$ $= 4(r \cos \theta) \ln (r \sin \theta) \Rightarrow \frac{\partial z}{\partial r} = (4 \cos \theta) \ln (r \sin \theta)$ $+ 4(r \cos \theta)\left(\frac{\sin \theta}{r \sin \theta}\right) = (4 \cos \theta) \ln (r \sin \theta) + 4 \cos \theta$;

   also $\frac{\partial z}{\partial \theta} = (-4r \sin \theta) \ln(r \sin \theta) + 4(r \cos \theta)\left(\frac{r \cos \theta}{r \sin \theta}\right)$ $= (-4r \sin \theta) \ln(r \sin \theta) + \frac{4r \cos^2 \theta}{\sin \theta}$

   b. At $\left(2, \frac{\pi}{4}\right)$: $\frac{\partial z}{\partial r} = 4 \cos \frac{\pi}{4} \ln\left(2 \sin \frac{\pi}{4}\right) + 4 \cos \frac{\pi}{4}$ $= 2\sqrt{2} \ln \sqrt{2} + 2\sqrt{2} = \sqrt{2}(\ln 2 + 2)$; $\frac{\partial z}{\partial \theta}$ $= (-4)(2)\sin \frac{\pi}{4} \ln \left(2 \sin \frac{\pi}{4}\right) + \frac{(4)(2)\left(\cos^2 \frac{\pi}{4}\right)}{\left(\sin \frac{\pi}{4}\right)}$ $= -4\sqrt{2} \ln \sqrt{2} + 4\sqrt{2} = -2\sqrt{2} \ln 2 + 4\sqrt{2}$

9. a. $\frac{\partial w}{\partial u} = \frac{\partial w}{\partial x}\frac{\partial x}{\partial u} + \frac{\partial w}{\partial y}\frac{\partial y}{\partial u} + \frac{\partial w}{\partial z}\frac{\partial z}{\partial u} = (y+z)(1)$ $+ (x+z)(1) + (y+x)(v) = x + y + 2z + v(y+x)$ $= (u+v) + (u-v) + 2uv + v(2u) = 2u + 4uv$;

   $\frac{\partial w}{\partial v} = \frac{\partial w}{\partial x}\frac{\partial x}{\partial v} + \frac{\partial w}{\partial y}\frac{\partial y}{\partial v} + \frac{\partial w}{\partial z}\frac{\partial z}{\partial v} = (y+z)(1)$ $+ (x+z)(-1) + (y+x)(u) = y - x + (y+x)u$ $= -2v + (2u)u = -2v + 2u^2$; $w = xy + yz + xz$ $= (u^2 - v^2) + (u^2v - uv^2) + (u^2v + uv^2) = u^2 - v^2$ $+ 2u^2v \Rightarrow \frac{\partial w}{\partial u} = 2u + 4uv$ and $\frac{\partial w}{\partial v} = -2v + 2u^2$

   b. At $\left(\frac{1}{2}, 1\right)$: $\frac{\partial w}{\partial u} = 2\left(\frac{1}{2}\right) + 4\left(\frac{1}{2}\right)(1) = 3$ and $\frac{\partial w}{\partial v} = -2(1)$ $+ 2\left(\frac{1}{2}\right)^2 = -\frac{3}{2}$

11. a. $\frac{\partial u}{\partial x} = \frac{\partial u}{\partial p}\frac{\partial p}{\partial x} + \frac{\partial u}{\partial q}\frac{\partial q}{\partial x} + \frac{\partial u}{\partial r}\frac{\partial r}{\partial x} = \frac{1}{q-r} + \frac{r-p}{(q-r)^2}$ $+ \frac{p-q}{(q-r)^2} = \frac{q - r + r - p + p - q}{(q-r)^2} = 0$;

   $\frac{\partial u}{\partial y} = \frac{\partial u}{\partial p}\frac{\partial p}{\partial y} + \frac{\partial u}{\partial q}\frac{\partial q}{\partial y} + \frac{\partial u}{\partial r}\frac{\partial r}{\partial y} = \frac{1}{q-r} - \frac{r-p}{(q-r)^2}$ $- \frac{p-q}{(q-r)^2} = \frac{q - r - r + p + p - q}{(q-r)^2} = \frac{2p - 2r}{(q-r)^2}$ $= \frac{(2x + 2y + 2z) - (2x + 2y - 2z)}{(2z - 2y)^2} = \frac{z}{(z-y)^2}$;

   $\frac{\partial u}{\partial z} = \frac{\partial u}{\partial p}\frac{\partial p}{\partial z} + \frac{\partial u}{\partial q}\frac{\partial q}{\partial z} + \frac{\partial u}{\partial r}\frac{\partial r}{\partial z} = \frac{1}{q-r} + \frac{r-p}{(q-r)^2}$ $- \frac{p-q}{(q-r)^2} = \frac{q - r + r - p - p + q}{(q-r)^2} = \frac{2q - 2p}{(q-r)^2}$ $= \frac{-4y}{(2z-2y)^2} = -\frac{y}{(z-y)^2}$; $u = \frac{p-q}{q-r} = \frac{2y}{2z-2y}$ $= \frac{y}{z-y} \Rightarrow \frac{\partial u}{\partial x} = 0$, $\frac{\partial u}{\partial y} = \frac{(z-y) - y(-1)}{(z-y)^2} = \frac{z}{(z-y)^2}$, and $\frac{\partial u}{\partial z} = \frac{(z-y)(0) - y(1)}{(z-y)^2} = -\frac{y}{(z-y)^2}$

   b. At $(\sqrt{3}, 2, 1)$: $\frac{\partial u}{\partial x} = 0$, $\frac{\partial u}{\partial y} = \frac{1}{(1-2)^2} = 1$, and $\frac{\partial u}{\partial z}$ $= \frac{-2}{(1-2)^2} = -2$

13. $\frac{dz}{dt} = \frac{\partial z}{\partial x}\frac{dx}{dt} + \frac{\partial z}{\partial y}\frac{dy}{dt}$

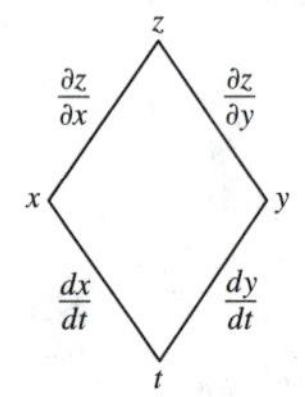

15. $\frac{\partial w}{\partial u} = \frac{\partial w}{\partial x}\frac{\partial x}{\partial u} + \frac{\partial w}{\partial y}\frac{\partial y}{\partial u} + \frac{\partial w}{\partial z}\frac{\partial z}{\partial u}$

$\frac{\partial w}{\partial v} = \frac{\partial w}{\partial x}\frac{\partial x}{\partial v} + \frac{\partial w}{\partial y}\frac{\partial y}{\partial v} + \frac{\partial w}{\partial z}\frac{\partial z}{\partial v}$

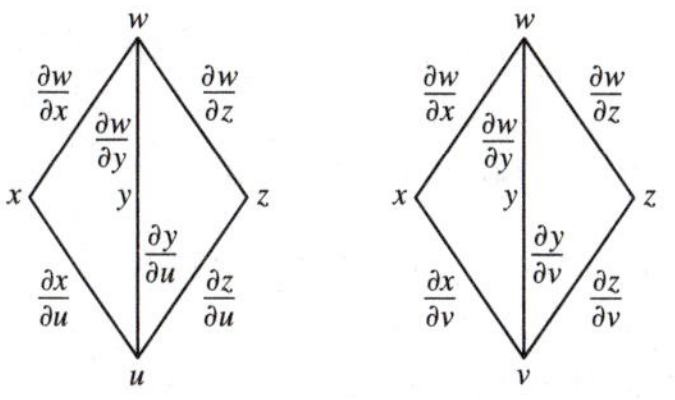

17. $\frac{\partial w}{\partial u} = \frac{\partial w}{\partial x}\frac{\partial x}{\partial u} + \frac{\partial w}{\partial y}\frac{\partial y}{\partial u}$

$\frac{\partial w}{\partial v} = \frac{\partial w}{\partial x}\frac{\partial x}{\partial v} + \frac{\partial w}{\partial y}\frac{\partial y}{\partial v}$

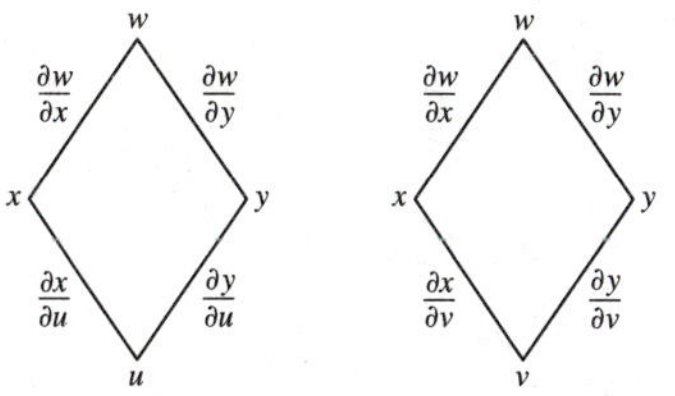

19. $\frac{\partial z}{\partial t} = \frac{\partial z}{\partial x}\frac{\partial x}{\partial t} + \frac{\partial z}{\partial y}\frac{\partial y}{\partial t}$

$\frac{\partial z}{\partial s} = \frac{\partial z}{\partial x}\frac{\partial x}{\partial s} + \frac{\partial z}{\partial y}\frac{\partial y}{\partial s}$

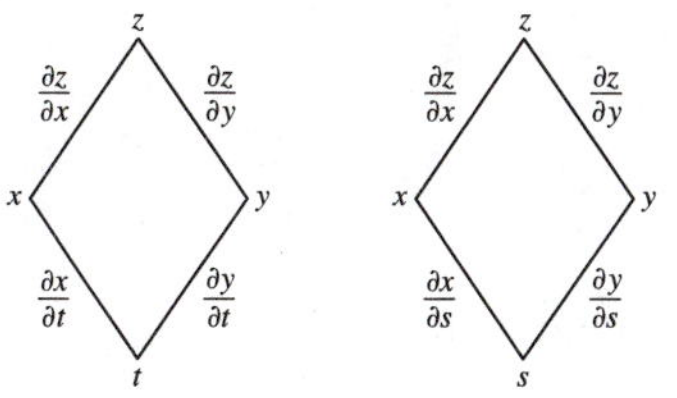

21. $\frac{\partial w}{\partial s} = \frac{dw}{du}\frac{\partial u}{\partial s}$

$\frac{\partial w}{\partial t} = \frac{dw}{du}\frac{\partial u}{\partial t}$

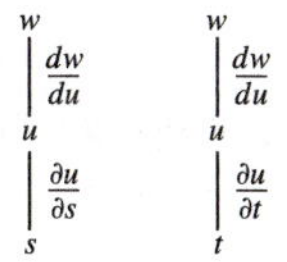

23. $\frac{\partial w}{\partial r} = \frac{\partial w}{\partial x}\frac{dx}{dr} + \frac{\partial w}{\partial y}\frac{dy}{dr} = \frac{\partial w}{\partial x}\frac{dx}{dr}$ since $\frac{dy}{dr} = 0$

$\frac{\partial w}{\partial s} = \frac{\partial w}{\partial x}\frac{dx}{ds} + \frac{\partial w}{\partial y}\frac{dy}{ds} = \frac{\partial w}{\partial y}\frac{dy}{ds}$ since $\frac{dx}{ds} = 0$

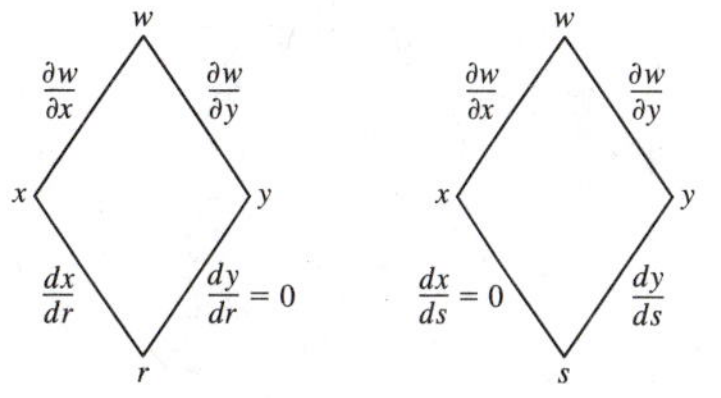

25. Let $F(x, y) = x^3 - 2y^2 + xy = 0 \Rightarrow F_x(x, y) = 3x^2 + y$ and $F_y(x, y) = -4y + x \Rightarrow \frac{dy}{dx} = -\frac{F_x}{F_y} = -\frac{3x^2 + y}{(-4y + x)}$ $\Rightarrow \frac{dy}{dx}(1, 1) = \frac{4}{3}$

27. Let $F(x, y) = x^2 + xy + y^2 - 7 = 0 \Rightarrow F_x(x, y) = 2x + y$ and $F_y(x, y) = x + 2y \Rightarrow \frac{dy}{dx} = -\frac{F_x}{F_y} = -\frac{2x + y}{x + 2y}$ $\Rightarrow \frac{dy}{dx}(1, 2) = -\frac{4}{5}$

29. $\frac{\partial w}{\partial r} = \frac{\partial w}{\partial x}\frac{\partial x}{\partial r} + \frac{\partial w}{\partial y}\frac{\partial y}{\partial r} + \frac{\partial w}{\partial z}\frac{\partial z}{\partial r} = 2(x + y + z)(1)$
$+ 2(x + y + z)[-\sin(r + s)] + 2(x + y + z)[\cos(r + s)]$
$= 2(x + y + z)[1 - \sin(r + s) + \cos(r + s)]$
$= 2[r - s + \cos(r + s) + \sin(r + s)][1 - \sin(r + s)$
$+ \cos(r + s)] \Rightarrow \left.\frac{\partial w}{\partial r}\right|_{r = 1,\, s = -1} = 2(3)(2) = 12$

31. $\frac{\partial w}{\partial v} = \frac{\partial w}{\partial x}\frac{\partial x}{\partial v} + \frac{\partial w}{\partial y}\frac{\partial y}{\partial v} = \left(2x - \frac{y}{x^2}\right)(-2) + \left(\frac{1}{x}\right)(1)$
$= \left[2(u - 2v + 1) - \frac{2u + v - 2}{(u - 2v + 1)^2}\right](-2) + \frac{1}{u - 2v + 1}$
$\Rightarrow \left.\frac{\partial w}{\partial v}\right|_{u = 0,\, v = 0} = -7$

33. $\frac{\partial z}{\partial u} = \frac{dz}{dx}\frac{\partial x}{\partial u} = \left(\frac{5}{1 + x^2}\right)e^u = \left[\frac{5}{1 + (e^u + \ln v)^2}\right]e^u$
$\Rightarrow \left.\frac{\partial z}{\partial u}\right|_{u = \ln 2,\, v = 1} = \left[\frac{5}{1 + (2)^2}\right](2) = 2;\ \frac{\partial z}{\partial v} = \frac{dz}{dx}\frac{\partial x}{\partial v} =$
$\left(\frac{5}{1 + x^2}\right)\left(\frac{1}{v}\right) = \left[\frac{5}{1 + (e^u + \ln v)^2}\right]\left(\frac{1}{v}\right) \Rightarrow \left.\frac{\partial z}{\partial v}\right|_{u = \ln 2,\, v = 1}$
$= \left[\frac{5}{1 + (2)^2}\right](1) = 1$

35. $\frac{dw}{dt} = \frac{\partial w}{\partial x}\frac{dx}{dt} + \frac{\partial w}{\partial y}\frac{dy}{dt} + \frac{\partial w}{\partial z}\frac{dz}{dt} = (2xe^{2y}\cos 3z)(-\sin t)$
$+ (2x^2e^{2y}\cos 3z)\left(\frac{1}{t + 2}\right) + (-3x^2e^{2y}\sin 3z)(1)$
$= -2xe^{2y}\cos 3z \sin t + \frac{2x^2e^{2y}\cos 3z}{t + 2} - 3x^2e^{2y}\sin 3z$; at the point on the curve $z = 0 \Rightarrow t = z = 0 \Rightarrow \left.\frac{dw}{dt}\right|_{(1,\, \ln 2,\, 0)}$
$= 0 + \frac{2(1)^2(4)(1)}{2} - 0 = 4$

37. a. $\frac{\partial T}{\partial x} = y$ and $\frac{\partial T}{\partial y} = x \Rightarrow \frac{dT}{dt} = \frac{\partial T}{\partial x}\frac{dx}{dt} + \frac{\partial T}{\partial y}\frac{dy}{dt}$
$= y(-2\sqrt{2}\sin t) + x(\sqrt{2}\cos t)$
$= (\sqrt{2}\sin t)(-2\sqrt{2}\sin t) + (2\sqrt{2}\cos t)(\sqrt{2}\cos t)$
$= -4\sin^2 t + 4\cos^2 t = -4\sin^2 t + 4(1 - \sin^2 t)$
$= 4 - 8\sin^2 t \Rightarrow \frac{d^2T}{dt^2} = -16\sin t\cos t = -8\sin 2t$;
$\frac{dT}{dt} = 0$
$\Rightarrow 4 \cdot 8\sin^2 t = 0 \Rightarrow \sin^2 t = \frac{1}{2} \Rightarrow \sin t = \pm\frac{1}{\sqrt{2}}$

$\Rightarrow t = \frac{\pi}{4}, \frac{3\pi}{4}, \frac{5\pi}{4}, \frac{7\pi}{4}$ on the interval $0 \le t \le 2\pi$;

$\left.\frac{d^2T}{dt^2}\right|_{t=\frac{\pi}{4}} = -8 \sin 2\left(\frac{\pi}{4}\right) = -8 \Rightarrow T$ has a maximum at $(x, y) = (2, 1)$; $\left.\frac{d^2T}{dt^2}\right|_{t=\frac{3\pi}{4}} = -8 \sin 2\left(\frac{3\pi}{4}\right) = 8 \Rightarrow T$ has a maximum at $(x, y) = (-2, 1)$; $\left.\frac{d^2T}{dt^2}\right|_{t=\frac{5\pi}{4}} = -8 \sin 2\left(\frac{5\pi}{4}\right) = -8 \Rightarrow T$ has a minimum at $(x, y) = (-2, -1)$; $\left.\frac{d^2T}{dt^2}\right|_{t=\frac{7\pi}{4}} = -8 \sin 2\left(\frac{7\pi}{4}\right) = 8 \Rightarrow T$ has a maximum at $(x, y) = (2, -1)$

b. $T = xy - 2 \Rightarrow \frac{\partial T}{\partial x} = y$ and $\frac{\partial T}{\partial y} = x$ so the extreme values occur at the four points found in part (a): $T(2, 1) = T(-2, -1) = 0$, the maximum and $T(-2, 1) = T(2, -1) = -4$, the minimum.

39. $\frac{\partial f}{\partial x} = \frac{\partial f}{\partial u}\frac{\partial u}{\partial x} + \frac{\partial f}{\partial v}\frac{\partial v}{\partial x} + \frac{\partial f}{\partial w}\frac{\partial w}{\partial x} = \frac{\partial f}{\partial u}(1) + \frac{\partial f}{\partial v}(0) + \frac{\partial f}{\partial w}(-1) = \frac{\partial f}{\partial u} - \frac{\partial f}{\partial w}$,

$\frac{\partial f}{\partial y} = \frac{\partial f}{\partial u}\frac{\partial u}{\partial y} + \frac{\partial f}{\partial v}\frac{\partial v}{\partial y} + \frac{\partial f}{\partial w}\frac{\partial w}{\partial y} = \frac{\partial f}{\partial u}(-1) + \frac{\partial f}{\partial v}(1) + \frac{\partial f}{\partial w}(0) = -\frac{\partial f}{\partial u} + \frac{\partial f}{\partial v}$, and

$\frac{\partial f}{\partial z} = \frac{\partial f}{\partial u}\frac{\partial u}{\partial z} + \frac{\partial f}{\partial v}\frac{\partial v}{\partial z} + \frac{\partial f}{\partial w}\frac{\partial w}{\partial z} = \frac{\partial f}{\partial u}(0) + \frac{\partial f}{\partial v}(-1) + \frac{\partial f}{\partial w}(1) = -\frac{\partial f}{\partial v} + \frac{\partial f}{\partial w} \Rightarrow \frac{\partial f}{\partial x} + \frac{\partial f}{\partial y} + \frac{\partial f}{\partial z} = 0$

41. $V = abc \Rightarrow \frac{dV}{dt} = \frac{\partial V}{\partial a}\frac{da}{dt} + \frac{\partial V}{\partial b}\frac{db}{dt} + \frac{\partial V}{\partial c}\frac{dc}{dt} = (bc)\frac{da}{dt} + (ac)\frac{db}{dt} + (ab)\frac{dc}{dt} \Rightarrow \left.\frac{dV}{dt}\right|_{a=1, b=2, c=3}$ $= (2 \text{ m})(3 \text{ m})(1 \text{ m/sec}) + (1 \text{ m})(3 \text{ m})(1 \text{ m/sec}) + (1 \text{ m})(2 \text{ m})(-3 \text{ m/sec}) = 3 \text{ m}^3\text{/sec}$ and the volume is increasing; $S = 2ab + 2ac + 2bc \Rightarrow \frac{dS}{dt} = \frac{\partial S}{\partial a}\frac{da}{dt} + \frac{\partial S}{\partial b}\frac{db}{dt} + \frac{\partial S}{\partial c}\frac{dc}{dt} = 2(b + c)\frac{da}{dt} + 2(a + c)\frac{db}{dt} + 2(a + b)\frac{dc}{dt} \Rightarrow$ $\left.\frac{dS}{dt}\right|_{a=1, b=2, c=3} = 2(5 \text{ m})(1 \text{ m/sec}) + 2(4 \text{ m})(1 \text{ m/sec}) + 2(3 \text{ m})(-3 \text{ m/sec}) = 0 \text{ m}^2\text{/sec}$ and the surface area is not changing; $D = \sqrt{a^2 + b^2 + c^2} \Rightarrow \frac{dD}{dt} = \frac{\partial D}{\partial a}\frac{da}{dt} + \frac{\partial D}{\partial b}\frac{db}{dt} + \frac{\partial D}{\partial c}\frac{dc}{dt} = \frac{1}{\sqrt{a^2 + b^2 + c^2}}\left(a\frac{da}{dt} + b\frac{db}{dt} + c\frac{dc}{dt}\right)$

$\Rightarrow \left.\frac{dD}{dt}\right|_{a=1, b=2, c=3} = \left(\frac{1}{\sqrt{14}} \text{ m}\right)[(1 \text{ m})(1 \text{ m/sec}) + (2 \text{ m})(1 \text{ m/sec}) + (3 \text{ m})(-3 \text{ m/sec})] = -\frac{6}{\sqrt{14}} \text{ m/sec}$ $< 0 \Rightarrow$ the diagonals are decreasing in length.

43. $\frac{\partial w}{\partial x} = f'(u)(1) + g'(v)(1) = f'(u) + g'(v)$

$\Rightarrow w_{xx} = f''(u)(1) + g''(v)(1) = f''(u) + g''(v)$;

$\frac{\partial w}{\partial y} = f'(u)(i) + g'(v)(-i) \Rightarrow w_{yy} = f''(u)(i^2) + g''(v)(i^2)$

$= -f''(u) - g''(v) \Rightarrow w_{xx} + w_{yy} = 0$

45. a. $\frac{\partial w}{\partial r} = f_x\frac{\partial x}{\partial r} + f_y\frac{\partial y}{\partial r} = f_x \cos\theta + f_y \sin\theta$ and $\frac{\partial w}{\partial \theta} = f_x(-r\sin\theta) + f_y(r\cos\theta) \Rightarrow \frac{1}{r}\frac{\partial w}{\partial \theta} = -f_x \sin\theta + f_y \cos\theta$

b. $\frac{\partial w}{\partial r}\sin\theta = f_x \sin\theta\cos\theta + f_y \sin^2\theta$ and $\left(\frac{\cos\theta}{r}\right)\frac{\partial w}{\partial \theta} = -f_x \sin\theta\cos\theta + f_y \cos^2\theta \Rightarrow f_y = (\sin\theta)\frac{\partial w}{\partial r} + \left(\frac{\cos\theta}{r}\right)\frac{\partial w}{\partial \theta}$; then $\frac{\partial w}{\partial r} = f_x\cos\theta + \left[(\sin\theta)\frac{\partial w}{\partial r} + \left(\frac{\cos\theta}{r}\right)\frac{\partial w}{\partial \theta}\right](\sin\theta) \Rightarrow f_x \cos\theta = \frac{\partial w}{\partial r} - (\sin^2\theta)\frac{\partial w}{\partial r} - \left(\frac{\sin\theta\cos\theta}{r}\right)\frac{\partial w}{\partial \theta} = (1 - \sin^2\theta)\frac{\partial w}{\partial r} - \left(\frac{\sin\theta\cos\theta}{r}\right)\frac{\partial w}{\partial \theta}$ $\Rightarrow f_x = (\cos\theta)\frac{\partial w}{\partial r} - \left(\frac{\sin\theta}{r}\right)\frac{\partial w}{\partial \theta}$

c. $(f_x)^2 = (\cos^2\theta)\left(\frac{\partial w}{\partial r}\right)^2 - \left(\frac{2\sin\theta\cos\theta}{r}\right)\left(\frac{\partial w}{\partial r}\frac{\partial w}{\partial \theta}\right) + \left(\frac{\sin^2\theta}{r^2}\right)\left(\frac{\partial w}{\partial \theta}\right)^2$ and $(f_y)^2 = (\sin^2\theta)\left(\frac{\partial w}{\partial r}\right)^2 + \left(\frac{2\sin\theta\cos\theta}{r}\right)\left(\frac{\partial w}{\partial r}\frac{\partial w}{\partial \theta}\right) + \left(\frac{\cos^2\theta}{r^2}\right)\left(\frac{\partial w}{\partial \theta}\right)^2$

$\Rightarrow (f_x)^2 + (f_y)^2 = \left(\frac{\partial w}{\partial r}\right)^2 + \frac{1}{r^2}\left(\frac{\partial w}{\partial \theta}\right)^2$

■ Section 13.6 Directional Derivatives, Gradient Vectors, and Tangent Planes (pp. 723–736)

Quick Review 13.6

1. $|\mathbf{u}| = \sqrt{2^2 + (-1)^2 + 1^2} = \sqrt{6}$

2. $\frac{\mathbf{u}}{|\mathbf{u}|} = \frac{2}{\sqrt{6}}\mathbf{i} - \frac{1}{\sqrt{6}}\mathbf{j} + \frac{1}{\sqrt{6}}\mathbf{k}$

3. $\mathbf{u}\cdot\mathbf{v} = (3)(-1) + (-1)(2) + (0)(-1) = -5$

4. $\mathbf{u}\times\mathbf{v} = \begin{vmatrix} \mathbf{i} & \mathbf{j} & \mathbf{k} \\ 3 & -1 & 0 \\ -1 & 2 & -1 \end{vmatrix} = \begin{vmatrix} -1 & 0 \\ 2 & -1 \end{vmatrix}\mathbf{i} - \begin{vmatrix} 3 & 0 \\ -1 & -1 \end{vmatrix}\mathbf{j} + \begin{vmatrix} 3 & -1 \\ -1 & 2 \end{vmatrix}\mathbf{k}$

$= (1 - 0)\mathbf{i} - (-3 - 0)\mathbf{j} + (6 - 1)\mathbf{k} = \mathbf{i} + 3\mathbf{j} + 5\mathbf{k}$

5. $\mathbf{u}\times\mathbf{v} = \mathbf{i} + 3\mathbf{j} + 5\mathbf{k}$

6. $\frac{\mathbf{u}\times\mathbf{v}}{|\mathbf{u}\times\mathbf{v}|} = \frac{1}{\sqrt{1^2 + 3^2 + 5^2}}(\mathbf{i} + 3\mathbf{j} + 5\mathbf{k})$

$= \frac{1}{\sqrt{35}}\mathbf{i} + \frac{3}{\sqrt{35}}\mathbf{j} + \frac{5}{\sqrt{35}}\mathbf{k}$

7. $\mathbf{u} = -\mathbf{i} + 2\mathbf{j} - \mathbf{k} + [(3\mathbf{i} + 2\mathbf{j}) - (-\mathbf{i} + 2\mathbf{j} - \mathbf{k})]t$
$= -\mathbf{i} + 2\mathbf{j} - \mathbf{k} + (4\mathbf{i} + \mathbf{k})t$
$= (-1 + 4t)\mathbf{i} + 2\mathbf{j} + (-1 + t)\mathbf{k}$
$\Rightarrow x = -1 + 4t, y = 2, z = -1 + t$

8. $\mathbf{u} = 2\mathbf{i} - \mathbf{j} + \mathbf{k} + (3\mathbf{i} - 2\mathbf{j} + \mathbf{k})t$
$= (2 + 3t)\mathbf{i} + (-1 - 2t)\mathbf{j} + (1 + t)\mathbf{k}$
$\Rightarrow x = 2 + 3t, y = -1 - 2t, z = 1 + t$

9. $\mathbf{u} = 2\mathbf{i} - 2\mathbf{k} + (\mathbf{i} - 2\mathbf{j} + 3\mathbf{k})t$
$= (2 + t)\mathbf{i} + (-2t)\mathbf{j} + (-2 + 3t)\mathbf{k}$
$\Rightarrow x = 2 + t, y = -2t, z = -2 + 3t$

10. $[\mathbf{u} - (\mathbf{i} - 2\mathbf{k})] \times (-\mathbf{i} + 3\mathbf{j} - 2\mathbf{k}) = 0$
$\Rightarrow \mathbf{u} \times (-\mathbf{i} + 3\mathbf{j} - 2\mathbf{k}) = (\mathbf{i} - 2\mathbf{k}) \times (-\mathbf{i} + 3\mathbf{j} - 2\mathbf{k})$
$\Rightarrow$ with $\mathbf{u} = x\mathbf{i} + y\mathbf{j} + z\mathbf{k}$:
$-x + 3y - 2z = (1)(-1) + (0)(3) + (-2)(-2)$
$\Rightarrow x - 3y + 2z = -3$

Section 13.6 Exercises

1. $\frac{\partial f}{\partial x} = -1, \frac{\partial f}{\partial y} = 1 \Rightarrow \nabla f = -\mathbf{i} + \mathbf{j}; f(2, 1) = -1$
$\Rightarrow -1 = y - x$ is the level curve

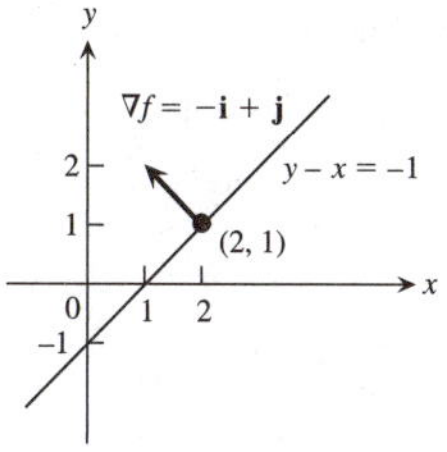

3. $\frac{\partial g}{\partial x} = -2x \Rightarrow \frac{\partial g}{\partial x}(-1, 0) = 2; \frac{\partial g}{\partial y} = 1 \Rightarrow \nabla g = 2\mathbf{i} + \mathbf{j};$
$g(-1, 0) = -1 \Rightarrow -1 = y - x^2$ is the level curve

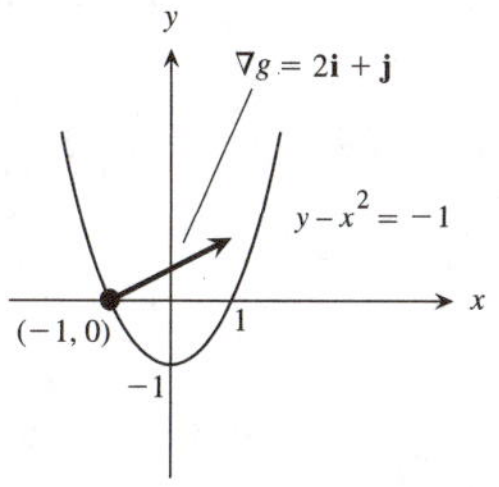

5. $\frac{\partial f}{\partial x} = 2x + \frac{z}{x} \Rightarrow \frac{\partial f}{\partial x}(1, 1, 1) = 3; \frac{\partial f}{\partial y} = 2y \Rightarrow \frac{\partial f}{\partial y}(1, 1, 1)$
$= 2; \frac{\partial f}{\partial z} = -4z + \ln x \Rightarrow \frac{\partial f}{\partial z}(1, 1, 1) = -4$; thus
$\nabla f = 3\mathbf{i} + 2\mathbf{j} - 4\mathbf{k}$

7. $\frac{\partial f}{\partial x} = -\frac{x}{(x^2 + y^2 + z^2)^{3/2}} + \frac{1}{x} \Rightarrow \frac{\partial f}{\partial x}(-1, 2, -2) = -\frac{26}{27};$
$\frac{\partial f}{\partial y} = -\frac{y}{(x^2 + y^2 + z^2)^{3/2}} + \frac{1}{y} \Rightarrow \frac{\partial f}{\partial y}(-1, 2, -2) = \frac{23}{54};$
$\frac{\partial f}{\partial z} = -\frac{z}{(x^2 + y^2 + z^2)^{3/2}} + \frac{1}{z} \Rightarrow \frac{\partial f}{\partial z}(-1, 2, -2) = -\frac{23}{54};$
thus $\nabla f = -\frac{26}{27}\mathbf{i} + \frac{23}{54}\mathbf{j} - \frac{23}{54}\mathbf{k}$

9. $\mathbf{u} = \frac{\mathbf{A}}{|\mathbf{A}|} = \frac{4\mathbf{i} + 3\mathbf{j}}{\sqrt{4^2 + 3^2}} = \frac{4}{5}\mathbf{i} + \frac{3}{5}\mathbf{j}; f_x(x, y) = 2y \Rightarrow f_x(5, 5)$
$= 10; f_y(x, y) = 2x - 6y \Rightarrow f_y(5, 5) = -20 \Rightarrow \nabla f$
$= 10\mathbf{i} - 20\mathbf{j} \Rightarrow (D_{\mathbf{u}}f)_{P_0} = \nabla f \cdot \mathbf{u} = 10\left(\frac{4}{5}\right) - 20\left(\frac{3}{5}\right) = -4$

11. $\mathbf{u} = \frac{\mathbf{A}}{|\mathbf{A}|} = \frac{12\mathbf{i} + 5\mathbf{j}}{\sqrt{12^2 + 5^2}} = \frac{12}{13}\mathbf{i} + \frac{5}{13}\mathbf{j}; g_x(x, y) = 1 + \frac{y^2}{x^2}$
$+ \frac{2y\sqrt{3}}{2xy\sqrt{4x^2y^2 - 1}} \Rightarrow g_x(1, 1) = 3; g_y(x, y) = -\frac{2y}{x}$
$+ \frac{2x\sqrt{3}}{2xy\sqrt{4x^2y^2 - 1}} \Rightarrow g_y(1, 1) = -1 \Rightarrow \nabla g = 3\mathbf{i} - \mathbf{j}$
$\Rightarrow (D_{\mathbf{u}}g)_{P_0} = \nabla g \cdot \mathbf{u} = \frac{36}{13} - \frac{5}{13} = \frac{31}{13}$

13. $\mathbf{u} = \frac{\mathbf{A}}{|\mathbf{A}|} = \frac{3\mathbf{i} + 6\mathbf{j} - 2\mathbf{k}}{\sqrt{3^2 + 6^2 + (-2)^2}} = \frac{3}{7}\mathbf{i} + \frac{6}{7}\mathbf{j} - \frac{2}{7}\mathbf{k};$
$f_x(x, y, z) = y + z \Rightarrow f_x(1, -1, 2) = 1; f_y(x, y, z) = x + z$
$\Rightarrow f_y(1, -1, 2) = 3; f_z(x, y, z) = y + x \Rightarrow f_z(1, -1, 2) = 0$
$\Rightarrow \nabla f = \mathbf{i} + 3\mathbf{j} \Rightarrow (D_{\mathbf{u}}f)_{P_0} = \nabla f \cdot \mathbf{u} = \frac{3}{7} + \frac{18}{7} = 3$

15. $\mathbf{u} = \frac{\mathbf{A}}{|\mathbf{A}|} = \frac{2\mathbf{i} + \mathbf{j} - 2\mathbf{k}}{\sqrt{2^2 + 1^2 + (-2)^2}} = \frac{2}{3}\mathbf{i} + \frac{1}{3}\mathbf{j} - \frac{2}{3}\mathbf{k}; g_x(x, y, z)$
$= 3e^x \cos yz; g_x(0, 0, 0) = 3; g_y(x, y, z) = -3ze^x \sin yz$
$\Rightarrow g_y(0, 0, 0) = 0; g_z(x, y, z) = -3ye^x \sin yz \Rightarrow g_z(0, 0, 0)$
$= 0 \Rightarrow \nabla g = 3\mathbf{i} \Rightarrow (D_{\mathbf{u}}g)_{P_0} = \nabla g \cdot \mathbf{u} = 2$

17. $\nabla f = (2x + y)\mathbf{i} + (x + 2y)\mathbf{j} \Rightarrow \nabla f(-1, 1) = -\mathbf{i} + \mathbf{j} \Rightarrow \mathbf{u}$
$= \frac{\nabla f}{|\nabla f|} = \frac{-\mathbf{i} + \mathbf{j}}{\sqrt{(-1)^2 + 1^2}} = -\frac{1}{\sqrt{2}}\mathbf{i} + \frac{1}{\sqrt{2}}\mathbf{j}$; f increases most rapidly in the direction $\mathbf{u} = -\frac{1}{\sqrt{2}}\mathbf{i} + \frac{1}{\sqrt{2}}\mathbf{j}$ and decreases most rapidly in the direction $-\mathbf{u} = \frac{1}{\sqrt{2}}\mathbf{i} - \frac{1}{\sqrt{2}}\mathbf{j}$; $(D_{\mathbf{u}}f)_{P_0}$
$= \nabla f \cdot \mathbf{u} = |\nabla f| = \sqrt{2}$ and $(D_{-\mathbf{u}}f)_{P_0} = -\sqrt{2}$

19. $\nabla f = \frac{1}{y}\mathbf{i} - \left(\frac{x}{y^2} + z\right)\mathbf{j} - y\mathbf{k} \Rightarrow \nabla f(4, 1, 1) = \mathbf{i} - 5\mathbf{j} - \mathbf{k}$
$\Rightarrow \mathbf{u} = \frac{\nabla f}{|\nabla f|} = \frac{\mathbf{i} - 5\mathbf{j} - \mathbf{k}}{\sqrt{1^2 + (-5)^2 + (-1)^2}} = \frac{1}{3\sqrt{3}}\mathbf{i} - \frac{5}{3\sqrt{3}}\mathbf{j}$
$- \frac{1}{3\sqrt{3}}\mathbf{k}$; f increases most rapidly in the direction of $\mathbf{u} = \frac{1}{3\sqrt{3}}\mathbf{i} - \frac{5}{3\sqrt{3}}\mathbf{j} - \frac{1}{3\sqrt{3}}\mathbf{k}$ and decreases most rapidly in the direction $-\mathbf{u} = -\frac{1}{3\sqrt{3}}\mathbf{i} + \frac{5}{3\sqrt{3}}\mathbf{j} + \frac{1}{3\sqrt{3}}\mathbf{k}$; $(D_{\mathbf{u}}f)_{P_0}$
$= \nabla f \cdot \mathbf{u} = |\nabla f| = 3\sqrt{3}$ and $(D_{-\mathbf{u}}f)_{P_0} = -3\sqrt{3}$

21. $\nabla f = \left(\frac{1}{x} + \frac{1}{x}\right)\mathbf{i} + \left(\frac{1}{y} + \frac{1}{y}\right)\mathbf{j} + \left(\frac{1}{z} + \frac{1}{z}\right)\mathbf{k} \Rightarrow \nabla f(1, 1, 1) =$
$2\mathbf{i} + 2\mathbf{j} + 2\mathbf{k} \Rightarrow \mathbf{u} = \frac{\nabla f}{|\nabla f|} = \frac{1}{\sqrt{3}}\mathbf{i} + \frac{1}{\sqrt{3}}\mathbf{j} + \frac{1}{\sqrt{3}}\mathbf{k}$; f increases most rapidly in the direction $\mathbf{u} = \frac{1}{\sqrt{3}}\mathbf{i} + \frac{1}{\sqrt{3}}\mathbf{j} + \frac{1}{\sqrt{3}}\mathbf{k}$ and decreases most rapidly in the direction
$-\mathbf{u} = -\frac{1}{\sqrt{3}}\mathbf{i} - \frac{1}{\sqrt{3}}\mathbf{j} - \frac{1}{\sqrt{3}}\mathbf{k}; (D_{\mathbf{u}}f)_{P_0} = \nabla f \cdot \mathbf{u} = |\nabla f| =$
$2\sqrt{3}$ and $(D_{-\mathbf{u}}f)_{P_0} = -2\sqrt{3}$

23. $\nabla f = \left(\frac{x}{x^2 + y^2 + z^2}\right)\mathbf{i} + \left(\frac{y}{x^2 + y^2 + z^2}\right)\mathbf{j} + \left(\frac{z}{x^2 + y^2 + z^2}\right)\mathbf{k}$
$\Rightarrow \nabla f(3, 4, 12) = \frac{3}{169}\mathbf{j} + \frac{4}{169}\mathbf{j} + \frac{12}{169}\mathbf{k}; \mathbf{u} = \frac{\mathbf{A}}{|\mathbf{A}|}$
$= \frac{3\mathbf{i} + 6\mathbf{j} - 2\mathbf{k}}{\sqrt{3^2 + 6^2 + (-2)^2}} = \frac{3}{7}\mathbf{i} + \frac{6}{7}\mathbf{j} - \frac{2}{7}\mathbf{k}$
$\Rightarrow \nabla f \cdot \mathbf{u} = \frac{9}{1183}$ and $df = (\nabla f \cdot \mathbf{u})ds = \left(\frac{9}{1183}\right)(0.1)$
≈ 0.000761

25. $\nabla g = (1 + \cos z)\mathbf{i} + (1 - \sin z)\mathbf{j} + (-x \sin z - y \cos z)\mathbf{k}$
$\Rightarrow \nabla g(2, -1, 0) = 2\mathbf{i} + \mathbf{j} + \mathbf{k}$; $\mathbf{A} = \overrightarrow{P_0P_1} = -2\mathbf{i} + 2\mathbf{j} + 2\mathbf{k} \Rightarrow \mathbf{u} = \frac{\mathbf{A}}{|\mathbf{A}|} = \frac{-2\mathbf{i} + 2\mathbf{j} + 2\mathbf{k}}{\sqrt{(-2)^2 + 2^2 + 2^2}} = -\frac{1}{\sqrt{3}}\mathbf{i} + \frac{1}{\sqrt{3}}\mathbf{j} + \frac{1}{\sqrt{3}}\mathbf{k} \Rightarrow \nabla g \cdot \mathbf{u} = 0$ and $dg = (\nabla g \cdot \mathbf{u})ds = (0)(0.2) = 0$

27. $\nabla f = 2x\mathbf{i} + 2y\mathbf{j} + 2z\mathbf{k} \Rightarrow \nabla f(1, 1, 1) = 2\mathbf{i} + 2\mathbf{j} + 2\mathbf{k}$
$\Rightarrow$ Tangent plane: $2(x - 1) + 2(y - 1) + 2(z - 1) = 0$
$2x - 2 + 2y - 2 + 2z - 2 = 0$
$\Rightarrow x + y + z = 3$; Normal line: $x = 1 + 2t$, $y = 1 + 2t$, $z = 1 + 2t$

29. $\nabla f = -2x\mathbf{i} + 2\mathbf{k} \Rightarrow \nabla f(2, 0, 2) = -4\mathbf{i} + 2\mathbf{k} \Rightarrow$ Tangent plane: $-4(x - 2) + 2(z - 2) = 0 \Rightarrow -4x + 2z + 4 = 0$
$\Rightarrow -2x + z + 2 = 0$; Normal line: $x = 2 - 4t$, $y = 0$, $z = 2 + 2t$

31. $\nabla f = (-\pi \sin \pi x - 2xy + ze^{xz})\mathbf{i} + (-x^2 + z)\mathbf{j} + (xe^{xz} + y)\mathbf{k} \Rightarrow \nabla f(0, 1, 2) = 2\mathbf{i} + 2\mathbf{j} + \mathbf{k} \Rightarrow$ Tangent plane: $2(x - 0) + 2(y - 1) + 1(z - 2) = 0 \Rightarrow 2x + 2y + z - 4 = 0$;
Normal line: $x = 2t$, $y = 1 + 2t$, $z = 2 + t$

33. $\nabla f = \mathbf{i} + \mathbf{j} + \mathbf{k}$ for all points $\Rightarrow \nabla f(0, 1, 0) = \mathbf{i} + \mathbf{j} + \mathbf{k} \Rightarrow$
Tangent plane: $1(x - 0) + 1(y - 1) + 1(z - 0) = 0 \Rightarrow x + y + z - 1 = 0$; Normal line: $x = t$, $y = 1 + t$, $z = t$

35. $z = f(x, y) = \ln(x^2 + y^2) \Rightarrow f_x(x, y) = \frac{2x}{x^2 + y^2}$ and $f_y(x, y) = \frac{2y}{x^2 + y^2} \Rightarrow f_x(1, 0) = 2$ and $f_y(1, 0) = 0$
$\Rightarrow$ from Eq. (10) the tangent plane at (1, 0, 0) is
$2(x - 1) - z = 0$ or $2x - z - 2 = 0$

37. $z = f(x, y) = \sqrt{y - x} \Rightarrow f_x(x, y) = -\frac{1}{2}(y - x)^{-1/2}$ and $f_y(x, y) = \frac{1}{2}(y - x)^{-1/2} \Rightarrow f_x(1, 2) = -\frac{1}{2}$ and $f_y(1, 2) = \frac{1}{2}$
$\Rightarrow$ from Eq. (10) the tangent plane at (1, 2, 1) is
$-\frac{1}{2}(x - 1) + \frac{1}{2}(y - 2) - (z - 1) = 0 \Rightarrow x - y + 2z - 1 = 0$

39. $\nabla f = 2x\mathbf{i} + 2y\mathbf{j} \Rightarrow \nabla f(\sqrt{2}, \sqrt{2}) = 2\sqrt{2}\mathbf{i} + 2\sqrt{2}\mathbf{j}$
$\Rightarrow$ Tangent line: $2\sqrt{2}(x - \sqrt{2}) + 2\sqrt{2}(y - \sqrt{2}) = 0$
$\Rightarrow \sqrt{2}x + \sqrt{2}y = 4$

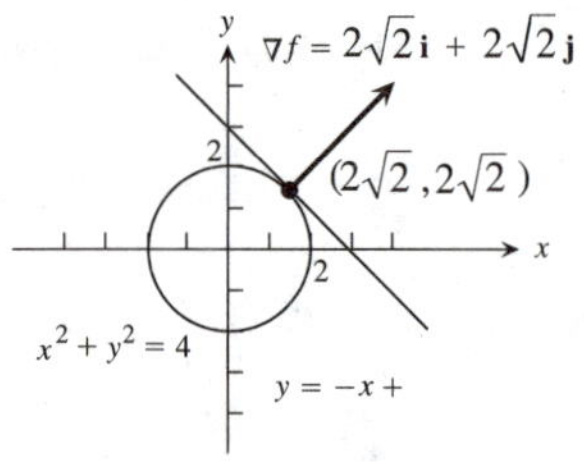

41. $\nabla f = y\mathbf{i} + x\mathbf{j} \Rightarrow \nabla f(2, -2) = -2\mathbf{i} + 2\mathbf{j} \Rightarrow$ Tangent line:
$-2(x - 2) + 2(y + 2) = 0 \Rightarrow y = x - 4$

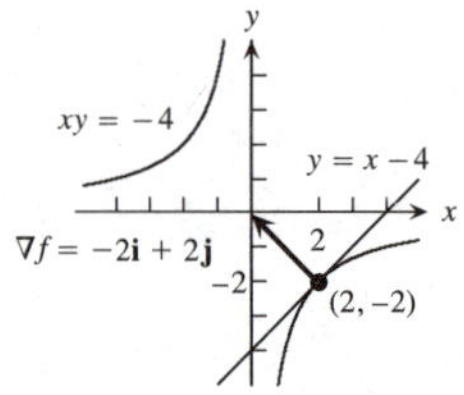

43. $\nabla f = \mathbf{i} + 2y\mathbf{j} + 2\mathbf{k} \Rightarrow \nabla f(1, 1, 1) = \mathbf{i} + 2\mathbf{j} + 2\mathbf{k}$ and $\nabla g = \mathbf{i}$ for all points; $\mathbf{v} = \nabla f \times \nabla g \Rightarrow \mathbf{v} = \begin{vmatrix} \mathbf{i} & \mathbf{j} & \mathbf{k} \\ 1 & 2 & 2 \\ 1 & 0 & 0 \end{vmatrix}$
$= 2\mathbf{j} - 2\mathbf{k} \Rightarrow$ Tangent line: $x = 1$, $y = 1 + 2t$, $z = 1 - 2t$

45. $\nabla f = 2x\mathbf{i} + 2\mathbf{j} + 2\mathbf{k} \Rightarrow \nabla f\left(1, 1, \frac{1}{2}\right) = 2\mathbf{i} + 2\mathbf{j} + 2\mathbf{k}$ and $\nabla g = \mathbf{j}$ for all points; $\mathbf{v} = \nabla f \times \nabla g$
$\Rightarrow \mathbf{v} = \begin{vmatrix} \mathbf{i} & \mathbf{j} & \mathbf{k} \\ 2 & 2 & 2 \\ 0 & 1 & 0 \end{vmatrix} = -2\mathbf{i} + 2\mathbf{k} \Rightarrow$ Tangent line:
$x = 1 - 2t$, $y = 1$, $z = \frac{1}{2} + 2t$

47. $\nabla f = (3x^2 + 6xy^2 + 4y)\mathbf{i} + (6x^2y + 3y^2 + 4x)\mathbf{j} - 2z\mathbf{k}$
$\Rightarrow \nabla f(1, 1, 3) = 13\mathbf{i} + 13\mathbf{j} - 6\mathbf{k}$; $\nabla g = 2x\mathbf{i} + 2y\mathbf{j} + 2z\mathbf{k}$
$\Rightarrow \nabla g(1, 1, 3) = 2\mathbf{i} + 2\mathbf{j} + 6\mathbf{k}$; $\mathbf{v} = \nabla f \times \nabla g$
$\Rightarrow \mathbf{v} = \begin{vmatrix} \mathbf{i} & \mathbf{j} & \mathbf{k} \\ 13 & 13 & -6 \\ 2 & 2 & 6 \end{vmatrix} = 90\mathbf{i} - 90\mathbf{j} \Rightarrow$ Tangent line:
$x = 1 + 90t$, $y = 1 - 90t$, $z = 3$

49. $\nabla f = y\mathbf{i} + (x + 2y)\mathbf{j} \Rightarrow \nabla f(3, 2) = 2\mathbf{i} + 7\mathbf{j}$; a vector orthogonal to ∇f is $\mathbf{A} = 7\mathbf{i} - 2\mathbf{j} \Rightarrow \mathbf{u} = \frac{\mathbf{A}}{|\mathbf{A}|} = \frac{7\mathbf{i} - 2\mathbf{j}}{\sqrt{7^2 + (-2)^2}} = \frac{7}{\sqrt{53}}\mathbf{i} - \frac{2}{\sqrt{53}}\mathbf{j}$ and $-\mathbf{u} = -\frac{7}{\sqrt{53}}\mathbf{i} + \frac{7}{\sqrt{53}}\mathbf{j}$ are the directions where the derivative is zero

51. $\nabla f = (2x - 3y)\mathbf{i} + (-3x + 8y)\mathbf{j} \Rightarrow \nabla f(1, 2) = -4\mathbf{i} + 13\mathbf{j}$
$\Rightarrow |\nabla f(1, 2)| = \sqrt{(-4)^2 + (13)^2} = \sqrt{185}$; no, the maximum rate of change is $\sqrt{185} < 14$

53. $\nabla f = f_x(1, 2)\mathbf{i} + f_y(1, 2)\mathbf{j}$ and $\mathbf{u}_1 = \frac{\mathbf{i}+\mathbf{j}}{\sqrt{1^2+1^2}}$

$= \frac{1}{\sqrt{2}}\mathbf{i} + \frac{1}{\sqrt{2}}\mathbf{j} \Rightarrow (D_{\mathbf{u}_1}f)(1, 2) = f_x(1, 2)\left(\frac{1}{\sqrt{2}}\right)$

$+ f_y(1, 2)\left(\frac{1}{\sqrt{2}}\right) = 2\sqrt{2} \Rightarrow f_x(1, 2) + f_y(1, 2) = 4;\ \mathbf{u}_2 = -\mathbf{j}$

$\Rightarrow (D_{\mathbf{u}_2}f)_{(1,2)} = f_x(1, 2)(0) + f_y(1, 2)(-1) = -3$

$\Rightarrow -f_y(1, 2) = -3 \Rightarrow f_y(1, 2) = 3$; then $f_x(1, 2) + 3 = 4$

$\Rightarrow f_x(1, 2) = 1$; thus $\nabla f(1, 2) = \mathbf{i} + 3\mathbf{j}$ and $\mathbf{u} = \frac{\mathbf{A}}{|\mathbf{A}|}$

$= \frac{-\mathbf{i}-2\mathbf{j}}{\sqrt{(-1)^2+(-2)^2}} = -\frac{1}{\sqrt{5}}\mathbf{i} - \frac{2}{\sqrt{5}}\mathbf{j} \Rightarrow (D_{\mathbf{u}}f)_{P_0} = \nabla f \cdot \mathbf{u}$

$= -\frac{1}{\sqrt{5}} - \frac{6}{\sqrt{5}} = -\frac{7}{\sqrt{5}}$

55. a. The unit tangent vector at $\left(\frac{1}{2}, \frac{\sqrt{3}}{2}\right)$ in the direction of motion is $\mathbf{u} = \frac{\sqrt{3}}{2}\mathbf{i} - \frac{1}{2}\mathbf{j}$; $\nabla T = (\sin 2y)\mathbf{i}$ $+ (2x \cos 2y)\mathbf{j} \Rightarrow \nabla T\left(\frac{1}{2}, \frac{\sqrt{3}}{2}\right) = (\sin \sqrt{3})\mathbf{i}$ $+ (\cos \sqrt{3})\mathbf{j} \Rightarrow D_{\mathbf{u}}T\left(\frac{1}{2}, \frac{\sqrt{3}}{2}\right) = \nabla T \cdot \mathbf{u}$

$= \frac{\sqrt{3}}{2} \sin \sqrt{3} - \frac{1}{2} \cos \sqrt{3} \approx 0.935°\text{C/m}$

b. At P, $\frac{dT}{dt} = \frac{dT}{ds} \cdot \frac{ds}{dt} \approx (0.935°\text{C/m})(2 \text{ m/s}) = 1.87°\text{C/s}$.

57. $\nabla f = 2x\mathbf{i} + 2y\mathbf{j} + 2z\mathbf{k} = (2 \cos t)\mathbf{i} + (2 \sin t)\mathbf{j} + 2t\mathbf{k}$ and $\mathbf{v}$ $= (-\sin t)\mathbf{i} + (\cos t)\mathbf{j} + \mathbf{k} \Rightarrow \mathbf{u} = \frac{\mathbf{v}}{|\mathbf{v}|}$

$= \frac{(-\sin t)\mathbf{i} + (\cos t)\mathbf{j} + \mathbf{k}}{\sqrt{(\sin t)^2 + (\cos t)^2 + 1^2}} = \left(\frac{-\sin t}{\sqrt{2}}\right)\mathbf{i} + \left(\frac{\cos t}{\sqrt{2}}\right)\mathbf{j} + \frac{1}{\sqrt{2}}\mathbf{k}$

$\Rightarrow (D_{\mathbf{u}}f)_{P_0} = \nabla f \cdot \mathbf{u} = (2 \cos t)\left(\frac{-\sin t}{\sqrt{2}}\right) + (2 \sin t)\left(\frac{\cos t}{\sqrt{2}}\right)$

$+ (2t)\left(\frac{1}{\sqrt{2}}\right) = \frac{2t}{\sqrt{2}} \Rightarrow (D_{\mathbf{u}}f)\left(\frac{-\pi}{4}\right) = \frac{-\pi}{2\sqrt{2}}$, $(D_{\mathbf{u}}f)(0) = 0$

and $(D_{\mathbf{u}}f)\left(\frac{\pi}{4}\right) = \frac{\pi}{2\sqrt{2}}$

59. If (x, y) is a point on the line, then $\mathbf{T}(x, y) = (x - x_0)\mathbf{i}$ $+ (y - y_0)\mathbf{j}$ is a vector parallel to the line $\Rightarrow \mathbf{T} \cdot \mathbf{N} = 0 \Rightarrow A(x - x_0) + B(y - y_0) = 0$, as claimed.

61. $x = g(t)$ and $y = h(t) \Rightarrow \mathbf{r} = g(t)\mathbf{i} + h(t)\mathbf{j} \Rightarrow \mathbf{v} = g'(t)\mathbf{i}$ $+ h'(t)\mathbf{j} \Rightarrow \mathbf{T} = \frac{\mathbf{v}}{|\mathbf{v}|} = \frac{g'(t)\mathbf{i} + h'(t)\mathbf{j}}{\sqrt{[g'(t)]^2 + [h'(t)]^2}}$; $z = f(x, y)$

$\Rightarrow \frac{df}{dt} = \frac{\partial f}{\partial x}\frac{dx}{dt} + \frac{\partial f}{\partial y}\frac{dy}{dt} = \frac{\partial f}{\partial x}g'(t) + \frac{\partial f}{\partial y}h'(t) = \nabla f \cdot \mathbf{T}$. If $f(g(t), h(t)) = c$, then $\frac{df}{dt} = 0 \Rightarrow \frac{\partial f}{\partial x}g'(t) + \frac{\partial f}{\partial y}h'(t) = 0$ $\Rightarrow \nabla f \cdot \mathbf{T} = 0 \Rightarrow \nabla f$ is normal to $\mathbf{T}$

63. The directional derivative is the scalar component. With ∇f evaluated at P_0, the scalar component of ∇f in the direction of $\mathbf{u}$ is $\nabla f \cdot \mathbf{u} = (D_{\mathbf{u}}f)P_0$.

65. a. $\nabla(kf) = \frac{\partial(kf)}{\partial x}\mathbf{i} + \frac{\partial(kf)}{\partial y}\mathbf{j} + \frac{\partial(kf)}{\partial z}\mathbf{k} = k\left(\frac{\partial f}{\partial x}\right)\mathbf{i} + k\left(\frac{\partial f}{\partial y}\right)\mathbf{j}$ $+ k\left(\frac{\partial f}{\partial z}\right)\mathbf{k} = k\left(\frac{\partial f}{\partial x}\mathbf{i} + \frac{\partial f}{\partial y}\mathbf{j} + \frac{\partial f}{\partial z}\mathbf{k}\right) = k\nabla f$

b. $\nabla(f + g) = \frac{\partial(f+g)}{\partial x}\mathbf{i} + \frac{\partial(f+g)}{\partial y}\mathbf{j} + \frac{\partial(f+g)}{\partial z}\mathbf{k} = \left(\frac{\partial f}{\partial x}\right.$ $\left.+ \frac{\partial g}{\partial x}\right)\mathbf{i} + \left(\frac{\partial f}{\partial y} + \frac{\partial g}{\partial y}\right)\mathbf{j} + \left(\frac{\partial f}{\partial z} + \frac{\partial g}{\partial z}\right)\mathbf{k} = \frac{\partial f}{\partial x}\mathbf{i} + \frac{\partial g}{\partial x}\mathbf{i}$

$+ \frac{\partial f}{\partial y}\mathbf{j} + \frac{\partial g}{\partial y}\mathbf{j} + \frac{\partial f}{\partial z}\mathbf{k} + \frac{\partial g}{\partial z}\mathbf{k} = \left(\frac{\partial f}{\partial x}\mathbf{i} + \frac{\partial f}{\partial y}\mathbf{j} + \frac{\partial f}{\partial z}\mathbf{k}\right)$

$+ \left(\frac{\partial g}{\partial x}\mathbf{i} + \frac{\partial g}{\partial y}\mathbf{j} + \frac{\partial g}{\partial z}\mathbf{k}\right) = \nabla f + \nabla g$

c. $\nabla(f - g) = \nabla f - \nabla g$ (Substitute $-g$ for g in part (b) above)

d. $\nabla(fg) = \frac{\partial(fg)}{\partial x}\mathbf{i} + \frac{\partial(fg)}{\partial y}\mathbf{j} + \frac{\partial(fg)}{\partial z}\mathbf{k} = \left(\frac{\partial f}{\partial x}g + \frac{\partial g}{\partial x}f\right)\mathbf{i}$

$+ \left(\frac{\partial f}{\partial y}g + \frac{\partial g}{\partial y}f\right)\mathbf{j} + \left(\frac{\partial f}{\partial z}g + \frac{\partial g}{\partial z}f\right)\mathbf{k} = \left(\frac{\partial f}{\partial x}g\right)\mathbf{i}$

$+ \left(\frac{\partial g}{\partial x}f\right)\mathbf{i} + \left(\frac{\partial f}{\partial y}g\right)\mathbf{j} + \left(\frac{\partial g}{\partial y}f\right)\mathbf{j} + \left(\frac{\partial f}{\partial z}g\right)\mathbf{k} + \left(\frac{\partial g}{\partial z}f\right)\mathbf{k}$

$= f\left(\frac{\partial g}{\partial x}\mathbf{i} + \frac{\partial g}{\partial y}\mathbf{j} + \frac{\partial g}{\partial z}\mathbf{k}\right) + g\left(\frac{\partial f}{\partial x}\mathbf{i} + \frac{\partial f}{\partial y}\mathbf{j} + \frac{\partial f}{\partial z}\mathbf{k}\right)$

$= f\nabla g + g\nabla f$

e. $\nabla\left(\frac{f}{g}\right) = \frac{\partial\left(\frac{f}{g}\right)}{\partial x}\mathbf{i} + \frac{\partial\left(\frac{f}{g}\right)}{\partial y}\mathbf{j} + \frac{\partial\left(\frac{f}{g}\right)}{\partial z}\mathbf{k} = \left(\frac{g\frac{\partial f}{\partial x} - f\frac{\partial g}{\partial x}}{g^2}\right)\mathbf{i}$

$+ \left(\frac{g\frac{\partial f}{\partial y} - f\frac{\partial g}{\partial y}}{g^2}\right)\mathbf{j} + \left(\frac{g\frac{\partial f}{\partial z} - f\frac{\partial g}{\partial z}}{g^2}\right)\mathbf{k}$

$= \left(\frac{g\frac{\partial f}{\partial x}\mathbf{i} + g\frac{\partial f}{\partial y}\mathbf{j} + g\frac{\partial f}{\partial z}\mathbf{k}}{g^2}\right) - \left(\frac{f\frac{\partial g}{\partial x}\mathbf{i} + f\frac{\partial g}{\partial y}\mathbf{j} + f\frac{\partial g}{\partial z}\mathbf{k}}{g^2}\right)$

$= \frac{g\left(\frac{\partial f}{\partial x}\mathbf{i} + \frac{\partial f}{\partial y}\mathbf{j} + \frac{\partial f}{\partial z}\mathbf{k}\right)}{g^2} - \frac{f\left(\frac{\partial g}{\partial x}\mathbf{i} + \frac{\partial g}{\partial y}\mathbf{j} + \frac{\partial g}{\partial z}\mathbf{k}\right)}{g^2}$

$= \frac{g\nabla f}{g^2} - \frac{f\nabla g}{g^2} = \frac{g\nabla f - f\nabla g}{g^2}$

■ Section 13.7 Extreme Values and Saddle Points (pp. 736–745)

Quick Review 13.7

1. $\frac{df}{dx} = 0 \Rightarrow \frac{(x^2+1)(1) - (x)(2x)}{(x^2+1)^2} = \frac{1 - x^2}{(x^2+1)^2} = 0$

$\Rightarrow x = 1$ or $x = -1 \Rightarrow$ critical points are

$(1, f(1)) = \left(1, \frac{1}{2}\right)$ and $(-1)(f(-1)) = \left(-1, -\frac{1}{2}\right)$. $\frac{df}{dx}$ is never undefined.

2. $\frac{d^2f}{dx^2} = \frac{(x^2+1)^2(-2x) - 4x(x^2+1)(1-x^2)}{(x^2+1)^4}$

$= \frac{2x(x^2-3)}{(x^2+1)^3} \Rightarrow \left.\frac{d^2f}{dx^2}\right|_{(-1, -\frac{1}{2})} = \frac{-2(-2)}{2^3} = \frac{1}{2} > 0$

$\Rightarrow f$ has a local minimum of $-\frac{1}{2}$ at $\left(-1, -\frac{1}{2}\right)$.

3. $\left.\frac{d^2f}{dx^2}\right|_{(1, \frac{1}{2})} = \frac{2(-2)}{2^3} = -\frac{1}{2} < 0 \Rightarrow f$ has a local maximum of $\frac{1}{2}$ at $\left(1, \frac{1}{2}\right)$.

4. Absolute minimum: $\left(-1, -\frac{1}{2}\right)$; absolute maximum: $\left(1, \frac{1}{2}\right)$.

5. $\frac{df}{dx} = 0 \Rightarrow 2 - 2x = 0 \Rightarrow x = 1 \Rightarrow$ critical point is $(1, f(1)) = (1, 4)$. $\frac{df}{dx}$ is never undefined.

6. $\frac{d^2f}{dx^2} = -2 < 0 \Rightarrow$ local minima are the endpoints of the parabolic segment: $(-2, f(-2)) = (-2, -5)$ and $(5, f(5)) = (5, -12)$.

7. f has a local maximum of 4 at $(1, 4)$.

8. Absolute minimum: $(5, -12)$; absolute maximum: $(1, 4)$.

9. $-2x + y = 2, x - 2y = 2 \Rightarrow x = 2y + 2$ $\Rightarrow -2(2y + 2) + y = 2 \Rightarrow y = -2, x = -2$

10. $3x - 4y = 5, 2x + y = 1 \Rightarrow y = 1 - 2x$ $\Rightarrow 3x - 4(1 - 2x) = 5 \Rightarrow x = \frac{9}{11}, y = -\frac{7}{11}$

Section 13.7 Exercises

1. $f_x(x, y) = 2x + y + 3 = 0$ and $f_y(x, y) = x + 2y - 3 = 0$ $\Rightarrow x = -3$ and $y = 3 \Rightarrow$ critical point is $(-3, 3)$; $f_{xx}(-3, 3) = 2, f_{yy}(-3, 3) = 2, f_{xy}(-3, 3) = 1 \Rightarrow f_{xx}f_{yy} - f_{xy}^2 = 3 > 0$ and $f_{xx} > 0 \Rightarrow$ local minimum of $f(-3, 3) = -5$

3. $f_x(x, y) = 2x + y + 3 = 0$ and $f_y(x, y) = x + 2 = 0 \Rightarrow x = -2$ and $y = 1 \Rightarrow$ critical point is $(-2, 1)$; $f_{xx}(-2, 1) = 2, f_{yy}(-2, 1) = 0, f_{xy}(-2, 1) = 1 \Rightarrow f_{xx}f_{yy} - f_{xy}^2 = -1 < 0 \Rightarrow$ saddle point

5. $f_x(x, y) = 5y - 14x + 3 = 0$ and $f_y(x, y) = 5x - 6 = 0$ $\Rightarrow x = \frac{6}{5}$ and $y = \frac{69}{25} \Rightarrow$ critical point is $\left(\frac{6}{5}, \frac{69}{25}\right)$; $f_{xx}\left(\frac{6}{5}, \frac{69}{25}\right) = -14, f_{yy}\left(\frac{6}{5}, \frac{69}{25}\right) = 0, f_{xy}\left(\frac{6}{5}, \frac{69}{25}\right) = 5 \Rightarrow f_{xx}f_{yy} - f_{xy}^2 = -25 < 0 \Rightarrow$ saddle point

7. $f_x(x, y) = 4x + 3y - 5 = 0$ and $f_y(x, y) = 3x + 8y + 2 = 0$ $\Rightarrow x = 2$ and $y = -1 \Rightarrow$ critical point is $(2, -1)$; $f_{xx}(2, -1) = 4, f_{yy}(2, -1) = 8, f_{xy}(2, -1) = 3 \Rightarrow f_{xx}f_{yy} - f_{xy}^2 = 23 > 0$ and $f_{xx} > 0 \Rightarrow$ local minimum of $f(2, -1) = -6$

9. $f_x(x, y) = 2x - 2 = 0$ and $f_y(x, y) = -2y + 4 = 0$ $\Rightarrow x = 1$ and $y = 2 \Rightarrow$ critical point is $(1, 2)$; $f_{xx}(1, 2) = 2$, $f_{yy}(1, 2) = -2, f_{xy}(1, 2) = 0 \Rightarrow f_{xx}f_{yy} - f_{xy}^2 = -4 < 0$ $\Rightarrow$ saddle point

11. $f_x(x, y) = 12x - 6x^2 + 6y = 0$ and $f_y(x, y) = 6y + 6x = 0$ $\Rightarrow x = 0$ and $y = 0$, or $x = 1$ and $y = -1 \Rightarrow$ critical points are $(0, 0)$ and $(1, -1)$; for $(0, 0)$: $f_{xx}(0, 0) = 12 - 12x|_{(0,0)} = 12, f_{yy}(0, 0) = 6, f_{xy}(0, 0) = 6 \Rightarrow f_{xx}f_{yy} - f_{xy}^2 = 36 > 0$ and $f_{xx} > 0 \Rightarrow$ local minimum of $f(0, 0) = 0$; for $(1, -1)$: $f_{xx}(1, -1) = 0, f_{yy}(1, -1) = 6, f_{xy}(1, -1) = 6 \Rightarrow f_{xx}f_{yy} - f_{xy}^2 = -36 < 0 \Rightarrow$ saddle point

13. $f_x(x, y) = 4y - 4x^3 = 0$ and $f_y(x, y) = 4x - 4y^3 = 0 \Rightarrow y = x^3 \Rightarrow x = x^9 \Rightarrow x = 0, 1, -1 \Rightarrow$ the critical points are $(0, 0)$, $(1, 1)$, and $(-1, -1)$; for $(0, 0)$: $f_{xx}(0, 0) = -12x^2|_{(0,0)} = 0, f_{yy}(0, 0) = -12y^2|_{(0,0)} = 0, f_{xy}(0, 0) = 4 \Rightarrow f_{xx}f_{yy} - f_{xy}^2 = -16 < 0 \Rightarrow$ saddle point; for $(1, 1)$: $f_{xx}(1, 1) = -12, f_{yy}(1, 1) = -12, f_{xy}(1, 1) = 4 \Rightarrow f_{xx}f_{yy} - f_{xy}^2 = 128 > 0$ and $f_{xx} < 0 \Rightarrow$ local maximum of $f(1, 1) = 2$; for $(-1, -1)$: $f_{xx}(-1, -1) = -12, f_{yy}(-1, -1) = -12, f_{xy}(-1, -1) = 4 \Rightarrow f_{xx}f_{yy} - f_{xy}^2 = 128 > 0$ and $f_{xx} < 0 \Rightarrow$ local maximum of $f(-1, -1) = 2$

15. $f_x(x, y) = \frac{-2x}{(x^2 + y^2 - 1)^2} = 0$ and $f_y(x, y) = \frac{-2y}{(x^2 + y^2 - 1)^2} = 0 \Rightarrow x = 0$ and $y = 0 \Rightarrow$ the critical point is $(0, 0)$; $f_{xx} = \frac{6x^2 - 2y^2 + 2}{(x^2 + y^2 - 1)^3}, f_{yy} = \frac{-2x^2 + 6y^2 + 2}{(x^2 + y^2 - 1)^3}$, $f_{xy} = \frac{8xy}{(x^2 + y^2 - 1)^3}$; $f_{xx}(0, 0) = -2, f_{yy}(0, 0) = -2$, $f_{xy}(0, 0) = 0 \Rightarrow f_{xx}f_{yy} - f_{xy}^2 = 4 > 0$ and $f_{xx} < 0 \Rightarrow$ local maximum of $f(0, 0) = -1$

17. $f_x(x, y) = y \cos x = 0$ and $f_y(x, y) = \sin x = 0 \Rightarrow x = n\pi$, n an integer, and $y = 0 \Rightarrow$ the critical points are $(n\pi, 0)$, n an integer (Note: $\cos x$ and $\sin x$ cannot both be 0 for the same x, so $\sin x$ must be 0 and $y = 0$); $f_{xx} = -y \sin x, f_{yy} = 0$, $f_{xy} = \cos x$; $f_{xx}(n\pi, 0) = 0, f_{yy}(n\pi, 0) = 0, f_{xy}(n\pi, 0) = 1$ if n is even and $f_{xy}(n\pi, 0) = -1$ if n is odd $\Rightarrow f_{xx}f_{yy} - f_{xy}^2 = -1 < 0 \Rightarrow$ saddle point; $f(n\pi, 0) = 0$ for every n

19. i. On OA, $f(x, y) = f(0, y) = y^2 - 4y + 1$ on $0 \le y \le 2$; $f'(0, y) = 2y - 4 = 0 \Rightarrow y = 2$; $f(0, 0) = 1$ and $f(0, 2) = -3$

ii. On AB, $f(x, y) = f(x, 2) = 2x^2 - 4x - 3$ on $0 \le x \le 1$; $f'(x, 2) = 4x - 4 = 0 \Rightarrow x = 1$; $f(0, 2) = -3$ and $f(1, 2) = -5$

iii. On OB, $f(x, y) = f(x, 2x) = 6x^2 - 12x + 1$ on $0 \le x \le 1$; endpoint values have been found above; $f'(x, 2x) = 12x - 12 = 0 \Rightarrow x = 1$ and $y = 2$, but $(1, 2)$ is not an interior point of OB

iv. For interior points of the triangular region, $f_x(x, y) = 4x - 4 = 0$ and $f_y(x, y) = 2y - 4 = 0 \Rightarrow x = 1$ and $y = 2$, but $(1, 2)$ is not an interior point of the region. Therefore, the absolute maximum is 1 at $(0, 0)$ and the absolute minimum is -5 at $(1, 2)$.

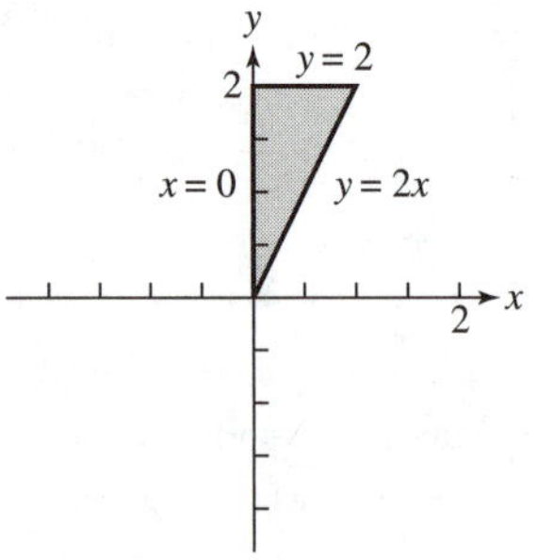

21. i. On OC, $T(x, y) = T(x, 0) = x^2 - 6x + 2$ on $0 \le x \le 5$; $T'(x, 0) = 2x - 6 = 0 \Rightarrow x = 3$ and $y = 0$; $T(3, 0) = -7$, $T(0, 0) = 2$, and $T(5, 0) = -3$

ii. On CB, $T(x, y) = T(5, y) = y^2 + 5y - 3$ on $-3 \le y \le 0$; $T'(5, y) = 2y + 5 = 0 \Rightarrow y = -\frac{5}{2}$ and $x = 5$; $T\left(5, -\frac{5}{2}\right) = -\frac{37}{4}$ and $T(5, -3) = -9$

iii. On AB, $T(x, y) = T(x, -3) = x^2 - 9x + 11$ on $0 \le x \le 5$; $T'(x, -3) = 2x - 9 = 0 \Rightarrow x = \frac{9}{2}$ and $y = -3$; $T\left(\frac{9}{2}, -3\right) = -\frac{37}{4}$ and $T(0, -3) = 11$

iv. On AO, $T(x, y) = T(0, y) = y^2 + 2$ on $-3 \le y \le 0$; $T'(0, y) = 2y = 0 \Rightarrow y = 0$ and $x = 0$, but $(0, 0)$ is not an interior point of AO

v. For interior points of the rectangular region, $T_x(x, y) = 2x + y - 6 = 0$ and $T_y(x, y) = x + 2y = 0 \Rightarrow x = 4$ and $y = -2$, an interior critical point with $T(4, -2) = -10$. Therefore the absolute maximum is 11 at $(0, -3)$ and the absolute minimum is -10 at $(4, -2)$.

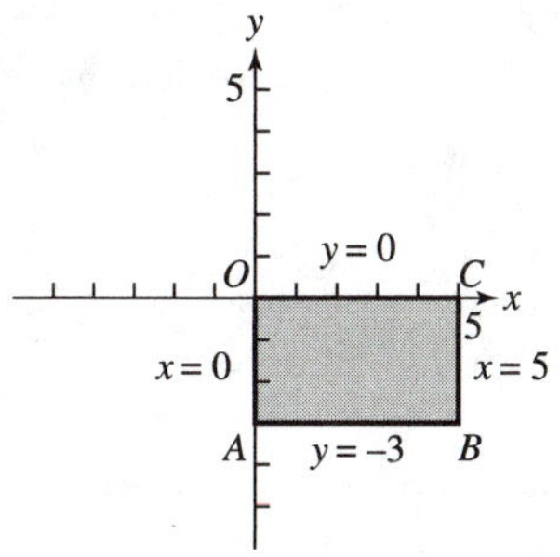

23. i. On AB, $f(x, y) = f(1, y) = 3 \cos y$ on $-\frac{\pi}{4} \le y \le \frac{\pi}{4}$; $f'(1, y) = -3 \sin y = 0 \Rightarrow y = 0$ and $x = 1$; $f(1, 0) = 3$, $f\left(1, -\frac{\pi}{4}\right) = \frac{3\sqrt{2}}{2}$, and $f\left(1, \frac{\pi}{4}\right) = \frac{3\sqrt{2}}{2}$

ii. On CD, $f(x, y) = f(3, y) = 3 \cos y$ on $-\frac{\pi}{4} \le y \le \frac{\pi}{4}$; $f'(3, y) = -3 \sin y = 0 \Rightarrow y = 0$ and $x = 3$, $f(3, 0) = 3$, $f\left(3, -\frac{\pi}{4}\right) = \frac{3\sqrt{2}}{2}$ and $f\left(3, \frac{\pi}{4}\right) = \frac{3\sqrt{2}}{2}$

iii. On BC, $f(x, y) = f\left(x, \frac{\pi}{4}\right) = \frac{\sqrt{2}}{2}(4x - x^2)$ on $1 \le x \le 3$; $f'\left(x, \frac{\pi}{4}\right) = \sqrt{2}(2 - x) = 0 \Rightarrow x = 2$ and $y = \frac{\pi}{4}$; $f\left(2, \frac{\pi}{4}\right) = 2\sqrt{2}$, $f\left(1, \frac{\pi}{4}\right) = \frac{3\sqrt{2}}{2}$, and $f\left(3, \frac{\pi}{4}\right) = \frac{3\sqrt{2}}{2}$

iv. On AD, $f(x, y) = f\left(x, -\frac{\pi}{4}\right) = \frac{\sqrt{2}}{2}(4x - x^2)$ on $1 \le x \le 3$; $f'\left(x, -\frac{\pi}{4}\right) = \sqrt{2}(2 - x) = 0 \Rightarrow x = 2$ and $y = -\frac{\pi}{4}$; $f\left(2, -\frac{\pi}{4}\right) = 2\sqrt{2}$, $f\left(1, -\frac{\pi}{4}\right) = \frac{3\sqrt{2}}{2}$, and $f\left(3, -\frac{\pi}{4}\right) = \frac{3\sqrt{2}}{2}$

v. For interior points of the region, $f_x(x, y) = (4 - 2x)\cos y = 0$ and $f_y(x, y) = -(4x - x^2)\sin y = 0 \Rightarrow x = 2$ and $y = 0$, which is an interior critical point with $f(2, 0) = 4$. Therefore the absolute maximum is 4 at $(2, 0)$ and the absolute minimum is $\frac{3\sqrt{2}}{2}$ at $\left(3, -\frac{\pi}{4}\right)$, $\left(3, \frac{\pi}{4}\right)$, $\left(1, -\frac{\pi}{4}\right)$, and $\left(1, \frac{\pi}{4}\right)$.

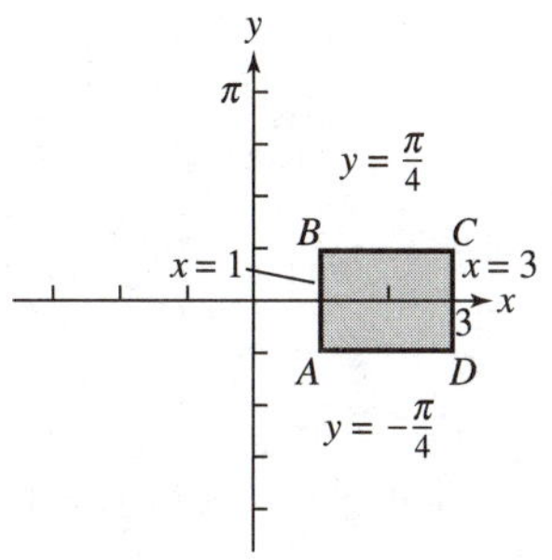

25. i. On the upper half-circle ABC we have $y = \sqrt{1 - x^2}$ with $-1 \le x \le 1$ so $f(x, y) = f(x, \sqrt{1 - x^2}) = x^3 + (1 - x^2)$. $f'(x, \sqrt{1 - x^2}) = 3x^2 - 2x = x(3x - 2)$. Setting $f'(x, \sqrt{1 - x^2}) = 0$ yields $x = 0 \Rightarrow y = 1$ and $x = \frac{2}{3} \Rightarrow y = \frac{\sqrt{5}}{3}$. $f(-1, 0) = -1$, $f(0, 1) = 1$, $f\left(\frac{2}{3}, \frac{\sqrt{5}}{3}\right) = \frac{23}{27}$, $f(1, 0) = 1$

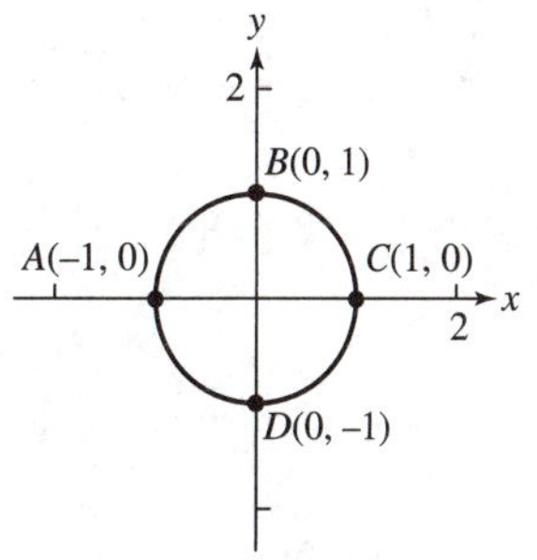

ii. On the lower half-circle ADC we have $y = -\sqrt{1 - x^2}$ with $-1 \le x \le 1$ so $f(x, y) = (x, -\sqrt{1 - x^2}) = x^3 + (1 - x^2)$. This is similar to (i) above so we only note that $f(0, -1) = 1$ and $f\left(\frac{2}{3}, -\frac{\sqrt{5}}{3}\right) = \frac{23}{27}$.

iii. For interior points of the disk, $f_x(x, y) = 3x^2$ and $f_y(x, y) = 2y$. Setting $f_x(x, y) = f_y(x, y) = 0$ yields $(x, y) = (0, 0)$. $f(0, 0) = 0$.

Therefore the absolute maximum is 1 and this occurs at $(0, 1)$, $(0, -1)$, and $(1, 0)$. The absolute minimum is -1 and this occurs at $(-1, 0)$.

27. Let $F(a, b) = \int_a^b (6 - x - x^2)dx$ where $a \le b$. The boundary of the domain of F is the line $a = b$ in the ab-plane, and $F(a, a) = 0$, so F is identically 0 on the boundary of its domain. For interior critical points we have $\frac{\partial F}{\partial a} = -(6 - a - a^2) = 0 \Rightarrow a = -3, 2$ and $\frac{\partial F}{\partial b} = (6 - b - b^2) = 0 \Rightarrow b = -3, 2$. Since $a \le b$, there is only one interior critical point $(-3, 2)$ and $F(-3, 2) = \int_{-3}^{2} (6 - x - x^2)dx$ gives the area under the parabola $y = 6 - x - x^2$ that is above the x-axis. Therefore, $a = -3$ and $b = 2$.

29. $T_x(x, y) = 2x - 1 = 0$ and $T_y(x, y) = 4y = 0 \Rightarrow x = \frac{1}{2}$ and $y = 0$ with $T\left(\frac{1}{2}, 0\right) = -\frac{1}{4}$; on the boundary $x^2 + y^2 = 1$: $T(x, y) = -x^2 - x + 2$ for $-1 \le x \le 1 \Rightarrow T'(x, y)$

$= -2x - 1 = 0 \Rightarrow x = -\frac{1}{2}$ and $y = \pm\frac{\sqrt{3}}{2}$; $T\left(-\frac{1}{2}, \frac{\sqrt{3}}{2}\right)$ $= \frac{9}{4}$, $T\left(-\frac{1}{2}, -\frac{\sqrt{3}}{2}\right) = \frac{9}{4}$, $T(-1, 0) = 2$, and $T(1, 0) = 0$ $\Rightarrow$ the hottest is $2\frac{1}{4}°$ at $\left(-\frac{1}{2}, \frac{\sqrt{3}}{2}\right)$ and $\left(-\frac{1}{2}, -\frac{\sqrt{3}}{2}\right)$; the coldest is $-\frac{1}{4}°$ at $\left(\frac{1}{2}, 0\right)$.

31. a. $f_x(x, y) = 2x - 4y = 0$ and $f_y(x, y) = 2y - 4x = 0$ $\Rightarrow x = 0$ and $y = 0$; $f_{xx}(0, 0) = 2$, $f_{yy}(0, 0) = 2$, $f_{xy}(0, 0) = -4 \Rightarrow f_{xx}f_{yy} - f_{xy}^2 = -12 < 0 \Rightarrow$ saddle point at $(0, 0)$

b. $f_x(x, y) = 2x - 2 = 0$ and $f_y(x, y) = 2y - 4 = 0$ $\Rightarrow x = 1$ and $y = 2$; $f_{xx}(1, 2) = 2$, $f_{yy}(1, 2) = 2$, $f_{xy}(1, 2) = 0 \Rightarrow f_{xx}f_{yy} - f_{xy}^2 = 4 > 0$ and $f_{xx} > 0$ $\Rightarrow$ local minimum at $(1, 2)$

c. $f_x(x, y) = 9x^2 - 9 = 0$ and $f_y(x, y) = 2y + 4 = 0$ $\Rightarrow x = \pm 1$ and $y = -2$; $f_{xx}(1, -2) = 18x\Big|_{(1,-2)}$ $= 18$, $f_{yy}(1, -2) = 2$, $f_{xy}(1, -2) = 0 \Rightarrow f_{xx}f_{yy} - f_{xy}^2$ $= 36 > 0$ and $f_{xx} > 0 \Rightarrow$ local minimum at $(1, -2)$; $f_{xx}(-1, -2) = -18$, $f_{yy}(-1, -2) = 2$, $f_{xy}(-1, -2) = 0$ $\Rightarrow f_{xx}f_{yy} - f_{xy}^{\,2} = -36 < 0 \Rightarrow$ saddle point at $(-1, -2)$

33. If $k = 0$, then $f(x, y) = x^2 + y^2 \Rightarrow f_x(x, y) = 2x = 0$ and $f_y(x, y) = 2y = 0 \Rightarrow x = 0$ and $y = 0 \Rightarrow (0, 0)$ is the only critical point. If $k \neq 0$, $f_x(x, y) = 2x + ky = 0 \Rightarrow y = -\frac{2}{k}x$; $f_y(x, y) = kx + 2y = 0 \Rightarrow kx + 2\left(-\frac{2}{k}x\right) = 0 \Rightarrow kx - \frac{4x}{k}$ $= 0 \Rightarrow \left(k - \frac{4}{k}\right)x = 0 \Rightarrow x = 0$ or $k = \pm 2 \Rightarrow y = \left(-\frac{2}{k}\right)(0)$ $= 0$ or $y = -x$; in any case $(0, 0)$ is a critical point.

35. a. No; for example $f(x, y) = xy$ has a saddle point at $(a, b) = (0, 0)$ where $f_x = f_y = 0$.

b. If $f_{xx}(a, b)$ and $f_{yy}(a, b)$ differ in sign, then $f_{xx}(a, b) \cdot f_{yy}(a, b) < 0$ so $f_{xx}f_{yy} - f_{xy}^2 < 0$. The surface must therefore have a saddle point at (a, b) by the second derivative test.

37. We want the point on $z = 10 - x^2 - y^2$ where the tangent plane is parallel to the plane $x + 2y + 3z = 0$. To find a normal vector to $z = 10 - x^2 - y^2$ let $w = z + x^2 + y^2 - 10$. Then $\nabla w = 2x\mathbf{i} + 2y\mathbf{j} + \mathbf{k}$ is normal to $z = 10 - x^2 - y^2$ at (x, y). The vector ∇w is parallel to $\mathbf{i} + 2\mathbf{j} + 3\mathbf{k}$ which is normal to the plane $x + 2y + 3z = 0$ if $6x\mathbf{i} + 6y\mathbf{j} + 3\mathbf{k} = \mathbf{i} + 2\mathbf{j} + 3\mathbf{k}$ or $x = \frac{1}{6}$ and $y = \frac{1}{3}$. Thus the point is $\left(\frac{1}{6}, \frac{1}{3}, 10 - \frac{1}{36} - \frac{1}{9}\right)$ or $\left(\frac{1}{6}, \frac{1}{3}, \frac{355}{36}\right)$.

39. No, because the domain $x \geq 0$ and $y \geq 0$ is unbounded since x and y can be as large as we please. Absolute extrema are guaranteed for continuous functions defined over closed *and bounded* domains in the plane. Since the domain is unbounded, the continuous function $f(x, y) = x + y$ need not have an absolute maximum (although, in this case, it does have an absolute minimum value of $f(0, 0) = 0$).

41. a. $\frac{df}{dt} = \frac{\partial f}{\partial x}\frac{dx}{dt} + \frac{\partial f}{\partial y}\frac{dy}{dt} = \frac{dx}{dt} + \frac{dy}{dt} = -2\sin t + 2\cos t$ $= 0 \Rightarrow \cos t = \sin t \Rightarrow x = y$

i. On the semicircle $x^2 + y^2 = 4$, $y \geq 0$, we have $t = \frac{\pi}{4}$ and $x = y = \sqrt{2} \Rightarrow f(\sqrt{2}, \sqrt{2}) = 2\sqrt{2}$. At the endpoints, $f(-2, 0) = -2$ and $f(2, 0) = 2$. Therefore the absolute minimum is $f(-2, 0) = -2$ when $t = \pi$; the absolute maximum is $f(\sqrt{2}, \sqrt{2}) = 2\sqrt{2}$ when $t = \frac{\pi}{4}$.

ii. On the quarter circle $x^2 + y^2 = 4$, $x \geq 0$ and $y \geq 0$, the endpoints give $f(0, 2) = 2$ and $f(2, 0) = 2$. Therefore the absolute minimum is $f(2, 0) = 2$ and $f(0, 2) = 2$ when $t = 0, \frac{\pi}{2}$ respectively; the absolute maximum is $f(\sqrt{2}, \sqrt{2}) = 2\sqrt{2}$ when $t = \frac{\pi}{4}$.

b. $\frac{dg}{dt} = \frac{\partial g}{\partial x}\frac{dx}{dt} + \frac{\partial g}{\partial y}\frac{dy}{dt} = y\frac{dx}{dt} + x\frac{dy}{dt} = -4\sin^2 t$ $+ 4\cos^2 t = 0 \Rightarrow \cos t = \pm\sin t \Rightarrow x = \pm y$.

i. On the semicircle $x^2 + y^2 = 4$, $y \geq 0$, we obtain $x = y = \sqrt{2}$ at $t = \frac{\pi}{4}$ and $x = -\sqrt{2}$, $y = \sqrt{2}$ at $t = \frac{3\pi}{4}$. Then $g(\sqrt{2}, \sqrt{2}) = 2$ and $g(-\sqrt{2}, \sqrt{2}) = -2$. At the endpoints, $g(-2, 0) = g(2, 0) = 0$. Therefore the absolute minimum is $g(-\sqrt{2}, \sqrt{2}) = -2$ when $t = \frac{3\pi}{4}$; the absolute maximum is $g(\sqrt{2}, \sqrt{2}) = 2$ when $t = \frac{\pi}{4}$.

ii. On the quarter circle $x^2 + y^2 = 4$, $x \geq 0$ and $y \geq 0$, the endpoints give $g(0, 2) = 0$ and $g(2, 0) = 0$. Therefore the absolute minimum is $g(2, 0) = 0$ and $g(0, 2) = 0$ when $t = 0, \frac{\pi}{2}$ respectively; the absolute maximum is $g(\sqrt{2}, \sqrt{2}) = 2$ when $t = \frac{\pi}{4}$.

c. $\frac{dh}{dt} = \frac{\partial h}{\partial x}\frac{dx}{dt} + \frac{\partial h}{\partial y}\frac{dy}{dt} = 4x\frac{dx}{dt} + 2y\frac{dy}{dt}$ $= (8\cos t)(-2\sin t) + (4\sin t)(2\cos t)$ $= -8\cos t\sin t = 0 \Rightarrow t = 0, \frac{\pi}{2}, \pi$ yielding the points $(2, 0)$, $(0, 2)$, and $(-2, 0)$, respectively.

i. On the semicircle $x^2 + y^2 = 4$, $y \geq 0$, we have $h(2, 0) = 8$, $h(0, 2) = 4$, and $h(-2, 0) = 8$. Therefore the absolute minimum is $h(0, 2) = 4$ when $t = \frac{\pi}{2}$; the absolute maximum is $h(2, 0) = 8$ and $h(-2, 0) = 8$ when $t = 0, \pi$ respectively.

ii. On the quartercircle $x^2 + y^2 = 4$, $x \geq 0$ and $y \geq 0$, the absolute minimum is $h(0, 2) = 4$ when $t = \frac{\pi}{2}$; the absolute maximum is $h(2, 0) = 8$ when $t = 0$.

43. $\frac{df}{dx} = \frac{\partial f}{\partial x}\frac{dx}{dt} + \frac{\partial f}{\partial y}\frac{dy}{dt} = y\frac{dx}{dt} + x\frac{dy}{dt}$

i. $x = 2t$ and $y = t + 1 \Rightarrow \frac{df}{dt} = (t + 1)(2) + (2t)(1) = 4t$ $+ 2 = 0 \Rightarrow t = -\frac{1}{2} \Rightarrow x = -1$ and $y = \frac{1}{2}$ with

$f\left(-1, \frac{1}{2}\right) = -\frac{1}{2} \cdot \frac{d^2f}{dt^2} = 4 > 0$; the absolute minimum is $f\left(-1, \frac{1}{2}\right) = -\frac{1}{2}$ when $t = -\frac{1}{2}$; there is no absolute maximum.

ii. For the endpoints: $t = -1 \Rightarrow x = -2$ and $y = 0$ with $f(-2, 0) = 0$; $t = 0 \Rightarrow x = 0$ and $y = 1$ with $f(0, 1) = 0$. The absolute minimum is $f\left(-1, \frac{1}{2}\right) = -\frac{1}{2}$ when $t = -\frac{1}{2}$; the absolute maximum is $f(0, 1) = 0$ and $f(-2, 0) = 0$ when $t = -1, 0$ respectively.

iii. There are no interior critical points. For the endpoints: $t = 0 \Rightarrow x = 0$ and $y = 1$ with $f(0, 1) = 0$; $t = 1 \Rightarrow x = 2$ and $y = 2$ with $f(2, 2) = 4$. The absolute minimum is $f(0, 1) = 0$ when $t = 0$; the absolute maximum is $f(2, 2) = 4$ when $t = 1$.

45–50. Example CAS commands:

Maple:

```
with(plots):
f:=(x,y)->2*x^4+y^4-2*x^2-2*y^2+3;
plot3d(f(x,y),x=-1..1,y=-1..1,
  axes=BOXED);
contourplot(f(x,y),x=-3/2..3/2,
  y=-3/2..3/2,axes=NORMAL);
exp1:=diff(f(x,y),x)=0;
  exp2:=diff(f(x,y),y)=0;
critical:=evalf(solve({exp1,exp2},
  {x,y}));
diff(diff(f(x,y),x),x):fxx:=unapply(%,
  (x,y));
diff(diff(f(x,y),x),y):fxy:=unapply(%,
  (x,y));
diff(diff(f(x,y),y),y):fyy:=unapply(%,
  (x,y));
fxx(x,y)*fyy(x,y)-(fxy(x,y))^2:disc:=
  unapply(%,(x,y));
subs(critical[1],[fxx(x,y),disc(x,y)]);
subs(critical[2],[fxx(x,y),disc(x,y)]);
subs(critical[3],[fxx(x,y),disc(x,y)]);
subs(critical[4],[fxx(x,y),disc(x,y)]);
subs(critical[5],[fxx(x,y),disc(x,y)]);
subs(critical[6],[fxx(x,y),disc(x,y)]);
```

Mathematica:

```
Clear[x,y]
SetOptions[ContourPlot, PlotPoints->25,
  Contours->20, ContourShading->False];
f[x_,y_]=2x^4+y^4-2x^2-2y^2+3
{xa,xb}={-3/2,3/2};
{ya,yb}={-3/2,3/2};
Plot3D[f[x,y],{x,xa,xb},{y,ya,yb}]
ContourPlot[f[x,y], {x,xa,xb},
  {y,ya,yb}]
fx=D[f[x,y],x]
fy=D[f[x,y],y]
crit=Solve[{fx==0,fy==0}]
fxx=D[fx,x]
fxy=D[fx,y]
fyy=D[fy,y]
disc=fxx fyy - fxy^2
{{x,y},disc,fxx} /. crit
```

Section 13.8 Lagrange Multipliers (pp. 746–756)

Quick Review 13.8

1. $f_x(x, y) = -10x + 2y + 4, f_y(x, y) = 2x - 4y$, so $f_x(x, y) = 0, f_y(x, y) = 0 \Rightarrow x = \frac{4}{9}, y = \frac{2}{9}$. $f_{xx} = -10, f_{yy} = -4, f_{xy} = 2 \Rightarrow f_{xx}f_{yy} - f_{xy}^2 = 36 > 0$, and $f_{xx} < 0$, so $f\left(\frac{4}{9}, \frac{2}{9}\right) = -\frac{28}{9}$ is a local maximum.

2. $f_x(x, y) = -4x - 2y + 2, f_y(x, y) = -2x - 2y + 2$, so $f_x(x, y) = 0, f_y(x, y) = 0 \Rightarrow x = 0, y = 1$. $f_{xx} = -4, f_{yy} = -2, f_{xy} = -2 \Rightarrow f_{xx}f_{yy} - f_{xy}^2 = 4 > 0$, and $f_{xx} < 0$, so $f(0, 1) = 4$ is a local maximum.

3. $f_x(x, y) = 2x + 3y - 6, f_y(x, y) = 3x + 6y + 3$, so $f_x(x, y) = 0, f_y(x, y) = 0 \Rightarrow x = 15, y = -8$. $f_{xx} = 2, f_{yy} = 6, f_{xy} = 3 \Rightarrow f_{xx}f_{yy} - f_{xy}^2 = 3 > 0$, and $f_{xx} > 0$, so $f(15, -8) = -63$ is a local minimum.

4. $f_x(x, y) = 24x^2 + 6y, f_y(x, y) = 3y^2 + 6x$, so $f_x(x, y) = 0$, $f_y(x, y) = 0 \Rightarrow$ $x = 0, y = 0$ or $x = -\frac{1}{2}, y = -1$. $f_{xx} = 48x, f_{yy} = 6y$, $f_{xy} = 6$. At $(0, 0)$: $f_{xx} = 0, f_{yy} = 0, f_{xy} = 6$ $\Rightarrow f_{xx}f_{yy} - f_{xy}^2 = -6 < 0 \Rightarrow$ saddle point. At $\left(-\frac{1}{2}, -1\right)$: $f_{xx} = -24, f_{yy} = -6, f_{xy} = 6 \Rightarrow f_{xx}f_{yy} - f_{xy}^2 = 108 > 0$, and $f_{xx} < 0$, so $f\left(-\frac{1}{2}, -1\right) = 1$ is a local maximum.

5. $\nabla g = \frac{\partial g}{\partial x}\mathbf{i} + \frac{\partial g}{\partial y}\mathbf{j} + \frac{\partial g}{\partial z}\mathbf{k} = 4y\mathbf{i} - 6z\mathbf{k}$

6. $\nabla k = \frac{\partial k}{\partial x}\mathbf{i} + \frac{\partial k}{\partial y}\mathbf{j}$
$= (9x^2y - 2y^3 + y \cos xy)\mathbf{i} + (3x^3 - 6xy^2 + x \cos xy)\mathbf{j}$

7. $\nabla h = \frac{\partial h}{\partial x}\mathbf{i} + \frac{\partial h}{\partial y}\mathbf{j} + \frac{\partial h}{\partial z}\mathbf{k} = 3\mathbf{i} - 2\mathbf{j} + 4z\mathbf{k}$

8. $\mathbf{u} \cdot \mathbf{v} = (2)(1) + (-3)(-2) + (4)(-2) = 0$, so the vectors are perpendicular.

9. $\mathbf{u} \cdot \mathbf{w} = (2)(-7) + (-3)(14) + (4)(14) = 0$, so the vectors are perpendicular.

10. $\mathbf{w} = -7\mathbf{v}$, so the vectors are parallel.

Section 13.8 Exercises

1. $\nabla f = y\mathbf{i} + x\mathbf{j}$ and $\nabla g = 2x\mathbf{i} + 4y\mathbf{j}$ so that $\nabla f = \lambda \nabla g \Rightarrow y\mathbf{i} + x\mathbf{j} = \lambda(2x\mathbf{i} + 4y\mathbf{j}) \Rightarrow y = 2x\lambda$ and $x = 4y\lambda \Rightarrow x = 8x\lambda^2$ $\Rightarrow \lambda = \pm\frac{\sqrt{2}}{4}$ or $x = 0$.

CASE 1: If $x = 0$, then $y = 0$. But $(0, 0)$ is not on the ellipse so $x \neq 0$.

CASE 2: $x \neq 0 \Rightarrow \lambda = \pm\frac{\sqrt{2}}{4} \Rightarrow x \pm\sqrt{2}y \Rightarrow (\pm\sqrt{2}y)^2 + 2y^2 = 1 \Rightarrow y = \pm\frac{1}{2}$.

Therefore f takes on its extreme values at $\left(\pm\sqrt{2}, \frac{1}{2}\right)$ and $\left(\pm\sqrt{2}, -\frac{1}{2}\right)$. The extreme values of f on the ellipse are $\pm\frac{\sqrt{2}}{2}$.

3. $\nabla f = -2x\mathbf{i} - 2y\mathbf{j}$ and $\nabla g = \mathbf{i} + 3\mathbf{j}$ so that $\nabla f = \lambda \nabla g$ $\Rightarrow -2x\mathbf{i} - 2y\mathbf{j} = \lambda(\mathbf{i} + 3\mathbf{j}) \Rightarrow x = -\frac{\lambda}{2}$ and $y = -\frac{3\lambda}{2}$ $\Rightarrow \left(-\frac{\lambda}{2}\right) + 3\left(-\frac{3\lambda}{2}\right) = 10 \Rightarrow \lambda = -2 \Rightarrow x = 1$ and $y = 3$ $\Rightarrow f$ takes on its extreme value at (1, 3) on the line. The extreme value is $f(1, 3) = 49 - 1 - 9 = 39$.

5. We optimize $f(x, y) = x^2 + y^2$, the square of the distance to the origin, subject to the constraint $g(x, y) = xy^2 - 54 = 0$. Thus $\nabla f = 2x\mathbf{i} + 2y\mathbf{j}$ and $\nabla g = y^2\mathbf{i} + 2xy\mathbf{j}$ so that $\nabla f = \lambda \nabla g \Rightarrow 2x\mathbf{i} + 2y\mathbf{j} = \lambda(y^2\mathbf{i} + 2xy\mathbf{j}) \Rightarrow 2x = \lambda y^2$ and $2y = 2\lambda xy$.

CASE 1: If $y = 0$, then $x = 0$. But (0, 0) does not satisfy the constraint $xy^2 = 54$ so $y \neq 0$.

CASE 2: If $y \neq 0$, then $2 = 2\lambda x \Rightarrow x = \frac{1}{\lambda} \Rightarrow 2\left(\frac{1}{\lambda}\right) = \lambda y^2$ $\Rightarrow y^2 = \frac{2}{\lambda^2}$. Then $xy^2 = 54 \Rightarrow \left(\frac{1}{\lambda}\right)\left(\frac{2}{\lambda^2}\right) = 54 \Rightarrow \lambda^3 = \frac{1}{27}$ $\Rightarrow \lambda = \frac{1}{3} \Rightarrow x = 3$. So $y^2 = 18 \Rightarrow x = 3$ and $y = \pm 3\sqrt{2}$.

Therefore $(3, \pm 3\sqrt{2})$ are the points on the curve $xy^2 = 54$ nearest the origin (since $xy^2 = 54$ has points increasingly far away as y gets close to 0, no points are farthest away).

7. a. $\nabla f = \mathbf{i} + \mathbf{j}$ and $\nabla g = y\mathbf{i} + x\mathbf{j}$ so that $\nabla f = \lambda \nabla g \Rightarrow \mathbf{i} + \mathbf{j} = \lambda(y\mathbf{i} + x\mathbf{j}) \Rightarrow 1 = \lambda y$ and $1 = \lambda x \Rightarrow y = \frac{1}{\lambda}$ and $x = \frac{1}{\lambda} \Rightarrow \frac{1}{\lambda^2} = 16 \Rightarrow \lambda = \pm\frac{1}{4}$. Use $\lambda = \frac{1}{4}$ since $x > 0$ and $y > 0$. Then $x = 4$ and $y = 4 \Rightarrow$ the minimum value is 8 at the point (4, 4). Now, $xy = 16$, $x > 0$, $y > 0$ is a branch of a hyperbola in the first quadrant with the x- and y-axes as asymptotes. The equations $x + y = c$ give a family of parallel lines with $m = -1$. As these lines move away from the origin, the number c increases. Thus the minimum value of c occurs where $x + y = c$ is tangent to the hyperbola's branch.

b. $\nabla f = y\mathbf{i} + x\mathbf{j}$ and $\nabla g = \mathbf{i} + \mathbf{j}$ so that $\nabla f = \lambda \nabla g \Rightarrow y\mathbf{i} + x\mathbf{j} = \lambda(\mathbf{i} + \mathbf{j}) \Rightarrow y = \lambda = x \Rightarrow y + y = 16 \Rightarrow y = 8$ $\Rightarrow x = 8 \Rightarrow f(8, 8) = 64$ is the maximum value. The equations $xy = c$ ($x > 0$ and $y > 0$ or $x < 0$ and $y < 0$ to get a maximum value) give a family of hyperbolas in the first and third quadrants with the x- and y-axes as asymptotes. The maximum value of c occurs where the hyperbola $xy = c$ is tangent to the line $x + y = 16$.

9. $V = \pi r^2 h = 16\pi \Rightarrow 16 = r^2 h \Rightarrow g(r, h) = r^2 h - 16$; $S = 2\pi rh + 2\pi r^2 \Rightarrow \nabla S = (2\pi h + 4\pi r)\mathbf{i} + 2\pi r\mathbf{j}$ and $\nabla g = 2rh\mathbf{i} + r^2\mathbf{j}$ so that $\nabla S = \lambda \nabla g \Rightarrow (2\pi h + 4\pi r)\mathbf{i} + 2\pi r\mathbf{j} = \lambda(2rh\mathbf{i} + r^2\mathbf{j}) \Rightarrow 2\pi h + 4\pi r = 2rh\lambda$ and $2\pi r = \lambda r^2 \Rightarrow r = 0$ or $\lambda = \frac{2\pi}{r}$. But $r = 0$ gives no physical can, so $r \neq 0 \Rightarrow \lambda = \frac{2\pi}{r} \Rightarrow 2\pi h + 4\pi r = 2rh\left(\frac{2\pi}{r}\right) \Rightarrow 2r = h \Rightarrow 16 = r^2(2r) \Rightarrow r = 2 \Rightarrow h = 4$, thus $r = 2$ cm and $h = 4$ cm give the only extreme surface area of 24π cm^2. Since $r = 4$ cm and $h = 1$ cm $\Rightarrow V = 16\pi$ cm^3 and $S = 40\pi$ cm^2, which is a larger surface area, then 24π cm^2 must be the minimum surface area.

11. $A = (2x)(2y) = 4xy$ subject to $g(x, y) = \frac{x^2}{16} + \frac{y^2}{9} - 1 = 0$; $\nabla A = 4y\mathbf{i} + 4x\mathbf{j}$ and $\nabla g = \frac{x}{8}\mathbf{i} + \frac{2y}{9}\mathbf{j}$ so that $\nabla A = \lambda \nabla g$ $\Rightarrow 4y\mathbf{i} + 4x\mathbf{j} = \lambda\left(\frac{x}{8}\mathbf{i} + \frac{2y}{9}\mathbf{j}\right) \Rightarrow 4y = \left(\frac{x}{8}\right)\lambda$ and $4x = \left(\frac{2y}{9}\right)\lambda$ $\Rightarrow \lambda = \frac{32y}{x}$ and $4x = \left(\frac{2y}{9}\right)\left(\frac{32y}{x}\right) \Rightarrow y = \pm\frac{3}{4}x \Rightarrow \frac{x^2}{16} + \frac{\left(\frac{\pm 3}{4}x\right)^2}{9} = 1 \Rightarrow x^2 = 8 \Rightarrow x = \pm 2\sqrt{2}$. We use $x = 2\sqrt{2}$ since x represents distance. Then $y = \frac{3}{4}(2\sqrt{2}) = \frac{3\sqrt{2}}{2}$, so the length is $2x = 4\sqrt{2}$ and the width is $2y = 3\sqrt{2}$.

13. $\nabla f = 2x\mathbf{i} + 2y\mathbf{j}$ and $\nabla g = (2x - 2)\mathbf{i} + (2y - 4)\mathbf{j}$ so that $\nabla f = \lambda \nabla g = 2x\mathbf{i} + 2y\mathbf{j} = \lambda[(2x - 2)\mathbf{i} + (2y - 4)\mathbf{j}]$ $\Rightarrow 2x = \lambda(2x - 2)$ and $2y = \lambda(2y - 4) \Rightarrow x = \frac{\lambda}{\lambda - 1}$ and $y = \frac{2\lambda}{\lambda - 1}$, $\lambda \neq 1 \Rightarrow y = 2x \Rightarrow x^2 - 2x + (2x)^2 - 4(2x) = 0 \Rightarrow x = 0$ and $y = 0$, or $x = 2$ and $y = 4$. Therefore $f(0, 0) = 0$ is the minimum value and $f(2, 4) = 20$ is the maximum value. (Note that $\lambda = 1$ gives $2x = 2x - 2$ or $0 = -2$, which is impossible.)

15. $\nabla T = (8x - 4y)\mathbf{i} + (-4x + 2y)\mathbf{j}$ and $g(x, y) = x^2 + y^2 - 25 = 0 \Rightarrow \nabla g = 2x\mathbf{i} + 2y\mathbf{j}$ so that $\nabla T = \lambda \nabla g$ $\Rightarrow (8x - 4y)\mathbf{i} + (-4x + 2y)\mathbf{j} = \lambda(2x\mathbf{i} + 2y\mathbf{j}) \Rightarrow 8x - 4y = 2\lambda x$ and $-4x + 2y = 2\lambda y \Rightarrow y = \frac{-2x}{\lambda - 1}$, $\lambda \neq 1$ $\Rightarrow 8x - 4\left(\frac{-2x}{\lambda - 1}\right) = 2\lambda x \Rightarrow x = 0$, or $\lambda = 0$, or $\lambda = 5$.

CASE 1: $x = 0 \Rightarrow y = 0$; but (0, 0) is not on $x^2 + y^2 = 25$ so $x \neq 0$.

CASE 2: $\lambda = 0 \Rightarrow y = 2x \Rightarrow x^2 + (2x)^2 = 25 \Rightarrow x = \pm\sqrt{5}$ and $y = 2x$.

CASE 3: $\lambda = 5 \Rightarrow y = \frac{-2x}{4} = -\frac{x}{2} \Rightarrow x^2 + \left(-\frac{x}{2}\right)^2 = 25$ $\Rightarrow x = \pm 2\sqrt{5} \Rightarrow x = 2\sqrt{5}$ and $y = -\sqrt{5}$, or $x = -2\sqrt{5}$ and $y = \sqrt{5}$.

Therefore $T(\sqrt{5}, 2\sqrt{5}) = 0° = T(-\sqrt{5}, -2\sqrt{5})$ is the minimum value and $T(2\sqrt{5}, -\sqrt{5}) = 125°$ $= T(-2\sqrt{5}, \sqrt{5})$ is the maximum value. (Note: $\lambda = 1 \Rightarrow x = 0$ from the equation $-4x + 2y = 2\lambda y$; but we found $x \neq 0$ in CASE 1.)

17. Let $f(x, y, z) = (x - 1)^2 + (y - 1)^2 + (z - 1)^2$ be the square of the distance from (1, 1, 1). Then $\nabla f = 2(x - 1)\mathbf{i} + 2(y - 1)\mathbf{j} + 2(z - 1)\mathbf{k}$ and $\nabla g = \mathbf{i} + 2\mathbf{j} + 3\mathbf{k}$ so that $\nabla f = \lambda \nabla g \Rightarrow 2(x - 1)\mathbf{i} + 2(y - 1)\mathbf{j} + 2(z - 1)\mathbf{k} = \lambda(\mathbf{i} + 2\mathbf{j} + 3\mathbf{k}) \Rightarrow 2(x - 1) = \lambda$, $2(y - 1) = 2\lambda$, $2(z - 1) = 3\lambda$ $\Rightarrow 2(y - 1) = 2[2(x - 1)]$ and $2(z - 1) = 3[2(x - 1)]$ $\Rightarrow x = \frac{y + 1}{2} \Rightarrow z + 2 = 3\left(\frac{y + 1}{2}\right)$ or $z = \frac{3y - 1}{2}$; thus $\frac{y + 1}{2} + 2y + 3\left(\frac{3y - 1}{2}\right) - 13 = 0 \Rightarrow y = 2 \Rightarrow x = \frac{3}{2}$ and $z = \frac{5}{2}$. Therefore the point $\left(\frac{3}{2}, 2, \frac{5}{2}\right)$ is closest (since no point on the plane is farthest from the point (1, 1, 1)).

19. Let $f(x, y, z) = x^2 + y^2 + z^2$ be the square of the distance from the origin. Then $\nabla f = 2x\mathbf{i} + 2y\mathbf{j} + 2z\mathbf{k}$ and $\nabla g = 2x\mathbf{i} - 2y\mathbf{j} - 2z\mathbf{k}$ so that $\nabla f = \lambda \nabla g \Rightarrow 2x\mathbf{i} + 2y\mathbf{j} + 2z\mathbf{k} = \lambda(2x\mathbf{i} - 2y\mathbf{j} - 2z\mathbf{k}) \Rightarrow 2x = 2x\lambda$, $2y = -2y\lambda$, and $2z = -2z\lambda \Rightarrow x = 0$ or $\lambda = 1$.

CASE 1: $\lambda = 1 \Rightarrow 2y = -2y \Rightarrow y = 0$; $2x = -2z \Rightarrow z = 0 \Rightarrow x^2 - 1 = 0 \Rightarrow x = \pm 1$ and $y = z = 0$.

CASE 2: $x = 0 \Rightarrow -y^2 - z^2 = 1$, which has no solution.

Therefore the points $(\pm 1, 0, 0)$ are closest to the origin $\Rightarrow$ the minimum distance from the surface to the origin is 1 (since there is no maximum distance from the surface to the origin).

21. Let $f(x, y, z) = x^2 + y^2 + z^2$ be the square of the distance to the origin. Then $\nabla f = 2x\mathbf{i} + 2y\mathbf{j} + 2z\mathbf{k}$ and $\nabla g = -y\mathbf{i} - x\mathbf{j} + 2z\mathbf{k}$ so that $\nabla f = \lambda \nabla g \Rightarrow 2x\mathbf{i} + 2y\mathbf{j} + 2z\mathbf{k} = \lambda(-y\mathbf{i} - x\mathbf{j} + 2z\mathbf{k}) \Rightarrow 2x = -y\lambda$, $2y = -x\lambda$, and $2z = 2z\lambda \Rightarrow \lambda = 1$ or $z = 0$.

CASE 1: $\lambda = 1 \Rightarrow 2x = -y$ and $2y = -x \Rightarrow y = 0$ and $x = 0 \Rightarrow z^2 - 4 = 0 \Rightarrow z = \pm 2$ and $x = y = 0$.

CASE 2: $z = 0 \Rightarrow -xy - 4 = 0 \Rightarrow y = -\frac{4}{x}$.
Then $2x = \frac{4}{x}\lambda \Rightarrow \lambda = \frac{x^2}{2}$, and $-\frac{8}{x} = -x\lambda$
$\Rightarrow -\frac{8}{x} = -x\left(\frac{x^2}{2}\right) \Rightarrow x^4 = 16$
$\Rightarrow x = \pm 2$. Thus, $x = 2$ and $y = -2$, or $x = -2$ and $y = 2$.

Therefore we get four points: $(2, -2, 0)$, $(-2, 2, 0)$, $(0, 0, 2)$ and $(0, 0, -2)$. But the points $(0, 0, 2)$ and $(0, 0, -2)$ are closest to the origin since they are 2 units away and the others are $2\sqrt{2}$ units away.

23. $\nabla f = \mathbf{i} - 2\mathbf{j} + 5\mathbf{k}$ and $\nabla g = 2x\mathbf{i} + 2y\mathbf{j} + 2z\mathbf{k}$ so that $\nabla f = \lambda \nabla g \Rightarrow \mathbf{i} - 2\mathbf{j} + 5\mathbf{k} = \lambda(2x\mathbf{i} + 2y\mathbf{j} + 2z\mathbf{k}) \Rightarrow 1 = 2x\lambda$, $-2 = 2y\lambda$, and $5 = 2z\lambda \Rightarrow x = \frac{1}{2\lambda}$, $y = -\frac{1}{\lambda} = -2x$, and $z = \frac{5}{2\lambda} = 5x \Rightarrow x^2 + (-2x)^2 + (5x)^2 = 30 \Rightarrow x = \pm 1$.
Thus, $x = 1$, $y = -2$, $z = 5$ or $x = -1$, $y = 2$, $z = -5$. Therefore $f(1, -2, 5) = 30$ is the maximum value and $f(-1, 2, -5) = -30$ is the minimum value.

25. Let $f(x, y, z) = x^2 + y^2 + z^2$ and $g(x, y, z) = x + y + z - 9 = 0 \Rightarrow \nabla f = 2x\mathbf{i} + 2y\mathbf{j} + 2z\mathbf{k}$ and $\nabla g = \mathbf{i} + \mathbf{j} + \mathbf{k}$ so that $\nabla f = \lambda \nabla g \Rightarrow 2x\mathbf{i} + 2y\mathbf{j} + 2z\mathbf{k} = \lambda(\mathbf{i} + \mathbf{j} + \mathbf{k}) \Rightarrow 2x = \lambda$ $2y = \lambda$, and $2z = \lambda \Rightarrow x = y = z \Rightarrow x + x + x - 9 = 0 \Rightarrow x = 3$, $y = 3$, and $z = 3$.

27. $V = 8xyz$ and $g(x, y, z) = x^2 + y^2 + z^2 - 1 = 0 \Rightarrow \nabla V = 8yz\mathbf{i} + 8xz\mathbf{j} + 8xy\mathbf{k}$ and $\nabla g = 2x\mathbf{i} + 2y\mathbf{j} + 2z\mathbf{k}$ so that $\nabla f = \lambda \nabla g \Rightarrow 4yz = \lambda x$, $4xz = \lambda y$, and $4xy = \lambda z \Rightarrow 4xyz = \lambda y^2$ and $4xyz = \lambda y^2 \Rightarrow y = \pm x \Rightarrow z = \pm x \Rightarrow x^2 + x^2 + x^2 = 1 \Rightarrow x = \frac{1}{\sqrt{3}}$ since $x > 0 \Rightarrow$ the dimensions of the box are $\frac{2}{\sqrt{3}}$ by $\frac{2}{\sqrt{3}}$ by $\frac{2}{\sqrt{3}}$ for maximum volume. (Note that there is no minimum volume since the box could be made arbitrarily thin.)

29. $\nabla T = 16x\mathbf{i} + 4z\mathbf{j} + (4y - 16)\mathbf{k}$ and $\nabla g = 8x\mathbf{i} + 2y\mathbf{j} + 8z\mathbf{k}$ so that $\nabla T = \lambda \nabla g \Rightarrow 16x\mathbf{i} + 4z\mathbf{j} + (4y - 16)\mathbf{k} = \lambda(8x\mathbf{i} + 2y\mathbf{j} + 8z\mathbf{k}) \Rightarrow 16x = 8x\lambda$, $4z = 2y\lambda$, and $4y - 16 = 8z\lambda \Rightarrow \lambda = 2$ or $x = 0$.

CASE 1: $\lambda = 2 \Rightarrow 4z = 2y(2) \Rightarrow z = y$. Then $4z - 16 = 16z \Rightarrow z = -\frac{4}{3} \Rightarrow y = -\frac{4}{3}$. Then $4x^2 + \left(-\frac{4}{3}\right)^2 + 4\left(-\frac{4}{3}\right)^2 = 16 \Rightarrow x = \pm\frac{4}{3}$.

CASE 2: $x = 0 \Rightarrow \lambda = \frac{2z}{y} \Rightarrow 4y - 16 = 8z\left(\frac{2z}{y}\right) \Rightarrow y^2 - 4y = 4z^2 \Rightarrow 4(0)^2 + y^2 + (y^2 - 4y) - 16 = 0 \Rightarrow y^2 - 2y - 8 = 0 \Rightarrow (y - 4)(y + 2) = 0 \Rightarrow y = 4$ or $y = -2$. Now $y = 4 \Rightarrow 4z^2 = 4^2 - 4(4) \Rightarrow z = 0$ and $y = -2 \Rightarrow 4z^2 = (-2)^2 - 4(-2) \Rightarrow z = \pm\sqrt{3}$.

The temperatures are $T\left(\pm\frac{4}{3}, -\frac{4}{3}, -\frac{4}{3}\right) = 642\frac{2}{3}°$, $T(0, 4, 0) = 600°$, $T(0, -2, \sqrt{3}) = (600 - 24\sqrt{3})° \approx 558.43°$, and $T(0, -2, -\sqrt{3}) = (600 + 24\sqrt{3})° \approx 641.6°$. Therefore $\left(\pm\frac{4}{3}, -\frac{4}{3}, -\frac{4}{3}\right)$ are the hottest points on the space probe.

31. $\nabla U = (y + 2)\mathbf{i} + x\mathbf{j}$ and $\nabla g = 2\mathbf{i} + \mathbf{j}$ so that $\nabla U = \lambda \nabla g \Rightarrow (y + 2)\mathbf{i} + x\mathbf{j} = \lambda(2\mathbf{i} + \mathbf{j}) \Rightarrow y + 2 = 2\lambda$ and $x = \lambda \Rightarrow y + 2 = 2x \Rightarrow y = 2x - 2 \Rightarrow 2x + (2x - 2) = 30 \Rightarrow x = 8$ and $y = 14$. Therefore $U(8, 14) = \$128$ is the maximum value of U under the constraint.

33. Let $g_1(x, y, z) = 2x - y = 0$ and $g_2(x, y, z) = y + z = 0 \Rightarrow \nabla g_1 = 2\mathbf{i} - \mathbf{j}$, $\nabla g_2 = \mathbf{j} + \mathbf{k}$, and $\nabla f = 2x\mathbf{i} + 2\mathbf{j} - 2z\mathbf{k}$ so that $\nabla f = \lambda \nabla g_1 + \mu \nabla g_2 \Rightarrow 2x\mathbf{i} + 2\mathbf{j} - 2z\mathbf{k} = \lambda(2\mathbf{i} - \mathbf{j}) + \mu(\mathbf{j} + \mathbf{k}) \Rightarrow 2x\mathbf{i} + 2\mathbf{j} - 2z\mathbf{k} = 2\lambda\mathbf{i} + (\mu - \lambda)\mathbf{j} + \mu\mathbf{k} \Rightarrow 2x = 2\lambda$, $2 = \mu - \lambda$, and $-2z = \mu \Rightarrow x = \lambda$. Then $2 = -2z - x \Rightarrow x = -2z - 2$ so that $2x - y = 0 \Rightarrow 2(-2z - 2) - y = 0 \Rightarrow -4z - 4 - y = 0$. This equation coupled with $y + z = 0$ implies $z = -\frac{4}{3}$ and $y = \frac{4}{3}$. Then $x = \frac{2}{3}$ so that $\left(\frac{2}{3}, \frac{4}{3}, -\frac{4}{3}\right)$ is the point that gives the maximum value $f\left(\frac{2}{3}, \frac{4}{3}, -\frac{4}{3}\right) = \left(\frac{2}{3}\right)^2 + 2\left(\frac{4}{3}\right) - \left(-\frac{4}{3}\right)^2 = \frac{4}{3}$

35. Let $f(x, y, z) = x^2 + y^2 + z^2$ be the square of the distance from the origin. We want to minimize $f(x, y, z)$ subject to the constraints $g_1(x, y, z) = y + 2z - 12 = 0$ and $g_2(x, y, z) = x + y - 6 = 0$. Thus $\nabla f = 2x\mathbf{i} + 2y\mathbf{j} + 2z\mathbf{k}$, $\nabla g_1 = \mathbf{j} + 2\mathbf{k}$, and $\nabla g_2 = \mathbf{i} + \mathbf{j}$ so that $\nabla f = \lambda \nabla g_1 + \mu \nabla g_2 \Rightarrow 2x = \mu$, $2y = \lambda + \mu$, $2z = 2\lambda$. Then $0 = y + 2z - 12 = \left(\frac{\lambda}{2} + \frac{\mu}{2}\right) + 2\lambda - 12 \Rightarrow \frac{5}{2}\lambda + \frac{1}{2}\mu = 12 \Rightarrow 5\lambda + \mu = 24$; $0 = x + y - 6 = \frac{\mu}{2} + \left(\frac{\lambda}{2} + \frac{\mu}{2}\right) - 6 \Rightarrow \frac{1}{2}\lambda + \mu = 6 \Rightarrow \lambda + 2\mu = 12$. Solving these two equations for λ and μ gives $\lambda = 4$ and $\mu = 4 \Rightarrow x = \frac{\mu}{2} = 2$, $y = \frac{\lambda + \mu}{2} = 4$, and $z = \lambda = 4$. The point $(2, 4, 4)$ on the line of intersection is closest to the origin. (There is no maximum distance from the origin since points on the line can be arbitrarily far away.)

37. Let $g_1(x, y, z) = z - 1 = 0$ and $g_2(x, y, z) = x^2 + y^2 + z^2 - 10 = 0 \Rightarrow \nabla g_1 = \mathbf{k}$, $\nabla g_2 = 2x\mathbf{i} + 2y\mathbf{j} + 2z\mathbf{k}$, and $\nabla f = 2xyz\mathbf{i} + x^2z\mathbf{j} + x^2y\mathbf{k}$ so that $\nabla f = \lambda \nabla g_1 + \mu \nabla g_2 \Rightarrow 2xyz\mathbf{i}$

$+ x^2z\mathbf{j} + x^2y\mathbf{k} = \lambda(\mathbf{k}) + \mu(2x\mathbf{i} + 2y\mathbf{j} + 2z\mathbf{k}) \Rightarrow 2xyz = 2x\mu$, $x^2z = 2y\mu$, and $x^2y = 2z\mu + \lambda \Rightarrow xyz = x\mu \Rightarrow x = 0$ or $yz = \mu \Rightarrow \mu = y$ since $z = 1$.

CASE 1: $x = 0$ and $z = 1 \Rightarrow y^2 - 9 = 0$ (from g_2) $\Rightarrow y = \pm 3$ yielding the points $(0, \pm 3, 1)$.

CASE 2: $\mu = y \Rightarrow x^2z = 2y^2 \Rightarrow x^2 = 2y^2$ (since $z = 1$) $\Rightarrow 2y^2 + y^2 + 1 - 10 = 0$ (from g_2) $\Rightarrow 3y^2 - 9 = 0 \Rightarrow y = \pm\sqrt{3} \Rightarrow x^2 = 2(\pm\sqrt{3})^2 \Rightarrow x = \pm\sqrt{6}$ yielding the points $(\pm\sqrt{6}, \pm\sqrt{3}, 1)$.

Now $f(0, \pm 3, 1) = 1$ and $f(\pm\sqrt{6}, \pm\sqrt{3}, 1) = 6(\pm\sqrt{3}) + 1 = 1 \pm 6\sqrt{3}$. Therefore the maximum of f is $1 + 6\sqrt{3}$ at $(\pm\sqrt{6}, \sqrt{3}, 1)$, and the minimum of f is $1 - 6\sqrt{3}$ at $(\pm\sqrt{6}, -\sqrt{3}, 1)$.

39. Let $g_1(x, y, z) = y - x = 0$ and $g_2(x, y, z) = x^2 + y^2 + z^2 - 4 = 0$. Then $\nabla f = y\mathbf{i} + x\mathbf{j} + 2z\mathbf{k}$, $\nabla g_1 = -\mathbf{i} + \mathbf{j}$, and $\nabla g_2 = 2x\mathbf{i} + 2y\mathbf{j} + 2z\mathbf{k}$ so that $\nabla f = \lambda \nabla g_1 + \mu \nabla g_2 \Rightarrow y\mathbf{i} + x\mathbf{j} + 2z\mathbf{k} = \lambda(-\mathbf{i} + \mathbf{j}) + \mu(2x\mathbf{i} + 2y\mathbf{j} + 2z\mathbf{k})$ $\Rightarrow y = -\lambda + 2x\mu$, $x = \lambda + 2y\mu$, and $2z = 2z\mu \Rightarrow z = 0$ or $\mu = 1$.

CASE 1: $z = 0 \Rightarrow x^2 + y^2 - 4 = 0 \Rightarrow 2x^2 - 4 = 0$ (since $x = y$) $\Rightarrow x = \pm\sqrt{2}$ and $y = \pm\sqrt{2}$ yielding the points $(\pm\sqrt{2}, \pm\sqrt{2}, 0)$.

CASE 2: $\mu = 1 \Rightarrow y = -\lambda + 2x$ and $x = \lambda + 2y \Rightarrow x + y = 2(x + y) \Rightarrow 2x = 2(2x)$ since $x = y \Rightarrow x = 0 \Rightarrow y = 0 \Rightarrow z^2 - 4 = 0 \Rightarrow z = \pm 2$ yielding the points $(0, 0, \pm 2)$.

Now, $f(0, 0, \pm 2) = 4$ and $f(\pm\sqrt{2}, \pm\sqrt{2}, 0) = 2$. Therefore the maximum value of f is 4 at $(0, 0, \pm 2)$ and the minimum value of f is 2 at $(\pm\sqrt{2}, \pm\sqrt{2}, 0)$.

41. $\nabla f = \mathbf{i} + \mathbf{j}$ and $\nabla g = y\mathbf{i} + x\mathbf{j}$ so that $\nabla f = \lambda \nabla g \Rightarrow \mathbf{i} + \mathbf{j} = \lambda(y\mathbf{i} + x\mathbf{j}) \Rightarrow 1 = y\lambda$ and $1 = x\lambda \Rightarrow y = x \Rightarrow y^2 = 16 \Rightarrow y = \pm 4 \Rightarrow (4, 4)$ and $(-4, -4)$ are candidates for the location of extreme values. But as $x \to \infty$, $y \to 0$ and $f(x, y) \to \infty$; as $x \to -\infty$, $y \to 0$ and $f(x, y) \to -\infty$. Therefore no maximum or minimum value exists subject to the constraint.

43 a. Maximize $f(a, b, c) = a^2b^2c^2$ subject to $a^2 + b^2 + c^2 = r^2$. Thus $\nabla f = 2ab^2c^2\mathbf{i} + 2a^2bc^2\mathbf{j} + 2a^2b^2c\mathbf{k}$ and $\nabla g = 2a\mathbf{i} + 2b\mathbf{j} + 2c\mathbf{k}$ so that $\nabla f = \lambda \nabla g \Rightarrow$ $2ab^2c^2 = 2a\lambda$, $2a^2bc^2 = 2b\lambda$, and $2a^2b^2c = 2c\lambda$ $\Rightarrow 2a^2b^2c^2 = 2a^2\lambda = 2b^2\lambda = 2c^2\lambda \Rightarrow \lambda = 0$ or $a^2 = b^2 = c^2$.

CASE 1: $\lambda = 0 \Rightarrow a^2b^2c^2 = 0$.

CASE 2: $a^2 = b^2 = c^2 \Rightarrow f(a, b, c) = a^2a^2a^2$ and $3a^2 = r^2 \Rightarrow f(a, b, c) = \left(\frac{r^2}{3}\right)^3$ is the maximum value.

b. The point $(\sqrt{a}, \sqrt{b}, \sqrt{c})$ is on the sphere if $a + b + c = r^2$. Moreover, by part (a), $abc = f(\sqrt{a}, \sqrt{b}, \sqrt{c}) \le \left(\frac{r^2}{3}\right)^3 \Rightarrow (abc)^{1/3} \le \frac{r^2}{3} = \frac{a + b + c}{3}$, as claimed.

45–49. Example CAS commands:

Maple:

```
f:=(x,y,z)->x*y+y*z;
g1:=(x,y,z)->x^2+y^2-2;
g2:=(x,y,z)->x^2+z^2-2;
lambda1:='lambda1':lambda2:='lambda2':
h:=(x,y,z)->f(x,y,z)-lambda1*g1(x,y,z)-
   lambda2*g2(x,y,z);
expn1:=diff(h(x,y,z),x)=0;
expn2:=diff(h(x,y,z),y)=0;
expn3:=diff(h(x,y,z),z)=0;
expn4:=diff(h(x,y,z),lambda1)=0;
expn5:=diff(h(x,y,z),lambda2)=0;
s:=evalf(solve({expn1,expn2,expn3,expn4,
   expn5},{x,y,z,lambda1,lambda2}));
subs(s[1],f(x,y,z));
subs(s[2],f(x,y,z));
```

Mathematica:

```
Clear[w,x,y,z,L,M]
f[x_,y_,z_] = x y + y z
g1[x_,y_,z_] = x^2 + y^2 - 2
g2[x_,y_,z_] = x^2 + z^2 - 2
h = f[x,y,z] - L g1[x,y,z] -
  M g2[x,y,z]
hx = D[h,x]
hy = D[h,y]
hz = D[h,z]
hL = D[h,L]
hM = D[h,M]
crit = NSolve[{hx==0,hy==0,hz==0,hM==0,
   hM==0}]
{{x,y,z},f[x,y,z]} /. crit
```

Chapter 13 Review Exercises

(pp. 756– 759)

1. Domain: All points in the xy-plane
Range: $z \ge 0$

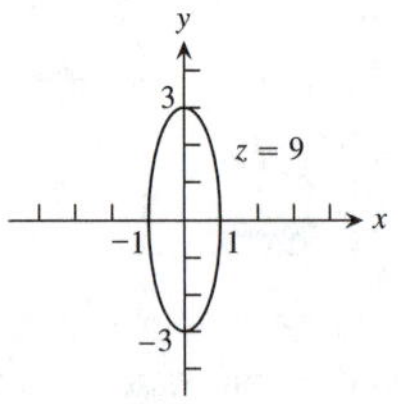

Level curves are ellipses with major axis along the y-axis and minor axis along the x-axis.

2. Domain: All points in the xy-plane
Range: $0 < z < \infty$

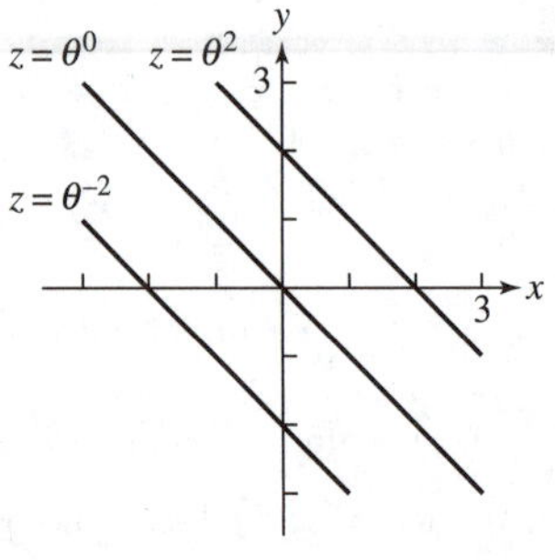

Level curves are the straight lines $x + y = \ln z$ with slope -1, and $z > 0$.

3. Domain: All (x, y) such that $x \neq 0$ and $y \neq 0$
Range: $z \neq 0$

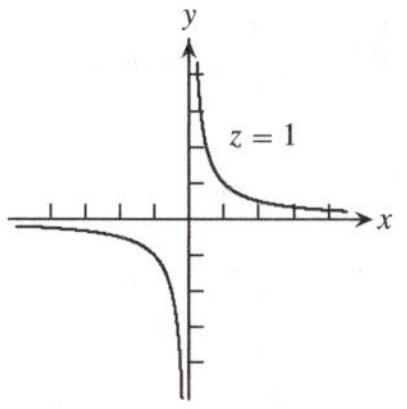

Level curves are hyperbolas with the x- and y-axes as asymptotes.

4. Domain: All (x, y) so that $x^2 - y \geq 0$
Range: $z \geq 0$

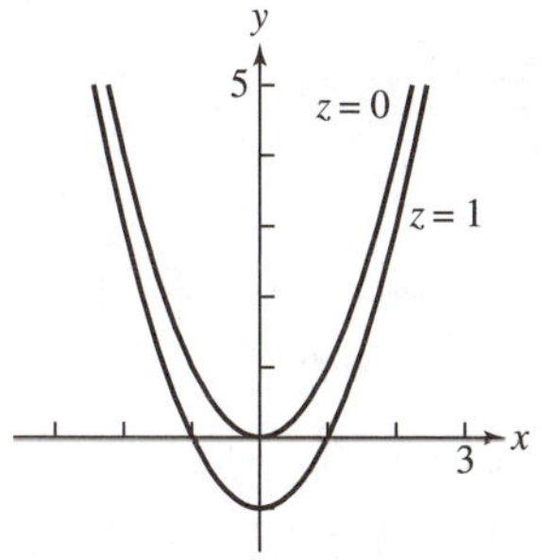

Level curves are the parabolas $y = x^2 - c,\ c \geq 0$.

5. Domain: All points (x, y, z) in space.
Range: All real numbers

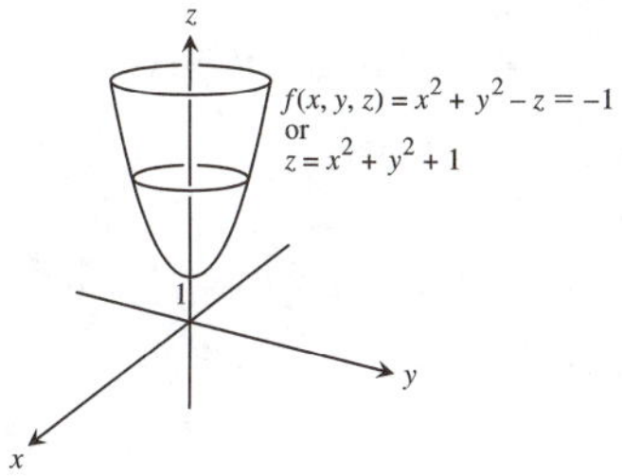

Level surfaces are paraboloids of revolution with the z-axis as axis.

6. Domain: All points (x, y, z) in space
Range: Nonnegative real numbers

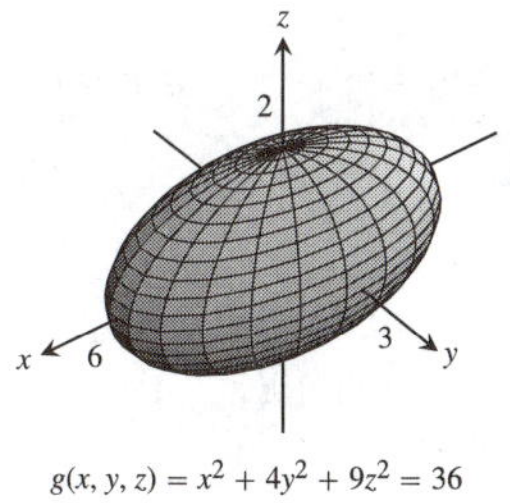

Level surfaces are ellipsoids with center $(0, 0, 0)$.

7. Domain: All (x, y, z) such that $(x, y, z) \neq (0, 0, 0)$
Range: Positive real numbers

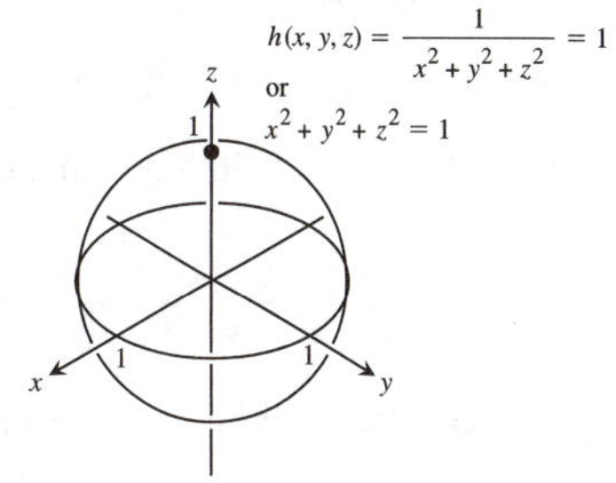

Level surfaces are spheres with center $(0, 0, 0)$ and radius $r > 0$.

8. Domain: All points (x, y, z) in space
Range: $(0, 1]$

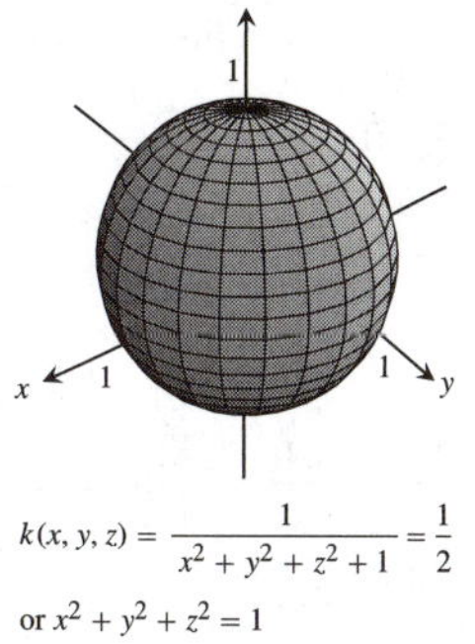

Level surfaces are spheres with center $(0, 0, 0)$ and radius $r > 0$.

9. $\displaystyle\lim_{(x, y) \to (\pi, \ln 2)} e^y \cos x = e^{\ln 2} \cos \pi = (2)(-1) = -2$

10. $\displaystyle\lim_{(x, y) \to (0, 0)} \frac{2 + y}{x + \cos y} = \frac{2 + 0}{0 + \cos 0} = 2$

11. $\displaystyle\lim_{\substack{(x, y) \to (1, 1) \\ x \neq \pm y}} \frac{x - y}{x^2 - y^2} = \lim_{\substack{(x, y) \to (1, 1) \\ x \neq \pm y}} \frac{x - y}{(x - y)(x + y)}$
$\displaystyle= \lim_{(x, y) \to (1, 1)} \frac{1}{x + y} = \frac{1}{1 + 1} = \frac{1}{2}$

12. $\displaystyle\lim_{(x, y) \to (1, 1)} \frac{x^3y^3 - 1}{xy - 1} = \lim_{(x, y) \to (1, 1)} \frac{(xy - 1)(x^2y^2 + xy + 1)}{xy - 1}$
$\displaystyle= \lim_{(x, y) \to (1, 1)} (x^2y^2 + xy + 1) = 1^2 \cdot 1^2 + 1 \cdot 1 + 1 = 3$

13. $\displaystyle\lim_{P \to (1, -1, e)} \ln |x + y + z| = \ln |1 + (-1) + e| = \ln e = 1$

14. $\displaystyle\lim_{P \to (1, -1, -1)} \tan^{-1}(x + y + z) = \tan^{-1}(1 + (-1) + (-1))$
$\displaystyle= \tan^{-1}(-1) = -\frac{\pi}{4}$

15. Let $y = kx^2,\ k \neq 1$. Then $\displaystyle\lim_{(x, y) \to (0, 0)} \frac{y}{x^2 - y}$
$\displaystyle= \lim_{x \to 0} \frac{kx^2}{x^2 - kx^2} = \frac{k}{1 - k^2}$ which gives different limits for different values of $k \Rightarrow$ the limit does not exist.

16. Let $y = kx$, $k \neq 0$. Then $\lim_{\substack{(x, y) \to (0, 0) \\ xy \neq 0}} \frac{x^2 + y^2}{xy}$
$= \lim_{x \to 0} \frac{x^2 + (kx)^2}{x(kx)} = \frac{1 + k^2}{k}$ which gives different limits for different values of $k \Rightarrow$ the limit does not exist.

17. a. Let $y = kx$. Then $\lim_{(x, y) \to (0, 0)} \frac{x^2 - y^2}{x^2 + y^2} = \lim_{x \to 0} \frac{x^2 - k^2x^2}{x^2 + k^2x^2}$
$= \frac{1 - k^2}{1 + k^2}$ which gives different limits for different values of $k \Rightarrow$ the limit does not exist so $f(0, 0)$ cannot be defined in a way that makes f continuous at the origin.

b. Along the x-axis, $y = 0$ and $\lim_{(x, y) \to (0, 0)} \frac{\sin(x - y)}{|x + y|}$
$= \lim_{x \to 0} \frac{\sin x}{|x|} = \begin{cases} 1, x > 0 \\ -1, x < 0 \end{cases}$, so the limit fails to exist
$\Rightarrow f$ is not continuous at $(0, 0)$.

18. a. $\lim_{r \to 0} \frac{\sin 6r}{6r} = \lim_{t \to 0} \frac{\sin t}{t} = 1$, where $t = 6r$

b. $f_r(0, 0) = \lim_{h \to 0} \frac{f(0 + h, 0) - f(0, 0)}{h} = \lim_{h \to 0} \frac{\left(\frac{\sin 6h}{6h}\right) - 1}{h}$
$= \lim_{h \to 0} \frac{\sin 6h - 6h}{6h^2} = \lim_{h \to 0} \frac{6 \cos 6h - 6}{12h}$
$= \lim_{h \to 0} \frac{-36 \sin 6h}{12} = 0$ (applying l'Hôpital's rule twice)

c. $f_\theta(r, \theta) = \lim_{h \to 0} \frac{f(r, \theta + h) - f(r, \theta)}{h}$
$= \lim_{h \to 0} \frac{\left(\frac{\sin 6r}{6r}\right) - \left(\frac{\sin 6r}{6r}\right)}{h} = \lim_{h \to 0} \frac{0}{h} = 0$

19. $\frac{\partial g}{\partial r} = \cos\theta + \sin\theta$, $\frac{\partial g}{\partial \theta} = -r\sin\theta + r\cos\theta$

20. $\frac{\partial f}{\partial x} = \frac{1}{2}\left(\frac{2x}{x^2 + y^2}\right) + \frac{\left(-\frac{y}{x^2}\right)}{1 + \left(\frac{y}{x}\right)^2} = \frac{x}{x^2 + y^2} - \frac{y}{x^2 + y^2}$
$= \frac{x - y}{x^2 + y^2}$, $\frac{\partial f}{\partial y} = \frac{1}{2}\left(\frac{2y}{x^2 + y^2}\right) + \frac{\left(\frac{1}{x}\right)}{1 + \left(\frac{y}{x}\right)^2} = \frac{y}{x^2 + y^2}$
$+ \frac{x}{x^2 + y^2} = \frac{x + y}{x^2 + y^2}$

21. $\frac{\partial f}{\partial R_1} = -\frac{1}{R_1^2}$, $\frac{\partial f}{\partial R_2} = -\frac{1}{R_2^2}$, $\frac{\partial f}{\partial R_3} = -\frac{1}{R_3^2}$

22. $h_x(x, y, z) = 2\pi \cos(2\pi x + y - 3z)$, $h_y(x, y, z)$
$= \cos(2\pi x + y - 3z)$, $h_z(x, y, z) = -3\cos(2\pi x + y - 3z)$

23. $\frac{\partial P}{\partial n} = \frac{RT}{V}$, $\frac{\partial P}{\partial R} = \frac{nT}{V}$, $\frac{\partial P}{\partial T} = \frac{nR}{V}$, $\frac{\partial P}{\partial V} = -\frac{nRT}{V^2}$

24. $f_r(r, \ell, T, w) = -\frac{1}{2r^2\ell}\sqrt{\frac{T}{\pi w}}$, $f_\ell(r, \ell, T, w) = -\frac{1}{2r\ell^2}\sqrt{\frac{T}{\pi w}}$,
$f_T(r, \ell, T, w) = \left(\frac{1}{2r\ell}\right)\left(\frac{1}{\sqrt{\pi w}}\right)\left(\frac{1}{2\sqrt{T}}\right) = \frac{1}{4r\ell}\sqrt{\frac{1}{T\pi w}}$
$= \frac{1}{4r\ell T}\sqrt{\frac{T}{\pi w}}$, $f_w(r, \ell, T, w) = \left(\frac{1}{2r\ell}\right)\sqrt{\frac{T}{\pi}}\left(-\frac{1}{2}w^{-3/2}\right)$
$= -\frac{1}{4r\ell w}\sqrt{\frac{T}{\pi w}}$

25. $\frac{\partial g}{\partial x} = \frac{1}{y}$, $\frac{\partial g}{\partial y} = 1 - \frac{x}{y^2}$, $\frac{\partial^2 g}{\partial x^2} = 0$, $\frac{\partial^2 g}{\partial y^2} = \frac{2x}{y^3}$, $\frac{\partial^2 g}{\partial y \partial x} = \frac{\partial^2 g}{\partial x \partial y}$
$= -\frac{1}{y^2}$

26. $g_x(x, y) = e^x + y\cos x$, $g_y(x, y) = \sin x \Rightarrow g_{xx}(x, y)$
$= e^x - y\sin x$, $g_{yy}(x, y) = 0$, $g_{xy}(x, y) = g_{xy}(x, y) = \cos x$

27. $\frac{\partial f}{\partial x} = 1 + y - 15x^2 + \frac{2x}{x^2 + 1}$, $\frac{\partial f}{\partial y} = x \Rightarrow \frac{\partial^2 f}{\partial x^2} = -30x$
$+ \frac{2 - 2x^2}{(x^2 + 1)^2}$, $\frac{\partial^2 f}{\partial y^2} = 0$, $\frac{\partial^2 f}{\partial y \partial x} = \frac{\partial^2 f}{\partial x \partial y} = 1$

28. $f_x(x, y) = -3y$, $f_y(x, y) = 2y - 3x - \sin y + 7e^y \Rightarrow f_{xx}(x, y)$
$= 0$, $f_{yy}(x, y) = 2 - \cos y + 7e^y$, $f_{xy}(x, y) = f_{yx}(x, y) = -3$

29. $f\left(\frac{\pi}{4}, \frac{\pi}{4}\right) = \frac{1}{2}$, $f_x\left(\frac{\pi}{4}, \frac{\pi}{4}\right) = \cos x \cos y|_{(\pi/4, \pi/4)} = \frac{1}{2}$, $f_y\left(\frac{\pi}{4}, \frac{\pi}{4}\right)$
$= -\sin x \sin y|_{(\pi/4, \pi/4)} = -\frac{1}{2} \Rightarrow L(x, y) = \frac{1}{2} + \frac{1}{2}\left(x - \frac{\pi}{4}\right)$
$- \frac{1}{2}\left(y - \frac{\pi}{4}\right) = \frac{1}{2} + \frac{1}{2}x - \frac{1}{2}y$; $f_{xx}(x, y) = -\sin x \cos y$,
$f_{yy}(x, y) = -\sin x \cos y$, and $f_{xy}(x, y) = -\cos x \sin y$. Thus an upper bound for E depends on the bound M used for $|f_{xx}|$, $|f_{xy}|$, and $|f_{yy}|$. With $M = \frac{\sqrt{2}}{2}$ we have $|E(x, y)|$
$\leq \frac{1}{2}\left(\frac{\sqrt{2}}{2}\right)\left(\left|x - \frac{\pi}{4}\right| + \left|y - \frac{\pi}{4}\right|\right)^2 \leq \frac{\sqrt{2}}{4}(0.2)^2 \leq 0.0142$;
with $M = 1$, $|E(x, y)| \leq \frac{1}{2}(1)\left(\left|x - \frac{\pi}{4}\right| + \left|y - \frac{\pi}{4}\right|\right)^2$
$= \frac{1}{2}(0.2)^2 = 0.02$.

30. $f(1, 1) = 0$, $f_x(1, 1) = y|_{(1, 1)} = 1$, $f_y(1, 1) = x - 6y|_{(1, 1)}$
$= -5 \Rightarrow L(x, y) = (x - 1) - 5(y - 1) = x - 5y + 4$;
$f_{xx}(x, y) = 0$, $f_{yy}(x, y) = -6$, and $f_{xy}(x, y) = 1 \Rightarrow$ maximum of $|f_{xx}|$, $|f_{yy}|$, and $|f_{xy}|$ is $6 \Rightarrow M = 6 \Rightarrow |E(x, y)|$
$\leq \frac{1}{2}(6)(|x - 1| + |y - 1|)^2 = \frac{1}{2}(6)(0.1 + 0.2)^2 = 0.27$

31. $f(1, 0, 0) = 0$, $f_x(1, 0, 0) = y - 3z|_{(1, 0, 0)} = 0$, $f_y(1, 0, 0)$
$= x + 2z|_{(1, 0, 0)} = 1$, $f_z(1, 0, 0) = 2y - 3x|_{(1, 0, 0)} = -3$
$\Rightarrow L(x, y, z) = 0(x - 1) + (y - 0) - 3(z - 0) = y - 3z$;
$f(1, 1, 0) = 1$, $f_x(1, 1, 0) = 1$, $f_y(1, 1, 0) = 1$, $f_z(1, 1, 0)$
$= -1 \Rightarrow L(x, y, z) = 1 + (x - 1) + (y - 1) - 1(z - 0)$
$= x + y - z - 1$

32. $f\left(0, 0, \frac{\pi}{4}\right) = 1, f_x\left(0, 0, \frac{\pi}{4}\right) = -\sqrt{2}\sin x \sin(y+z)\Big|_{\left(0,0,\frac{\pi}{4}\right)}$ $= 0, f_y\left(0, 0, \frac{\pi}{4}\right) = \sqrt{2}\cos x\cos(y+z)\Big|_{\left(0,0,\frac{\pi}{4}\right)} = 1,$ $f_z\left(0, 0, \frac{\pi}{4}\right) = \sqrt{2}\cos x\cos(y+z)\Big|_{\left(0,0,\frac{\pi}{4}\right)} = 1 \Rightarrow L(x, y, z)$ $= 1 + 1(y-0) + 1\left(z - \frac{\pi}{4}\right) = 1 + y + z - \frac{\pi}{4}; f\left(\frac{\pi}{4}, \frac{\pi}{4}, 0\right)$ $= \frac{\sqrt{2}}{2}, f_x\left(\frac{\pi}{4}, \frac{\pi}{4}, 0\right) = -\frac{\sqrt{2}}{2}, f_y\left(\frac{\pi}{4}, \frac{\pi}{4}, 0\right) = \frac{\sqrt{2}}{2}, f_z\left(\frac{\pi}{4}, \frac{\pi}{4}, 0\right)$ $= \frac{\sqrt{2}}{2} \Rightarrow L(x, y, z) = \frac{\sqrt{2}}{2} - \frac{\sqrt{2}}{2}\left(x - \frac{\pi}{4}\right) + \frac{\sqrt{2}}{2}\left(y - \frac{\pi}{4}\right)$ $+ \frac{\sqrt{2}}{2}(z - 0) = \frac{\sqrt{2}}{2} - \frac{\sqrt{2}}{2}x + \frac{\sqrt{2}}{2}y + \frac{\sqrt{2}}{2}z$

33. $V = \pi r^2 h \Rightarrow dV = 2\pi rh\,dr + \pi r^2\,dh \Rightarrow dV\big|_{(1.5,\,5280)}$ $= 2\pi(1.5)(5280)\,dr + \pi(1.5)^2\,dh = 15{,}840\pi\,dr$ $+ 2.25\pi\,dh$. You should be more careful with the diameter since it has a greater effect on dV.

34. $df = (2x - y)dx + (-x + 2y)dy \Rightarrow df\big|_{(1,2)} = 3\,dy \Rightarrow f$ is more sensitive to changes in y; in fact, near the point (1, 2) a change in x does not change f.

35. $dI = \frac{1}{R}\,dV - \frac{V}{R^2}\,dR \Rightarrow dI\big|_{(24,\,100)} = \frac{1}{100}\,dV - \frac{24}{100^2}\,dR$

$\Rightarrow dI\big|_{dV=-1,\,dR=-20} = -0.01 + (480)(0.0001) = 0.038,$

or increases by 0.038 amps; % change in $V = (100)\left(-\frac{1}{24}\right)$

$\approx -4.17\%$; % change in $R = \left(-\frac{20}{100}\right)(100) = -20\%$;

$I = \frac{24}{100} = 0.24 \Rightarrow$ estimated % change in $I = \frac{dI}{I} \times 100$

$= \frac{0.038}{0.24} \times 100 \approx 15.83\%$

36. $A = \pi ab \Rightarrow dA = \pi b\,da + \pi a\,db \Rightarrow dA\big|_{(10,\,16)} = 16\pi\,da$ $+ 10\pi\,db$; $da = \pm 0.1$ and $db = \pm 0.1 \Rightarrow dA = \pm 26\pi(0.1)$ $= \pm 2.6\pi$ and $A = \pi(10)(16) = 160\pi \Rightarrow \left|\frac{dA}{A} \times 100\right|$

$= \left|\frac{2.6\pi}{160\pi} \times 100\right| \approx 1.625\%$

37. a. $y = uv \Rightarrow dy = v\,du + u\,dv$; percentage change in $u \le 2\% \Rightarrow |du| \le 0.02$, and percentage change in $v \le 3\% \Rightarrow |dv| \le 0.03$; $\frac{dy}{y} = \frac{v\,du + u\,dv}{uv} = \frac{du}{u} + \frac{dv}{v}$

$\Rightarrow \left|\frac{dy}{y} \times 100\right| = \left|\frac{du}{u} \times 100 + \frac{dv}{v} \times 100\right|$

$\le \left|\frac{du}{u} \times 100\right| + \left|\frac{dv}{v} \times 100\right| \le 2\% + 3\% = 5\%$

b. $z = u + v \Rightarrow \frac{dz}{z} = \frac{du + dv}{u + v} = \frac{du}{u + v} + \frac{dv}{u + v}$

$\le \frac{du}{u} + \frac{dv}{v}$ (since $u > 0, v > 0$) $\Rightarrow \left|\frac{dz}{z} \times 100\right|$

$\le \left|\frac{du}{u} \times 100 + \frac{dv}{v} \times 100\right| = \left|\frac{dy}{y} \times 100\right|$

38. $C = \frac{7}{71.84w^{0.425}h^{0.725}} \Rightarrow C_w = \frac{(-0.425)(7)}{71.84w^{1.425}h^{0.725}}$ and $C_h = \frac{(-0.725)(7)}{71.84w^{0.425}h^{1.725}} \Rightarrow dC = \frac{-2.975}{71.84w^{1.425}h^{0.725}}\,dw$ $+ \frac{-5.075}{71.84w^{0.425}h^{1.725}}\,dh$; thus when $w = 70$ and $h = 180$ we have $dC\big|_{(70,\,180)} \approx -(0.00000225)dw - (0.00000149)\,dh$

$\Rightarrow$ 1 kg error in weight has more effect

39. $\frac{\partial w}{\partial x} = y\cos(xy + \pi), \frac{\partial w}{\partial y} = x\cos(xy + \pi), \frac{dx}{dt} = e^t, \frac{dy}{dt} =$ $\frac{1}{t+1} \Rightarrow \frac{dw}{dt} = [y\cos(xy + \pi)]e^t$ $+ [x\cos(xy + \pi)]\left(\frac{1}{t+1}\right)$; $t = 0 \Rightarrow x = 1$ and $y = 0$ $\Rightarrow \frac{dw}{dt}\Big|_{t=0} = 0 \cdot 1 + [1 \cdot (-1)]\left(\frac{1}{0+1}\right) = -1$

40. $\frac{\partial w}{\partial u} = \frac{dw}{dx}\frac{\partial x}{\partial u} = \left(\frac{x}{1 + x^2} - \frac{1}{x^2 + 1}\right)(2e^u\cos v)$; $u = v = 0$ $\Rightarrow x = 2 \Rightarrow \frac{\partial w}{\partial u}\Big|_{(0,0)} = \left(\frac{2}{5} - \frac{1}{5}\right)(2) = \frac{2}{5}; \frac{\partial w}{\partial v} = \frac{dw}{dx}\frac{\partial x}{\partial v}$ $= \left(\frac{x}{1 + x^2} - \frac{1}{x^2 + 1}\right)(-2e^u\sin v) \Rightarrow \frac{\partial w}{\partial v}\Big|_{(0,0)} = \left(\frac{2}{5} - \frac{1}{5}\right)(0)$ $= 0$

41. $\frac{\partial f}{\partial x} = y + z, \frac{\partial f}{\partial y} = x + z, \frac{\partial f}{\partial z} = y + x, \frac{dx}{dt} = -\sin t,$ $\frac{dy}{dt} = \cos t, \frac{dz}{dt} = -2\sin 2t \Rightarrow \frac{df}{dt} = -(y + z)(\sin t)$ $+ (x + z)(\cos t) - 2(y + x)(\sin 2t)$; $t = 1 \Rightarrow x = \cos 1,$ $y = \sin 1$, and $z = \cos 2 \Rightarrow \frac{df}{dt}\Big|_{t=1}$ $= -(\sin 1 + \cos 2)(\sin 1) + (\cos 1 + \cos 2)(\cos 1)$ $- 2(\sin 1 + \cos 1)(\sin 2)$

42. $\frac{\partial w}{\partial x} = \frac{dw}{ds}\frac{\partial s}{\partial x} = (5)\frac{dw}{ds}$ and $\frac{\partial w}{\partial y} = \frac{dw}{ds}\frac{\partial s}{\partial y} = (1)\frac{dw}{ds}$ $\Rightarrow \frac{\partial w}{\partial x} - 5\frac{\partial w}{\partial y} = 5\frac{dw}{ds} - 5\frac{dw}{ds} = 0$

43. $F(x, y) = 1 - x - y^2 - \sin xy \Rightarrow F_x = -1 - y\cos xy$ and $F_y = -2y - x\cos xy \Rightarrow \frac{dy}{dx} = -\frac{F_x}{F_y} = -\frac{-1 - y\cos xy}{-2y - x\cos xy}$ $= \frac{1 + y\cos xy}{-2y - x\cos xy} \Rightarrow$ at $(x, y) = (0, 1)$ we have $\frac{dy}{dx}\Big|_{(0,1)}$ $= \frac{1 + 1}{-2} = -1$

44. $F(x, y) = 2xy + e^{x+y} - 2 \Rightarrow F_x = 2y + e^{x+y}$ and $F_y = 2x + e^{x+y} \Rightarrow \frac{dy}{dx} = -\frac{F_x}{F_y} = -\frac{2y + e^{x+y}}{2x + e^{x+y}} \Rightarrow$ at (x, y) $= (0, \ln 2)$ we have $\frac{dy}{dx}\Big|_{(0,\ln 2)} = -\frac{2\ln 2 + 2}{0 + 2}$ $= -(\ln 2 + 1)$

45. $\nabla f = (-\sin x\cos y)\mathbf{i} - (\cos x\sin y)\mathbf{j} \Rightarrow \nabla f\Big|_{\left(\frac{\pi}{4},\frac{\pi}{4}\right)} = -\frac{1}{2}\mathbf{i}$ $- \frac{1}{2}\mathbf{j} \Rightarrow |\nabla f| = \sqrt{\left(-\frac{1}{2}\right)^2 + \left(-\frac{1}{2}\right)^2} = \frac{1}{\sqrt{2}} = \frac{\sqrt{2}}{2}; \mathbf{u} = \frac{\nabla f}{|\nabla f|}$ $= -\frac{\sqrt{2}}{2}\mathbf{i} - \frac{\sqrt{2}}{2}\mathbf{j} \Rightarrow f$ increases most rapidly in the direction $\mathbf{u} = -\frac{\sqrt{2}}{2}\mathbf{i} - \frac{\sqrt{2}}{2}\mathbf{j}$ and decreases most rapidly in the direction $-\mathbf{u} = \frac{\sqrt{2}}{2}\mathbf{i} + \frac{\sqrt{2}}{2}\mathbf{j}$; and $(D_{\mathbf{u}}f)_{P_0} = |\nabla f|$ $= \frac{\sqrt{2}}{2}; (D_{-\mathbf{u}}f)_{P_0} = -\frac{\sqrt{2}}{2}; \mathbf{u}_1 = \frac{\mathbf{A}}{|\mathbf{A}|} = \frac{3\mathbf{i} + 4\mathbf{j}}{\sqrt{3^2 + 4^2}} = \frac{3}{5}\mathbf{i}$ $+ \frac{4}{5}\mathbf{j} \Rightarrow (D_{\mathbf{u}_1}f)_{P_0} = \nabla f \cdot \mathbf{u}_1 = \left(-\frac{1}{2}\right)\left(\frac{3}{5}\right) + \left(-\frac{1}{2}\right)\left(\frac{4}{5}\right) = -\frac{7}{10}$

46. $\nabla f = 2xe^{-2y}\mathbf{i} - 2x^2e^{-2y}\mathbf{j} \Rightarrow \nabla f\big|_{(1,\,0)} = 2\mathbf{i} - 2\mathbf{j} \Rightarrow |\nabla f|$ $= \sqrt{2^2 + (-2)^2} = 2\sqrt{2}$; $\mathbf{u} = \frac{\nabla f}{|\nabla f|} = \frac{1}{\sqrt{2}}\mathbf{i} - \frac{1}{\sqrt{2}}\mathbf{j} \Rightarrow f$ increases most rapidly in the direction $\mathbf{u} = \frac{1}{\sqrt{2}}\mathbf{i} - \frac{1}{\sqrt{2}}\mathbf{j}$ and decreases most rapidly in the direction $-\mathbf{u} = -\frac{1}{\sqrt{2}}\mathbf{i}$ $+ \frac{1}{\sqrt{2}}\mathbf{j}$; $(D_{\mathbf{u}}f)_{P_0} = |\nabla f| = 2\sqrt{2}$ and $(D_{-\mathbf{u}}f)_{P_0} = -2\sqrt{2}$; $\mathbf{u}_1 = \frac{\mathbf{A}}{|\mathbf{A}|} = \frac{2\mathbf{i} - 2\mathbf{j}}{\sqrt{2^2 + 2^2}} = \frac{1}{\sqrt{2}}\mathbf{i} - \frac{1}{\sqrt{2}}\mathbf{j} \Rightarrow (D_{\mathbf{u}_1}f)_{P_0} = \nabla f \cdot \mathbf{u}_1$ $= (2)\left(\frac{1}{\sqrt{2}}\right) + (-2)\left(\frac{1}{\sqrt{2}}\right) = 0$

47. $\nabla f = \left(\frac{2}{2x + 3y + 6z}\right)\mathbf{i} + \left(\frac{3}{2x + 3y + 6z}\right)\mathbf{j} + \left(\frac{6}{2x + 3y + 6z}\right)\mathbf{k}$ $\Rightarrow \nabla f\big|_{(-1,\,-1,\,1)} = 2\mathbf{i} + 3\mathbf{j} + 6\mathbf{k}$; $\mathbf{u} = \frac{\nabla f}{|\nabla f|}$ $= \frac{2\mathbf{i} + 3\mathbf{j} + 6\mathbf{k}}{\sqrt{2^2 + 3^2 + 6^2}} = \frac{2}{7}\mathbf{i} + \frac{3}{7}\mathbf{j} + \frac{6}{7}\mathbf{k} \Rightarrow f$ increases most rapidly in the direction $\mathbf{u} = \frac{2}{7}\mathbf{i} + \frac{3}{7}\mathbf{j} + \frac{6}{7}\mathbf{k}$ and decreases most rapidly in the direction $-\mathbf{u} = -\frac{2}{7}\mathbf{i} - \frac{3}{7}\mathbf{j} - \frac{6}{7}\mathbf{k}$; $(D_{\mathbf{u}}f)_{P_0} = |\nabla f| = 7$, $(D_{-\mathbf{u}}f)_{P_0} = -7$; $\mathbf{u}_1 = \frac{\mathbf{A}}{|\mathbf{A}|} = \frac{2}{7}\mathbf{i} + \frac{3}{7}\mathbf{j}$ $+ \frac{6}{7}\mathbf{k} \Rightarrow (D_{\mathbf{u}_1}f)_{P_0} = (D_{\mathbf{u}}f)_{P_0} = 7$

48. $\nabla f = (2x + 3y)\mathbf{i} + (3x + 2)\mathbf{j} + (1 - 2z)\mathbf{k} \Rightarrow \nabla f\big|_{(0,\,0,\,0)}$ $= 2\mathbf{j} + \mathbf{k}$; $\mathbf{u} = \frac{\nabla f}{|\nabla f|} = \frac{2}{\sqrt{5}}\mathbf{j} + \frac{1}{\sqrt{5}}\mathbf{k} \Rightarrow f$ increases most rapidly in the direction $\mathbf{u} = \frac{2}{\sqrt{5}}\mathbf{j} + \frac{1}{\sqrt{5}}\mathbf{k}$ and decreases most rapidly in the direction $-\mathbf{u} = -\frac{2}{\sqrt{5}}\mathbf{j} - \frac{1}{\sqrt{5}}\mathbf{k}$; $(D_{\mathbf{u}}f)_{P_0} = |\nabla f| = \sqrt{5}$ and $(D_{-\mathbf{u}}f)_{P_0} = -\sqrt{5}$; $\mathbf{u}_1 = \frac{\mathbf{A}}{|\mathbf{A}|}$ $= \frac{\mathbf{i} + \mathbf{j} + \mathbf{k}}{\sqrt{1^2 + 1^2 + 1^2}} = \frac{1}{\sqrt{3}}\mathbf{i} + \frac{1}{\sqrt{3}}\mathbf{j} + \frac{1}{\sqrt{3}}\mathbf{k} \Rightarrow (D_{\mathbf{u}_1}f)_{P_0}$ $= \nabla f \cdot \mathbf{u}_1 = (0)\left(\frac{1}{\sqrt{3}}\right) + (2)\left(\frac{1}{\sqrt{3}}\right) + (1)\left(\frac{1}{\sqrt{3}}\right) = \frac{3}{\sqrt{3}} = \sqrt{3}$

49. $\mathbf{r}(t) = (\cos 3t)\mathbf{i} + (\sin 3t)\mathbf{j} + 3t\mathbf{k} \Rightarrow \mathbf{v}(t) = (-3 \sin 3t)\mathbf{i}$ $+ (3 \cos 3t)\mathbf{j} + 3\mathbf{k} \Rightarrow \mathbf{v}\left(\frac{\pi}{3}\right) = -3\mathbf{j} + 3\mathbf{k} \Rightarrow \mathbf{u} = -\frac{1}{\sqrt{2}}\mathbf{j}$ $+ \frac{1}{\sqrt{2}}\mathbf{k}$; $f(x, y, z) = xyz \Rightarrow \nabla f = yz\mathbf{i} + xz\mathbf{j} + xy\mathbf{k}$; $t = \frac{\pi}{3}$ yields the point on the helix $(-1, 0, \pi) \Rightarrow \nabla f\big|_{(1,\,0,\,\pi)}$ $= -\pi\mathbf{j} \Rightarrow \nabla f \cdot \mathbf{u} = (-\pi\mathbf{j}) \cdot \left(-\frac{1}{\sqrt{2}}\mathbf{j} + \frac{1}{\sqrt{2}}\mathbf{k}\right) = \frac{\pi}{\sqrt{2}}$

50. $f(x, y, z) = xyz \Rightarrow \nabla f = yz\mathbf{i} + xz\mathbf{j} + xy\mathbf{k}$; at (1, 1, 1) we get $\nabla f = \mathbf{i} + \mathbf{j} + \mathbf{k} \Rightarrow$ the maximum value of $D_{\mathbf{u}}f\big|_{(1,\,1,\,1)}$ $= |\nabla f| = \sqrt{3}$

51. a. Let $\nabla f = a\mathbf{i} + b\mathbf{j}$ at (1, 2). The direction toward (2, 2) is determined by $\mathbf{v}_1 = (2 - 1)\mathbf{i} + (2 - 2)\mathbf{j} = \mathbf{i} = \mathbf{u}$ so that $\nabla f \cdot \mathbf{u} = 2 \Rightarrow a = 2$. The direction toward (1, 1) is determined by $\mathbf{v}_2 + (1 - 1)\mathbf{i} + (1 - 2)\mathbf{j} = -\mathbf{j} = \mathbf{u}$ so that $\nabla f \cdot \mathbf{u} = -2 \Rightarrow -b = -2 \Rightarrow b = 2$. Therefore ∇f $= 2\mathbf{i} + 2\mathbf{j}$.

b. The direction toward (4, 6) is determined by $\mathbf{v}_3$ $= (4 - 1)\mathbf{i} + (6 - 2)\mathbf{j} = 3\mathbf{i} + 4\mathbf{j} \Rightarrow \mathbf{u} = \frac{3}{5}\mathbf{i} + \frac{4}{5}\mathbf{j}$ $\Rightarrow \nabla f \cdot \mathbf{u} = \frac{14}{5}$.

52. a. True

b. False

c. True

d. True

53. $\nabla f = 2x\mathbf{i} + \mathbf{j} + 2z\mathbf{k} \Rightarrow \nabla f\big|_{(0,\,-1,\,-1)} = \mathbf{j} - 2\mathbf{k}$, $\nabla f\big|_{(0,\,0,\,0)} = \mathbf{j}$, $\nabla f\big|_{(0,\,-1,\,1)} = \mathbf{j} + 2\mathbf{k}$

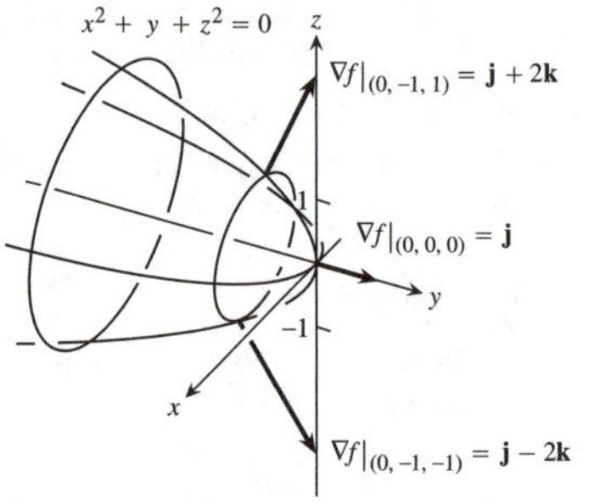

54. $\nabla f = 2y\mathbf{j} + 2z\mathbf{k} \Rightarrow \nabla f\big|_{(2,\,2,\,0)} = 4\mathbf{j}$, $\nabla f\big|_{(2,\,-2,\,0)} = -4\mathbf{j}$, $\nabla f\big|_{(2,\,0,\,2)} = 4\mathbf{k}$, $\nabla f\big|_{(2,\,0,\,-2)} = -4\mathbf{k}$

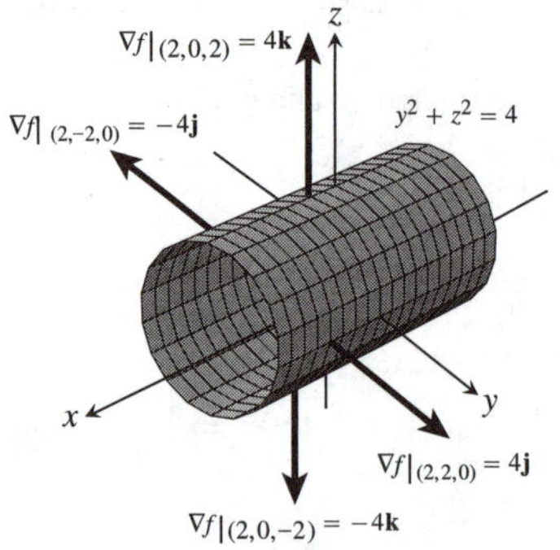

55. $\nabla f = 2x\mathbf{i} - \mathbf{j} - 5\mathbf{k} \Rightarrow \nabla f\big|_{(2,\,-1,\,1)} = 4\mathbf{i} - \mathbf{j} - 5\mathbf{k}$ $\Rightarrow$ Tangent Plane: $4(x - 2) - (y + 1) - 5(z - 1) = 0$ $\Rightarrow 4x - y - 5z = 4$; Normal Line: $x = 2 + 4t$, $y = -1$ $- t$, $z = 1 - 5t$

56. $\nabla f = 2x\mathbf{i} + 2y\mathbf{j} + \mathbf{k} \Rightarrow \nabla f\big|_{(1,\,1,\,2)} = 2\mathbf{i} + 2\mathbf{j} + \mathbf{k}$ $\Rightarrow$ Tangent Plane: $2(x - 1) + 2(y - 1) + (z - 2) = 0$ $\Rightarrow 2x + 2y + z - 6 = 0$; Normal Line: $x = 1 + 2t$, $y = 1$ $+ 2t$, $z = 2 + t$

57. $\frac{\partial z}{\partial x} = \frac{2x}{x^2 + y^2} \Rightarrow \frac{\partial z}{\partial x}\Big|_{(0,\,1,\,0)} = 0$ and $\frac{\partial z}{\partial y} = \frac{2y}{x^2 + y^2}$ $\Rightarrow \frac{\partial z}{\partial y}\Big|_{(0,\,1,\,0)} = 2$; thus the tangent plane is $2(y - 1)$ $- (z - 0) = 0$ or $2y - z - 2 = 0$

58. $\frac{\partial z}{\partial x} = -2x(x^2 + y^2)^{-2} \Rightarrow \frac{\partial z}{\partial x}\Big|_{\left(1,\,1,\,\frac{1}{2}\right)} = -\frac{1}{2}$ and $\frac{\partial z}{\partial y}$ $= -2y(x^2 + y^2)^{-2} \Rightarrow \frac{\partial z}{\partial y}\Big|_{\left(1,\,1,\,\frac{1}{2}\right)} = -\frac{1}{2}$; thus the tangent plane is $-\frac{1}{2}(x-1) - \frac{1}{2}(y-1) - \left(z - \frac{1}{2}\right) = 0$ or $x + y + 2z - 3 = 0$

59. $\nabla f = (-\cos x)\mathbf{i} + \mathbf{j} \Rightarrow \nabla f\big|_{(\pi, 1)} = \mathbf{i} + \mathbf{j} \Rightarrow$ the tangent line is $(x - \pi) + (y - 1) = 0 \Rightarrow x + y = \pi + 1$; the normal line is $y - 1 = 1(x - \pi) \Rightarrow y = x - \pi + 1$

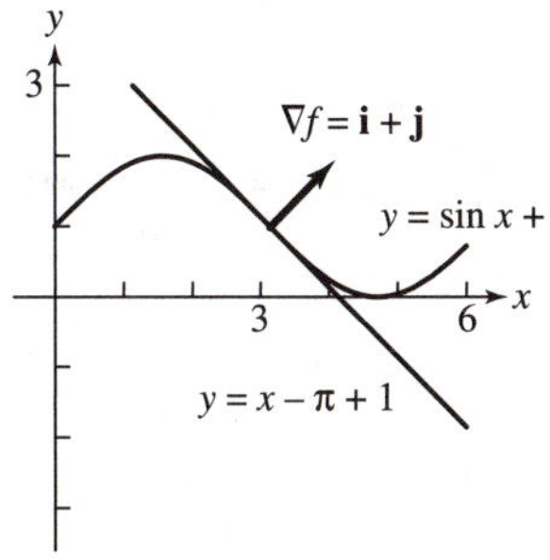

60. $\nabla f = -x\mathbf{i} + y\mathbf{j} \Rightarrow \nabla f\big|_{(1, 2)} = -\mathbf{i} + 2\mathbf{j} \Rightarrow$ the tangent line is $-(x - 1) + 2(y - 2) = 0 \Rightarrow y = \frac{1}{2}x + \frac{3}{2}$; the normal line is $y - 2 = -2(x - 1) \Rightarrow y = -2x + 4$

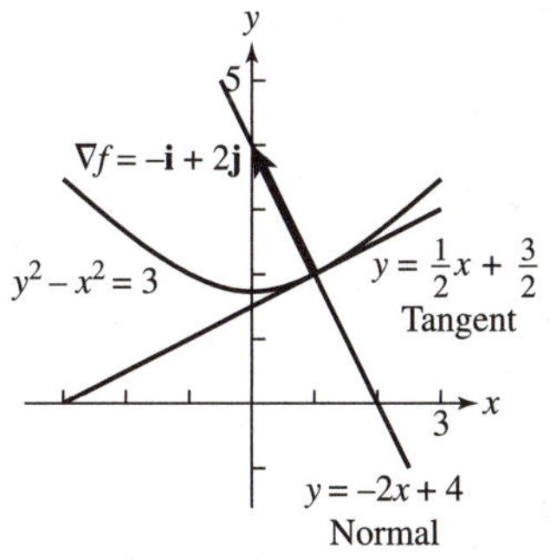

61. Let $f(x, y, z) = x^2 + 2y + 2z - 4$ and $g(x, y, z) = y - 1$. Then $\nabla f = 2x\mathbf{i} + 2\mathbf{j} + 2\mathbf{k}\big|_{\left(1, 1, \frac{1}{2}\right)} = 2\mathbf{i} + 2\mathbf{j} + 2\mathbf{k}$ and $\nabla g = \mathbf{j} \Rightarrow \nabla f \times \nabla g = \begin{vmatrix} \mathbf{i} & \mathbf{j} & \mathbf{k} \\ 2 & 2 & 2 \\ 0 & 1 & 0 \end{vmatrix} = -2\mathbf{i} + 2\mathbf{k} \Rightarrow$ the line is $x = 1 - 2t,\ y = 1,\ z = \frac{1}{2} + 2t$

62. Let $f(x, y, z) = x + y^2 + z - 2$ and $g(x, y, z) = y - 1$. Then $\nabla f = \mathbf{i} + 2y\mathbf{j} + \mathbf{k}\big|_{\left(\frac{1}{2}, 1, \frac{1}{2}\right)} = \mathbf{i} + 2\mathbf{j} + \mathbf{k}$ and $\nabla g = \mathbf{j}$ $\Rightarrow \nabla f \times \nabla g = \begin{vmatrix} \mathbf{i} & \mathbf{j} & \mathbf{k} \\ 1 & 2 & 1 \\ 0 & 1 & 0 \end{vmatrix} = -\mathbf{i} + \mathbf{k} \Rightarrow$ the line is $x = \frac{1}{2} - t,\ y = 1,\ z = \frac{1}{2} + t$

63. $f_x(x, y) = 2x - y + 2 = 0$ and $f_y(x, y) = -x + 2y + 2 = 0$ $\Rightarrow x = -2$ and $y = -2 \Rightarrow (-2, -2)$ is the critical point; $f_{xx}(-2, -2) = 2, f_{yy}(-2, -2) = 2, f_{xy}(-2, -2) = -1$ $\Rightarrow f_{xx}f_{yy} - f_{xy}^2 = 3 > 0$ and $f_{xx} > 0 \Rightarrow$ local minimum value of $f(-2, -2) = -8$

64. $f_x(x, y) = 10x + 4y + 4 = 0$ and $f_y(x, y) = 4x - 4y - 4 = 0 \Rightarrow x = 0$ and $y = -1 \Rightarrow (0, -1)$ is the critical point; $f_{xx}(0, -1) = 10, f_{yy}(0, -1) = -4, f_{xy}(0, -1) = 4 \Rightarrow f_{xx}f_{yy} - f_{xy}^2 = -56 < 0 \Rightarrow$ saddle point with $f(0, -1) = 2$

65. $f_x(x, y) = 6x^2 + 3y = 0$ and $f_y(x, y) = 3x + 6y^2 = 0 \Rightarrow y = -2x^2$ and $3x + 6(4x^4) = 0 \Rightarrow x(1 + 8x^3) = 0 \Rightarrow x = 0$ and $y = 0$, or $x = -\frac{1}{2}$ and $y = -\frac{1}{2} \Rightarrow$ the critical points are $(0, 0)$ and $\left(-\frac{1}{2}, -\frac{1}{2}\right)$. For $(0, 0)$: $f_{xx}(0, 0) = 12x\big|_{(0, 0)} = 0$, $f_{yy}(0, 0) = 12y\big|_{(0, 0)} = 0, f_{xy}(0, 0) = 3 \Rightarrow f_{xx}f_{yy} - f_{xy}^2 = -9 < 0 \Rightarrow$ saddle point with $f(0, 0) = 0$. For $\left(-\frac{1}{2}, -\frac{1}{2}\right)$: $f_{xx} = -6, f_{yy} = -6, f_{xy} = 3 \Rightarrow f_{xx}f_{yy} - f_{xy}^2 = 27 > 0$ and $f_{xx} < 0 \Rightarrow$ local maximum value of $f\left(-\frac{1}{2}, -\frac{1}{2}\right) = \frac{1}{4}$

66. $f_x(x, y) = 3x^2 - 3y = 0$ and $f_y(x, y) = 3y^2 - 3x = 0 \Rightarrow y = x^2$ and $x^4 - x = 0 \Rightarrow x(x^3 - 1) = 0 \Rightarrow$ the critical points are $(0, 0)$ and $(1, 1)$. For $(0, 0)$: $f_{xx}(0, 0) = 6x\big|_{(0, 0)} = 0$, $f_{yy}(0, 0) = 6y\big|_{(0, 0)} = 0, f_{xy}(0, 0) = -3 \Rightarrow f_{xx}f_{yy} - f_{xy}^2 = -9 < 0 \Rightarrow$ saddle point with $f(0, 0) = 15$. For $(1, 1)$: $f_{xx}(1, 1) = 6, f_{yy}(1, 1) = 6, f_{xy}(1, 1) = -3 \Rightarrow f_{xx}f_{yy} - f_{xy}^2 = 27 > 0$ and $f_{xx} > 0 \Rightarrow$ local maximum value of $f(1, 1) = 14$

67. $f_x(x, y) = 3x^2 + 6x = 0$ and $f_y(x, y) = 3y^2 - 6y = 0$ $\Rightarrow x(x + 2) = 0$ and $y(y - 2) = 0 \Rightarrow x = 0$ or $x = -2$ and $y = 0$ or $y = 2 \Rightarrow$ the critical points are $(0, 0)$, $(0, 2)$, $(-2, 0)$, $(-2, 2)$. For $(0, 0)$: $f_{xx}(0, 0) = 6x + 6\big|_{(0, 0)} = 6$, $f_{yy}(0, 0) = 6y - 6\big|_{(0, 0)} = -6, f_{xy}(0, 0) = 0 \Rightarrow f_{xx}f_{yy} - f_{xy}^2 = -36 < 0 \Rightarrow$ saddle point with $f(0, 0) = 0$. For $(0, 2)$: $f_{xx}(0, 2) = 6, f_{yy}(0, 2) = 6, f_{xy}(0, 2) = 0 \Rightarrow f_{xx}f_{yy} - f_{xy}^2 = 36 > 0$ and $f_{xx} > 0 \Rightarrow$ local minimum value of $f(0, 2) = -4$. For $(-2, 0)$: $f_{xx}(-2, 0) = -6, f_{yy}(-2, 0) = -6$, $f_{xy}(-2, 0) = 0 \Rightarrow f_{xx}f_{yy} - f_{xy}^2 = 36 > 0$ and $f_{xx} < 0$ $\Rightarrow$ local maximum value of $f(-2, 0) = 4$. For $(-2, 2)$: $f_{xx}(-2, 2) = -6, f_{yy}(-2, 2) = 6, f_{xy}(-2, 2) = 0 \Rightarrow f_{xx}f_{yy} - f_{xy}^2 = -36 < 0 \Rightarrow$ saddle point with $f(-2, 2) = 0$.

68. $f_x(x, y) = 4x^3 - 16x = 0 \Rightarrow 4x(x^2 - 4) = 0 \Rightarrow x = 0, 2, -2$; $f_y(x, y) = 6y - 6 = 0 \Rightarrow y = 1$. Therefore the critical points are $(0, 1)$, $(2, 1)$, and $(-2, 1)$. For $(0, 1)$: $f_{xx}(0, 1) = 12x^2 - 16\big|_{(0, 1)} = -16, f_{yy}(0, 1) = 6, f_{xy}(0, 1) = 0$ $\Rightarrow f_{xx}f_{yy} - f_{xy}^2 = -96 < 0 \Rightarrow$ saddle point with $f(0, 1) = -3$. For $(2, 1)$: $f_{xx}(2, 1) = 32, f_{yy}(2, 1) = 6, f_{xy}(2, 1) = 0$ $\Rightarrow f_{xx}f_{yy} - f_{xy}^2 = 192 > 0$ and $f_{xx} > 0 \Rightarrow$ local minimum value of $f(2, 1) = -19$. For $(-2, 1)$: $f_{xx}(-2, 1) = 32$, $f_{yy}(-2, 1) = 6, f_{xy}(-2, 1) = 0 \Rightarrow f_{xx}f_{yy} - f_{xy}^2 = 192 > 0$ and $f_{xx} > 0 \Rightarrow$ local minimum value of $f(-2, 1) = -19$.

69. **i.** On OA, $f(x, y) = f(0, y) = y^2 + 3y$ for $0 \le y \le 4$ $\Rightarrow f'(0, y) = 2y + 3 = 0 \Rightarrow y = -\frac{3}{2}$. But $\left(0, -\frac{3}{2}\right)$ is not in the region. Endpoints: $f(0, 0) = 0$ and $f(0, 4) = 28$.

ii. On AB, $f(x, y) = f(x, -x + 4) = x^2 - 10x + 28$ for $0 \le x \le 4 \Rightarrow f'(x, -x + 4) = 2x - 10 = 0 \Rightarrow x = 5$, $y = -1$. But $(5, -1)$ is not in the region. Endpoints: $f(4, 0) = 4$ and $f(0, 4) = 28$.

iii. On OB, $f(x, y) = f(x, 0) = x^2 - 3x$ for $0 \le x \le 4$ $\Rightarrow f'(x, 0) = 2x - 3 \Rightarrow x = \frac{3}{2}$ and $y = 0$. $\left(\frac{3}{2}, 0\right)$ is a critical point with $f\left(\frac{3}{2}, 0\right) = -\frac{9}{4}$.
Endpoints: $f(0, 0) = 0$ and $f(4, 0) = 4$.

iv. For the interior of the triangular region, $f_x(x, y) = 2x + y - 3 = 0$ and $f_y(x, y) = x + 2y + 3 = 0 \Rightarrow x = 3$ and $y = -3$. But $(3, -3)$ is not in the region. Therefore the absolute maximum is 28 at $(0, 4)$ and the absolute minimum is $-\frac{9}{4}$ at $\left(\frac{3}{2}, 0\right)$.

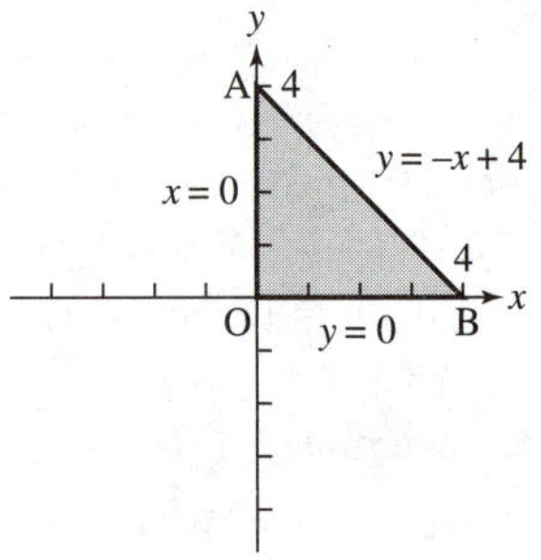

70. i. On OA, $f(x, y) = f(0, y) = -y^2 + 4y + 1$ for $0 \le y \le 2 \Rightarrow f'(0, y) = -2y + 4 = 0 \Rightarrow y = 2$ and $x = 0$. But $(0, 2)$ is not in the interior of OA.
Endpoints: $f(0, 0) = 1$ and $f(0, 2) = 5$.

ii. On AB, $f(x, y) = f(x, 2) = x^2 - 2x + 5$ for $0 \le x \le 4$ $\Rightarrow f'(x, 2) = 2x - 2 = 0 \Rightarrow x = 1$ and $y = 2 \Rightarrow (1, 2)$ is an interior critical point of AB with $f(1, 2) = 4$.
Endpoints: $f(4, 2) = 13$ and $f(0, 2) = 5$.

iii. On BC, $f(x, y) = f(4, y) = -y^2 + 4y + 9$ for $0 \le y \le 2 \Rightarrow f'(4, y) = -2y + 4 = 0 \Rightarrow y = 2$ and $x = 4$. But $(4, 2)$ is not in the interior of BC.
Endpoints: $f(4, 0) = 9$ and $f(4, 2) = 13$.

iv. On OC, $f(x, y) = f(x, 0) = x^2 - 2x + 1$ for $0 \le x \le 4$ $\Rightarrow f'(x, 0) = 2x - 2 = 0 \Rightarrow x = 1$ and $y = 0 \Rightarrow (1, 0)$ is an interior critical point of OC with $f(1, 0) = 0$.
Endpoints: $f(0, 0) = 1$ and $f(4, 0) = 9$.

v. For the interior of the triangular region, $f_x(x, y) = 2x - 2 = 0$ and $f_y(x, y) = -2y + 4 = 0 \Rightarrow x = 1$ and $y = 2$. But $(1, 2)$ is not in the interior of the region. Therefore the absolute maximum is 13 at $(4, 2)$ and the absolute minimum is 0 at $(1, 0)$.

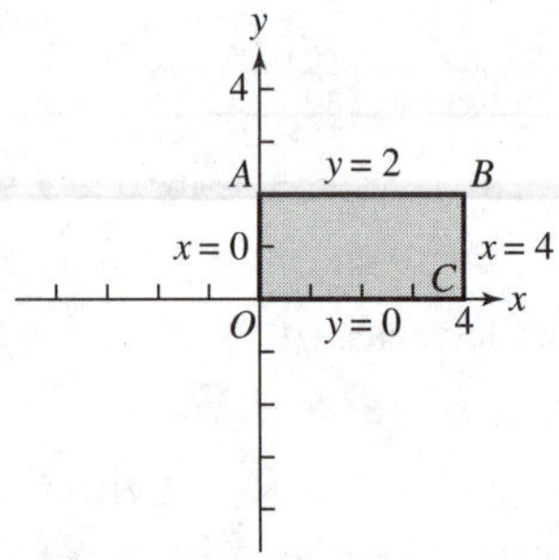

71. i. On AB, $f(x, y) = f(-2, y) = y^2 - y - 4$ for $-2 \le y \le 2 \Rightarrow f'(-2, y) = 2y - 1 \Rightarrow y = \frac{1}{2}$ and $x = -2$ $\Rightarrow \left(-2, \frac{1}{2}\right)$ is an interior critical point in AB with $f\left(-2, \frac{1}{2}\right) = -\frac{17}{4}$.
Endpoints: $f(-2, -2) = 2$ and $f(-2, 2) = -2$.

ii. On BC, $f(x, y) = f(x, 2) = -2$ for $-2 \le x \le 2$ $\Rightarrow f'(x, 2) = 0 \Rightarrow$ no critical points in the interior of BC.
Endpoints: $f(-2, 2) = -2$ and $f(2, 2) = -2$.

iii. On CD, $f(x, y) = f(2, y) = y^2 - 5y + 4$ for $-2 \le y \le 2 \Rightarrow f'(2, y) = 2y - 5 = 0 \Rightarrow y = \frac{5}{2}$ and $x = 2$. But $\left(2, \frac{5}{2}\right)$ is not in the region.
Endpoints: $f(2, -2) = 18$ and $f(2, 2) = -2$.

iv. On AD, $f(x, y) = f(x, -2) = 4x + 10$ for $-2 \le x \le 2$ $\Rightarrow f'(x, -2) = 4 \Rightarrow$ no critical points in the interior of AD.
Endpoints: $f(-2, -2) = 2$ and $f(2, -2) = 18$.

v. For the interior of the square, $f_x(x, y) = -y + 2 = 0$ and $f_y(x, y) = 2y - x - 3 = 0 \Rightarrow y = 2$ and $x = 1$ $\Rightarrow (1, 2)$ is an interior critical point of the square with $f(1, 2) = -2$. Therefore the absolute maximum is 18 at $(2, -2)$ and the absolute minimum is $-\frac{17}{4}$ at $\left(-2, \frac{1}{2}\right)$.

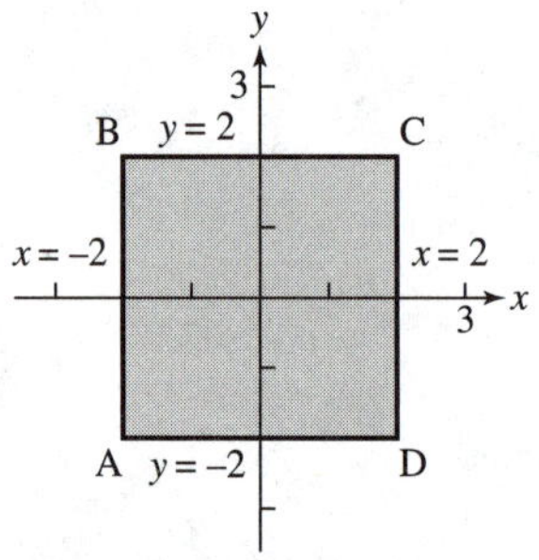

72. i. On OA, $f(x, y) = f(0, y) = 2y - y^2$ for $0 \le y \le 2$ $\Rightarrow f'(0, y) = 2 - 2y = 0 \Rightarrow y = 1$ and $x = 0 \Rightarrow (0, 1)$ is an interior critical point of OA with $f(0, 1) = 1$.
Endpoints: $f(0, 0) = 0$ and $f(0, 2) = 0$.

ii. On AB, $f(x, y) = f(x, 2) = 2x - x^2$ for $0 \le x \le 2$ $\Rightarrow f'(x, 2) = 2 - 2x = 0 \Rightarrow x = 1$ and $y = 2 \Rightarrow (1, 2)$ is an interior critical point of AB with $f(1, 2) = 1$.
Endpoints: $f(0, 2) = 0$ and $f(2, 2) = 0$.

iii. On BC, $f(x, y) = f(2, y) = 2y - y^2$ for $0 \le y \le 2$ $\Rightarrow f'(2, y) = 2 - 2y = 0 \Rightarrow y = 1$ and $x = 2 \Rightarrow (2, 1)$ is an interior critical point of BC with $f(2, 1) = 1$.
Endpoints: $f(2, 0) = 0$ and $f(2, 2) = 0$.

iv. On OC, $f(x, y) = f(x, 0) = 2x - x^2$ for $0 \le x \le 2$ $\Rightarrow f'(x, 0) = 2 - 2x = 0 \Rightarrow x = 1$ and $y = 0 \Rightarrow (1, 0)$ is an interior critical point of OC with $f(1, 0) = 1$.
Endpoints: $f(0, 0) = 0$ and $f(0, 2) = 0$.

v. For the interior of the rectangular region, $f_x(x, y) = 2 - 2x = 0$ and $f_y(x, y) = 2 - 2y = 0 \Rightarrow x = 1$ and $y = 1 \Rightarrow (1, 1)$ is an interior critical point of the square with $f(1, 1) = 2$. Therefore the absolute maximum is 2 at $(1, 1)$ and the absolute minimum is 0 at the four corners $(0, 0)$, $(0, 2)$, $(2, 2)$, and $(2, 0)$.

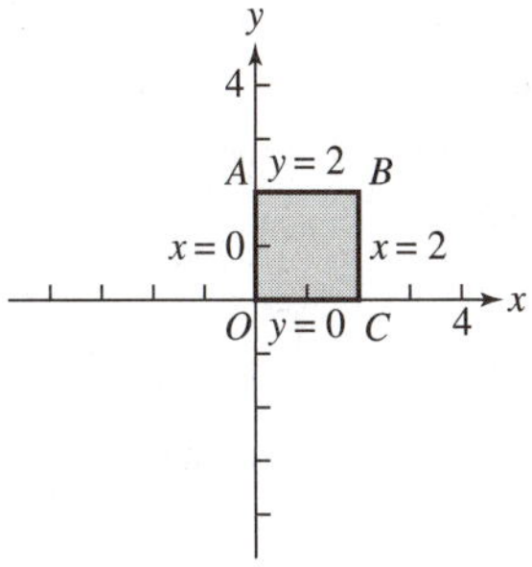

73. i. On AB, $f(x, y) = f(x, x + 2) = -2x + 4$ for $-2 \le x \le 2 \Rightarrow f'(x, x + 2) = -2 = 0 \Rightarrow$ no critical points in the interior of AB.
Endpoints: $f(-2, 0) = 8$ and $f(2, 4) = 0$.

ii. On BC, $f(x, y) = f(2, y) = -y^2 + 4y$ for $0 \le y \le 4 \Rightarrow f'(2, y) = -2y + 4 = 0 \Rightarrow y = 2$ and $x = 2 \Rightarrow (2, 2)$ is an interior critical point of BC with $f(2, 2) = 4$. Endpoints: $f(2, 0) = 0$ and $f(2, 4) = 0$.

iii. On AC, $f(x, y) = f(x, 0) = x^2 - 2x$ for $-2 \le y \le 2 \Rightarrow f'(x, 0) = 2x - 2 \Rightarrow x = 1$ and $y = 0 \Rightarrow (1, 0)$ is an interior critical point of AC with $f(1, 0) = -1$.
Endpoints: $f(-2, 0) = 8$ and $f(2, 0) = 0$.

iv. For the interior of the triangular region, $f_x(x, y) = 2x - 2 = 0$ and $f_y(x, y) = -2y + 4 = 0 \Rightarrow x = 1$ and $y = 2 \Rightarrow (1, 2)$ is an interior critical point of the region with $f(1, 2) = 3$. Therefore the absolute maximum is 8 at $(-2, 0)$ and the absolute minimum is -1 at $(1, 0)$.

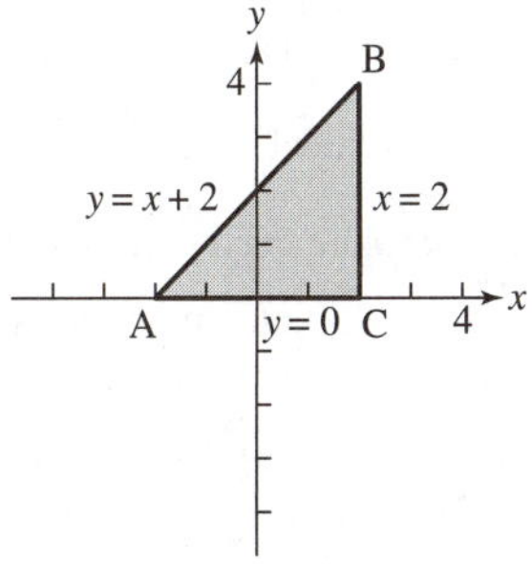

74. i. On AB, $f(x, y) = f(x, x) = 4x^2 - 2x^4 + 16$ for $-2 \le x \le 2 \Rightarrow f'(x, x) = 8x - 8x^3 = 0 \Rightarrow x = 0$ and $y = 0$, or $x = 1$ and $y = 1$, or $x = -1$ and $y = -1 \Rightarrow (0, 0)$, $(1, 1)$, $(-1, -1)$ are all interior points of AB with $f(0, 0) = 16$, $f(1, 1) = 18$, and $f(-1, -1) = 18$.
Endpoints: $f(-2, -2) = 0$ and $f(2, 2) = 0$.

ii. On BC, $f(x, y) = f(2, y) = 8y - y^4$ for $-2 \le y \le 2 \Rightarrow f'(2, y) = 8 - 4y^3 = 0 \Rightarrow y = \sqrt[3]{2}$ and $x = 2 \Rightarrow (2, \sqrt[3]{2})$ is an interior critical point of BC with $f(2, \sqrt[3]{2}) = 6\sqrt[3]{2}$.
Endpoints: $f(2, -2) = -32$ and $f(2, 2) = 0$.

iii. On AC, $f(x, y) = f(x, -2) = -8x - x^4$ for $-2 \le y \le 2 \Rightarrow f'(x, -2) = -8 - 4x^3 = 0 \Rightarrow x = \sqrt[3]{-2}$ and $y = -2 \Rightarrow (\sqrt[3]{-2}, -2)$ is an interior critical point of AC with $f(\sqrt[3]{-2}, -2) = 6\sqrt[3]{2}$.
Endpoints: $f(-2, -2) = 0$ and $f(2, -2) = -32$.

iv. For the interior of the triangular region, $f_x(x, y) = 4y - 4x^3 = 0$ and $f_y(x, y) = 4x - 4y^3 = 0 \Rightarrow x = 0$ and $y = 0$, or $x = 1$ and $y = 1$, or $x = -1$ and $y = -1$. But none of the points $(0, 0)$, $(1, 1)$ and $(-1, -1)$ are interior to the region. Therefore the absolute maximum is 18 at $(1, 1)$ and $(-1, -1)$, and the absolute minimum is -32 at $(2, -2)$.

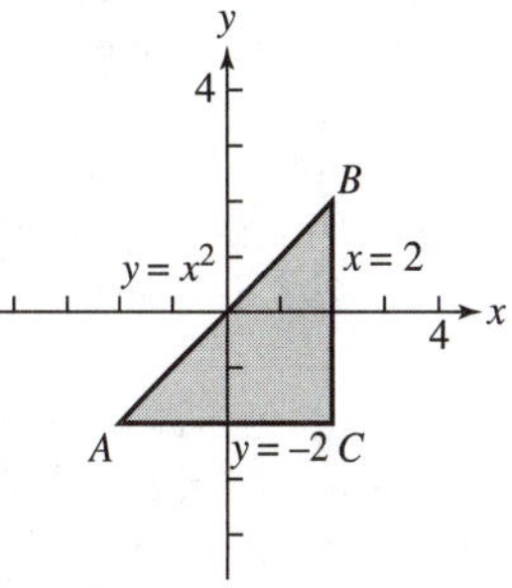

75. i. On AB, $f(x, y) = f(-1, y) = y^3 - 3y^2 + 2$ for $-1 \le y \le 1 \Rightarrow f'(-1, y) = 3y^2 - 6y = 0 \Rightarrow y = 0$ and $x = -1$, or $y = 2$ and $x = -1 \Rightarrow (-1, 0)$ is an interior critical point of AB with $f(-1, 0) = 2$; $(-1, 2)$ is outside the boundary.
Endpoints: $f(-1, -1) = -2$ and $f(-1, 1) = 0$.

ii. On BC, $f(x, y) = f(x, 1) = x^3 + 3x^2 - 2$ for $-1 \le x \le 1 \Rightarrow f'(x, 1) = 3x^2 + 6x = 0 \Rightarrow x = 0$ and $y = 1$, or $x = -2$ and $y = 1 \Rightarrow (0, 1)$ is an interior critical point of BC with $f(0, 1) = -2$; $(-2, 1)$ is outside the boundary. Endpoints: $f(-1, 1) = 0$ and $f(1, 1) = 2$.

iii. On CD, $f(x, y) = f(1, y) = y^3 - 3y^2 + 4$ for $-1 \le y \le 1 \Rightarrow f'(1, y) = 3y^2 - 6y = 0 \Rightarrow y = 0$ and $x = 1$, or $y = 2$ and $x = 1 \Rightarrow (1, 0)$ is an interior critical point of CD with $f(1, 0) = 4$; $(1, 2)$ is outside the boundary.
Endpoints: $f(1, 1) = 2$ and $f(1, -1) = 0$.

iv. On AD, $f(x, y) = f(x, -1) = x^3 + 3x^2 - 4$ for $-1 \le x \le 1 \Rightarrow f'(x, -1) = 3x^2 + 6x = 0 \Rightarrow x = 0$ and $y = -1$, or $x = -2$ and $y = -1 \Rightarrow (0, -1)$ is an interior point of AD with $f(0, -1) = -4$; $(-2, -1)$ is outside the boundary.
Endpoints: $f(-1, -1) = -2$ and $f(1, -1) = 0$.

v. For the interior of the square, $f_x(x, y) = 3x^2 + 6x = 0$ and $f_y(x, y) = 3y^2 - 6y = 0 \Rightarrow x = 0$ or $x = -2$, and $y = 0$ or $y = 2 \Rightarrow (0, 0)$ is an interior critical point of the square region with $f(0, 0) = 0$; the points $(0, 2)$, $(-2, 0)$, and $(-2, 2)$ are outside the region. Therefore the absolute maximum is 4 at $(1, 0)$ and the absolute minimum is -4 at $(0, -1)$.

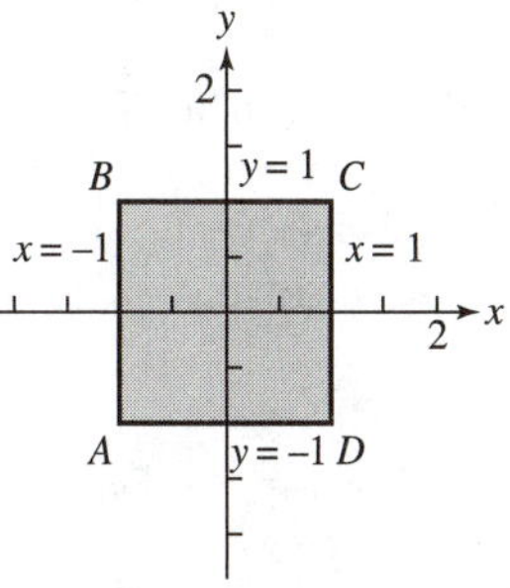

76. i. On AB, $f(x, y) = f(-1, y) = y^3 - 3y$ for $-1 \le y \le 1$ $\Rightarrow f'(-1, y) = 3y^2 - 3 = 0 \Rightarrow y = \pm 1$ and $x = -1$, yielding the corner points $(-1, -1)$ and $(-1, 1)$ with $f(-1, -1) = 2$ and $f(-1, 1) = -2$

ii. On BC, $f(x, y) = f(x, 1) = x^3 + 3x + 2$ for $-1 \le x \le 1 \Rightarrow f'(x, 1) = 3x^2 + 3 = 0 \Rightarrow$ no solution. Endpoints: $f(-1, 1) = -2$ and $f(1, 1) = 6$.

iii. On CD, $f(x, y) = f(1, y) = y^3 + 3y + 2$ for $-1 \le y \le 1 \Rightarrow f'(1, y) = 3y^2 + 3 \Rightarrow$ no solution. Endpoints: $f(1, 1) = 6$ and $f(1, -1) = -2$.

iv. On AD, $f(x, y) = f(x, -1) = x^3 - 3x$ for $-1 \le x \le 1$ $\Rightarrow f'(x, -1) = 3x^2 - 3 = 0 \Rightarrow x = \pm 1$ and $y = -1$ yielding the corner points $(-1, -1)$ and $(1, -1)$ with $f(-1, -1) = 2$ and $f(1, -1) = -2$

v. For the interior of the square, $f_x(x, y) = 3x^2 + 3y = 0$ and $f_y(x, y) = 3y^2 + 3x = 0 \Rightarrow y = -x^2$ and $x^4 + x = 0 \Rightarrow x = 0$ or $x = -1 \Rightarrow (0, 0)$ is an interior critical point of the square region with $f(0, 0) = 1$; $(-1, -1)$ is on the boundary. Therefore the absolute maximum is 6 at $(1, 1)$ and the absolute minimum is -2 at $(1, -1)$ and $(-1, 1)$.

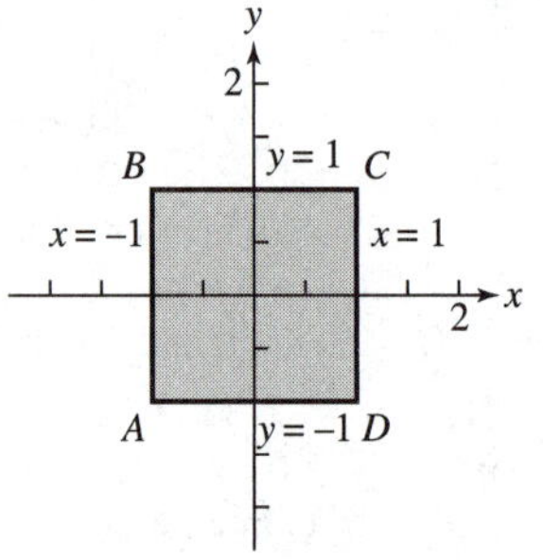

77. $\nabla f = 3x^2\mathbf{i} + 2y\mathbf{j}$ and $\nabla g = 2x\mathbf{i} + 2y\mathbf{j}$ so that $\nabla f = \lambda \nabla g$ $\Rightarrow 3x^2\mathbf{i} + 2y\mathbf{j} = \lambda(2x\mathbf{i} + 2y\mathbf{j}) \Rightarrow 3x^2 = 2x\lambda$ and $2y = 2y\lambda$ $\Rightarrow \lambda = 1$ or $y = 0$.

CASE 1: $\lambda = 1 \Rightarrow 3x^2 = 2x \Rightarrow x = 0$ or $x = \frac{2}{3}$; $x = 0 \Rightarrow y = \pm 1$ yielding the points $(0, 1)$ and $(0, -1)$; $x = \frac{2}{3} \Rightarrow y = \pm\frac{\sqrt{5}}{3}$ yielding the points $\left(\frac{2}{3}, \frac{\sqrt{5}}{3}\right)$ and $\left(\frac{2}{3}, -\frac{\sqrt{5}}{3}\right)$.

CASE 2: $y = 0 \Rightarrow x^2 - 1 = 0 \Rightarrow x = \pm 1$ yielding the points $(1, 0)$ and $(-1, 0)$.

Evaluations give $f(0, \pm 1) = 1$, $f\left(\frac{2}{3}, \frac{\sqrt{5}}{3}\right) = \frac{23}{27}$, $f(1, 0) = 1$, and $f(-1, 0) = -1$. Therefore the absolute maximum is 1 at $(0, \pm 1)$ and $(1, 0)$, and the absolute minimum is -1 at $(-1, 0)$.

78. $\nabla f = y\mathbf{i} + x\mathbf{j}$ and $\nabla g = 2x\mathbf{i} + 2y\mathbf{j}$ so that $\nabla f = \lambda \nabla g \Rightarrow y\mathbf{i} + x\mathbf{j} = \lambda(2x\mathbf{i} + 2y\mathbf{j}) \Rightarrow y = 2\lambda x$ and $x = 2\lambda y \Rightarrow x = 2\lambda(2\lambda x) = 4\lambda^2 x \Rightarrow x = 0$ or $4\lambda^2 = 1$.

CASE 1: $x = 0 \Rightarrow y = 0$ but $(0, 0)$ does not lie on the circle, so no solution.

CASE 2: $4\lambda^2 = 1 \Rightarrow \lambda = \frac{1}{2}$ or $\lambda = -\frac{1}{2}$. For $\lambda = \frac{1}{2}$, $y = x$ $\Rightarrow 1 = x^2 + y^2 = 2x^2 \Rightarrow x = y = \pm\frac{1}{\sqrt{2}}$ yielding the points $\left(\frac{1}{\sqrt{2}}, \frac{1}{\sqrt{2}}\right)$ and $\left(-\frac{1}{\sqrt{2}}, -\frac{1}{\sqrt{2}}\right)$. For $\lambda = -\frac{1}{2}$, $y = -x \Rightarrow 1 = x^2 + y^2 = 2x^2 \Rightarrow x = \pm\frac{1}{\sqrt{2}}$ and $y = -x$ yielding the points $\left(-\frac{1}{\sqrt{2}}, \frac{1}{\sqrt{2}}\right)$ and $\left(\frac{1}{\sqrt{2}}, -\frac{1}{\sqrt{2}}\right)$. Evaluations give the absolute maximum value $f\left(\frac{1}{\sqrt{2}}, \frac{1}{\sqrt{2}}\right) = f\left(-\frac{1}{\sqrt{2}}, -\frac{1}{\sqrt{2}}\right) = \frac{1}{2}$ and the absolute minimum value $f\left(-\frac{1}{\sqrt{2}}, \frac{1}{\sqrt{2}}\right) = f\left(\frac{1}{\sqrt{2}}, -\frac{1}{\sqrt{2}}\right) = -\frac{1}{2}$.

79. i. $f(x, y) = x^2 + 3y^2 + 2y$ on $x^2 + y^2 = 1 \Rightarrow \nabla f = 2x\mathbf{i} + (6y + 2)\mathbf{j}$ and $\nabla g = 2x\mathbf{i} + 2y\mathbf{j}$ so that $\nabla f = \lambda \nabla g$ $\Rightarrow 2x\mathbf{i} + (6y + 2)\mathbf{j} = \lambda(2x\mathbf{i} + 2y)\mathbf{j} \Rightarrow 2x = 2x\lambda$ and $6y + 2 = 2y\lambda \Rightarrow \lambda = 1$ or $x = 0$.

CASE 1: $\lambda = 1 \Rightarrow 6y + 2 = 2y \Rightarrow y = -\frac{1}{2}$ and $x = \pm\frac{\sqrt{3}}{2}$ yielding the points $\left(\pm\frac{\sqrt{3}}{2}, -\frac{1}{2}\right)$.

CASE 2: $x = 0 \Rightarrow y^2 = 1 \Rightarrow y = \pm 1$ yielding the points $(0, \pm 1)$.

Evaluations give $f\left(\pm\frac{\sqrt{3}}{2}, -\frac{1}{2}\right) = \frac{1}{2}$, $f(0, 1) = 5$, and $f(0, -1) = 1$. Therefore $\frac{1}{2}$ and 5 are the extreme values on the boundary of the disk.

ii. For the interior of the disk, $f_x(x, y) = 2x = 0$ and $f_y(x, y) = 6y + 2 = 0 \Rightarrow x = 0$ and $y = -\frac{1}{3}$ $\Rightarrow \left(0, -\frac{1}{3}\right)$ is an interior critical point with $f\left(0, -\frac{1}{3}\right) = -\frac{1}{3}$. Therefore the absolute maximum of f on the disk is 5 at $(0, 1)$ and the absolute minimum of f on the disk is $-\frac{1}{3}$ at $\left(0, -\frac{1}{3}\right)$.

80. i. $f(x, y) = x^2 + y^2 - 3x - xy$ on $x^2 + y^2 = 9 \Rightarrow \nabla f = (2x - 3 - y)\mathbf{i} + (2y - x)\mathbf{j}$ and $\nabla g = 2x\mathbf{i} + 2y\mathbf{j}$ so that $\nabla f = \lambda \nabla g \Rightarrow (2x - 3 - y)\mathbf{i} + (2y - x)\mathbf{j} = \lambda(2x\mathbf{i} + 2y\mathbf{j}) \Rightarrow 2x - 3 - y = 2x\lambda$ and $2y - x = 2y\lambda$ $\Rightarrow 2x(1 - \lambda) - y = 3$ and $-x + 2y(1 - \lambda) = 0 \Rightarrow 1 - \lambda = \frac{x}{2y}$ and $(2x)\left(\frac{x}{2y}\right) - y = 3 \Rightarrow x^2 - y^2 = 3y \Rightarrow x^2 = y^2 + 3y$. Thus, $9 = x^2 + y^2 = y^2 + 3y + y^2 \Rightarrow 2y^2 + 3y - 9 = 0 \Rightarrow (2y - 3)(y + 3) = 0 \Rightarrow y = -3, \frac{3}{2}$. For $y = -3$, $x^2 + y^2 = 9 \Rightarrow x = 0$ yielding the point $(0, -3)$. For $y = \frac{3}{2}$, $x^2 + y^2 = 9 \Rightarrow x^2 + \frac{9}{4} = 9 \Rightarrow x^2 = \frac{27}{4} \Rightarrow x = \pm\frac{3\sqrt{3}}{2}$. Evaluations give $f(0, -3) = 9$, $f\left(-\frac{3\sqrt{3}}{2}, \frac{3}{2}\right) = 9 + \frac{27\sqrt{3}}{4} \approx 20.691$. and $f\left(\frac{3\sqrt{3}}{2}, \frac{3}{2}\right) = 9 - \frac{27\sqrt{3}}{4} \approx -2.691$.

ii. For the interior of the disk, $f_x(x, y) = 2x - 3 - y = 0$ and $f_y(x, y) = 2y - x = 0 \Rightarrow x = 2$ and $y = 1 \Rightarrow (2, 1)$ is an interior critical point of the disk with $f(2, 1) = -3$. Therefore, the absolute maximum of f on the disk is $9 + \frac{27\sqrt{3}}{4}$ at $\left(-\frac{3\sqrt{3}}{2}, \frac{3}{2}\right)$ and the absolute minimum of f on the disk is -3 at $(2, 1)$.

81. $\nabla f = \mathbf{i} - \mathbf{j} + \mathbf{k}$ and $\nabla g = 2x\mathbf{i} + 2y\mathbf{j} + 2z\mathbf{k}$ so that $\nabla f = \lambda \nabla g \Rightarrow \mathbf{i} - \mathbf{j} + \mathbf{k} = \lambda(2x\mathbf{i} + 2y\mathbf{j} + 2z\mathbf{k}) \Rightarrow 1 = 2x\lambda$, $-1 = 2y\lambda$, $1 = 2z\lambda \Rightarrow x = -y = z = \frac{1}{2\lambda}$. Thus $x^2 + y^2 + z^2 = 1 \Rightarrow 3x^2 = 1 \Rightarrow x = \pm\frac{1}{\sqrt{3}}$ yielding the points $\left(\frac{1}{\sqrt{3}}, -\frac{1}{\sqrt{3}}, \frac{1}{\sqrt{3}}\right)$ and $\left(-\frac{1}{\sqrt{3}}, \frac{1}{\sqrt{3}}, -\frac{1}{\sqrt{3}}\right)$. Evaluations give the absolute maximum value of $f\left(\frac{1}{\sqrt{3}}, -\frac{1}{\sqrt{3}}, \frac{1}{\sqrt{3}}\right) = \frac{3}{\sqrt{3}} = \sqrt{3}$ and the absolute minimum value of $f\left(-\frac{1}{\sqrt{3}}, \frac{1}{\sqrt{3}}, -\frac{1}{\sqrt{3}}\right) = -\sqrt{3}$.

82. Let $f(x, y, z) = x^2 + y^2 + z^2$ be the square of the distance to the origin and $g(x, y, z) = z^2 - xy - 4$. Then $\nabla f = 2x\mathbf{i} + 2y\mathbf{j} + 2z\mathbf{k}$ and $\nabla g = -y\mathbf{i} - x\mathbf{j} + 2z\mathbf{k}$ so that $\nabla f = \lambda \nabla g \Rightarrow 2x = -\lambda y$, $2y = -\lambda x$, and $2z = 2\lambda z \Rightarrow z = 0$ or $\lambda = 1$.

CASE 1: $z = 0 \Rightarrow xy = -4 \Rightarrow x = -\frac{4}{y}$ and $y = -\frac{4}{x} \Rightarrow 2\left(-\frac{4}{y}\right) = -\lambda y$ and $2\left(-\frac{4}{x}\right) = -\lambda x$, $\Rightarrow \frac{8}{\lambda} = y^2$ and $\frac{8}{\lambda} = x^2 \Rightarrow y^2 = x^2 \Rightarrow y = \pm x$. But $y = x \Rightarrow x^2 = -4$ leads to no solution, so $y = -x \Rightarrow x^2 = 4 \Rightarrow x = \pm 2$ yielding the points $(-2, 2, 0)$ and $(2, -2, 0)$.

CASE 2: $\lambda = 1 \Rightarrow 2x = -y$ and $2y = -x \Rightarrow 2y = -\left(-\frac{y}{2}\right) \Rightarrow 4y = y \Rightarrow y = 0 \Rightarrow x = 0 \Rightarrow z^2 - 4 = 0 \Rightarrow z = \pm 2$ yielding the points $(0, 0, -2)$ and $(0, 0, 2)$.

Evaluations give $f(-2, 2, 0) = f(2, -2, 0) = 8$ and $f(0, 0, -2) = f(0, 0, 2) = 4$. Thus the points $(0, 0, -2)$ and $(0, 0, 2)$ on the surface are closest to the origin.

83. The cost is $f(x, y, z) = 2axy + 2bxz + 2cyz$ subject to the constraint $xyz = V$. Then $\nabla f = \lambda \nabla g \Rightarrow 2ay + 2bz = \lambda yz$, $2ax + 2cz = \lambda xz$, and $2bx + 2cy = \lambda xy \Rightarrow 2axy + 2bxz = \lambda xyz$, $2axy + 2cyz = \lambda xyz$, and $2bxz + 2cyz = \lambda xyz \Rightarrow 2axy + 2bxz = 2axy + 2cyz \Rightarrow y = \left(\frac{b}{c}\right)x$. Also $2axy + 2bxz = 2bxz + 2cyz \Rightarrow z = \left(\frac{a}{c}\right)x$. Then $x\left(\frac{b}{c}x\right)\left(\frac{a}{c}x\right) = V \Rightarrow x^3 = \frac{c^2V}{ab} \Rightarrow$ Width $= x = \left(\frac{c^2V}{ab}\right)^{1/3}$, Depth $= y = \left(\frac{b}{c}\right)\left(\frac{c^2V}{ab}\right)^{1/3} = \left(\frac{b^2V}{ac}\right)^{1/3}$, and Height $= z = \left(\frac{a}{c}\right)\left(\frac{c^2V}{ab}\right)^{1/3} = \left(\frac{a^2V}{bc}\right)^{1/3}$.

84. The volume of the pyramid in the first octant formed by the plane is $V(a, b, c) = \frac{1}{3}\left(\frac{1}{2}ab\right)c = \frac{1}{6}abc$. The point $(2, 1, 2)$ on the plane $\Rightarrow \frac{2}{a} + \frac{1}{b} + \frac{2}{c} = 1$. We want to maximize V subject to the constraint $2bc + ac + 2ab = abc$. Thus, $\nabla V = \frac{bc}{6}\mathbf{i} + \frac{ac}{6}\mathbf{j} + \frac{ab}{6}\mathbf{k}$ and $\nabla g = (c + 2b - bc)\mathbf{i} + (2c + 2a - ac)\mathbf{j} + (2b + a - ab)\mathbf{k}$ so that $\nabla V = \lambda \nabla g \Rightarrow \frac{bc}{6} = \lambda(c + 2b - bc)$, $\frac{ac}{6} = \lambda(2c + 2a - ac)$, and $\frac{ab}{6} = \lambda(2b + a - ab) \Rightarrow \frac{abc}{6} = \lambda(ac + 2ab - abc)$, $\frac{abc}{6} = \lambda(2bc + 2ab - abc)$, $\frac{abc}{6} = \lambda(2bc + ac - abc) \Rightarrow \lambda ac = 2\lambda bc$ and $2\lambda ab = 2\lambda bc$. Now $\lambda \neq 0$ since $a \neq 0$, $b \neq 0$, and $c \neq 0 \Rightarrow ac = 2bc$ and $ab = bc \Rightarrow a = 2b = c$. Substituting into the constraint equation gives $\frac{2}{a} + \frac{2}{a} + \frac{2}{a} = 1 \Rightarrow a = 6 \Rightarrow b = 3$ and $c = 6$. Therefore the desired plane is $\frac{x}{6} + \frac{y}{3} + \frac{z}{6} = 1$ or $x + 2y + z = 6$.

85. $\nabla f = (y + z)\mathbf{i} + x\mathbf{j} + x\mathbf{k}$, $\nabla g = 2x\mathbf{i} + 2y\mathbf{j}$, and $\nabla h = z\mathbf{i} + x\mathbf{k}$ so that $\nabla f = \lambda \nabla g + \mu \nabla h \Rightarrow (y + z)\mathbf{i} + x\mathbf{j} + x\mathbf{k} = \lambda(2x\mathbf{i} + 2y\mathbf{j}) + \mu(z\mathbf{i} + x\mathbf{k}) \Rightarrow y + z = 2\lambda x + \mu z$, $x = 2\lambda y$, $x = \mu x \Rightarrow x = 0$ or $\mu = 1$.

CASE 1: $x = 0$ which is impossible since $xz = 1$.

CASE 2: $\mu = 1 \Rightarrow y + z = 2\lambda x + z \Rightarrow y = 2\lambda x$ and $x = 2\lambda y \Rightarrow y = (2\lambda)(2\lambda y) \Rightarrow y = 0$ or $4\lambda^2 = 1$. If $y = 0$, then $x^2 = 1 \Rightarrow x = \pm 1$ so with $xz = 1$ we obtain the points $(1, 0, 1)$ and $(-1, 0, -1)$. If $4\lambda^2 = 1$, then $\lambda = \pm\frac{1}{2}$. For $\lambda = -\frac{1}{2}$, $y = -x$ so $x^2 + y^2 = 1 \Rightarrow x^2 = \frac{1}{2} \Rightarrow x = \pm\frac{1}{\sqrt{2}}$ with $xz = 1 \Rightarrow z = \pm\sqrt{2}$, and we obtain the points $\left(\frac{1}{\sqrt{2}}, -\frac{1}{\sqrt{2}}, \sqrt{2}\right)$ and $\left(-\frac{1}{\sqrt{2}}, \frac{1}{\sqrt{2}}, -\sqrt{2}\right)$. For $\lambda = \frac{1}{2}$, $y = x \Rightarrow x^2 = \frac{1}{2} \Rightarrow x = \pm\frac{1}{\sqrt{2}}$ with $xz = 1 \Rightarrow z = \pm\sqrt{2}$, and we obtain the points $\left(\frac{1}{\sqrt{2}}, \frac{1}{\sqrt{2}}, \sqrt{2}\right)$ and $\left(-\frac{1}{\sqrt{2}}, -\frac{1}{\sqrt{2}}, -\sqrt{2}\right)$.

Evaluations give $f(1, 0, 1) = 1$, $f(-1, 0, -1) = 1$, $f\left(\frac{1}{\sqrt{2}}, -\frac{1}{\sqrt{2}}, \sqrt{2}\right) = \frac{1}{2}$, $f\left(-\frac{1}{\sqrt{2}}, \frac{1}{\sqrt{2}}, -\sqrt{2}\right) = \frac{1}{2}$, $f\left(\frac{1}{\sqrt{2}}, \frac{1}{\sqrt{2}}, \sqrt{2}\right) = \frac{3}{2}$, and $f\left(-\frac{1}{\sqrt{2}}, -\frac{1}{\sqrt{2}}, -\sqrt{2}\right) = \frac{3}{2}$. Therefore the absolute maximum is $\frac{3}{2}$ at $\left(\frac{1}{\sqrt{2}}, \frac{1}{\sqrt{2}}, \sqrt{2}\right)$ and $\left(-\frac{1}{\sqrt{2}}, -\frac{1}{\sqrt{2}}, -\sqrt{2}\right)$, and the absolute minimum is $\frac{1}{2}$ at $\left(-\frac{1}{\sqrt{2}}, \frac{1}{\sqrt{2}}, -\sqrt{2}\right)$ and $\left(\frac{1}{\sqrt{2}}, -\frac{1}{\sqrt{2}}, \sqrt{2}\right)$.

86. Let $f(x, y, z) = x^2 + y^2 + z^2$ be the squre of the distance to the origin. Then $\nabla f = 2x\mathbf{i} + 2y\mathbf{j} + 2z\mathbf{k}$, $\nabla g = \mathbf{i} + \mathbf{j} + \mathbf{k}$, and $\nabla h = 4x\mathbf{i} + 4y\mathbf{j} - 2z\mathbf{k}$ so that $\nabla f = \lambda \nabla g + \mu \nabla h \Rightarrow 2x = \lambda + 4x\mu$, $2y = \lambda + 4y\mu$, and $2z = \lambda - 2z\mu \Rightarrow \lambda = 2x(1 - 2\mu) = 2y(1 - 2\mu) = 2z(1 + \mu) \Rightarrow x = y$ or $\mu = \frac{1}{2}$.

CASE 1: $x = y \Rightarrow z^2 = 4x^2 \Rightarrow z = \pm 2x$ so that $x + y + z = 1 \Rightarrow x + x + 2x = 1$ or $x + x - 2x = 1$ (impossible) $\Rightarrow x = \frac{1}{4} \Rightarrow y = \frac{1}{4}$ and $z = \frac{1}{2}$ yielding the point $\left(\frac{1}{4}, \frac{1}{4}, \frac{1}{2}\right)$.

CASE 2: $\mu = \frac{1}{2} \Rightarrow \lambda = 0 \Rightarrow 0 = 2z\left(1 + \frac{1}{2}\right) \Rightarrow z = 0$ so that $2x^2 + 2y^2 = 0 \Rightarrow x = y = 0$. But the origin $(0, 0, 0)$ fails to satisfy the first constraint $x + y + z = 1$.

Therefore, the point $\left(\frac{1}{4}, \frac{1}{4}, \frac{1}{2}\right)$ on the curve of intersection is closest to the origin.

87. Note that $x = r\cos\theta$ and $y = r\sin\theta \Rightarrow r = \sqrt{x^2+y^2}$ and $\theta = \tan^{-1}\left(\frac{y}{x}\right)$. Thus, $\frac{\partial w}{\partial x} = \frac{\partial w}{\partial r}\frac{\partial r}{\partial x} + \frac{\partial w}{\partial \theta}\frac{\partial \theta}{\partial x}$ $= \left(\frac{\partial w}{\partial r}\right)\left(\frac{x}{\sqrt{x^2+y^2}}\right) + \left(\frac{\partial w}{\partial \theta}\right)\left(\frac{-y}{x^2+y^2}\right) = (\cos\theta)\frac{\partial w}{\partial r}$ $- \left(\frac{\sin\theta}{r}\right)\frac{\partial w}{\partial \theta}$; $\frac{\partial w}{\partial y} = \frac{\partial w}{\partial r}\frac{\partial r}{\partial y} + \frac{\partial w}{\partial \theta}\frac{\partial \theta}{\partial y} = \left(\frac{\partial w}{\partial r}\right)\left(\frac{y}{\sqrt{x^2+y^2}}\right)$ $+ \left(\frac{\partial w}{\partial \theta}\right)\left(\frac{x}{x^2+y^2}\right) = (\sin\theta)\frac{\partial w}{\partial r} + \left(\frac{\cos\theta}{r}\right)\frac{\partial w}{\partial \theta}$

88. $z_x = f_u\frac{\partial u}{\partial x} + f_v\frac{\partial v}{\partial x} = af_u + af_v$, and $z_y = f_u\frac{\partial u}{\partial y} + f_v\frac{\partial v}{\partial y}$ $= bf_u - bf_v$

89. $\frac{\partial w}{\partial x} = \frac{2x}{x^2+y^2+2z} = \frac{2(r+s)}{(r+s)^2+(r-s)^2+4rs}$ $= \frac{2(r+s)}{2(r^2+2rs+s^2)} = \frac{1}{r+s}$, $\frac{\partial w}{\partial y} = \frac{2y}{x^2+y^2+2z} = \frac{2(r-s)}{2(r+s)^2}$ $= \frac{r-s}{(r+s)^2}$, and $\frac{\partial w}{\partial z} = \frac{2}{x^2+y^2+2z} = \frac{1}{(r+s)^2} \Rightarrow \frac{\partial w}{\partial r}$ $= \frac{\partial w}{\partial x}\frac{\partial x}{\partial r} + \frac{\partial w}{\partial y}\frac{\partial y}{\partial r} + \frac{\partial w}{\partial z}\frac{\partial z}{\partial r} = \frac{1}{r+s} + \frac{r-s}{(r+s)^2}$ $+ \left[\frac{1}{(r+s)^2}\right](2s) = \frac{2r+2s}{(r+s)^2} = \frac{2}{r+s}$ and $\frac{\partial w}{\partial s} = \frac{\partial w}{\partial x}\frac{\partial x}{\partial s}$ $+ \frac{\partial w}{\partial y}\frac{\partial y}{\partial s} + \frac{\partial w}{\partial z}\frac{\partial z}{\partial s} = \frac{1}{r+s} - \frac{r-s}{(r+s)^2} + \left[\frac{1}{(r+s)^2}\right](2r)$ $= \frac{2}{r+s}$

90. $e^u\cos v - x = 0 \Rightarrow (e^u\cos v)\frac{\partial u}{\partial x} - (e^u\sin v)\frac{\partial v}{\partial x} = 1$; $e^u\sin v - y = 0 \Rightarrow (e^u\sin v)\frac{\partial u}{\partial x} + (e^u\cos v)\frac{\partial v}{\partial x} = 0$. Solving this system yields $\frac{\partial u}{\partial x} = e^{-u}\cos v$ and $\frac{\partial v}{\partial x}$ $= -e^{-u}\sin v$. Similarly, $e^u\cos v - x = 0 \Rightarrow (e^u\cos v)\frac{\partial u}{\partial y}$ $- (e^u\sin v)\frac{\partial v}{\partial y} = 0$ and $e^u\sin v - y = 0 \Rightarrow (e^u\sin v)\frac{\partial u}{\partial y}$ $+ (e^u\cos v)\frac{\partial v}{\partial y} = 1$. Solving this second system yields $\frac{\partial u}{\partial y}$ $= e^{-u}\sin v$ and $\frac{\partial v}{\partial y} = e^{-u}\cos v$. Therefore $\left(\frac{\partial u}{\partial x}\mathbf{i} + \frac{\partial u}{\partial y}\mathbf{j}\right)$ $\cdot\left(\frac{\partial v}{\partial x}\mathbf{i} + \frac{\partial v}{\partial y}\mathbf{j}\right) = [(e^{-u}\cos v)\mathbf{i} + (e^{-u}\sin v)\mathbf{j}]$ $\cdot[(-e^{-u}\sin v)\mathbf{i} + (e^{-u}\cos v)\mathbf{j}] = 0 \Rightarrow$ the vectors are orthogonal $\Rightarrow$ the angle between the vectors is the constant $\frac{\pi}{2}$.

91. $(y+z)^2 + (z-x)^2 = 16 \Rightarrow \nabla f = -2(z-x)\mathbf{i} + 2(y+z)\mathbf{j}$ $+ 2(y+2z-x)\mathbf{k}$; if the normal line is parallel to the yz-plane, then x is constant $\Rightarrow \frac{\partial f}{\partial x} = 0 \Rightarrow -2(z-x) = 0$ $\Rightarrow z = x \Rightarrow (y+z)^2 + (z-z)^2 = 16 \Rightarrow y + z = \pm 4$. Let $x = t \Rightarrow z = t \Rightarrow y = -t \pm 4$. Therefore the points are $(t, -t \pm 4, t)$, t a real number.

92. Let $f(x, y, z) = xy + yz + zx - x - z^2 = 0$. If the tangent plane is to be parallel to the xy-plane, then ∇f is perpendicular to the xy-plane $\Rightarrow \nabla f \cdot \mathbf{i} = 0$ and $\nabla f \cdot \mathbf{j} = 0$. Now $\nabla f = (y + z - 1)\mathbf{i} + (x + z)\mathbf{j} + (y + x - 2z)\mathbf{k}$ so that $\nabla f \cdot \mathbf{i} = y + z - 1 = 0 \Rightarrow y + z = 1 \Rightarrow y = 1 - z$, and $\nabla f \cdot \mathbf{j} = x + z = 0 \Rightarrow x = -z$. Then $-z(1-z)$ $+ (1-z)z + z(-z) - (-z) - z^2 = 0 \Rightarrow z - 2z^2 = 0$ $\Rightarrow z = \frac{1}{2}$ or $z = 0$. Now $z = \frac{1}{2} \Rightarrow x = -\frac{1}{2}$ and $y = \frac{1}{2}$ $\Rightarrow \left(-\frac{1}{2}, \frac{1}{2}, \frac{1}{2}\right)$ is one desired point; $z = 0 \Rightarrow x = 0$ and $y = 1 \Rightarrow (0, 1, 0)$ is a second desired point.

93. It is possible to travel from $(0, 0, a)$ to $(0, 0, -a)$ along a circular arc centered at the origin, and since ∇f is everywhere normal to such an arc, the value of f does not change along the arc.

94. $\left(\frac{df}{ds}\right)_{\mathbf{u},(0,0,0)} = \lim\limits_{s\to 0}\frac{f(0+su_1, 0+su_2, 0+su_3) - f(0,0,0)}{s}$, $= \lim\limits_{s\to 0}\frac{\sqrt{s^2u_1^2 + s^2u_2^2 + s^2u_3^2} - 0}{s}$ $= \lim\limits_{s\to 0}\frac{s\sqrt{u_1^2+u_2^2+u_3^2}}{s} = \lim\limits_{s\to 0}|\mathbf{u}| = 1$; however, $\nabla f = \frac{x}{\sqrt{x^2+y^2+z^2}}\mathbf{i} + \frac{y}{\sqrt{x^2+y^2+z^2}}\mathbf{k}$ $+ \frac{z}{\sqrt{x^2+y^2+z^2}}\mathbf{k}$ fails to exist at the origin $(0, 0, 0)$

95. Let $f(x, y, z) = xy + z - 2 \Rightarrow \nabla f = y\mathbf{i} + x\mathbf{j} + \mathbf{k}$. At $(1, 1, 1)$, we have $\nabla f = \mathbf{i} + \mathbf{j} + \mathbf{k} \Rightarrow$ the normal line is $x = 1 + t$, $y = 1 + t$, $z = 1 + t$ so at $t = -1 \Rightarrow x = 0$, $y = 0$, $z = 0$ and the normal line passes through the origin.

96. **a.**

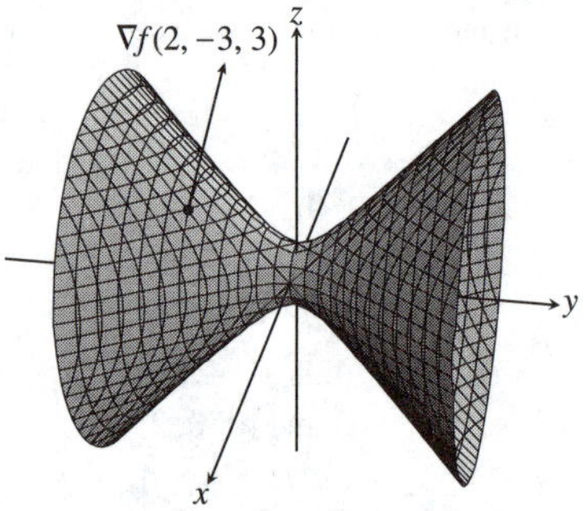

$x^2 - y^2 + z^2 = 4$

b. $f(x, y, z) = x^2 - y^2 + z^2 = 4 \Rightarrow \nabla f = 2x\mathbf{i} - 2y\mathbf{j} + 2z\mathbf{k}$ $\Rightarrow$ at $(2, -3, 3)$ the gradient is $\nabla f = 4\mathbf{i} + 6\mathbf{j} + 6\mathbf{k}$ which is normal to the surface

c. Tangent plane: $4x + 6y + 6z = 8$ or $2x + 3y + 3z = 4$
Normal line: $x = 2 + 4t$, $y = -3 + 6t$, $z = 3 + 6t$

Chapter 14
Multiple Integrals

Section 14.1 Double Integrals (pp.761–773)

Chapter Opener

The kinetic energy of the body will be given by

$$KE = \frac{1}{2} I\,\omega^2$$

where $I = mr^2$, m is the mass, r is the radius and ω is the angular velocity.

Since a unit of measure for KE is Joules and J = 1 Joule = kg m^2/sec^2, we will convert the disk's given measurements to match.

$$\text{density} = \delta = 8.89\,\frac{\text{gm}}{\text{cm}^3} = 8890\,\frac{\text{kg}}{\text{m}^3}$$

volume $= \pi(0.06)^2 \cdot (0.005) = 1.8 \times 10^{-5}\pi\, m^3$

mass = density · volume $= 1.8 \times 10^{-5}(8890) =$ 0.16002kg

$I = (0.16002)(.06)^2 = 5.76072 \times 10^{-4}$ kg m^2

$$KE = \left(\frac{1}{2}\right) \cdot 5.76072 \times 10^{-4} \cdot 1800^2 = 933.2366 \text{ J}$$

Quick Review 14.1

1. $\int_0^3 x^3\,dx$

2. $\int_{-1}^{1} \sqrt[3]{2 - x^2}\,dx$

3. $e^{\sin x} + C$

4. $-2\ln(|x - 1|) + C$

5. $e^x + \frac{5}{x} + C$

6. $\sqrt{2} - 1$

7. $\frac{11}{2}$

8. ≈ 1.89

9. 4

10. $\frac{512\pi}{15} \approx 107.23$

Section 14.1 Exercises

1. $\int_0^3 \int_0^2 (4-y^2)\,dy\,dx = \int_0^3 \left[4y - \frac{y^3}{3}\right]_0^2 dx = \frac{16}{3}\int_0^3 dx = 16$

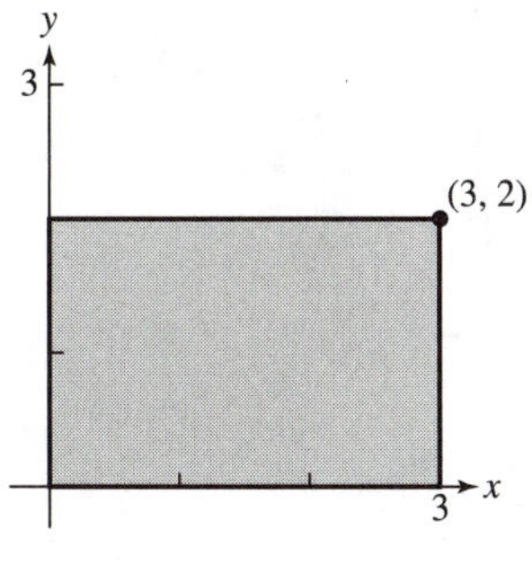

3. $\int_{-1}^{0} \int_{-1}^{1} (x + y + 1)\,dx\,dy = \int_{-1}^{0} \left[\frac{x^2}{2} + yx + x\right]_{-1}^{1} dy$

$= \int_{-1}^{0} (2y + 2)\,dy$

$= \left[y^2 + 2y\right]_{-1}^{0} = 1$

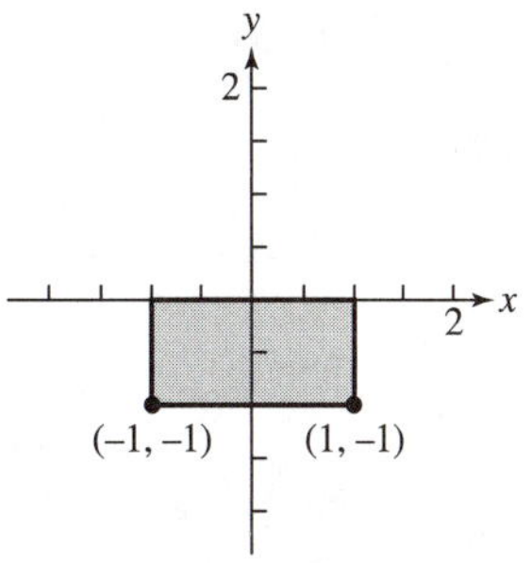

5. $\int_0^{\pi} \int_0^{x} (x \sin y)\,dy\,dx = \int_0^{\pi} [-x \cos y]_0^x \; dx$

$= \int_0^{\pi} (x - x\cos x)dx = \left[\frac{x^2}{2} - (\cos x + x \sin x)\right]_0^{\pi}$

$= \frac{\pi^2}{2} + 2$

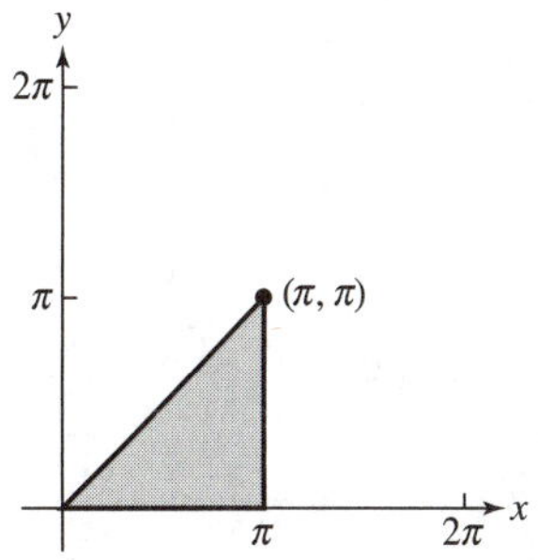

7. $\int_1^{\ln 8} \int_0^{\ln y} e^{x+y}\,dx\,dy = \int_1^{\ln 8} [e^{x+y}]_0^{\ln y}\,dy$

$= \int_1^{\ln 8} (ye^y - e^y)\,dy = [(y-1)e^y - e^y)]_0^{\ln 8}$

$= 8(\ln 8 - 1) - 8 + e = 8\ln 8 - 16 + e$

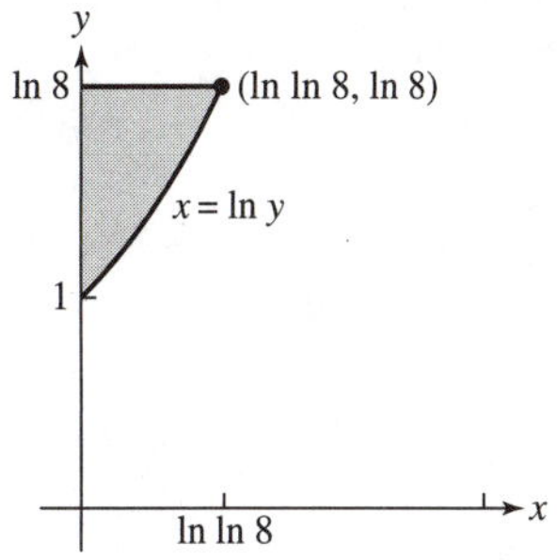

9. $\int_0^1 \int_0^{y^2} 3y^3 e^{xy}\,dx\,dy = \int_0^1 [3y^2e^{xy}] \quad dy$

$= \int_0^1 (3y^2e^{y^3} - 3y^2)\,dy = [e^{y^3} - y^3] \quad = e - 2$

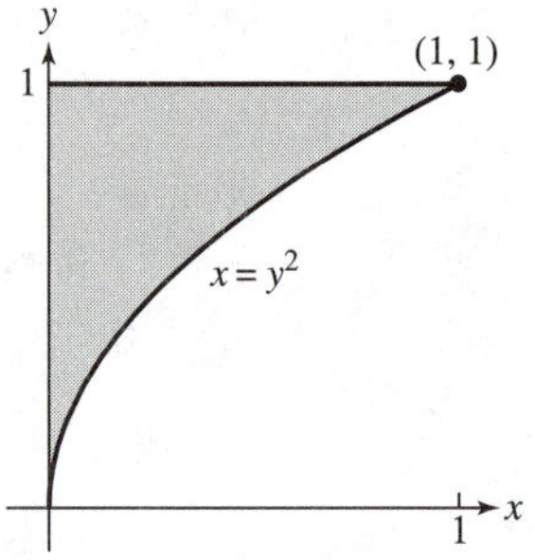

11. $\int_1^2 \int_x^{2x} \frac{x}{y}\,dy\,dx = \int_1^2 [x \ln y]\,dx = (\ln 2)\int_1^2 x\,dx = \frac{3}{2}\ln 2$

13. $\int_0^1 \int_0^{1-x} (x^2 + y^2)\,dy\,dx = \int_0^1 \left[x^2y + \frac{y^3}{3}\right]_0^{1-x} dx$

$= \int_0^1 \left[x^2(1-x) + \frac{(1-x)^3}{3}\right] dx$

$= \int_0^1 \left[x^2 - x^3 + \frac{(1-x)^3}{3}\right] dx$

$= \left[\frac{x^3}{3} - \frac{x^4}{4} - \frac{(1-x)^4}{12}\right]_1^4$

$= \left(\frac{1}{3} - \frac{1}{4} - 0\right) - \left(0 - 0 - \frac{1}{12}\right) = \frac{1}{6}$

15. $\int_0^1 \int_0^{1-u} (v - \sqrt{u})\,dv\,du = \int_0^1 \left[\frac{v^2}{2} - x\sqrt{u}\right]_0^{1-u} du$

$= \int_0^1 \left[\frac{1 - 2u + u^2}{2} - \sqrt{u}(1-u)\right] du$

$= \int_0^1 \left(\frac{1}{2} - u + \frac{u^2}{2} - u^{1/2} + u^{3/2}\right) du$

$= \left[\frac{u}{2} - \frac{u^2}{2} + \frac{u^3}{6} - \frac{2}{3}u^{3/2} + \frac{2}{5}u^{5/2}\right]_0^1$

$= \frac{1}{2} - \frac{1}{2} + \frac{1}{6} - \frac{2}{3} + \frac{2}{5} = -\frac{1}{2} + \frac{2}{5} = -\frac{1}{10}$

17. $\int_{-2}^0 \int_v^{-v} 2\,dp\,dv = 2\int_{-2}^0 [p]_v^{-v}\,dv = 2\int_{-2}^0 -2v\,dv$

$= -2[v^2]_{-2}^0 = 8$

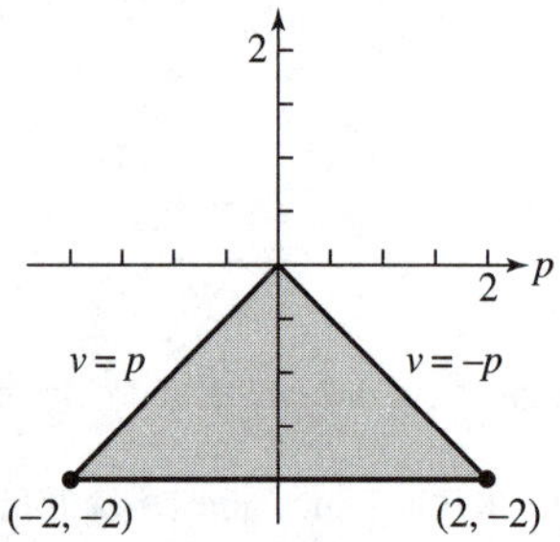

19. $\int_{-\pi/3}^{\pi/3} \int_0^{\sec t} 3\cos t\,du\,dt = \int_{-\pi/3}^{\pi/3} [(3\cos t)u]_0^{\sec t}$

$= \int_{-\pi/3}^{\pi/3} 3\,dt = 2\pi$

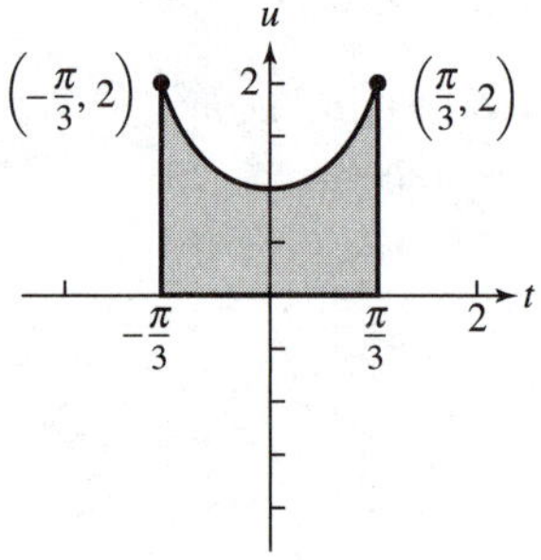

21. $\int_2^4 \int_0^{(4-y)/2} dx\,dy$

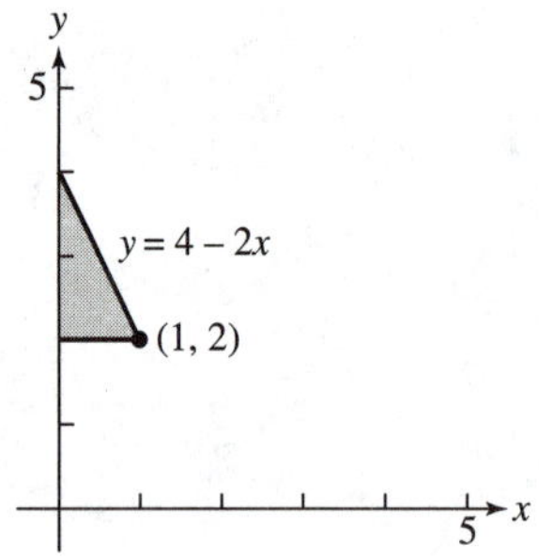

23. $\int_0^1 \int_{x^2}^x dy\,dx$

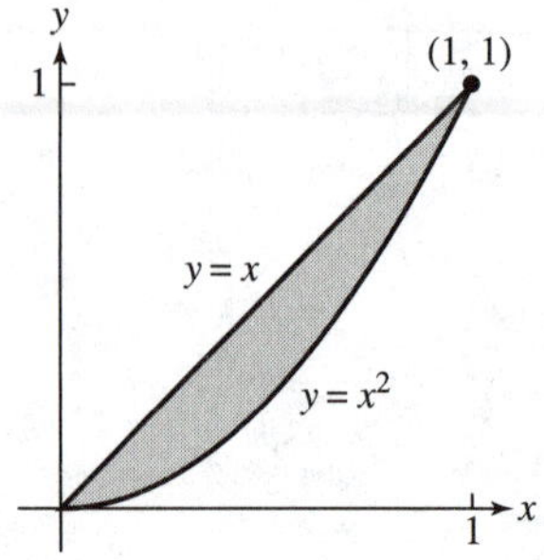

25. $\int_1^e \int_{\ln y}^1 dx\,dy$

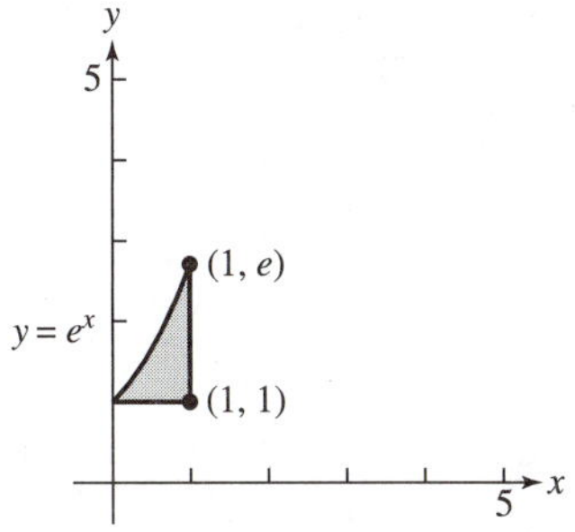

27. $\int_0^9 \int_0^{1/2\sqrt{9-y}} 16x\,dx\,dy$

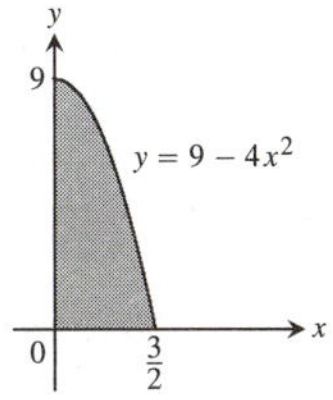

29. $\int_{-1}^1 \int_0^{\sqrt{1-x^2}} 3y\,dy\,dx$

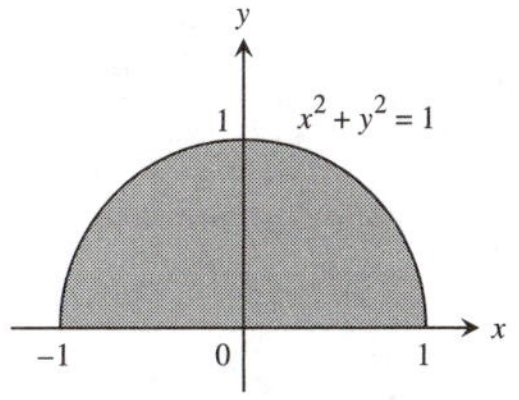

31. $\int_0^\pi \int_x^\pi \frac{\sin y}{y}\,dy\,dx = \int_0^\pi \int_0^y \frac{\sin y}{y}\,dx\,dy = \int_0^\pi \sin y\,dy = 2$

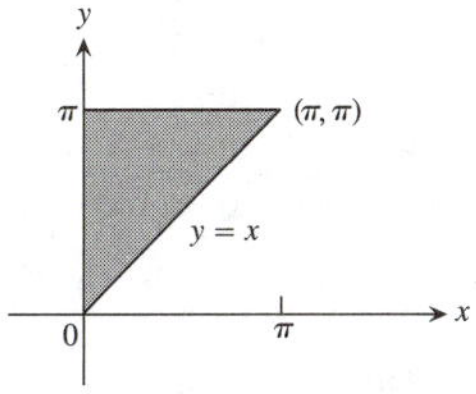

33. $\int_0^1 \int_y^1 x^2 e^{xy}\,dx\,dy = \int_0^1 \int_0^x x^2 e^{xy}\,dy\,dx = \int_0^1 [xe^{xy}]\,dx$

$= \int_0^1 (xe^{x^2} - x)\,dx = \left[\frac{1}{2}e^{x^2} - \frac{x^2}{2}\right]_0^1 = \frac{e-2}{2}$

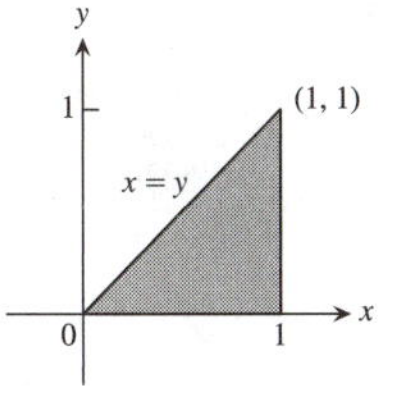

35. $\int_0^{2\sqrt{\ln 3}} \int_{y/2}^{\sqrt{\ln 3}} e^{x^2}\,dx\,dy = \int_0^{\sqrt{\ln 3}} \int_0^{2x} e^{x^2}\,dy\,dx$

$= \int_0^{\sqrt{\ln 3}} 2xe^{x^2}\,dx = [e^{x^2}]_0^{\sqrt{\ln 3}} = e^{\ln 3} - 1 = 2$

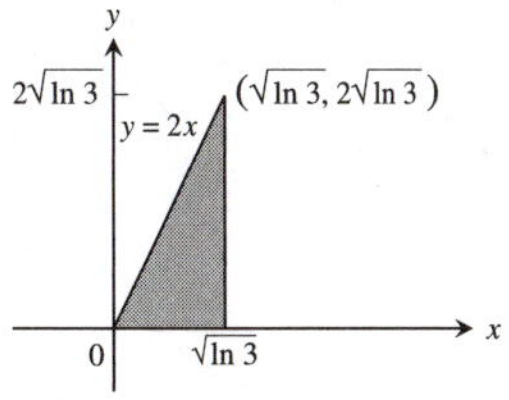

37. $\int_0^{1/16} \int_{y^{1/4}}^{1/2} \cos(16\pi x^5)\,dx\,dy = \int_0^{1/2} \int_0^{x^4} \cos(16\pi x^5)\,dy\,dx$

$= \int_0^{1/2} x^4 \cos(16\pi x^5)\,dx = \left[\frac{\sin(16\pi x^5)}{80\pi}\right]_0^{1/2} = \frac{1}{80\pi}$

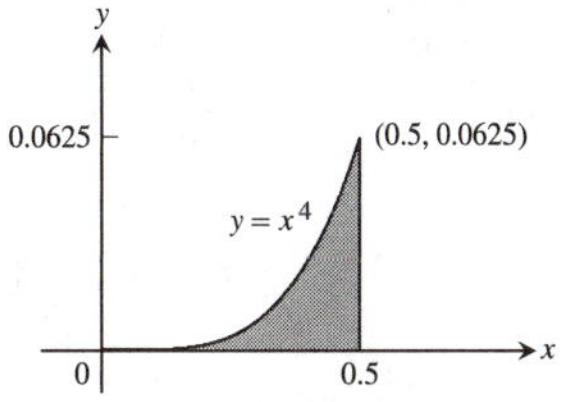

39. $\int\int_R (y - 2x^2)\,dA = \int_{-1}^0 \int_{-x-1}^{x+1} (y - 2x^2)\,dy\,dx$

$+ \int_0^1 \int_{x-1}^{1-x} (y - 2x^2)\,dy\,dx$

$= \int_{-1}^0 \left[\frac{1}{2}y^2 - 2x^2y\right]_{-x-1}^{x+1} dx + \int_0^1 \left[\frac{1}{2}y^2 - 2x^2y\right]_{x-1}^{1-x} dx$

$= \int_{-1}^0 \left[\frac{1}{2}(x+1)^2 - 2x^2(x+1) - \frac{1}{2}(-x-1)^2\right.$

$\left. + 2x^2(-x-1)\right] dx + \int_0^1 \left[\frac{1}{2}(1-x)^2 - 2x^2(1-x)\right.$

$\left. - \frac{1}{2}(x-1)^2 + 2x^2(x-1)\right] dx$

$= 4\int_{-1}^0 (x^3 + x^2)\,dx + 4\int_0^1 (x^3 - x^2)\,dx$

$= 4\left[\frac{x^4}{4} + \frac{x^3}{3}\right]_{-1}^0 + 4\left[\frac{x^4}{4} - \frac{x^3}{3}\right]_0^1$

$= -4\left[\frac{(-1)^4}{4} + \frac{(-1)^3}{3}\right] + 4\left(\frac{1}{4} - \frac{1}{3}\right)$

$= 8\left(\frac{3}{12} - \frac{4}{12}\right) = -\frac{8}{12} = -\frac{2}{3}$

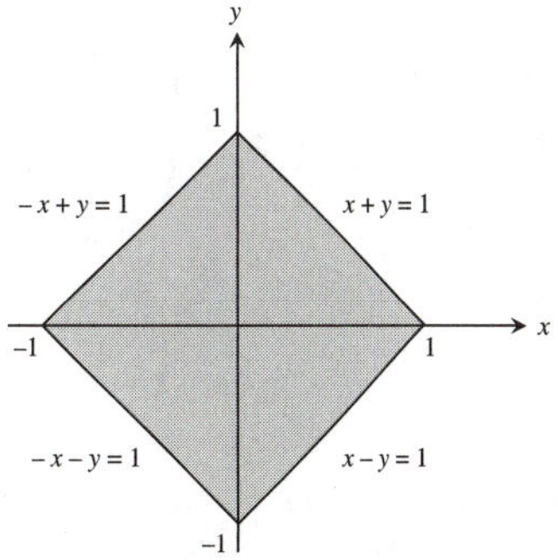

41. $V = \int_0^1 \int_x^{2-x} (x^2 + y^2)\, dy\, dx = \int_0^1 \left[x^2y + \frac{y^3}{3}\right]_x^{2-x} dx$

$= \int_0^1 \left[2x^2 - \frac{7x^3}{3} + \frac{(2-x)^3}{3}\right] dx$

$= \left[\frac{2x^3}{3} - \frac{7x^4}{12} - \frac{(2-x)^4}{12}\right]_0^1$

$= \left(\frac{2}{3} - \frac{7}{12} - \frac{1}{12}\right) - \left(0 - 0 - \frac{16}{12}\right)$

$= \frac{4}{3}$

43. $V = \int_{-4}^1 \int_{3x}^{4-x^2} (x + 4)\, dy\, dx = \int_{-4}^1 [xy + 4y]_{3x}^{4-x^2} dx$

$= \int_{-4}^1 [x(4 - x^2) + 4(4 - x^2) - 3x^2 - 12x]\, dx$

$= \int_{-4}^1 (-x^3 - 7x^2 - 8x + 16)\, dx$

$= \left[-\frac{1}{4}x^4 - \frac{7}{3}x^3 - 4x^2 + 16x\right]_{-4}^1$

$= \left(-\frac{1}{4} - \frac{7}{3} + 12\right) - \left(\frac{64}{3} - 64\right)$

$= \frac{157}{3} - \frac{1}{4} = \frac{625}{12}$

45. $V = \int_0^2 \int_0^3 (4 - y^2)\, dx\, dy = \int_0^2 [4x - y^2x]_0^3\, dy$

$= \int_0^2 (12 - 3y^2)\, dy = [12y - y^3]_0^2$

$= 24 - 8 = 16$

47. $V = \int_0^2 \int_0^{2-x} (12 - 3y^2)\, dy\, dx = \int_0^2 [12y - y^3]_0^{2-x} dx$

$= \int_0^2 [24 - 12x - (2 - x)^3]\, dx$

$= \left[24x - 6x^2 + \frac{(2-x)^4}{4}\right]_0^2 = 20$

49. $V = \int_1^2 \int_{-1/x}^{1/x} (x + 1)\, dy\, dx = \int_1^2 [xy + y]_{-1/x}^{1/x} dx$

$= \int_1^2 \left[1 + \frac{1}{x} - \left(-1 - \frac{1}{x}\right)\right] = 2\int_1^2 \left(1 + \frac{1}{x}\right) dx$

$= 2[x + \ln x]_1^2 = 2(1 + \ln 2)$

51. $\int_1^\infty \int_{e^{-x}}^1 \frac{1}{x^3y}\, dy\, dx = \int_1^\infty \left[\frac{\ln y}{x^3}\right]_{e^{-x}}^1 dx = \int_1^\infty -\left(\frac{-x}{x^3}\right) dx$

$= -\lim_{b\to\infty}\left[\frac{1}{x}\right]_1^b = -\lim_{b\to\infty}\left(\frac{1}{b} - 1\right) = 1$

53. $\int_{-\infty}^\infty \int_{-\infty}^\infty \frac{1}{(x^2+1)(y^2+1)}\, dx\, dy$

$= 2\int_0^\infty \left(\frac{2}{y^2+1}\right)\left(\lim_{b\to\infty} \tan^{-1} b - \tan^{-1} 0\right) dy$

$= 2\pi \lim_{b\to\infty} \int_0^b \frac{1}{y^2+1}\, dy$

$= 2\pi\left(\lim_{b\to\infty} \tan^{-1} b - \tan^{-1} 0\right) = (2\pi)\left(\frac{\pi}{2}\right) = \pi^2$

55. $\int\int_R f(x, y)\, dA \approx \frac{1}{4}f\left(-\frac{1}{2}, 0\right) + \frac{1}{8}f(0, 0) + \frac{1}{8}f\left(\frac{1}{4}, 0\right)$

$+ \frac{1}{4}f\left(\frac{1}{2}, 0\right) + \frac{1}{4}f\left(-\frac{1}{2}, \frac{1}{2}\right) + \frac{1}{8}f\left(0, \frac{1}{2}\right)$

$+ \frac{1}{8}f\left(\frac{1}{4}, \frac{1}{2}\right)$

$= \frac{1}{4}\left(-\frac{1}{2} + \frac{1}{2} + 0\right) + \frac{1}{8}\left(0 + \frac{1}{4} + \frac{1}{2} + \frac{3}{4}\right)$

$= \frac{3}{16}$

57. The ray $\theta = \frac{\pi}{6}$ meets the circle $x^2 + y^2 = 4$ at the point $(\sqrt{3}, 1) \Rightarrow$ the ray is represented by the line $y = \frac{x}{\sqrt{3}}$.

Thus $\int\int_R f(x, y)\, dA = \int_0^{\sqrt{3}} \int_{x/\sqrt{3}}^{\sqrt{4-x^2}} \sqrt{4 - x^2}\, dy\, dx$

$= \int_0^{\sqrt{3}} \left[(4 - x^2) - \frac{x}{\sqrt{3}}\sqrt{4 - x^2}\right] dx$

$= \left[4x - \frac{x^3}{3} + \frac{(4 - x^2)^{3/2}}{3\sqrt{3}}\right]_0^{\sqrt{3}}$

$= \frac{20\sqrt{3}}{9}$

59.

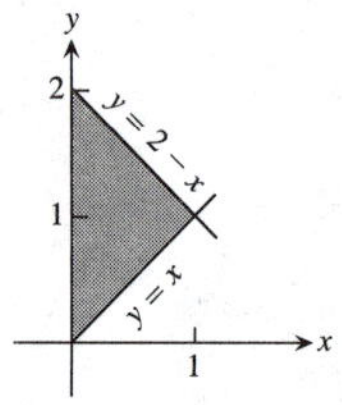

$V = \int_0^1 \int_x^{2-x} (x^2 + y^2)\, dy\, dx = \int_0^1 \left[x^2y + \frac{y^3}{3}\right]_x^{2-x} dx$

$= \int_0^1 \left[2x^2 - \frac{7x^3}{3} + \frac{(2-x)^3}{3}\right] dx$

$= \left[\frac{2x^3}{3} - \frac{7x^4}{12} - \frac{(2-x)^4}{12}\right]_0^1$

$= \left(\frac{2}{3} - \frac{7}{12} - \frac{1}{12}\right) - \left(0 - 0 - \frac{16}{12}\right)$

$= \left(\frac{2}{3} + \frac{8}{12}\right) = \frac{4}{3}$

61. To maximize the integral, we want the domain to include all points where the integrand is positive and to exclude all points where the integrand is negative. These criteria are met by the points (x, y) such that $4 - x^2 - 2y^2 \geq 0$ or $x^2 + 2y^2 \leq 4$, which is the ellipse $x^2 + 2y^2 = 4$ together with is interior.

63. No, it is not all right. By Fubini's theorem, the two orders of integration must give the same result.

65. $\int_{-b}^{b}\int_{-b}^{b} e^{-x^2-y^2}\,dx\,dy = \int_{-b}^{b}\int_{-b}^{b} e^{-y^2}e^{-x^2}\,dx\,dy$

$= \int_{-b}^{b} e^{-y^2}\left(\int_{-b}^{b} e^{-x^2}\,dx\right)dy$

$= \left(\int_{-b}^{b} e^{-x^2}\,dx\right)\left(\int_{-b}^{b} e^{-y^2}\,dy\right)$

$= \left(\int_{-b}^{b} e^{-x^2}\,dx\right)^2 = \left(2\int_{0}^{b} e^{-x^2}\,dx\right)^2$

$= 4\left(\int_{0}^{b} e^{-x^2}\,dx\right)^2;$

taking limits as $b \to \infty$ gives the stated result.

67. $\int_{1}^{3}\int_{1}^{x} \frac{1}{xy}\,dy\,dx \approx 0.603$

69. $\int_{0}^{1}\int_{0}^{1} \tan^{-1} xy\,dy\,dx \approx 0.233$

■ Section 14.2 Areas, Moments, and Centers of Mass (pp 773-783)

Quick Review 14.2

1. $e - 1$

2. 1

3. $\int_{-2}^{2} (2 - x^2 + 2)\,dx = \frac{32}{3}$

4. $\int_{-2}^{2} 2\sqrt{2 - y}\,dy = \frac{32}{3}$

5. $\frac{2}{\pi} \approx 0.64$

6. $\frac{32}{3}$

9. $\bar{x} = 0, \bar{y} = \frac{12}{5}$

10. $\bar{x} = \frac{4}{3}, \bar{y} = 0$

Section 14.2 Exercises

1. $\int_{0}^{2}\int_{0}^{2-x} dy\,dx = \int_{0}^{2} (2 - x)\,dx = \left[2x - \frac{x^2}{2}\right]_0^2 = 2,$

or $\int_{0}^{2}\int_{0}^{2-y} dx\,dy = \int_{0}^{2} (2 - y)\,dy = 2$

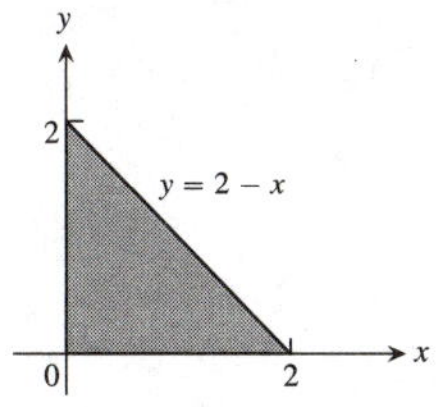

3. $\int_{-2}^{1}\int_{y-2}^{-y^2} dx\,dy = \int_{-2}^{1} (-y^2 - y + 2)\,dy$

$= \left[-\frac{y^3}{3} - \frac{y^2}{2} + 2y\right]_{-2}^{1}$

$= \left(-\frac{1}{3} - \frac{1}{2} + 2\right) - \left(\frac{8}{3} - 2 - 4\right) = \frac{9}{2}$

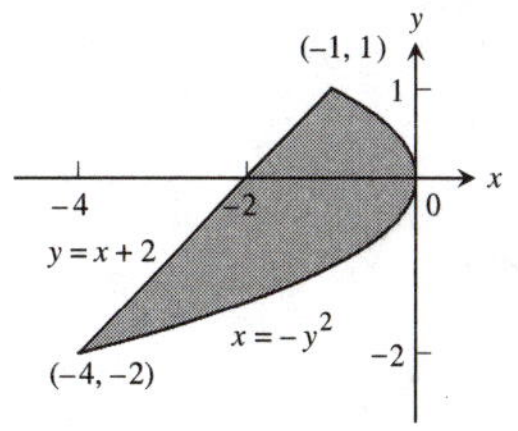

5. $\int_{0}^{\ln 2}\int_{0}^{e^x} dy\,dx = \int_{0}^{\ln 2} e^x\,dx = [e^x]_0^{\ln 2} = 2 - 1 = 1$

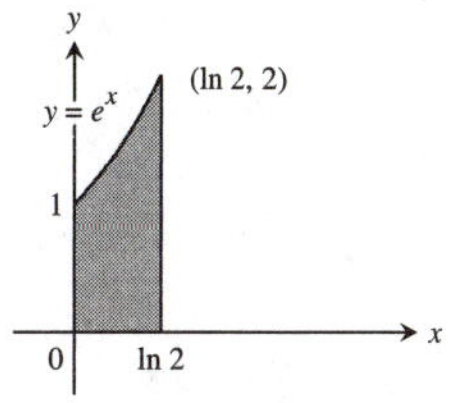

7. $\int_{0}^{1}\int_{y^2}^{2y-y^2} dx\,dy = \int_{0}^{1} (2y - 2y^2)\,dy = \left[y^2 - \frac{2}{3}y^3\right]_0^1 = \frac{1}{3}$

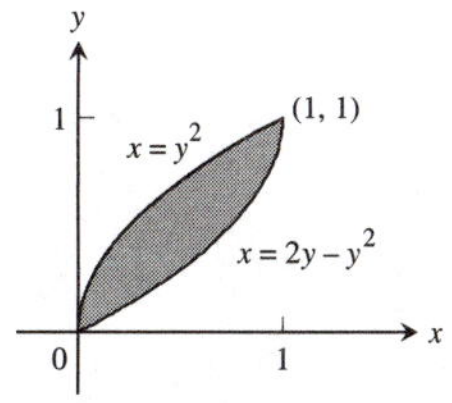

9. $\int_{0}^{6}\int_{y^2/3}^{2y} dx\,dy = \int_{0}^{6} \left(2y - \frac{y^2}{3}\right)dy = \left[y^2 - \frac{y^3}{9}\right]_0^6$

$= 35 - \frac{216}{9} = 12$

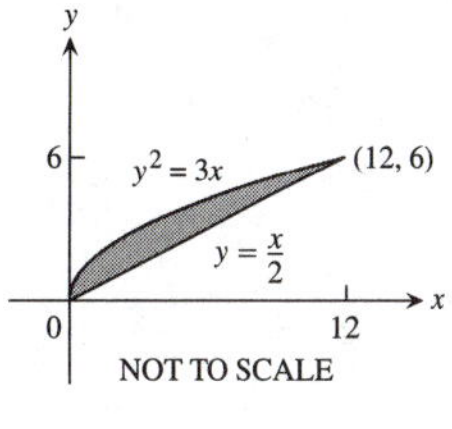

11. $\int_0^{\pi/4}\int_{\sin x}^{\cos x} dy\,dx = \int_0^{\pi/4} (\cos x - \sin x)\,dx$

$= [\sin x + \cos x]_0^{\pi/4}$

$= \left(\frac{\sqrt{2}}{2} + \frac{\sqrt{2}}{2}\right) - (0 + 1) = \sqrt{2} - 1$

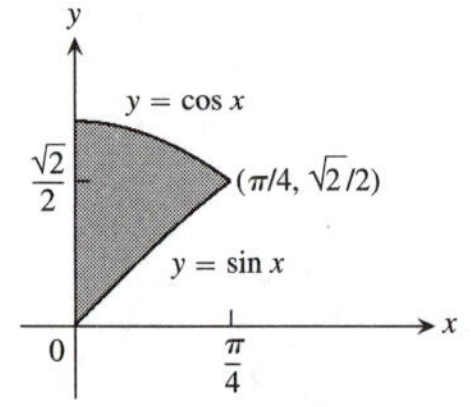

13. $\int_{-1}^{0}\int_{-2x}^{1-x} dy\,dx + \int_0^2\int_{-x/2}^{1-x} dy\,dx = \int_{-1}^{0} (1 + x)\,dx + \int_0^2 \left(1 - \frac{x}{2}\right) dx$

$= \left[x + \frac{x^2}{2}\right]_{-1}^{0} + \left[x - \frac{x^2}{4}\right]_0^2$

$= -\left(-1 + \frac{1}{2}\right) + (2 - 1) = \frac{3}{2}$

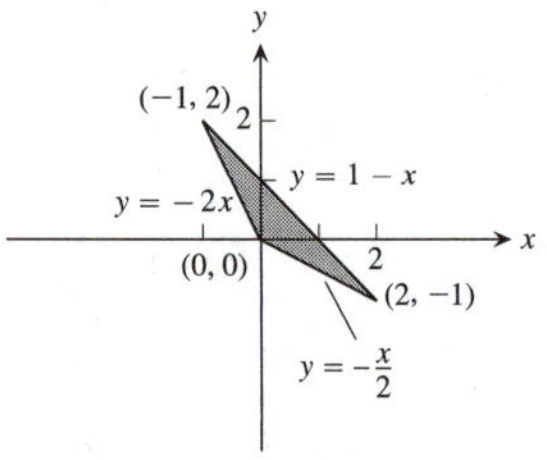

15. a. average $= \frac{1}{\pi^2}\int_0^{\pi}\int_0^{\pi} \sin(x + y)\,dy\,dx$

$= \frac{1}{\pi^2}\int_0^{\pi} [-\cos(x + y)]_0^{\pi}\,dx$

$= \frac{1}{\pi^2}\int_0^{\pi} [-\cos(x + \pi) + \cos x]\,dx$

$= \frac{1}{\pi^2} [-\sin(x + \pi) + \sin x]_0^{\pi}$

$= \frac{1}{\pi^2} [(-\sin 2\pi + \sin \pi) - (-\sin \pi + \sin 0)]$

$= 0$

b. average $= \frac{1}{\left(\frac{\pi^2}{2}\right)}\int_0^{\pi}\int_0^{\pi/2} \sin(x + y)\,dy\,dx$

$= \frac{2}{\pi^2}\int_0^{\pi} [-\cos(x + y)]_0^{\pi/2}\,dx$

$= \frac{2}{\pi^2}\int_0^{\pi} \left[-\cos\left(x + \frac{\pi}{2}\right) + \cos x\right] dx$

$= \frac{2}{\pi^2}\left[-\sin\left(x + \frac{\pi}{2}\right) + \sin x\right]_0^{\pi}$

$= \frac{2}{\pi^2}\left[\left(-\sin\frac{3\pi}{2} + \sin \pi\right) - \left(-\sin\frac{\pi}{2} + \sin 0\right)\right]$

$= \frac{4}{\pi^2}$

17. average height $= \frac{1}{4}\int_0^2\int_0^2 (x^2 + y^2)\,dy\,dx$

$= \frac{1}{4}\int_0^2 \left[x^2 y + \frac{y^3}{3}\right]_0^2 dx$

$= \frac{1}{4}\int_0^2 \left(2x^2 + \frac{8}{3}\right) dx$

$= \frac{1}{2}\left[\frac{x^3}{3} + \frac{4x}{3}\right]_0^2 = \frac{8}{3}$

19. $M = \int_0^1\int_x^{2-x^2} 3\,dy\,dx = 3\int_0^1 (2 - x^2 - x)\,dx = \frac{7}{2};$

$M_y = \int_0^1\int_x^{2-x^2} 3x\,dy\,dx = 3\int_0^1 [xy]_x^{2-x^2}\,dx$

$= 3\int_0^1 (2x - x^3 - x^2)\,dx = \frac{5}{4};$

$M_x = \int_0^1\int_x^{2-x^2} 3y\,dy\,dx = \frac{3}{2}\int_0^1 [y^2]_x^{2-x^2}\,dx$

$= \frac{3}{2}\int_0^1 (4 - 5x^2 + x^4)\,dx$

$= \frac{19}{5} \Rightarrow \bar{x} = \frac{5}{14}$ and $\bar{y} = \frac{38}{35}$

21. $M = \int_0^2\int_{y^2/2}^{4-y} dx\,dy = \int_0^2 \left(4 - y - \frac{y^2}{2}\right) dy = \frac{14}{3};$

$M_y = \int_0^2\int_{y^2/2}^{4-y} x\,dx\,dy = \frac{1}{2}\int_0^2 [x^2]_{y^2/2}^{4-y}\,dy$

$= \frac{1}{2}\int_0^2 \left(16 - 8y + y^2 - \frac{y^4}{4}\right) dy = \frac{128}{15};$

$M_x = \int_0^2\int_{y^2/2}^{4-y} y\,dx\,dy = \int_0^2 \left(4y - y^2 - \frac{y^3}{2}\right) dy$

$= \frac{10}{3} \Rightarrow \bar{x} = \frac{64}{35}$ and $\bar{y} = \frac{5}{7}$

23. $M = 2\int_0^1\int_0^{\sqrt{1-x^2}} dy\,dx = 2\int_0^1 \sqrt{1 - x^2}\,dx$

$= 2\left(\frac{\pi}{4}\right) = \frac{\pi}{2};$

$M_x = 2\int_0^1\int_0^{\sqrt{1-x^2}} y\,dy\,dx = \int_0^1 [y^2]_0^{\sqrt{1-x^2}}\,dx$

$= \int_0^1 (1 - x^2)\,dx = \left[x - \frac{x^3}{3}\right]_0^1 = \frac{2}{3}$

$\Rightarrow \bar{y} = \frac{4}{3\pi}$ and $\bar{x} = 0$, by symmetry

25. $M = \int_0^a\int_0^{\sqrt{a^2-x^2}} dy\,dx = \frac{\pi a^2}{4};$

$M_y = \int_0^a\int_0^{\sqrt{a^2-x^2}} x\,dy\,dx = \int_0^a [xy]_0^{\sqrt{a^2-x^2}}\,dx$

$= \int_0^a x\sqrt{a^2-x^2}\,dx = \frac{a^3}{3} \Rightarrow \bar{x} = \bar{y} = \frac{4a}{3\pi}$, by symmetry

27. $M = \int_0^{\pi}\int_0^{\sin x} dy\,dx = \int_0^{\pi} \sin x\,dx = 2;$

$M_x = \int_0^{\pi}\int_0^{\sin x} y\,dy\,dx = \frac{1}{2}\int_0^{\pi} [y^2] \quad dx$

$= \frac{1}{2}\int_0^{\pi} \sin^2 x\,dx = \frac{1}{4}\int_0^{\pi} (1 - \cos 2x)\,dx$

$= \frac{\pi}{4} \Rightarrow \bar{x} = \frac{\pi}{8}$ and $\bar{y} = \frac{\pi}{8}$

29. $M = \int_{-\infty}^{0}\int_0^{e^x} dy\,dx = \int_{-\infty}^{0} e^x\,dx = \lim_{b\to-\infty}\int_b^0 e^x\,dx$

$= 1 - \lim_{b\to-\infty} e^b = 1;$

$M_y = \int_{-\infty}^{0}\int_0^{e^x} x\,dy\,dx = \int_{-\infty}^{0} xe^x\,dx = \lim_{b\to-\infty}\int_b^0 xe^x\,dx$

$= \lim_{b\to-\infty} [xe^x - e^x]_b^0 = -1 - \lim_{b\to-\infty} (be^b - e^b)$

$= -1;$

$M_x = \int_{-\infty}^{0}\int_0^{e^x} y\,dy\,dx = \frac{1}{2}\int_{-\infty}^{0} e^{2x}\,dx = \frac{1}{4}\lim_{b\to-\infty}\int_b^0 e^{2x}\,dx$

$= \frac{1}{4} \Rightarrow \bar{x} = -1$ and $\bar{y} = \frac{1}{4}$

31. $M = \int_0^2\int_{-y}^{y-y^2} (x + y)\,dx\,dy = \int_0^2 \left[\frac{x^2}{2} + xy\right]_{-y}^{y-y^2} dy$

$= \int_0^2 \left(\frac{y^4}{2} - 2y^3 + 2y^2\right) dy = \left[\frac{y^5}{10} - \frac{y^4}{2} + \frac{2y^3}{3}\right]_0^2 = \frac{8}{15};$

$I_x = \int_0^2\int_{-y}^{y-y^2} y^2(x + y)\,dx\,dy = \int_0^2 \left[\frac{x^2y^2}{2} + xy^3\right]_{-y}^{y-y^2} dy$

$= \int_0^2 \left(\frac{y^6}{2} - 2y^5 + 2y^4\right) dy = \frac{64}{105};$

$R_x = \sqrt{\frac{I_x}{M}} = \sqrt{\frac{8}{7}} = 2\sqrt{\frac{2}{7}}$

33. $M = \int_0^1\int_x^{2-x} (6x + 3y + 3)\,dy\,dx$

$= \int_0^1 \left[6xy + \frac{3}{2}y^2 + 3y\right]_x^{2-x} dx$

$= \int_0^1 (12 - 12x^2)\,dx = 8;$

$M_y = \int_0^1\int_x^{2-x} x(6x + 3y + 3)\,dy\,dx$

$= \int_0^1 (12x - 12x^3)\,dx = 3;$

$M_x = \int_0^1\int_x^{2-x} y(6x + 3y + 3)\,dy\,dx$

$= \int_0^1 (14 - 6x - 6x^2 - 2x^3)\,dx$

$= \frac{17}{2} \Rightarrow \bar{x} = \frac{3}{8}$ and $\bar{y} = \frac{17}{16}$

35. $M = \int_0^1\int_0^6 (x + y + 1)\,dx\,dy = \int_0^1 (6y + 24)\,dy = 27;$

$M_x = \int_0^1\int_0^6 y(x + y + 1)\,dx\,dy = \int_0^1 y(6y + 24)\,dy = 14;$

$M_y = \int_0^1\int_0^6 x(x + y + 1)\,dx\,dy = \int_0^1 (18y + 90)\,dy$

$= 99 \Rightarrow \bar{x} = \frac{11}{3}$ and $\bar{y} = \frac{14}{27};$

$I_y = \int_0^1\int_0^6 x^2(x + y + 1)\,dx\,dy$

$= 216\int_0^1 \left(\frac{y}{3} + \frac{11}{6}\right) dy = 432;$

$R_y = \sqrt{\frac{I_y}{M}} = 4$

37. $M = \int_{-1}^1\int_0^{x^2} (7y + 1)\,dy\,dx = \int_{-1}^1 \left(\frac{7x^4}{2} + x^2\right) dx = \frac{31}{15};$

$M_x = \int_{-1}^1\int_0^{x^2} y(7y + 1)\,dy\,dx = \int_{-1}^1 \left(\frac{7x^6}{3} + \frac{x^4}{2}\right) dx = \frac{13}{15};$

$M_y = \int_{-1}^1\int_0^{x^2} x(7y + 1)\,dy\,dx$

$= \int_{-1}^1 \left(\frac{7x^5}{2} + x^3\right) dx = 0 \Rightarrow \bar{x} = 0$ and $\bar{y} = \frac{13}{31};$

$I_y = \int_{-1}^1\int_0^{x^2} x^2(7y + 1)\,dy\,dx = \int_{-1}^1 \left(\frac{7x^6}{2} + x^4\right) dx = \frac{7}{5};$

$R_y = \sqrt{\frac{I_y}{M}} = \sqrt{\frac{21}{31}}$

39. $M = \int_0^1\int_{-y}^{y} (y + 1)\,dx\,dy = \int_0^1 (2y^2 + 2y)\,dy = \frac{5}{3};$

$M_x = \int_0^1\int_{-y}^{y} y(y + 1)\,dx\,dy = 2\int_0^1 (y^3 + y^2)\,dy = \frac{7}{6};$

$M_y = \int_0^1\int_{-y}^{y} x(y + 1)\,dx\,dy = \int_0^1 0\,dy$

$= 0 \Rightarrow \bar{x} = 0$ and $\bar{y} = \frac{7}{10};$

$I_x = \int_0^1\int_{-y}^{y} y^2(y + 1)\,dx\,dy = \int_0^1 (2y^4 + 2y^3)\,dy$

$= \frac{9}{10} \Rightarrow R_x = \sqrt{\frac{I_x}{M}} = \frac{3\sqrt{6}}{10};$

$I_y = \int_0^1\int_{-y}^{y} x^2(y + 1)\,dx\,dy = \frac{1}{3}\int_0^1 (2y^4 + 2y^3)\,dy$

$= \frac{3}{10} \Rightarrow R_y = \sqrt{\frac{I_y}{M}} = \frac{3\sqrt{2}}{10};$

$I_o = I_x + I_y = \frac{6}{5} \Rightarrow R_o = \sqrt{\frac{I_o}{M}} = \frac{3\sqrt{2}}{5}$

41. $\int_{-5}^{5}\int_{-2}^{0} \frac{10{,}000e^y}{1+\frac{|x|}{2}}\,dy\,dx = 10{,}000(1-e^{-2})\int_{-5}^{5}\frac{dx}{1+\frac{|x|}{2}}$

$= 10{,}000(1-e^{-2})\left[\int_{-5}^{0}\frac{dx}{1-\frac{x}{2}} + \int_{0}^{5}\frac{dx}{1+\frac{x}{2}}\right]$

$= 10{,}000(1-e^{-2})\left[-2\ln\left(1-\frac{x}{2}\right)\right]_{-5}^{0}$

$+ 10{,}000(1-e^{-2})\left[2\ln\left(1+\frac{x}{2}\right)\right]_{0}^{5}$

$= 10{,}000(1-e^{-2})\left[2\ln\left(1+\frac{5}{2}\right)\right]$

$+ 10{,}000(1-e^{-2})\left[2\ln\left(1+\frac{5}{2}\right)\right]$

$= 40{,}000(1-e^{-2})\ln\left(\frac{7}{2}\right) \approx 43{,}329$

43. $M = \int_{-1}^{1}\int_{0}^{a(1-x^2)} dy\,dx = 2a\int_{0}^{1}(1-x^2)\,dx$

$= 2a\left[x-\frac{x^3}{3}\right]_0^1 = \frac{4a}{3};$

$M_x = \int_{-1}^{1}\int_{0}^{a(1-x^2)} y\,dy\,dx$

$= \frac{2a^2}{2}\int_0^1 (1-2x^2+x^4)\,dx = a^2\left[x-\frac{2x^3}{3}+\frac{x^5}{5}\right]_0^1$

$= \frac{8a^2}{15} \Rightarrow \bar{y} = \frac{M_x}{M} = \frac{\left(\frac{8a^2}{15}\right)}{\left(\frac{4a}{3}\right)} = \frac{2a}{5}.$

The angle θ between the x–axis and the line segment from the fulcrum to the center of mass on the y–axis plus 45° must be no more than 90° if the center of mass is to lie on the left side of the line $x = 1 \Rightarrow \theta + \frac{\pi}{4} \le \frac{\pi}{2} \Rightarrow \tan^{-1}\left(\frac{2a}{5}\right) \le \frac{\pi}{4} \Rightarrow a \le \frac{5}{2}$. Thus, if $0 < a \le \frac{5}{2}$, then the appliance will have to be tipped more than 45° to fall over.

45. $M = \int_0^1 \int_{-1/\sqrt{1-x^2}}^{1/\sqrt{1-x^2}} dy\,dx = \int_0^1 \frac{2}{\sqrt{1-x^2}}\,dx$

$= [2\sin^{-1}x]_0^1 = 2\left(\frac{\pi}{2}-0\right) = \pi;$

$M_y = \int_0^1 \int_{-1/\sqrt{1-x^2}}^{1/\sqrt{1-x^2}} x\,dy\,dx = \int_0^1 \frac{2x}{\sqrt{1-x^2}}\,dx$

$= [-2(1-x^2)^{1/2}]_0^1 = 2 \Rightarrow \bar{x} = \frac{2}{\pi}$ and $\bar{y} = 0$ by symmetry

47. a. $\frac{1}{2} = M = \int_0^1\int_{y^2}^{2y-y^2} \delta\,dx\,dy = 2\delta\int_0^1 (y-y^2)\,dy$

$= 2\delta\left[\frac{y^2}{2}-\frac{y^3}{3}\right]_0^1 = 2\delta\left(\frac{1}{6}\right) = \frac{\delta}{3} \Rightarrow \delta = \frac{3}{2}$

b. average value $= \dfrac{\int_0^1\int_{y^2}^{2y-y^2}(y+1)\,dx\,dy}{\int_0^1\int_{y^2}^{2y-y^2}dx\,dy}$

$= \frac{\left(\frac{1}{2}\right)}{\left(\frac{1}{3}\right)} = \frac{3}{2} = \delta$

so the values are the same.

49. a. $\bar{x} = \frac{M_y}{M} = 0 \Rightarrow M_y = \iint_R x\delta(x, y)\,dy\,dx = 0$

b. $I_L = \iint_R (x-h)^2\,\delta(x, y)\,dA$

$= \iint_R x^2\delta(x, y)\,dA - \iint_R 2hx\delta(x, y)\,dA$

$+ \iint_R h^2\delta(x, y)\,dA$

$= I_y - 0 + h^2\iint_R \delta(x, y)\,dA = I_{c.m.} + mh^2$

51. $M_{x_{P_1 \cup P_2}} = \iint_{R_1} y\,dA_1 + \iint_{R_2} y\,dA_2$

$= M_{x_1} + M_{x_2} \Rightarrow \bar{x} = \frac{M_{x_1}+M_{x_2}}{m_1+m_2};$

likewise, $\bar{y} = \frac{M_{y_1}+M_{y_2}}{m_1+m_2}$; thus $\mathbf{c} = \bar{x}\mathbf{i} + \bar{y}\mathbf{j}$

$= \frac{1}{m_1+m_2}\left[\left(M_{x_1}+M_{x_2}\right)\mathbf{i} + \left(M_{y_1}+M_{y_2}\right)\mathbf{j}\right]$

$= \frac{1}{m_1+m_2}\left[(m_1x_1+m_2x_2)\mathbf{i} + (m_1y_1+m_2y_2)\mathbf{j}\right]$

$= \frac{1}{m_1+m_2}\left[m_1(x_1\mathbf{i}+y_1\mathbf{j}) + m_2(x_2\mathbf{i}+y_2\mathbf{j})\right]$

$= \frac{m_1\mathbf{c}_1 + m_2\mathbf{c}_2}{m_1+m_2}$

53. a. $\mathbf{c} = \frac{8(\mathbf{i}+3\mathbf{j}) + 2(3\mathbf{i}+3.5\mathbf{j})}{8+2} = \frac{14\mathbf{i}+31\mathbf{j}}{10}$

$\Rightarrow \bar{x} = \frac{7}{5}$ and $\bar{y} = \frac{31}{10}$

b. $\mathbf{c} = \frac{8(\mathbf{i}+3\mathbf{j}) + 6(5\mathbf{i}+2\mathbf{j})}{14} = \frac{38\mathbf{i}+36\mathbf{j}}{14}$

$\Rightarrow \bar{x} = \frac{19}{7}$ and $\bar{y} = \frac{18}{7}$

c. $\mathbf{c} = \frac{2(3\mathbf{i}+3.5\mathbf{j}) + 6(5\mathbf{i}+2\mathbf{j})}{8} = \frac{36\mathbf{i}+19\mathbf{j}}{8}$

$\Rightarrow \bar{x} = \frac{9}{2}$ and $\bar{y} = \frac{19}{8}$

d. $\mathbf{c} = \frac{8(\mathbf{i}+3\mathbf{j}) + 2(3\mathbf{i}+3.5\mathbf{j}) + 6(5\mathbf{i}+2\mathbf{j})}{16} = \frac{44\mathbf{i}+43\mathbf{j}}{16}$

$\Rightarrow \bar{x} = \frac{11}{4}$ and $\bar{y} = \frac{43}{16}$

55. Place the midpoint of the triangle's base at the origin and above the semicircle. Then the center of mass of the triangle is $\left(0, \frac{h}{3}\right)$, and the center of mass of the disk is $\left(0, -\frac{4a}{3\pi}\right)$ from Exercise 25. From Pappus's formula,

$$\mathbf{c} = \frac{(ah)\left(\frac{h}{3}\mathbf{j}\right) + \left(\frac{\pi a^2}{2}\right)\left(-\frac{4a}{3\pi}\mathbf{j}\right)}{\left(ah + \frac{\pi a^2}{2}\right)} = \frac{\left(\frac{ah^2 - 2a^3}{3}\right)\mathbf{j}}{\left(ah + \frac{\pi a^2}{2}\right)},$$ so the centroid

is on the boundary if $ah^2 - 2a^3 = 0 \Rightarrow h^2 = 2a^2 \Rightarrow h = a\sqrt{2}$. In order for the center of mass to be inside T we must have $ah^2 - 2a^3 > 0$ or $h > a\sqrt{2}$.

Section 14.3 Double Integrals in Polar Form (pp.784-790)

Quick Review 14.3

1. $r^2 = 9$
2. $r \cos\theta = 3$
3. $r = -4 \sin\theta$
4. $(x - 2)^2 + y^2 = 4$
5. $x = 1$
6. $r \le 2, \frac{\pi}{2} < \theta < \pi$
7. $r \le 2, \pi < \theta < 2\pi$
8. 4π
9. ≈ 3.39
10. $4(\pi - 1)$

Section 14.3 Exercises

1. $\int_{-1}^{1}\int_{0}^{\sqrt{1-x^2}} dy\, dx = \int_0^{\pi}\int_0^1 r\, dr\, d\theta = \frac{1}{2}\int_0^{\pi} d\theta = \frac{\pi}{2}$

3. $\int_0^1\int_0^{\sqrt{1-y^2}} (x^2 + y^2)\, dx\, dy = \int_0^{\pi/2}\int_0^1 r^3\, dr\, d\theta$

$= \frac{1}{4}\int_0^{\pi/2} d\theta = \frac{\pi}{8}$

5. $\int_{-a}^{a}\int_{-\sqrt{a^2-x^2}}^{\sqrt{a^2-x^2}} dy\, dx = \int_0^{2\pi}\int_0^a r\, dr\, d\theta = \frac{a^2}{2}\int_0^{2\pi} d\theta = \pi a^2$

7. $\int_0^6\int_0^y x\, dx\, dy = \int_{\pi/4}^{\pi/2}\int_0^{6\csc\theta} r^2\cos\theta\, dr\, d\theta$

$= 72\int_{\pi/4}^{\pi/2} \cot\theta \csc^2\theta\, d\theta = 36\,[\cot^2\theta]_{\pi/4}^{\pi/2}$

$= 36$

9. $\int_{-1}^{0}\int_{-\sqrt{1-x^2}}^{0} \frac{2}{1 + \sqrt{x^2 + y^2}}\, dy\, dx = \int_{\pi}^{3\pi/2}\int_0^1 \frac{2r}{1 + r}\, dr\, d\theta$

$= 2\int_{\pi}^{3\pi/2}\int_0^1 \left(1 - \frac{1}{1 + r}\right) dr\, d\theta$

$= 2\int_{\pi}^{3\pi/2} (1 - \ln 2)\, d\theta$

$= (1 - \ln 2)\pi$

11. $\int_0^{\ln 2}\int_0^{\sqrt{(\ln 2)^2 - y^2}} e^{\sqrt{x^2+y^2}}\, dx\, dy = \int_0^{\pi/2}\int_0^{\ln 2} re^r\, dr\, d\theta$

$= \int_0^{\pi/2} (2\ln 2 - 1)\, d\theta$

$= \frac{\pi}{2}(2\ln 2 - 1)$

13. $\int_0^2\int_0^{\sqrt{1-(x-1)^2}} \frac{x + y}{x^2 + y^2}\, dy\, dx$

$= \int_0^{\pi/2}\int_0^{2\cos\theta} \frac{r(\cos\theta + \sin\theta)}{r^2}\, r\, dr\, d\theta$

$= \int_0^{\pi/2} (2\cos^2\theta + 2\sin\theta\cos\theta)\, d\theta$

$= \left[\theta + \frac{\sin 2\theta}{2} + \sin^2\theta\right]_0^{\pi/2} = \frac{\pi + 2}{2} = \frac{\pi}{2} + 1$

15. $\int_{-1}^{1}\int_{-\sqrt{1-y^2}}^{\sqrt{1-y^2}} \ln(x^2 + y^2 + 1)\, dx\, dy$

$= 4\int_0^{\pi/2}\int_0^1 \ln(r^2 + 1)\, r\, dr\, d\theta$

$= 2\int_0^{\pi/2} (\ln 4 - 1)\, d\theta = \pi(\ln 4 - 1)$

17. $\int_0^{\pi/2}\int_0^{2\sqrt{2-\sin 2\theta}} r\, dr\, d\theta = 2\int_0^{\pi/2} (2 - \sin 2\theta)\, d\theta$

$= 2(\pi - 1)$

19. $A = 2\int_0^{\pi/6}\int_0^{12\cos 3\theta} r\, dr\, d\theta = 144\int_0^{\pi/6} \cos^2 3\theta\, d\theta = 12\pi$

21. $A = \int_0^{\pi/2}\int_0^{1+\sin\theta} r\, dr\, d\theta$

$= \frac{1}{2}\int_0^{\pi/2}\left(\frac{3}{2} + 2\sin\theta - \frac{\cos 2\theta}{2}\right) d\theta$

$= \frac{3\pi}{8} + 1$

23. $M_x = \int_0^{\pi}\int_0^{1-\cos\theta} 3r^2\sin\theta\, dr\, d\theta$

$= \int_0^{\pi} (1 - \cos\theta)^3\sin\theta\, d\theta$

$= 4$

25. $M = 2\int_{\pi/6}^{\pi/2}\int_3^{6\sin\theta} dr\, d\theta$

$= 2\int_{\pi/6}^{\pi/2} (6\sin\theta - 3)\, d\theta$

$= 6[-2\cos\theta - \theta]_{\pi/6}^{\pi/2}$

$= 6\sqrt{3} - 2\pi$

27. $M = 2\int_0^\pi \int_0^{1+\cos\theta} r\, dr\, d\theta = \int_0^\pi (1+\cos\theta)^2\, d\theta = \frac{3\pi}{2}$;

$$M_y = 2\int_0^{2\pi}\int_0^{1+\cos\theta} r^2\cos\theta\, dr\, d\theta$$

$$= \int_0^{2\pi}\left(\frac{4\cos\theta}{3} + \frac{15}{24} + \cos 2\theta\right.$$

$$\left. -\sin^2\theta\cos\theta + \frac{\cos 4\theta}{4}\right) d\theta$$

$$= \frac{5\pi}{4} \Rightarrow \bar{x} = \frac{5}{6} \text{ and } \bar{y} = 0, \text{ by symmetry}$$

29. average $= \frac{4}{\pi a^2}\int_0^{\pi/2}\int_0^a r\sqrt{a^2-r^2}\, dr\, d\theta$

$$= \frac{4}{3\pi a^2}\int_0^{\pi/2} a^3\, d\theta = \frac{2a}{3}$$

31. average $= \frac{1}{\pi a^2}\int_{-a}^a \int_{-\sqrt{a^2-x^2}}^{\sqrt{a^2-x^2}} \sqrt{x^2+y^2}\, dy\, dx$

$$= \frac{1}{\pi a^2}\int_0^{2\pi}\int_0^a r^2\, dr\, d\theta$$

$$= \frac{a}{3\pi}\int_0^{2\pi} d\theta$$

$$= \frac{2a}{3}$$

33. $\int_0^{2\pi}\int_1^{\sqrt{e}}\left(\frac{\ln r^2}{r}\right) r\, dr\, d\theta = \int_0^{2\pi}\int_1^{\sqrt{e}} 2\ln r\, dr\, d\theta$

$$= 2\int_0^{2\pi}\left[r\ln r - r\right]_1^{e^{1/2}} d\theta$$

$$= 2\int_0^{2\pi}\sqrt{e}\left[\left(\frac{1}{2}-1\right)+1\right] d\theta$$

$$= 2\pi\sqrt{e}$$

35. $V = 2\int_0^{\pi/2}\int_1^{1+\cos\theta} r^2\cos\theta\, dr\, d\theta$

$$= \frac{2}{3}\int_0^{\pi/2}(3\cos^2\theta + 3\cos^3\theta + \cos^4\theta)\, d\theta$$

$$= \frac{2}{3}\left[\frac{15\theta}{8} + \sin 2\theta + 3\sin\theta - \sin^3\theta + \frac{\sin 4\theta}{32}\right]_0^{\pi/2}$$

$$= \frac{4}{3} + \frac{5\pi}{8}$$

37. a. $I^2 = \int_0^\infty \int_0^\infty e^{-(x^2+y^2)}\, dx\, dy$

$$= \int_0^{\pi/2}\int_0^\infty (e^{-r^2})\, r\, dr\, d\theta$$

$$= \int_0^{\pi/2}\left[\lim_{b\to\infty}\int_0^b re^{-r^2}\, dr\right] d\theta$$

$$= -\frac{1}{2}\int_0^{\pi/2}\lim_{b\to\infty}(e^{-b^2}-1)\, d\theta$$

$$= \frac{1}{2}\int_0^{\pi/2} d\theta = \frac{\pi}{4}$$

$$\Rightarrow I = \frac{\sqrt{\pi}}{2}$$

b. $\lim_{x\to\infty}\int_0^x \frac{2e^{-t^2}}{\sqrt{\pi}}\, dt = \frac{2}{\sqrt{\pi}}\int_0^\infty e^{-t^2}\, dt = \left(\frac{2}{\sqrt{\pi}}\right)\left(\frac{\sqrt{\pi}}{2}\right) = 1$,

from part (a)

39. Over the disk $x^2 + y^2 \le \frac{3}{4}$: $\int\int_R \frac{1}{1-x^2-y^2}\, dA$

$$= \int_0^{2\pi}\int_0^{\sqrt{3}/2} \frac{r}{1-r^2}\, dr\, d\theta$$

$$= \int_0^{2\pi}\left[-\frac{1}{2}\ln(1-r^2)\right]_0^{\sqrt{3}/2} d\theta$$

$$= \int_0^{2\pi}\left(-\frac{1}{2}\ln\frac{1}{4}\right) d\theta$$

$$= (\ln 2)\int_0^{2\pi} d\theta$$

$$= \pi\ln 4$$

Over the disk $x^2 + y^2 \le 1$: $\int\int_R \frac{1}{1-x^2-y^2}\, dA$

$$= \int_0^{2\pi}\int_0^1 \frac{r}{1-r^2}\, dr\, d\theta$$

$$= \int_0^{2\pi}\left[\lim_{a\to 1^-}\int_0^a \frac{r}{1-r^2}\, dr\right] d\theta$$

$$= \int_0^{2\pi}\lim_{a\to 1^-}\left[-\frac{1}{2}\ln(1-a^2)\right] d\theta$$

$$= 2\pi\cdot\lim_{a\to 1^-}\left[-\frac{1}{2}\ln(1-a^2)\right]$$

$$= 2\pi\cdot\infty, \text{ so the integral does not exist over } x^2 + y^2 \le 1$$

41. average $= \frac{1}{\pi a^2}\int_0^{2\pi}\int_0^a [(r\cos\theta - h)^2$

$$+ r^2\sin^2\theta]\, r\, dr\, d\theta$$

$$= \frac{1}{\pi a^2}\int_0^{2\pi}\int_0^a (r^3 - 2r^2 h\cos\theta + rh^2)\, dr\, d\theta$$

$$= \frac{1}{\pi a^2}\int_0^{2\pi}\left(\frac{a^4}{4} - \frac{2a^3 h\cos\theta}{3} + \frac{a^2 h^2}{2}\right) d\theta$$

$$= \frac{1}{\pi}\int_0^{2\pi}\left(\frac{a^2}{4} - \frac{2ah\cos\theta}{3} + \frac{h^2}{2}\right) d\theta$$

$$= \frac{1}{\pi}\left[\frac{a^2\theta}{4} - \frac{2ah\sin\theta}{3} + \frac{a^2 h^2\theta}{2}\right]$$

$$= \frac{1}{2}(a^2 + 2h^2)$$

43.–45. Example CAS commands for Exercise 45:

Maple:

```
with (plots): y='y'; x:='x';
bdy1:= y= 0; bdy2:= x= 2 - y;
  bdy3:= y= x;
implicitplot({bdy1, bdy2, bdy3},
x=0..2,y=0..1,
  scaling=CONSTRAINED,title=`ORIGINAL
  PLOT`);
X:= r*cos(theta): Y:= r*sin(theta);
r1:= solve(Y=0, r);
theta1:=evalf(solve(Y=0,theta));
r2:=solve(Y=2-X,r);
theta2:=solve(Y=2-X; theta);
r3:=solve(Y=X,r);
theta3:=solve(Y=X,theta);
trbdy1:= theta=theta1; trbdy2:= r = r2;
  trbdy3:= theta=theta3;
implicitplot({trbdy1,trbdy2,trbdy3},
theta=0..1, r=0..2,
  title= `TRANSFORMED PLOT`);
f:=(x,y)->sqrt(x+y);
subs(x=X, y=Y, f(x, y));
g:= unapply(%,(r,theta));
int(int(g(r,theta)*r,
  r=0..r2),theta=0..theta3);
evalf(%);
```

Mathematica:

```
Clear[x,y,r,t]
topolar = {x -> r Cos[t],
  y -> r Sin[t]}
<< Graphics `ImplicitPlot`
f = Sqrt[x+y]
bdy1 = x == y
bdy2 = x == 2-y
ImplicitPlot[{bdy1,bdy2},{x,0,1},
  {y,0,2}]
bdy3 = y == 0
bdy1 /. topolar
```

Note: Mathematica cannot solve this directly, so we need to help by dividing the equation by the right–hand side:

```
%[[1]]/%[[2]] == 1
Solve[ %, t ]
t1 = t /. First[%]
bdy2 /. topolar
Solve[ %, t]
t2 = t/. First[%]
r1 = 0 bdy2/.topolar
Solve[ % , r ]
r2 =r/.%
ImplicitPlot[{r==r1,r==r2,t==t1,
  t==t2},{t,0,1}{r,0,2}]
f /. topolar
f = Simplify[%]
Integrate[ f r, {t,0,t1}, {r,r1,r2} ]
N[%]
```

■ Section 14.4 Triple Integrals in Rectangular Coordinates (pp. 790-799)

Quick Review 14.4

1. 4

2. $\frac{13}{2}$

5. 0

6. 4

7. 10

8. $2r + 2s$

9. $\frac{2}{\pi}$

10. $-\frac{4}{\pi^2}$

Section 14.4 Exercises

1. $$\int_0^1\int_0^{1-z}\int_0^2 dx\,dy\,dz = 2\int_0^1\int_0^{1-z} dy\,dz = 2\int_0^1 (1-z)\,dz = 2\left[z - \frac{z^2}{2}\right]_0^1 = 2\left(1 - \frac{1}{2}\right) = 1$$

3. $$\int_0^1\int_0^{2-2x}\int_0^{3-3x-3y/2} dz\,dy\,dx = \int_0^1\int_0^{2-2x}\left(3 - 3x - \frac{3}{2}y\right)dy\,dx = \int_0^1\left[3(1-x)\cdot 2(1-x) - \frac{3}{4}\cdot 4(1-x)^2\right]dx = 3\int_0^1 (1-x)^2\,dx = [-(1-x)^3]_0^1 = 1,$$

$$\int_0^2\int_0^{1-y/2}\int_0^{3-3x-3y/2} dz\,dx\,dy,$$

$$\int_0^1\int_0^{3-3x}\int_0^{2-2x-2z/3} dy\,dz\,dx,$$

$$\int_0^3\int_0^{1-z/3}\int_0^{2-2x-2z/3} dy\,dx\,dz,$$

$$\int_0^2\int_0^{3-3y/2}\int_0^{1-y/2-z/3} dx\,dz\,dy,$$

$$\int_0^3\int_0^{2-2z/3}\int_0^{1-y/2-z/3} dx\,dy\,dz$$

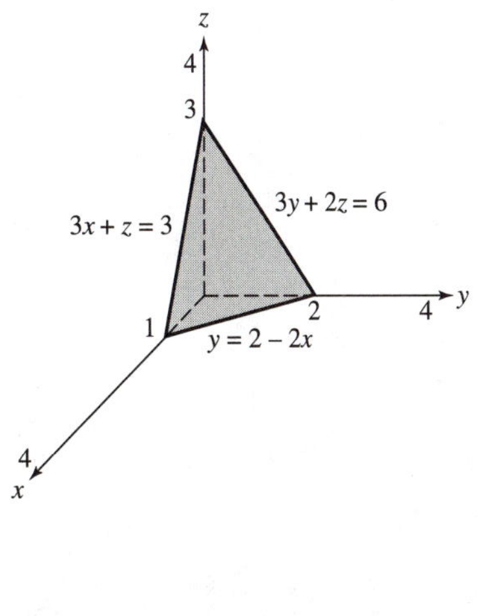

5. $$\int_{-2}^{2}\int_{-\sqrt{4-x^2}}^{\sqrt{4-x^2}}\int_{x^2+y^2}^{8-x^2-y^2} dz\,dy\,dx$$

$$= 4\int_0^2\int_0^{\sqrt{4-x^2}}\int_{x^2+y^2}^{8-x^2-y^2} dz\,dy\,dx$$

$$= 4\int_0^2\int_0^{\sqrt{4-x^2}} [8-2(x^2+y^2)]\,dy\,dx$$

$$= 8\int_0^2\int_0^{\sqrt{4-x^2}} (4-x^2-y^2)\,dydx$$

$$= 8\int_0^{\pi/2}\int_0^2 (4-r^2)\,r\,dr\,d\theta$$

$$= 8\int_0^{\pi/2}\left[2r^2-\frac{r^4}{4}\right]_0^2 d\theta = 32\int_0^{\pi/2} d\theta = 32\left(\frac{\pi}{2}\right) = 16\pi,$$

$$\int_{-2}^{2}\int_{-\sqrt{4-y^2}}^{\sqrt{4-y^2}}\int_{x^2+y^2}^{8-x^2-y^2} dz\,dx\,dy,$$

$$\int_{-2}^{2}\int_{y^2}^{4}\int_{-\sqrt{z-y^2}}^{\sqrt{z-y^2}} dx\,dz\,dy + \int_{-2}^{2}\int_{4}^{8-y^2}\int_{-\sqrt{8-z-y^2}}^{\sqrt{8-z-y^2}} dx\,dz\,dy,$$

$$\int_0^4\int_{-\sqrt{z}}^{\sqrt{z}}\int_{-\sqrt{z-y^2}}^{\sqrt{z-y^2}} dx\,dy\,dz$$

$$+\int_4^8\int_{-\sqrt{8-z}}^{\sqrt{8-z}}\int_{-\sqrt{8-z-y^2}}^{\sqrt{8-z-y^2}} dx\,dy\,dz,$$

$$\int_{-2}^{2}\int_{x^2}^{4}\int_{-\sqrt{z-x^2}}^{\sqrt{z-x^2}} dy\,dz\,dx + \int_{-2}^{2}\int_{4}^{8-x^2}\int_{-\sqrt{8-z-x^2}}^{\sqrt{8-z-x^2}} dy\,dz\,dx,$$

$$\int_0^4\int_{-\sqrt{z}}^{\sqrt{z}}\int_{-\sqrt{z-x^2}}^{\sqrt{z-x^2}} dy\,dx\,dz$$

$$+\int_4^8\int_{-\sqrt{8-z}}^{\sqrt{8-z}}\int_{-\sqrt{8-z-x^2}}^{\sqrt{8-z-x^2}} dy\,dx\,dz$$

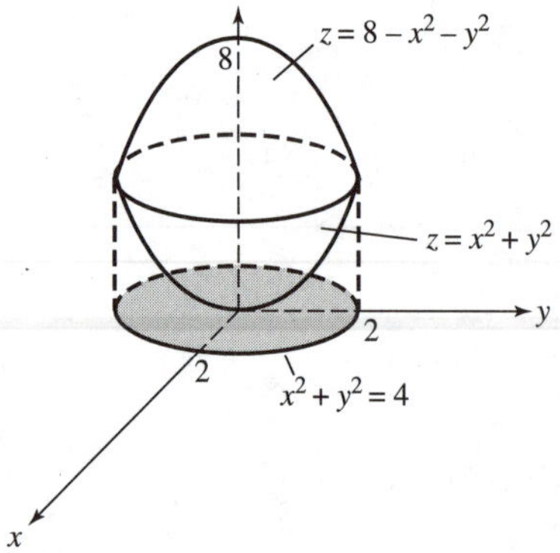

7. $$\int_0^1\int_0^1\int_0^1 (x^2+y^2+z^2)\,dz\,dy\,dx$$

$$= \int_0^1\int_0^1 \left(x^2+y^2+\frac{1}{3}\right) dy\,dx$$

$$= \int_0^1 \left(x^2+\frac{2}{3}\right) dx = 1$$

9. $$\int_1^e\int_1^e\int_1^e \frac{1}{xyz}\,dx\,dy\,dz = \int_1^e\int_1^e \left[\frac{\ln x}{yz}\right]_1^e dy\,dz$$

$$= \int_1^e\int_1^e \frac{1}{yz}\,dy\,dz$$

$$= \int_1^e \left[\frac{\ln y}{z}\right]_1^e dz$$

$$= \int_1^e \frac{1}{z}\,dz = 1$$

11. $$\int_0^1\int_0^\pi\int_0^\pi y\sin z\,dx\,dy\,dz = \int_0^1\int_0^\pi \pi y\sin z\,dy\,dz$$

$$= \frac{\pi^3}{2}\int_0^1 \sin z\,dz$$

$$= \frac{\pi^3}{2}(1-\cos 1)$$

13. $$\int_0^3\int_0^{\sqrt{9-x^2}}\int_0^{\sqrt{9-x^2}} dz\,dy\,dx$$

$$= \int_0^3\int_0^{\sqrt{9-x^2}} \sqrt{9-x^2}\,dy\,dx$$

$$= \int_0^3 (9-x^2)\,dx = \left[9x-\frac{x^3}{3}\right]_0^3 = 18$$

15. $$\int_0^1\int_0^{2-x}\int_0^{2-x-y} dz\,dy\,dx = \int_0^1\int_0^{2-x} (2-x-y)\,dy\,dx$$

$$= \int_0^1 \left[(2-x)^2-\frac{1}{2}(2-x)^2\right] dx$$

$$= \frac{1}{2}\int_0^1 (2-x)^2\,dx$$

$$= \left[-\frac{1}{6}(2-x)^3\right]_0^1$$

$$= -\frac{1}{6}+\frac{8}{6} = \frac{7}{6}$$

17. $$\int_0^\pi\int_0^\pi\int_0^\pi \cos(u+v+w)\,du\,dv\,dw$$

$$= \int_0^\pi\int_0^\pi [\sin(w+v+\pi)-\sin(w+v)]\,dv\,dw$$

$$= \int_0^\pi [(-\cos(w+2\pi)+\cos(w+\pi))$$

$$+ (\cos(w+\pi)-\cos w)]\,dw$$

$$= [-\sin(w+2\pi)+\sin(w+\pi)-\sin w$$

$$+ \sin(w+\pi)]_0^\pi$$

$$= 0$$

19. $$\int_0^{\pi/4}\int_0^{\ln\sec v}\int_{-\infty}^{2t} e^x\,dx\,dt\,dv$$

$$= \int_0^{\pi/4}\int_0^{\ln\sec v} \lim_{b\to-\infty}(e^{2t}-e^b)\,dt\,dv$$

$$= \int_0^{\pi/4}\int_0^{\ln\sec v} e^{2t}\,dt\,dv$$

$$= \int_0^{\pi/4} \frac{1}{2}e^{2\ln\sec v}\,dv$$

$$= \int_0^{\pi/4} \frac{\sec^2 v}{2}\,dv = \left[\frac{\tan v}{2}\right]_0^{\pi/4} = \frac{1}{2}$$

21. a. $\displaystyle\int_{-1}^{1}\int_{0}^{1-x^2}\int_{x^2}^{1-z} dy\,dz\,dx$

b. $\displaystyle\int_{0}^{1}\int_{-\sqrt{1-z}}^{\sqrt{1-z}}\int_{x^2}^{1-z} dy\,dx\,dz$

c. $\displaystyle\int_{0}^{1}\int_{0}^{1-z}\int_{-\sqrt{y}}^{\sqrt{y}} dx\,dy\,dz$

d. $\displaystyle\int_{0}^{1}\int_{0}^{1-y}\int_{-\sqrt{y}}^{\sqrt{y}} dx\,dz\,dy$

e. $\displaystyle\int_{0}^{1}\int_{-\sqrt{y}}^{\sqrt{y}}\int_{0}^{1-y} dz\,dx\,dy$

23. $\displaystyle V=\int_0^1\int_{-1}^1\int_0^{y^2} dz\,dy\,dx=\int_0^1\int_{-1}^1 y^2\,dy\,dx=\frac{2}{3}\int_0^1 dx=\frac{2}{3}$

25. $\displaystyle V=\int_0^4\int_0^{\sqrt{4-x}}\int_0^{2-y} dz\,dy\,dx=\int_0^4\int_0^{\sqrt{4-x}}(2-y)\,dy\,dx$

$\displaystyle =\int_0^4\left[2\sqrt{4-x}-\left(\frac{4-x}{2}\right)\right]dx$

$\displaystyle =\left[-\frac{4}{3}(4-x)^{3/2}+\frac{1}{4}(4-x)^2\right]_0^4$

$\displaystyle =\frac{4}{3}(4)^{3/2}-\frac{1}{4}(16)=\frac{32}{3}-4=\frac{20}{3}$

27. $\displaystyle V=\int_0^1\int_0^{2-2x}\int_0^{3-3x-3y/2} dz\,dy\,dx$

$\displaystyle =\int_0^1\int_0^{2-2x}\left(3-3x-\frac{3}{2}y\right)dy\,dx$

$\displaystyle =\int_0^1\left[6(1-x)^2-\frac{3}{4}\cdot 4(1-x)^2\right]dx$

$\displaystyle =\int_0^1 3(1-x)^2\,dx=[-(1-x)^3]_0^1=1$

29. $\displaystyle V=8\int_0^1\int_0^{\sqrt{1-x^2}}\int_0^{\sqrt{1-x^2}} dz\,dy\,dx$

$\displaystyle =8\int_0^1\int_0^{\sqrt{1-x^2}}\sqrt{1-x^2}\,dy\,dx$

$\displaystyle =8\int_0^1(1-x^2)\,dx=\frac{16}{3}$

31. $\displaystyle V=\int_0^4\int_0^{(\sqrt{16-y^2})/2}\int_0^{4-y} dx\,dz\,dy$

$\displaystyle =\int_0^4\int_0^{(\sqrt{16-y^2})/2}(4-y)\,dz\,dy$

$\displaystyle =\int_0^4\frac{\sqrt{16-y^2}}{2}(4-y)\,dy$

$\displaystyle =\int_0^4 2\sqrt{16-y^2}\,dy-\frac{1}{2}\int_0^4 y\sqrt{16-y^2}\,dy$

$\displaystyle =\left[y\sqrt{16-y^2}+16\sin^{-1}\frac{y}{4}\right]_0^4+\left[\frac{1}{6}(16-y^2)^{3/2}\right]_0^4$

$\displaystyle =16\left(\frac{\pi}{2}\right)-\frac{1}{6}(16)^{3/2}=8\pi-\frac{32}{3}$

33. $\displaystyle\int_0^2\int_0^{2-x}\int_{(2-x-y)/2}^{4-2x-2y} dz\,dy\,dx$

$\displaystyle =\int_0^2\int_0^{2-x}\left(3-\frac{3x}{2}-\frac{3y}{2}\right)dy\,dx$

$\displaystyle =\int_0^2\left[3\left(1-\frac{x}{2}\right)(2-x)-\frac{3}{4}(2-x)^2\right]dx$

$\displaystyle =\int_0^2\left[6-6x+\frac{3x^2}{2}-\frac{3(2-x)^2}{4}\right]dx$

$\displaystyle =\left[6x-3x^2+\frac{x^3}{2}+\frac{(2-x)^3}{4}\right]_0^2$

$\displaystyle =(12-12+4+0)-\frac{2^3}{4}=2$

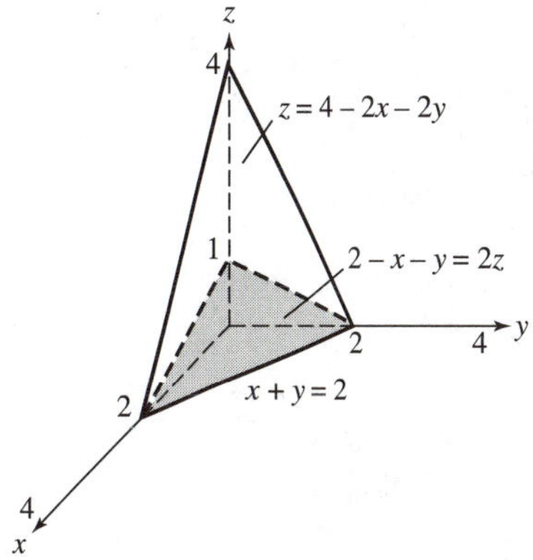

35. $\displaystyle V=2\int_{-2}^2\int_0^{\sqrt{4-x^2}/2}\int_0^{x+2} dz\,dy\,dx$

$\displaystyle =\int_{-2}^2\int_0^{\sqrt{4-x^2}/2}(x+2)\,dy\,dx$

$\displaystyle =\int_{-2}^2(x+2)\sqrt{4-x^2}\,dx$

$\displaystyle =\int_{-2}^2 2\sqrt{4-x^2}\,dx+\int_{-2}^2 x\sqrt{4-x^2}\,dx$

$\displaystyle =\left[x\sqrt{4-x^2}+4\sin^{-1}\frac{x}{2}\right]_{-2}^2+\left[-\frac{1}{3}(4-x^2)^{3/2}\right]_{-2}^2$

$\displaystyle =4\left(\frac{\pi}{2}\right)-4\left(\frac{\pi}{2}\right)=4\pi$

37. $\displaystyle \text{average}=\frac{1}{8}\int_0^2\int_0^2\int_0^2(x^2+9)\,dz\,dy\,dx$

$\displaystyle =\frac{1}{8}\int_0^2\int_0^2(2x^2+18)\,dy\,dx$

$\displaystyle =\frac{1}{8}\int_0^2(4x^2+36)\,dx=\frac{31}{3}$

39. $\displaystyle \text{average}=\int_0^1\int_0^1\int_0^1(x^2+y^2+z^2)\,dz\,dy\,dx$

$\displaystyle =\int_0^1\int_0^1\left(x^2+y^2+\frac{1}{3}\right)dy\,dx$

$\displaystyle =\int_0^1\left(x^2+\frac{2}{3}\right)dx=1$

41. $\int_0^4 \int_0^1 \int_{2y}^2 \frac{4\cos(x^2)}{2\sqrt{z}}\,dx\,dy\,dz$

$= \int_0^4 \int_0^2 \int_0^{x/2} \frac{4\cos(x^2)}{2\sqrt{z}}\,dy\,dx\,dz$

$= \int_0^4 \int_0^2 \frac{x\cos(x^2)}{\sqrt{z}}\,dx\,dz$

$= \int_0^4 \left(\frac{\sin 4}{2}\right) z^{-1/2}\,dz = [(\sin 4)z^{1/2}]_0^4$

$= 2\sin 4$

43. $\int_0^1 \int_{\sqrt[3]{z}}^1 \int_0^{\ln 3} \frac{\pi e^{2x}\sin(\pi y^2)}{y^2}\,dx\,dy\,dz$

$= \int_0^1 \int_{\sqrt[3]{z}}^1 \frac{4\pi\sin(\pi y^2)}{y^2}\,dy\,dz$

$= \int_0^1 \int_0^{y^3} \frac{4\pi\sin(\pi y^2)}{y^2}\,dz\,dy$

$= \int_0^1 4\pi y\sin(\pi y^2)\,dy = [-2\cos(\pi y^2)]_0^1$

$= -2(-1) + 2(1) = 4$

45. $\int_0^1 \int_0^{4-a-x^2} \int_a^{4-x^2-y} dz\,dy\,dx = \frac{4}{15}$

$\Rightarrow \int_0^1 \int_0^{4-a-x^2} (4 - x^2 - y - a)\,dy\,dx = \frac{4}{15}$

$\Rightarrow \int_0^1 \left[\left(4 - a - x^2\right)^2 - \frac{1}{2}(4 - a - x^2)^2\right] dx = \frac{4}{15}$

$\Rightarrow \frac{1}{2}\int_0^1 (4 - a - x^2)^2\,dx = \frac{4}{15}$

$\Rightarrow \int_0^1 [(4 - a)^2 - 2x^2(4 - a) + x^4]\,dx = \frac{8}{15}$

$\Rightarrow \left[(4 - a)^2 x - \frac{2}{3}x^3(4 - a) + \frac{x^5}{5}\right]_0^1 = \frac{8}{15}$

$\Rightarrow (4 - a)^2 - \frac{2}{3}(4 - a) + \frac{1}{5} = \frac{8}{15}$

$\Rightarrow 15(4 - a)^2 - 10(4 - a) - 5 = 0$

$\Rightarrow 3(4 - a)^2 - 2(4 - a) - 1 = 0$

$\Rightarrow [3(4 - a) + 1][(4 - a) - 1] = 0$

$\Rightarrow 4 - a = -\frac{1}{3}$ or $4 - a = 1 \Rightarrow a = \frac{13}{3}$ or $a = 3$

47. To minimize the integral, we want the domain to include all points where the integrand is negative and to exclude all points where it is positive. These criteria are met by the points (x, y, z) such that $4x^2 + 4y^2 + z^2 - 4 \le 0$ or $4x^2 + 4y^2 + z^2 \le 4$, which is an ellipsoid.

49.–52. Example CAS commands:

Maple:

```
int(int(int(x^2*y^2*z, z=0..1),
  y=-sqrt(1-x^2)..sqrt(1-x^2)),
  x=-1..1);
evalf(%);
```

Mathematica:

```
Clear [x, y, z]
Integrate[Integrate[Integrate[x^2 x^2 z,
 {z, 0, 1}],
 {y, -Sqrt[1-x^2], Sqrt[1-x^2]},
 {x, -1, 1}]
N[%]
```

■ Section 14.5 Masses and Moments in Three Dimension (pp.799-804)

Quick Review 14.5

1. $\frac{13}{3}$

2. $\frac{158}{15}$

3. $\frac{3}{2}$

4. $\bar{x} = \frac{9}{26}, \bar{y} = \frac{158}{65}$

5. $\frac{6091}{210}$

6. $\frac{23}{30}$

7. $\frac{1042}{35}$

8. $\sqrt{\frac{6091}{910}}$

9. $\sqrt{\frac{23}{130}}$

10. $\sqrt{\frac{3126}{455}}$

Section 14.5 Exercises

1. $I_x = \int_{-c/2}^{c/2} \int_{-b/2}^{b/2} \int_{-a/2}^{a/2} (y^2 + z^2)\,dx\,dy\,dz$

$= a\int_{-c/2}^{c/2} \int_{-b/2}^{b/2} (y^2 + z^2)\,dy\,dz$

$= a\int_{-c/2}^{c/2} \left[\frac{y^3}{3} + yz^2\right]_{-b/2}^{b/2} dz = a\int_{-c/2}^{c/2} \left(\frac{b^3}{12} + bz^2\right) dz$

$= ab\left[\frac{b^2}{12}z + \frac{z^3}{3}\right]_{-c/2}^{c/2} = ab\left(\frac{b^2c}{12} + \frac{c^3}{12}\right) = \frac{abc}{12}(b^2 + c^2)$

$= \frac{M}{12}(b^2 + c^2);$

$R_x = \sqrt{\frac{b^2 + c^2}{12}}$; likewise $R_y = \sqrt{\frac{a^2 + c^2}{12}}$ and $R_z = \sqrt{\frac{a^2 + b^2}{12}}$, by symmetry

3. $I_x = \int_0^a \int_0^b \int_0^c (y^2 + z^2)\,dz\,dy\,dx$

$= \int_0^a \int_0^b \left(cy^2 + \frac{c^3}{3}\right) dy\,dx$

$= \int_0^a \left(\frac{cb^3}{3} + \frac{c^3b}{3}\right) dx = \frac{abc(b^2 + c^2)}{3}$

$= \frac{M}{3}(b^2 + c^2)$ where $M = abc$;

$I_y = \frac{M}{3}(a^2 + c^2)$ and $I_z = \frac{M}{3}(a^2 + b^2)$, by symmetry

5. $M = 4\int_0^1\int_0^1\int_{4y^2}^4 dz\,dy\,dx = 4\int_0^1\int_0^1 (4-4y^2)\,dy\,dx$

$= 16\int_0^1 \frac{2}{3}\,dx = \frac{32}{3};$

$M_{xy} = 4\int_0^1\int_0^1\int_{4y^2}^4 z\,dz\,dy\,dx$

$= 2\int_0^1\int_0^1 (16-16y^4)\,dy\,dx = \frac{128}{5}\int_0^1 dx = \frac{128}{5}$

$\Rightarrow \bar{z} = \frac{12}{5}$, and $\bar{x} = \bar{y} = 0$, by symmetry;

$I_x = 4\int_0^1\int_0^1\int_{4y^2}^4 (y^2+z^2)\,dz\,dy\,dx$

$= 4\int_0^1\int_0^1 \left[\left(4y^2+\frac{64}{3}\right) - \left(4y^4+\frac{64y^6}{3}\right)\right] dy\,dx$

$= 4\int_0^1 \frac{1976}{105}\,dx = \frac{7904}{105};$

$I_y = 4\int_0^1\int_0^1\int_{4y^2}^4 (x^2+z^2)\,dz\,dy\,dx$

$= 4\int_0^1\int_0^1 \left[\left(4x^2+\frac{64}{3}\right) - \left(4x^2y^2+\frac{64y^6}{3}\right)\right] dy\,dx$

$= 4\int_0^1 \left(\frac{8}{3}x^2+\frac{128}{7}\right) dx = \frac{4832}{63};$

$I_z = 4\int_0^1\int_0^1\int_{4y^2}^4 (x^2+y^2)\,dz\,dy\,dx$

$= 16\int_0^1\int_0^1 (x^2 - x^2y^2 + y^2 - y^4)\,dy\,dx$

$= 16\int_0^1 \left(\frac{2x^2}{3}+\frac{2}{15}\right) dx = \frac{256}{45}$

7. a. $M = 4\int_0^2\int_0^{\sqrt{4-x^2}}\int_{x^2+y^2}^4 dz\,dy\,dx$

$= 4\int_0^{\pi/2}\int_0^2\int_{r^2}^4 r\,dz\,dr\,d\theta$

$= 4\int_0^{\pi/2}\int_0^2 (4r - r^3)\,dr\,d\theta = 4\int_0^{\pi/2} 4\,d\theta = 8\pi;$

$M_{xy} = \int_0^{2\pi}\int_0^2\int_{r^2}^4 zr\,dz\,dr\,d\theta$

$= \int_0^{2\pi}\int_0^2 \frac{r}{2}(16 - r^4)\,dr\,d\theta = \frac{32}{3}\int_0^{2\pi} d\theta = \frac{64\pi}{3}$

$\Rightarrow \bar{z} = \frac{8}{3}$, and $\bar{x} = \bar{y} = 0$, by symmetry

b. $M = 8\pi \Rightarrow 4\pi = \int_0^{2\pi}\int_0^{\sqrt{c}}\int_{r^2}^c r\,dz\,dr\,d\theta$

$= \int_0^{2\pi}\int_0^{\sqrt{c}} (cr - r^3)\,dr\,d\theta = \int_0^{2\pi} \frac{c^2}{4}\,d\theta = \frac{c^2\pi}{2}$

$\Rightarrow c^2 = 8 \Rightarrow c = 2\sqrt{2}$, since $c > 0$

9. The plane $y + 2z = 2$ is the top of the wedge

$\Rightarrow I_L = \int_{-2}^2\int_{-2}^4\int_{-1}^{(2-y)/2} [(y-6)^2 + z^2]\,dz\,dy\,dx$

$= \int_{-2}^2\int_{-2}^4 \left[\frac{(y-6)^2(4-y)}{2} + \frac{(2-y)^3}{24} + \frac{1}{3}\right] dy\,dx;$

let $t = 2 - y \Rightarrow I_L = 4\int_{-2}^4 \left(\frac{13t^3}{24} + 5t^2 + 16t + \frac{49}{3}\right) dt$

$= 1386;$

$M = \frac{1}{2}(3)(6)(4) = 36 \Rightarrow R_L = \sqrt{\frac{I_L}{M}} = \sqrt{\frac{77}{2}}$

11. $M = 8;$

$I_L = \int_0^4\int_0^2\int_0^1 [z^2 + (y-2)^2]dz\,dy\,dx$

$= \int_0^4\int_0^2 \left(y^2 - 4y + \frac{13}{3}\right) dy\,dx = \frac{10}{3}\int_0^4 dx = \frac{40}{3}$

$\Rightarrow R_L = \sqrt{\frac{I_L}{M}} = \sqrt{\frac{5}{3}}$

13. a. $M = \int_0^2\int_0^{2-x}\int_0^{2-x-y} 2x\,dz\,dy\,dx$

$= \int_0^2\int_0^{2-x} (4x - 2x^2 - 2xy)\,dy\,dx$

$= \int_0^2 (x^3 - 4x^2 + 4x)\,dx = \frac{4}{3}$

b. $M_{xy} = \int_0^2\int_0^{2-x}\int_0^{2-x-y} 2xz\,dz\,dy\,dx$

$= \int_0^2\int_0^{2-x} x(2-x-y)^2 dy\,dx = \int_0^2 \frac{x(2-x)^3}{3}\,dx$

$= \frac{8}{15};$

$M_{xz} = \frac{8}{15}$ by symmetry;

$M_{yz} = \int_0^2\int_0^{2-x}\int_0^{2-x-y} 2x^2\,dz\,dy\,dx$

$= \int_0^2\int_0^{2-x} 2x^2(2-x-y)\,dy\,dx = \int_0^2 (2x - x^2)^2\,dx$

$= \frac{16}{15} \Rightarrow \bar{x} = \frac{4}{5}$, and $= \bar{z} = \frac{2}{5}$

15. a. $M = \int_0^1\int_0^1\int_0^1 (x+y+z+1)\,dz\,dy\,dx$

$= \int_0^1\int_0^1 \left(x+y+\frac{3}{2}\right) dy\,dx = \int_0^1 (x+2)\,dx = \frac{5}{2}$

b. $M_{xy} = \int_0^1\int_0^1\int_0^1 z(x+y+z+1)\,dz\,dy\,dx$

$= \frac{1}{2}\int_0^1\int_0^1 \left(x+y+\frac{5}{3}\right) dy\,dx = \frac{1}{2}\int_0^1 \left(x+\frac{13}{6}\right) dx$

$= \frac{4}{3} \Rightarrow M_{xy} = M_{yz} = M_{xz} = \frac{4}{3}$, by symmetry

$\Rightarrow \bar{x} = \bar{y} = \bar{z} = \frac{8}{15}$

c. $I_z = \int_0^1 \int_0^1 \int_0^1 (x^2 + y^2)(x + y + z + 1)\, dz\, dy\, dx$

$= \int_0^1 \int_0^1 (x^2 + y^2)\left(x + y + \frac{3}{2}\right) dy\, dx$

$= \int_0^1 \left(x^3 + 2x^2 + \frac{1}{3}x + \frac{3}{4}\right) dx = \frac{11}{6}$

$\Rightarrow I_x = I_y = I_z = \frac{11}{6}$, by symmetry

d. $R_x = R_y = R_z = \sqrt{\frac{I_z}{M}} = \sqrt{\frac{11}{15}}$

17. $M = \int_0^1 \int_{z-1}^{1-z} \int_0^{\sqrt{z}} (2y + 5)\, dy\, dx\, dz$

$= \int_0^1 \int_{z-1}^{1-z} (z + 5\sqrt{z}) dx\, dz = \int_0^1 2(z + 5\sqrt{z})(1 - z)\, dz$

$= 2\int_0^1 (5z^{1/2} + z - 5z^{3/2} - z^2)\, dz$

$= 2\left[\frac{10}{3}z^{3/2} + \frac{1}{2}z^2 - 2z^{5/2} - \frac{1}{3}z^3\right]_0^1 = 2\left(\frac{9}{3} - \frac{3}{2}\right) = 3$

19. a. Let ΔV_i be the volume of the ith piece, and let (x_i, y_i, z_i) be a point in the ith piece. Then the work done by gravity in moving the ith piece to the xy–plane is approximately $W_i = m_i g z_i = (x_i + y_i + z_i + 1)\, g\, \Delta V_i\, z_i$ $\Rightarrow$ the total work done is the triple integral

$W = \int_0^1 \int_0^1 \int_0^1 (x + y + z + 1)\, gz\, dz\, dy\, dx$

$= g\int_0^1 \int_0^1 \left[\frac{1}{2}xz^2 + \frac{1}{2}yz^2 + \frac{1}{3}z^3 + \frac{1}{2}z^2\right]_0^1 dy\, dx$

$= g\int_0^1 \int_0^1 \left(\frac{1}{2}x + \frac{1}{2}y + \frac{5}{6}\right) dy\, dx$

$= g\int_0^1 \left[\frac{1}{2}xy + \frac{1}{4}y^2 + \frac{5}{6}y\right]_0^1 dx = g\int_0^1 \left(\frac{1}{2}x + \frac{13}{12}\right) dx$

$= g\left[\frac{x^2}{4} + \frac{13}{12}x\right]_0^1 = g\left(\frac{16}{12}\right) = \frac{4}{3}g$

b. From Exercise 15 the center of mass is $\left(\frac{8}{15}, \frac{8}{15}, \frac{8}{15}\right)$ and the mass of the liquid is $\frac{5}{2}$ $\Rightarrow$ the work done by gravity in moving the center of mass to the xy–plane is $W = mgd = \left(\frac{5}{2}\right)(g)\left(\frac{8}{15}\right) = \frac{4}{3}g$, which is the same as the work done in part (a).

21. a. $\bar{x} = \frac{M_{yz}}{M} = 0 \Rightarrow \iiint_R x\delta(x, y, z)\, dx\, dy\, dz = 0$

$\Rightarrow M_{yz} = 0$

b. $I_L = \iiint_R |\mathbf{v} - h\mathbf{i}|^2\, dm$

$= \iiint_R |(x - h)\,\mathbf{i} + y\mathbf{j}|^2\, dm$

$= \iiint_R (x^2 - 2xh + h^2 + y^2)\, dm$

$= \iiint_R (x^2 + y^2)\, dm - 2h\iiint_R x\, dm$

$+ h^2\iiint_R dm = I_x - 0 + h^2 m = I_{c.m.} + h^2 m$

23. a. $(\bar{x}, \bar{y}, \bar{z}) = \left(\frac{a}{2}, \frac{b}{2}, \frac{c}{2}\right) \Rightarrow I_z = I_{c.m.} + abc\left(\sqrt{\frac{a^2}{4} + \frac{b^2}{4}}\right)^2$

$\Rightarrow I_{c.m.} = I_z - \frac{abc(a^2 + b^2)}{4}$

$= \frac{abc(a^2 + b^2)}{3} - \frac{abc(a^2 + b^2)}{4} = \frac{abc(a^2 + b^2)}{12}$;

$R_{c.m.} = \sqrt{\frac{I_{c.m}}{M}} = \sqrt{\frac{a^2 + b^2}{12}}$

b. $I_L = I_{c.m.} + abc\left(\sqrt{\frac{a^2}{4} + \left(\frac{b}{2} - 2b\right)^2}\right)^2$

$= \frac{abc(a^2 + b^2)}{12} + \frac{abc(a^2 + 9b^2)}{4}$

$= \frac{abc(4a^2 + 28b^2)}{12} = \frac{abc(a^2 + 7b^2)}{3}$;

$R_L = \sqrt{\frac{I_L}{M}} = \sqrt{\frac{a^2 + 7b^2}{3}}$

25. $M_{yz_{B_1 \cup B_2}} = \iiint_{B_1} x\, dV_1 + \iiint_{B_2} x\, dV_2$

$= M_{(yz)_1} + M_{(yz)_2} \Rightarrow \bar{x} = \frac{M_{(yz)_1} + M_{(yz)_2}}{m_1 + m_2}$; similarly,

$\bar{y} = \frac{M_{(xz)_1} + M_{(xz)_2}}{m_1 + m_2}$ and $\bar{z} = \frac{M_{(xy)_1} + M_{(xy)_2}}{m_1 + m_2}$

$\Rightarrow \mathbf{c} = \bar{x}\mathbf{i} + \bar{y}\mathbf{j} + \bar{z}\mathbf{k}$

$= \frac{1}{m_1 + m_2}[(M_{(yz)_1} + M_{(yz)_2})\mathbf{i} + (M_{(xz)_1} + M_{(xz)_2})\mathbf{j}$
$+ (M_{(xy)_1} + M_{(xy)_2})\mathbf{k}]$

$= \frac{1}{m_1 + m_2}[(m_1\bar{x}_1 + m_2\bar{x}_2)\mathbf{i} + (m_1\bar{y}_1 + m_2\bar{y}_2)\mathbf{j}$
$+ (m_1\bar{z}_1 + m_2\bar{z}_2)\mathbf{k}]$

$= \frac{1}{m_1 + m_2}[m_1(\bar{x}_1\mathbf{i} + \bar{y}_1\mathbf{j} + \bar{z}_1\mathbf{k}) + m_2(\bar{x}_2\mathbf{i} + \bar{y}_2\mathbf{j} + \bar{z}_2\mathbf{k})]$

$= \frac{m_1\mathbf{c}_1 + m_2\mathbf{c}_2}{m_1 + m_2}$

27. a. $\mathbf{c} = \frac{\left(\frac{\pi a^2 h}{3}\right)\left(\frac{h}{4}\mathbf{k}\right) + \left(\frac{2\pi a^3}{3}\right)\left(-\frac{3a}{8}\mathbf{k}\right)}{m_1 + m_2}$

$= \frac{\left(\frac{a^2\pi}{3}\right)\left(\frac{h^2 - 3a^2}{4}\mathbf{k}\right)}{m_1 + m_2}$, where $m_1 = \frac{\pi a^2 h}{3}$ and

$m_2 = \frac{2\pi a^3}{3}$; if $\frac{h^2 - 3a^2}{4} = 0$, or $h = a\sqrt{3}$, then the centroid is on the common base.

b. See the solution to Exercise 55, Section 13.2, to see that $h = a\sqrt{2}$.

Section 14.6 Triple Integrals in Cylindrical and Spherical Coordinates (pp. 805-816)

Quick Review 14.6

1. $z = r$, $1 \le r \le 2$
2. $z^2 - x^2 - y^2 = 1$
3. Right circular cylinder parallel to the z-axis generated by the circle $r = 2\sin\theta$ in the $r\theta$-plane

4. Cylinder of lines parallel to the z-axis generated by the cardioid $r = 1 + \cos\theta$
5. $z = x^2$
6. $z = y^2$
7. The origin (0, 0)
8. $z = |x|$
9. $z = |y|$
10. Circle with radius 2 and center (0, 0, 2) in the plane $z = 2$

Section 14.6 Exercises

	Rectangular	Spherical
1.	(0, 0, 0)	(0, 0, 0)
3.	(0, 0, 1)	(1, 0, 0)
5.	$(0, -2\sqrt{2}, 0)$	$\left(2\sqrt{2}, \frac{\pi}{2}, \frac{3\pi}{2}\right)$

7. $r = 0 \Rightarrow$ rectangular, $x^2 + y^2 = 0$; spherical, $\phi = 0$ or $\phi = \pi$; the z–axis

9. $z = 0 \Rightarrow$ cylindrical, $z = 0$; spherical, $\phi = \frac{\pi}{2}$; the xy–plane

11. $z = \sqrt{x^2 + y^2}$, $z \le 1 \Rightarrow$ cylindrical, $z = r$, $0 \le r \le 1$;

spherical, $\phi = \tan^{-1}\frac{\sqrt{x^2+y^2}}{z} = \tan^{-1} 1 = \frac{\pi}{4}$,

$\rho = \sqrt{x^2 + y^2 + z^2} = \sqrt{x^2 + y^2 + x^2 + y^2}$

$= \sqrt{2(x^2 + y^2)} = \sqrt{2z^2} = \sqrt{2}\,|z| \le \sqrt{2} \Rightarrow 0 \le \rho \le \sqrt{2}$;

a (finite) cone

13. $\rho \sin\phi \cos\theta = 0 \Rightarrow$ rectangular, $x = 0$;
cylindrical $\theta = \frac{\pi}{2}$; the yz–plane

15. $\rho = 5\cos\phi \Rightarrow$ rectangular, $\sqrt{x^2 + y^2 + z^2}$

$= 5\cos\left(\cos^{-1}\left(\frac{z}{\sqrt{x^2+y^2+z^2}}\right)\right)$

$= \sqrt{x^2 + y^2 + z^2} = \frac{5z}{\sqrt{x^2+y^2+z^2}} \Rightarrow x^2 + y^2 + z^2 = 5z$

$\Rightarrow x^2 + y^2 + z^2 - 5z + \frac{25}{4} = \frac{25}{4} \Rightarrow x^2 + y^2 + \left(z - \frac{5}{2}\right)^2$

$= \frac{25}{4}$; cylindrical, $r^2 + \left(z - \frac{5}{2}\right)^2 = \frac{25}{4} \Rightarrow r^2 + z^2 = 5z$,

a sphere of radius $\frac{5}{2}$ centered at $\left(0, 0, \frac{5}{2}\right)$(rectangular)

17. $x^2 + y^2 + (z - 1)^2 = 1$, $z \le 1 \Rightarrow$ cylindrical,
$r^2 + (z - 1)^2 = 1 \Rightarrow r^2 + z^2 - 2z + 1 = 1$
$\Rightarrow r^2 + z^2 = 2z$, $z \le 1$; spherical $x^2 + y^2 + z^2 - 2z = 0$
$\Rightarrow \rho^2 - 2\rho\cos\phi = 0 \Rightarrow \rho(\rho - 2\cos\phi) = 0$

$\Rightarrow \rho = 2\cos\phi$, $\frac{\pi}{4} \le \phi \le \frac{\pi}{2}$ since $\rho \ne 0$,

the lower half(hemisphere) of the sphere of radius 1 centered at (0, 0, 1) (rectangular)

19. $\phi = \frac{3\pi}{4}$, $0 \le \rho \le \sqrt{2} \Rightarrow$ rectangular, $\cos\frac{3\pi}{4} = \cos\phi$

$= \frac{z}{\sqrt{x^2+y^2+z^2}} \Rightarrow \frac{-1}{\sqrt{2}} = \frac{z}{\sqrt{x^2+y^2+z^2}}$

$\Rightarrow \sqrt{x^2 + y^2 + z^2} = -\sqrt{2}z \Rightarrow x^2 + y^2 + z^2 = 2z^2$

$\Rightarrow x^2 + y^2 - z^2 = 0$ with $z \le 0 \Rightarrow z = -\sqrt{x^2 + y^2}$ and

$0 \ge z \ge \sqrt{2}\cos\frac{3\pi}{4} \Rightarrow z = -\sqrt{x^2 + y^2}$ and $-1 \le z \le 0$;

cylindrical $x^2 + y^2 - z^2 = 0 \Rightarrow r^2 - z^2 = 0 \Rightarrow r = -z$ or $r = z$, but $r \ge 0$ and $z \le 0 \Rightarrow r = -z$, a cone with vertex at the origin and base the circle $x^2 + y^2 = 1$ in the plane $z = -1$

21. Cardioid of revolution symmetric about the y–axis, cusp at the origin pointing down

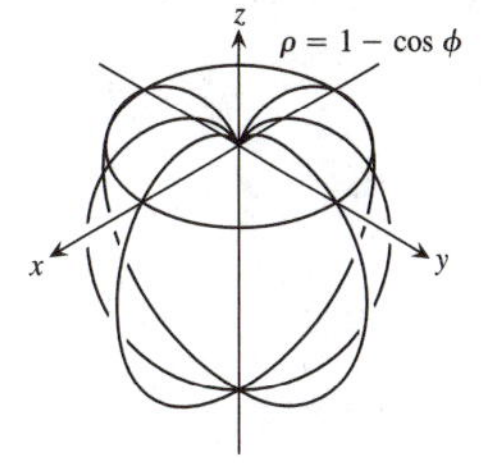

23. $x^2 + y^2 = a^2 \Rightarrow (\rho\sin\phi\cos\theta)^2 + (\rho\sin\phi\sin\theta)^2$
$= a^2 \Rightarrow (\rho^2\sin^2\phi)(\cos^2\theta + \sin^2\theta) = a^2 \Rightarrow \rho^2\sin^2\phi$
$= a^2 \Rightarrow \rho\sin\phi = a$ or $\rho\sin\phi = -a \Rightarrow \rho\sin\phi = a$ or
$\rho = a\csc\phi$, since $0 \le \phi \le \pi$ and $\rho \ge 0$

25. $\int_0^{2\pi}\int_0^1\int_r^{\sqrt{2-r^2}} dz\, r\, dr\, d\theta$

$= \int_0^{2\pi}\int_0^1 [r(2 - r^2)^{1/2} - r^2]\, dr\, d\theta$

$= \int_0^{2\pi}\left[-\frac{1}{3}(2 - r^2)^{3/2} - \frac{r^3}{3}\right]_0^1 d\theta = \int_0^{2\pi}\left(\frac{2^{3/2}}{3} - \frac{2}{3}\right) d\theta$

$= \frac{4\pi(\sqrt{2} - 1)}{3}$

27. $\int_0^{2\pi}\int_0^1\int_r^{(2-r^2)^{-1/2}} 3\, dz\, r\, dr\, d\theta$

$= 3\int_0^{2\pi}\int_0^1 [r(2 - r^2)^{-1/2} - r^2]\, dr\, d\theta$

$= 3\int_0^{2\pi}\left[-(2 - r^2)^{1/2} - \frac{r^3}{3}\right]_0^1 d\theta$

$= 3\int_0^{2\pi}\left(\sqrt{2} - \frac{4}{3}\right) d\theta = \pi(6\sqrt{2} - 8)$

29. $\int_0^{2\pi}\int_0^3\int_0^{z/3} r^3\, dr\, dz\, d\theta = \int_0^{2\pi}\int_0^3 \frac{z^4}{324}\, dz\, d\theta$

$= \int_0^{2\pi}\frac{3}{20}\, d\theta = \frac{3\pi}{10}$

31. $\int_0^1\int_0^{\sqrt{z}}\int_0^{2\pi} (r^2\cos^2\theta + z^2)\, r\, d\theta\, dr\, dz$

$= \int_0^1\int_0^{\sqrt{z}}\left[\frac{r^2\theta}{2} + \frac{r^2\sin 2\theta}{4} + z^2\theta\right]_0^{2\pi} r\, dr\, dz$

$= \int_0^1\int_0^{\sqrt{z}} (\pi r^3 + 2\pi r z^2)\, dr\, dz$

$= \int_0^1\left[\frac{\pi r^4}{4} + \pi r^2 z^2\right]_0^{\sqrt{z}} dz = \int_0^1\left(\frac{\pi z^2}{4} + \pi z^3\right) dz$

$= \left[\frac{\pi z^3}{12} + \frac{\pi z^4}{4}\right]_0^1 = \frac{\pi}{3}$

33. a. $\int_0^{2\pi}\int_0^1\int_0^{\sqrt{4-r^2}} dz\; r\; dr\; d\theta$

b. $\int_0^{2\pi}\int_0^{\sqrt{3}}\int_0^1 r\; dr\; dz\; d\theta + \int_0^{2\pi}\int_{\sqrt{3}}^2\int_0^{\sqrt{4-z^2}} r\; dr\; dz\; d\theta$

c. $\int_0^1\int_0^{\sqrt{4-r^2}}\int_0^{2\pi} r\; d\theta\; dz\; dr$

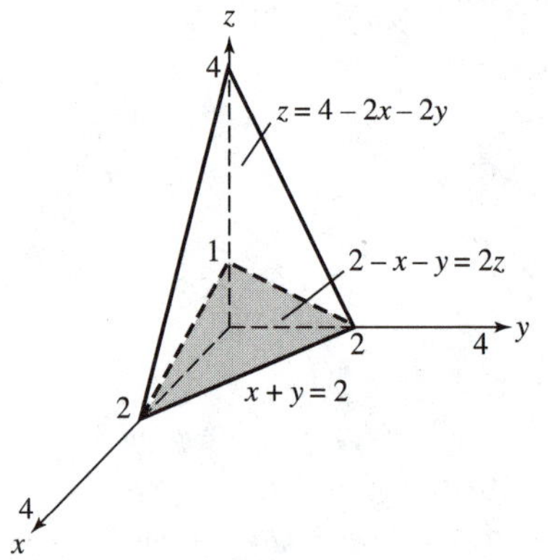

35. $\int_{-\pi/2}^{\pi/2}\int_0^{\cos\theta}\int_0^{3r^2} f(r, \theta, z)\; dz\; r\; dr\; d\theta$

37. $\int_0^{\pi}\int_0^{2\sin\theta}\int_0^{4-r\sin\theta} f(r, \theta, z)\; dz\; r\; dr\; d\theta$

39. $\int_{-\pi/2}^{\pi/2}\int_1^{1+\cos\theta}\int_0^{4} f(r, \theta, z)\; dz\; r\; dr\; d\theta$

41. $\int_0^{\pi/4}\int_0^{\sec\theta}\int_0^{2-r\sin\theta} f(r, \theta, z)\; dz\; r\; dr\; d\theta$

43. $\int_0^{\pi}\int_0^{\pi}\int_0^{2\sin\phi} \rho^2\sin\phi\; d\rho\; d\phi\; d\theta = \frac{8}{3}\int_0^{\pi}\int_0^{\pi}\sin^4\phi\; d\phi\; d\theta$

$= \frac{8}{3}\int_0^{\pi}\left(\left[-\frac{\sin^3\phi\cos\phi}{4}\right]_0^{\pi} + \frac{3}{4}\int_0^{\pi}\sin^2\phi\; d\phi\right) d\theta$

$= 2\int_0^{\pi}\int_0^{\pi}\sin^2\phi\; d\phi\; d\theta = \int_0^{\pi}\left[\theta - \frac{\sin 2\theta}{2}\right]_0^{\pi} d\theta$

$= \int_0^{\pi}\pi\; d\theta = \pi^2$

45. $\int_0^{2\pi}\int_0^{\pi/3}\int_{\sec\phi}^{2} 3\rho^2\sin\phi\; d\rho\; d\phi\; d\theta$

$= \int_0^{2\pi}\int_0^{\pi/3}(8 - \sec^3\phi)\sin\phi\; d\phi\; d\theta$

$= \int_0^{2\pi}\left[-8\cos\phi - \frac{1}{2}\sec^2\phi\right]_0^{\pi/3} d\theta$

$= \int_0^{2\pi}\left[(-4 - 2) - \left(-8 - \frac{1}{2}\right)\right] d\theta = \frac{5}{2}\int_0^{2\pi} d\theta = 5\pi$

47. $\int_0^2\int_{-\pi}^0\int_{\pi/4}^{\pi/2} \rho^3\sin 2\phi\; d\phi\; d\theta\; d\rho$

$= \int_0^2\int_{-\pi}^0 \rho^3\left[-\frac{\cos 2\phi}{2}\right]_{\pi/4}^{\pi/2} d\theta\; d\rho = \int_0^2\int_{-\pi}^0 \frac{\rho^3}{2}\; d\theta\; d\rho$

$= \int_0^2 \frac{\rho^3\pi}{2}\; d\rho = \left[\frac{\pi\rho^4}{8}\right]_0^2 = 2\pi$

49. $\int_0^1\int_0^{\pi}\int_0^{\pi/4} 12\,\rho\sin^3\phi\; d\phi\; d\theta\; d\rho$

$= \int_0^1\int_0^{\pi}\left(12\rho\left[\frac{-\sin^2\phi\cos\phi}{3}\right]_0^{\pi/4} + 8\rho\int_0^{\pi/4}\sin\phi\; d\phi\right) d\theta\; d\rho$

$= \int_0^1\int_0^{\pi}\left(-\frac{2\rho}{\sqrt{2}} - 8\rho[\cos\phi]_0^{\pi/4}\right) d\theta\; d\rho$

$= \int_0^1\int_0^{\pi}\left(8\rho - \frac{10\rho}{\sqrt{2}}\right) d\theta\; d\rho = \pi\int_0^1\left(8\rho - \frac{10\rho}{\sqrt{2}}\right) d\rho$

$= \pi\left[4\rho^2 - \frac{5\rho^2}{\sqrt{2}}\right]_0^1 = \frac{(4\sqrt{2} - 5)\pi}{\sqrt{2}}$

51. a. $x^2 + y^2 = 1 \Rightarrow \rho^2\sin^2\phi = 1$, and $\rho\sin\phi = 1$

$\Rightarrow \rho = \csc\phi$; thus $\int_0^{2\pi}\int_0^{\pi/6}\int_0^{2} \rho^2\sin\phi\; d\rho\; d\phi\; d\theta$

$+ \int_0^{2\pi}\int_{\pi/6}^{\pi/2}\int_0^{\csc\phi} \rho^2\sin\phi\; d\rho\; d\phi\; d\theta$

b. $\int_0^{2\pi}\int_1^{2}\int_{\pi/6}^{\sin^{-1}(1/\rho)} \rho^2\sin\phi\; d\phi\; d\rho\; d\theta$

$+ \int_0^{2\pi}\int_0^{2}\int_0^{\pi/6} \rho^2\sin\phi\; d\phi\; d\rho\; d\theta$

53. $V = \int_0^{2\pi}\int_0^{\pi/2}\int_{\cos\phi}^{2} \rho^2\sin\phi\; d\rho\; d\phi\; d\theta$

$= \frac{1}{3}\int_0^{2\pi}\int_0^{\pi/2}(8 - \cos^3\phi)\sin\phi\; d\phi\; d\theta$

$= \frac{1}{3}\int_0^{2\pi}\left[-8\cos\phi + \frac{\cos^4\phi}{4}\right]_0^{\pi/2} d\theta$

$= \frac{1}{3}\int_0^{2\pi}\left(8 - \frac{1}{4}\right) d\theta = \left(\frac{31}{12}\right)(2\pi) = \frac{31\pi}{6}$

55. $V = \int_0^{2\pi}\int_0^{\pi}\int_0^{1-\cos\phi} \rho^2\sin\phi\; d\rho\; d\phi\; d\theta$

$= \frac{1}{3}\int_0^{2\pi}\int_0^{\pi}(1 - \cos\phi)^3\sin\phi\; d\phi\; d\theta$

$= \frac{1}{3}\int_0^{2\pi}\left[\frac{(1 - \cos\phi)^4}{4}\right]_0^{\pi} d\theta = \frac{1}{12}(2)^4\int_0^{2\pi} d\theta$

$= \frac{4}{3}(2\pi) = \frac{8\pi}{3}$

57. $V = \int_0^{2\pi}\int_{\pi/4}^{\pi/2}\int_0^{2\cos\phi} \rho^2\sin\phi\; d\rho\; d\phi\; d\theta$

$= \frac{8}{3}\int_0^{2\pi}\int_{\pi/4}^{\pi/2}\cos^3\phi\sin\phi\; d\phi\; d\theta$

$= \frac{8}{3}\int_0^{2\pi}\left[-\frac{\cos^4\phi}{4}\right]_{\pi/4}^{\pi/2} d\theta = \left(\frac{8}{3}\right)\left(\frac{1}{16}\right)\int_0^{2\pi} d\theta$

$= \frac{1}{6}(2\pi) = \frac{\pi}{3}$

59. a. $8\int_0^{\pi/2}\int_0^{\pi/2}\int_0^{2} \rho^2\sin\phi\; d\rho\; d\phi\; d\theta$

b. $8\int_0^{\pi/2}\int_0^{2}\int_0^{\sqrt{4-r^2}} dz\; r\; dr\; d\theta$

c. $8\int_0^{2}\int_0^{\sqrt{4-x^2}}\int_0^{\sqrt{4-x^2-y^2}} dz\; dy\; dx$

61. a. $V = \int_0^{2\pi}\int_0^{\pi/3}\int_{\sec\phi}^{2} \rho^2 \sin\phi \, d\rho \, d\phi \, d\theta$

b. $V = \int_0^{2\pi}\int_0^{\sqrt{3}}\int_1^{\sqrt{4-r^2}} dz \, r \, dr \, d\theta$

c. $V = \int_{-\sqrt{3}}^{\sqrt{3}}\int_{-\sqrt{3-x^2}}^{\sqrt{3-x^2}}\int_1^{\sqrt{4-x^2-y^2}} dz \, dy \, dx$

d. $V = \int_0^{2\pi}\int_0^{\sqrt{3}} [r(4-r^2)^{1/2} - r] \, dr \, d\theta$
$= \int_0^{2\pi}\left[-\frac{(4-r^2)^{3/2}}{3} - \frac{r^2}{2}\right]_0^{\sqrt{3}} d\theta$
$= \int_0^{2\pi}\left(-\frac{1}{3} - \frac{3}{2} + \frac{4^{3/2}}{3}\right) d\theta = \frac{5}{6}\int_0^{2\pi} d\theta = \frac{5\pi}{3}$

63. $V = 4\int_0^{\pi/2}\int_0^1\int_{r^4-1}^{4-4r^2} dz \, r \, dr \, d\theta$
$= 4\int_0^{\pi/2}\int_0^1 (5r - 4r^3 - r^5) \, dr \, d\theta$
$= 4\int_0^{\pi/2}\left(\frac{5}{2} - 1 - \frac{1}{6}\right) d\theta = 4\int_0^{\pi/2}\frac{8}{6} \, d\theta = \frac{8\pi}{3}$

65. $V = \int_{3\pi/2}^{2\pi}\int_0^{3\cos\theta}\int_0^{-r\sin\theta} dz \, r \, dr \, d\theta$
$= \int_{3\pi/2}^{2\pi}\int_0^{3\cos\theta} -r^2 \sin\theta \, dr \, d\theta$
$= \int_{3\pi/2}^{2\pi} (-9\cos^3\theta)(\sin\theta) \, d\theta = \left[\frac{9}{4}\cos^4\theta\right]_{3\pi/2}^{2\pi}$
$= \frac{9}{4} - 0 = \frac{9}{4}$

67. $V = \int_0^{\pi/2}\int_0^{\sin\theta}\int_0^{\sqrt{1-r^2}} dz \, r \, dr \, d\theta$
$= \int_0^{\pi/2}\int_0^{\sin\theta} r\sqrt{1-r^2} \, dr \, d\theta$
$= \int_0^{\pi/2}\left[-\frac{1}{3}(1-r^2)^{3/2}\right]_0^{\sin\theta} d\theta$
$= -\frac{1}{3}\int_0^{\pi/2} [(1-\sin^2\theta)^{3/2} - 1] \, d\theta$
$= -\frac{1}{3}\int_0^{\pi/2} (\cos^3\theta - 1) \, d\theta$
$= -\frac{1}{3}\left(\left[\frac{\cos^2\theta\sin\theta}{3}\right]_0^{\pi/2} + \frac{2}{3}\int_0^{\pi/2}\cos\theta \, d\theta\right) + \left[\frac{\theta}{3}\right]_0^{\pi/2}$
$= -\frac{2}{9}[\sin\theta]_0^{\pi/2} + \frac{\pi}{6} = \frac{-4+3\pi}{18}$

69. $V = \int_0^{2\pi}\int_{\pi/3}^{2\pi/3}\int_0^a \rho^2 \sin\phi \, d\rho \, d\phi \, d\theta$
$= \int_0^{2\pi}\int_{\pi/3}^{2\pi/3}\frac{a^3}{3}\sin\phi \, d\phi \, d\theta = \frac{a^3}{3}\int_0^{2\pi}[-\cos\phi]_{\pi/3}^{2\pi/3} \, d\theta$
$= \frac{a^3}{3}\int_0^{2\pi}\left(\frac{1}{2} + \frac{1}{2}\right) d\theta = \frac{2\pi a^3}{3}$

71. $V = 4\int_0^{\pi/2}\int_0^{\pi/4}\int_{\sec\phi}^{2\sec\phi} \rho^2 \sin\phi \, d\rho \, d\phi \, d\theta$
$= \frac{4}{3}\int_0^{\pi/2}\int_0^{\pi/4} (8\sec^3\phi - \sec^3\phi)\sin\phi \, d\phi \, d\theta$
$= \frac{28}{3}\int_0^{\pi/2}\int_0^{\pi/4} \sec^3\phi \sin\phi \, d\phi \, d\theta$
$= \frac{28}{3}\int_0^{\pi/2}\int_0^{\pi/4} \tan\phi \sec^2\phi \, d\phi \, d\theta$
$= \frac{28}{3}\int_0^{\pi/2}\left[\frac{1}{2}\tan^2\phi\right]_0^{\pi/4} d\theta = \frac{14}{3}\int_0^{\pi/2} d\theta = \frac{7\pi}{3}$

73. $V = 8\int_0^{\pi/2}\int_0^{\sqrt{2}}\int_0^r dz \, r \, dr \, d\theta = 8\int_0^{\pi/2}\int_1^{\sqrt{2}} r^2 \, dr \, d\theta$
$= 8\left(\frac{2\sqrt{2}-1}{3}\right)\int_0^{\pi/2} d\theta = \frac{4\pi(2\sqrt{2}-1)}{3}$

75. $V = \int_0^{2\pi}\int_0^2\int_0^{4-r\sin\theta} dz \, r \, dr \, d\theta$
$= \int_0^{2\pi}\int_0^2 (4r - r^2\sin\theta) \, dr \, d\theta = 8\int_0^{2\pi}\left(1 - \frac{\sin\theta}{3}\right) d\theta$
$= 16\pi$

77. average $= \frac{1}{2\pi}\int_0^{2\pi}\int_0^1\int_{-1}^1 r^2 \, dz \, dr \, d\theta$
$= \frac{1}{2\pi}\int_0^{2\pi}\int_0^1 2r^2 \, dr \, d\theta = \frac{1}{3\pi}\int_0^{2\pi} d\theta = \frac{2}{3}$

79. $M = 4\int_0^{\pi/2}\int_0^1\int_0^r dz \, r \, dr \, d\theta = 4\int_0^{\pi/2}\int_0^1 r^2 \, dr \, d\theta$
$= \frac{4}{3}\int_0^{\pi/2} d\theta = \frac{2\pi}{3};$

$M_{xy} = \int_0^{2\pi}\int_0^1\int_0^r z \, dz \, r \, dr \, d\theta = \frac{1}{2}\int_0^{2\pi}\int_0^1 r^3 \, dr \, d\theta$
$= \frac{1}{8}\int_0^{2\pi} d\theta = \frac{\pi}{4} \Rightarrow \bar{z} = \frac{M_{xy}}{M} = \left(\frac{\pi}{4}\right)\left(\frac{3}{2\pi}\right) = \frac{3}{8},$
and $\bar{x} = \bar{y} = 0$, by symmetry

81. $M = \frac{8\pi}{3};$

$M_{xy} = \int_0^{2\pi}\int_{\pi/3}^{\pi/2}\int_0^2 z\rho^2 \sin\phi \, d\rho \, d\phi \, d\theta$
$= \int_0^{2\pi}\int_{\pi/3}^{\pi/2}\int_0^2 \rho^3 \cos\phi \sin\phi \, d\rho \, d\phi \, d\theta$
$= 4\int_0^{2\pi}\int_{\pi/3}^{\pi/2} \cos\phi \sin\phi \, d\phi \, d\theta$
$= 4\int_0^{2\pi}\left[\frac{\sin^2\phi}{2}\right]_{\pi/3}^{\pi/2} d\theta$
$= 4\int_0^{2\pi}\left(\frac{1}{2} - \frac{3}{8}\right) d\theta = \frac{1}{2}\int_0^{2\pi} d\theta = \pi \Rightarrow \bar{z} = \frac{M_{xy}}{M}$
$= (\pi)\left(\frac{3}{8\pi}\right) = \frac{3}{8}$, and $\bar{x} = \bar{y} = 0$, by symmetry

83. $M = \int_0^{2\pi}\int_0^4\int_0^{\sqrt{r}} dz \, r \, dr \, d\theta = \int_0^{2\pi}\int_0^4 r^{3/2} \, dr \, d\theta$
$= \frac{64}{5}\int_0^{2\pi} d\theta = \frac{128\pi}{5};$

$M_{xy} = \int_0^{2\pi}\int_0^4\int_0^{\sqrt{r}} z \, dz \, r \, dr \, d\theta = \frac{1}{2}\int_0^{2\pi}\int_0^4 r^2 \, dr \, d\theta$
$= \frac{32}{3}\int_0^{2\pi} d\theta = \frac{64\pi}{3} \Rightarrow \bar{z} = \frac{M_{xy}}{M} = \frac{5}{6}$, and $\bar{x} = \bar{y} = 0$,
by symmetry

85. $I_z = \int_0^{2\pi}\int_0^a \int_{-\sqrt{a^2-r^2}}^{\sqrt{a^2-r^2}} r^3\, dz\, dr\, d\theta$

$= \int_0^{2\pi}\int_0^a 2r^3\sqrt{a^2-r^2}\, dr\, d\theta$

$= 2\int_0^{2\pi}\left[\left(-\frac{r^2}{5}-\frac{2a^2}{15}\right)(a^2-r^2)^{3/2}\right]_0^a d\theta = 2\int_0^{2\pi}\frac{2}{15}a^5\, d\theta$

$= \frac{8\pi a^5}{15}$

87. The mass of the plant's atmosphere to an altitude h above the surface of the planet is the triple integral

$$M(h) = \int_0^{2\pi}\int_0^{\pi}\int_R^h \mu_0 e^{-c(\rho-R)}\rho^2 \sin\phi\, d\rho\, d\phi\, d\theta$$
$$= \int_R^h\int_0^{2\pi}\int_0^{\pi} \mu_0 e^{-c(\rho-R)}\,\rho^2 \sin\phi\, d\phi\, d\theta\, d\rho$$
$$= \int_R^h\int_0^{2\pi} [\mu_0 e^{-c(\rho-R)}\rho^2(-\cos\phi)]_0^{\pi}\, d\theta\, d\rho$$
$$= 2\int_R^h\int_0^{2\pi} \mu_0 e^{cR}e^{-c\rho}\rho^2\, d\theta\, d\rho$$
$$= 4\pi\mu_0 e^{cR}\int_R^h e^{-c\rho}\,\rho^2\, d\rho$$
$$= 4\pi\mu_0 e^{cR}\left[-\frac{\rho^2 e^{-c\rho}}{c}-\frac{2\rho e^{-c\rho}}{c^2}-\frac{2e^{-c\rho}}{c^3}\right]_R^h \text{ (by parts)}$$
$$= 4\pi\mu_0 e^{cR}\left(-\frac{h^2e^{-ch}}{c}-\frac{2he^{-ch}}{c^2}-\frac{2e^{-ch}}{c^3}+\frac{R^2e^{-cR}}{c}+\frac{2Re^{-cR}}{c^2}+\frac{2e^{-cR}}{c^3}\right).$$

The mass of the planet's atmosphere is therefore

$$M = \lim_{h\to\infty} M(h) = 4\pi\mu_0\left(\frac{R^2}{c}+\frac{2R}{c^2}+\frac{2}{c^3}\right).$$

Section 14.7 Substitutions in Multiple Integrals (pp. 816 – 825)

Quick Review 14.7

1. 17

2. 9

3. $x = 2,\ y = -1$

4. $x = -2,\ y = 1,\ z = 3$

5. $x = \frac{u+2v}{5},\ y = \frac{u-3v}{5}$

6. $x = u + v,\ y = 2v$

7. $x = \frac{u-v}{3},\ y = \frac{2u+v}{3}$

8. $x = \frac{u+4v-w-3}{10},\ y = \frac{-u+2v-w-3}{2},$
$z = \frac{-3u+8v-7w-11}{10}$

9. $x = u + v,\ y = 2v,\ z = 3w$

10. $\frac{\partial x}{\partial u} = \cos(vw),\ \frac{\partial x}{\partial v} = -uw\sin(vw),\ \frac{\partial x}{\partial w} = -uv\sin(vw)$

Section 14.7 Exercises

1. a. $x - y = u$ and $2x + y = v \Rightarrow 3x = u + v$

and $y = x - u \Rightarrow x = \frac{1}{3}(u + v)$ and $y = \frac{1}{3}(-2u + v)$;

$$\frac{\partial(x,y)}{\partial(u,v)} = \begin{vmatrix} \frac{1}{3} & \frac{1}{3} \\ -\frac{2}{3} & \frac{1}{3} \end{vmatrix} = \frac{1}{9}+\frac{2}{9} = \frac{1}{3}$$

b. The line segment $y = x$ from (0, 0) to (1, 1) is $x - y = 0 \Rightarrow u = 0$; the line segment $y = -2x$ from (0, 0) to (1, −2) is $2x + y = 0 \Rightarrow v = 0$; the line segment $x = 1$ from (1, 1) to (1, −2) is $(x - y) + (2x + y) = 3 \Rightarrow u + v = 3$. The transformed region is sketched below.

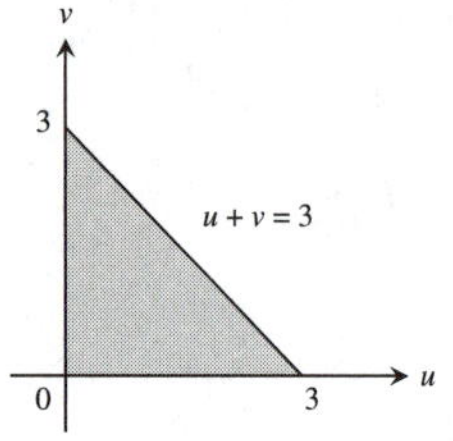

3. a. $3x + 2y = u$ and $x + 4y = v \Rightarrow -5x = -2u + v$ and $y = \frac{1}{2}(u - 3x) \Rightarrow x = \frac{1}{5}(2u - v)$ and $y = \frac{1}{10}(3v - u)$;

$$\frac{\partial(x,y)}{\partial(u,v)} = \begin{vmatrix} \frac{2}{5} & -\frac{1}{5} \\ -\frac{1}{10} & \frac{3}{10} \end{vmatrix} = \frac{6}{50}-\frac{1}{50} = \frac{1}{10}$$

b. The x–axis $y = 0 \Rightarrow u = 3v$; the y–axis $x = 0 \Rightarrow v = 2u$;

the line $x + y = 1 \Rightarrow \frac{1}{5}(2u - v) + \frac{1}{10}(3v - u) = 1 \Rightarrow 2(2u - v) + (3v - u) = 10 \Rightarrow 3u + v = 10$. The transformed region is sketched below.

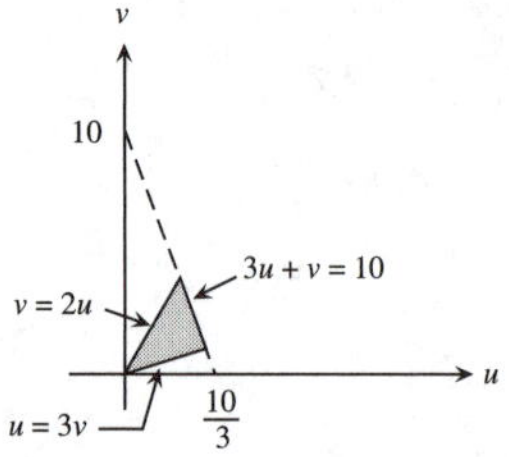

5. a. $x = u\cos v$ and $y = u\sin v \Rightarrow \frac{\partial(x,y)}{\partial(u,v)}$

$$= \begin{vmatrix} \cos v & -u\sin v \\ \sin v & u\cos v \end{vmatrix} = u\cos^2 v + u\sin^2 v = u$$

b. $x = u\sin v$ and $y = u\cos v \Rightarrow \frac{\partial(x,y)}{\partial(u,v)}$

$$= \begin{vmatrix} \sin v & u\cos v \\ \cos v & -u\sin v \end{vmatrix} = -u\sin^2 v - u\cos^2 v = -u$$

7. $\int_0^4\int_{y/2}^{(y/2)+1}\left(x-\frac{y}{2}\right)dx\,dy = \int_0^4\left[\frac{x^2}{2}-\frac{xy}{2}\right]_{y/2}^{y/2+1} dy$

$= \frac{1}{2}\int_0^4\left[\left(\frac{y}{2}+1\right)^2-\left(\frac{y}{2}\right)^2-\left(\frac{y}{2}+1\right)y+\left(\frac{y}{2}\right)y\right]dy$

$= \frac{1}{2}\int_0^4 (y + 1 - y)\, dy = \frac{1}{2}\int_0^4 dy = \frac{1}{2}(4) = 2$

9. $\int\int_R (3x^2 + 14xy + 8y^2)\,dx\,dy$

$= \int\int_R (3x + 2y)(x + 4y)\,dx\,dy$

$= \int\int_G uv \left|\frac{\partial(x, y)}{\partial(u, v)}\right| du\,dv$

$= \frac{1}{10}\int\int_G uv\,du\,dv;$

We find the boundaries of G from the boundaries of R, shown in the accompanying figure:

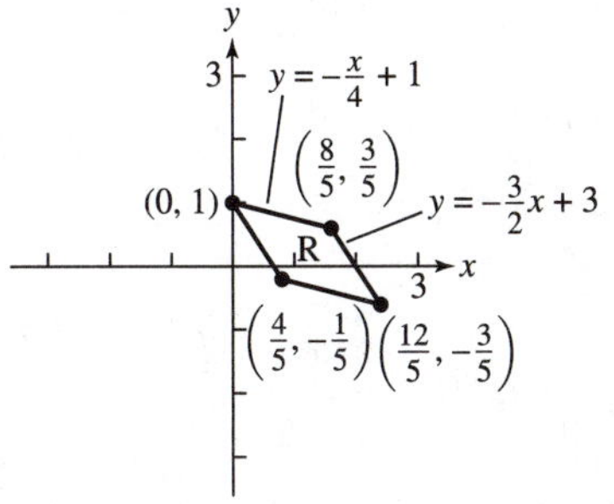

xy-equations for the boundary of R	Corresponding uv-equations for the boundary of G	Simplified uv-equations
$y = -\frac{3}{2}x + 1$	$\frac{1}{10}(3v - u) = -\frac{3}{10}(2u - v) + 1$	$u = 2$
$y = -\frac{3}{2}x + 3$	$\frac{1}{10}(3v - u) = -\frac{3}{10}(2u - v) + 3$	$u = 6$
$y = -\frac{1}{4}x$	$\frac{1}{10}(3v - u) = -\frac{1}{20}(2u - v)$	$v = 0$
$y = -\frac{1}{4}x + 1$	$\frac{1}{10}(3v - u) = -\frac{1}{20}(2u - v) + 1$	$v = 4$

$\Rightarrow \frac{1}{10}\int\int_G uv\,du\,dv = \frac{1}{10}\int_2^6\int_0^4 uv\,dv\,du$

$= \frac{1}{10}\int_2^6 u\left[\frac{v^2}{2}\right]_0^4 du = \frac{4}{5}\int_2^6 u\,du = \left(\frac{4}{5}\right)\left[\frac{u^2}{2}\right]_2^6$

$= \left(\frac{4}{5}\right)(18 - 2) = \frac{64}{5}$

11. $x = \frac{u}{v}$ and $y = uv \Rightarrow \frac{y}{x} = v^2$ and $xy = u^2$; $\frac{\partial(x, y)}{\partial(u, v)}$

$= J(u,v) = \begin{vmatrix} v^{-1} & -uv^{-2} \\ v & u \end{vmatrix} = v^{-1}u + v^{-1}u = \frac{2u}{v};$

$y = x \Rightarrow uv = \frac{u}{v} \Rightarrow v = 1$, and $y = 4x \Rightarrow v = 2$; $xy = 1$

$\Rightarrow u = 1$, and $xy = 9 \Rightarrow u = 3$;

thus $\int\int_R \left(\sqrt{\frac{y}{x}} + \sqrt{xy}\right) dx\,dy$

$= \int_1^3\int_1^2 (v + u)\left(\frac{2u}{v}\right) dv\,du = \int_1^3\int_1^2 \left(2u + \frac{2u^2}{v}\right) dv\,du$

$= \int_1^3 \left[2uv + 2u^2 \ln v\right]_1^2 du = \int_1^3 (2u + 2u^2 \ln 2)\,du$

$= \left[u^2 + \frac{2}{3}u^3 \ln 2\right]_1^3 = 8 + \frac{2}{3}(26)(\ln 2) = 8 + \frac{52}{3}(\ln 2)$

13. $x = ar\cos\theta$ and $y = ar\sin\theta \Rightarrow \frac{\partial(x, y)}{\partial(r, \theta)} = J(r, \theta)$

$= \begin{vmatrix} a\cos\theta & -ar\sin\theta \\ b\sin\theta & br\cos\theta \end{vmatrix} = abr\cos^2\theta + abr\sin^2\theta = abr;$

$I_0 = \int\int_R (x^2 + y^2)\,dA$

$= \int_0^{2\pi}\int_0^1 r^2(a^2\cos^2\theta + b^2\sin^2\theta)\,|J(r, \theta)|\,dr\,d\theta$

$= \int_0^{2\pi}\int_0^1 abr^3(a^2\cos^2\theta + b^2\sin^2\theta)\,dr\,d\theta$

$= \frac{ab}{4}\int_0^{2\pi} (a^2\cos^2\theta + b^2\sin^2\theta)\,d\theta$

$= \frac{ab}{4}\left[\frac{a^2\theta}{2} + \frac{a^2\sin 2\theta}{4} + \frac{b^2\theta}{2} + \frac{b^2\sin 2\theta}{4}\right]_0^{2\pi}$

$= \frac{ab\pi(a^2 + b^2)}{4}$

15. The region of integration R in the xy–plane is sketched in the figure below. The boundaries of the image G are obtained as follows, with G sketched below:

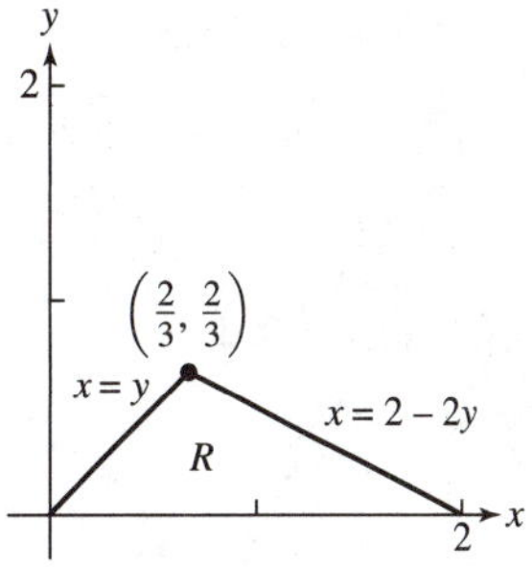

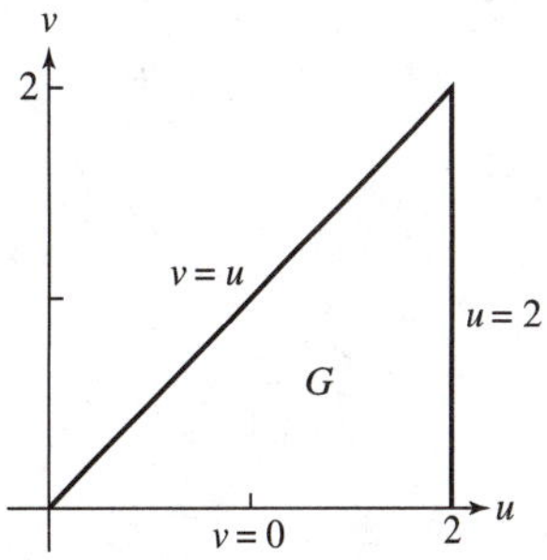

xy-equations for the boundary of R	Corresponding uv-equations for the boundary of G	Simplified uv-equations
$x = y$	$\frac{1}{3}(u + 2v) = \frac{1}{3}(u - v)$	$v = 0$
$x = 2 - 2y$	$\frac{1}{3}(u + 2v) = 2 - \frac{2}{3}(u - v)$	$u = 2$
$y = 0$	$0 = \frac{1}{3}(u - v)$	$v = u$

Also from Exercise 2, $\frac{\partial(x, y)}{\partial(u, v)} = J(u, v) = -\frac{1}{3}$

$\Rightarrow \int_0^{2/3}\int_y^{2-2y} (x + 2y)e^{(y-x)}\,dx\,dy$

$= \int_0^2\int_0^u ue^{-v}\left|-\frac{1}{3}\right| dv\,du = \frac{1}{3}\int_0^2 u\left[-e^{-v}\right]_0^u du$

$= \frac{1}{3}\int_0^2 u(1 - e^{-u})\,du = \frac{1}{3}\left[u(u + e^{-u}) - \frac{u^2}{2} + e^{-u}\right]_0^2$

$= \frac{1}{3}[2(2 + e^{-2}) - 2 + e^{-2} - 1] = \frac{1}{3}(3e^{-2} + 1) \approx 0.4687$

17. $\begin{vmatrix} \sin\phi\cos\theta & \rho\cos\phi\cos\theta & -\rho\sin\phi\sin\theta \\ \sin\phi\sin\theta & \rho\cos\phi\sin\theta & \rho\sin\phi\cos\theta \\ \cos\phi & -\rho\sin\phi & 0 \end{vmatrix}$

$= (\cos\phi)\begin{vmatrix} \rho\cos\phi\cos\theta & -\rho\sin\phi\sin\theta \\ \rho\cos\phi\sin\theta & \rho\sin\phi\cos\theta \end{vmatrix}$

$+ (\rho\sin\phi)\begin{vmatrix} \sin\phi\cos\theta & -\rho\sin\phi\sin\theta \\ \sin\phi\sin\theta & \rho\sin\phi\cos\theta \end{vmatrix}$

$= (\rho^2\cos\phi)(\sin\phi\cos\phi\cos^2\theta + \sin\phi\cos\phi\sin^2\theta)$
$\quad + (\rho^2\sin\phi)(\sin^2\phi\cos^2\theta + \sin^2\phi\sin^2\theta)$
$= \rho^2\sin\phi\cos^2\phi + \rho^2\sin^3\phi$
$= (\rho^2\sin\phi)(\cos^2\phi + \sin^2\phi) = \rho^2\sin\phi$

19. $J(u, v, w) = \begin{vmatrix} a & 0 & 0 \\ 0 & b & 0 \\ 0 & 0 & c \end{vmatrix} = abc$; the transformation takes the ellipsoid region $\frac{x^2}{a^2} + \frac{y^2}{b^2} + \frac{z^2}{c^2} \le 1$ in xyz–space into the spherical region $u^2 + v^2 + w^2 \le 1$ in the uvw–space (which has volume $V = \frac{4}{3}\pi$) $\Rightarrow V = \iiint_R dx\,dy\,dz$
$= \iiint_G abc\,du\,dv\,dw = \frac{4\pi abc}{3}$

21. $u = x$, $v = xy$, and $w = 3z \Rightarrow x = u$, $y = \frac{v}{u}$, and $z = \frac{1}{3}w \Rightarrow$

$J(u, v, w) = \begin{vmatrix} 1 & 0 & 0 \\ -\frac{v}{u^2} & \frac{1}{u} & 0 \\ 0 & 0 & \frac{1}{3} \end{vmatrix} = \frac{1}{3u}$;

$\iiint_R (x^2y + 3xyz)\,dx\,dy\,dz$
$= \iiint_G \left[u^2\left(\frac{v}{u}\right) + 3u\left(\frac{v}{u}\right)\left(\frac{w}{3}\right)\right] |J(u, v, w)|\,du\,dv\,dw$
$= \frac{1}{3}\int_0^3\int_0^2\int_1^2 \left(v + \frac{vw}{u}\right) du\,dv\,dw$
$= \frac{1}{3}\int_0^3\int_0^2 (v + vw\ln 2)\,dv\,dw$
$= \frac{1}{3}\int_0^3 (1 + w\ln 2)\left[\frac{v^2}{2}\right]_0^2 dw = \frac{2}{3}\int_0^3 (1 + w\ln 2)\,dw$
$= \frac{2}{3}\left[w + \frac{w^2}{2}\ln 2\right]_0^3 = \frac{2}{3}\left(3 + \frac{9}{2}\ln 2\right) = 2 + 3\ln 2$
$= 2 + \ln 8$

23. Let $u = g(x) \Rightarrow J(x) = \frac{du}{dx} = g'(x) \Rightarrow \int_a^b f(u)\,du$
$= \int_{g(a)}^{g(b)} f(g(x))g'(x)\,dx$ in accordance with formula (1) in Section 4.8. Note that $g'(x)$ represents the Jacobian of the transformation $u = g(x)$ or $x = g^{-1}(u)$. Several examples are presented in Section 4.8.

■ Chapter 14 Review Exercises

(pp. 825-827)

1. $\int_1^{10}\int_0^{1/y} ye^{xy}\,dx\,dy = \int_1^{10} [e^{xy}]_0^{1/y}\,dy = \int_1^{10} (e - 1)\,dy$
$= 9e - 9$

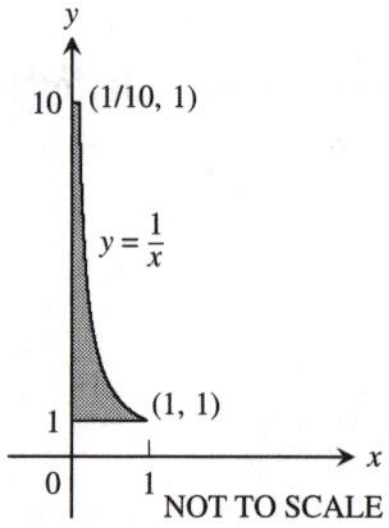

2. $\int_0^1\int_0^{x^3} e^{y/x}\,dy\,dx = \int_1^1 x[e^{y/x}]_0^{x^3}\,dx = \int_0^1 (xe^{x^2} - x)\,dx$
$= \left[\frac{1}{2}e^{x^2} - \frac{x^2}{2}\right]_0^1 = \frac{e-2}{2}$

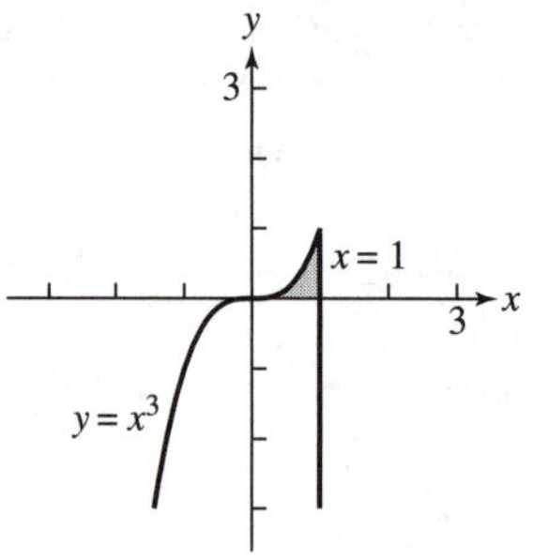

3. $\int_0^{3/2}\int_{-\sqrt{9-4t^2}}^{\sqrt{9-4t^2}} t\,ds\,dt = \int_0^{3/2} [ts]_{-\sqrt{9-4t^2}}^{\sqrt{9-4t^2}}\,dt$
$= \int_0^{3/2} 2t\sqrt{9 - 4t^2}\,dt$
$= \left[-\frac{1}{6}(9 - 4t^2)^{3/2}\right]_0^{3/2}$
$= -\frac{1}{6}(0^{3/2} - 9^{3/2}) = \frac{27}{6} = \frac{9}{2}$

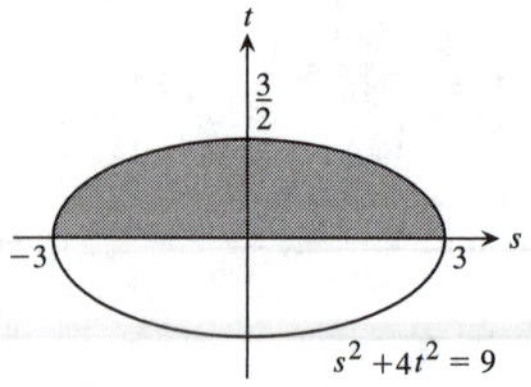

4. $\int_0^1\int_{\sqrt{y}}^{2-\sqrt{y}} xy\,dx\,dy = \int_0^1 y\left[\frac{x^2}{2}\right]_{\sqrt{y}}^{2-\sqrt{y}}\,dy$
$= \frac{1}{2}\int_0^1 y(4 - 4\sqrt{y} + y - y)\,dy$
$= \int_0^1 (2y - 2y^{3/2})\,dy = \left[y^2 - \frac{4y^{5/2}}{5}\right]_0^1$
$= \frac{1}{5}$

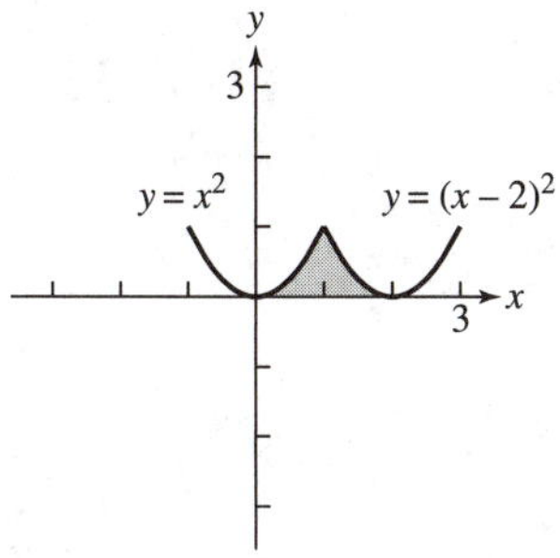

5. $\int_{-2}^{0}\int_{2x+4}^{4-x^2} dy\,dx = \int_{-2}^{0}(-x^2-2x)\,dx = \left[-\frac{x^3}{3}-x^2\right]_{-2}^{0}$

$= -\left(\frac{8}{3}-4\right)=\frac{4}{3}$

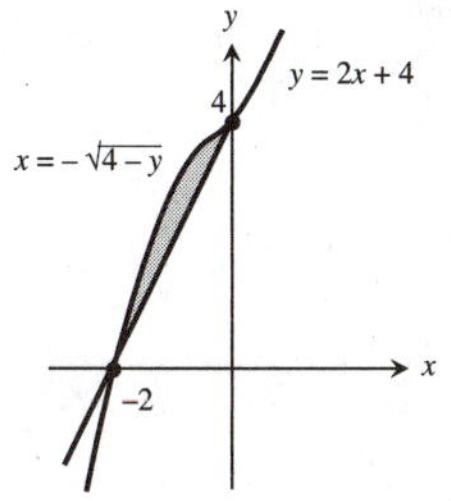

6. $\int_0^1\int_y^{\sqrt{y}}\sqrt{x}\,dx\,dy = \int_0^1\left[\frac{2}{3} x^{3/2}\right]_y^{\sqrt{y}}dy = \frac{2}{3}\int_0^1 (y^{3/4}-y^{3/2})dy$

$= \frac{2}{3}\left[\frac{4}{7} y^{7/4}-\frac{2}{5} y^{5/2}\right]_0^1$

$= \frac{2}{3}\left(\frac{4}{7}-\frac{2}{5}\right)=\frac{4}{35}$

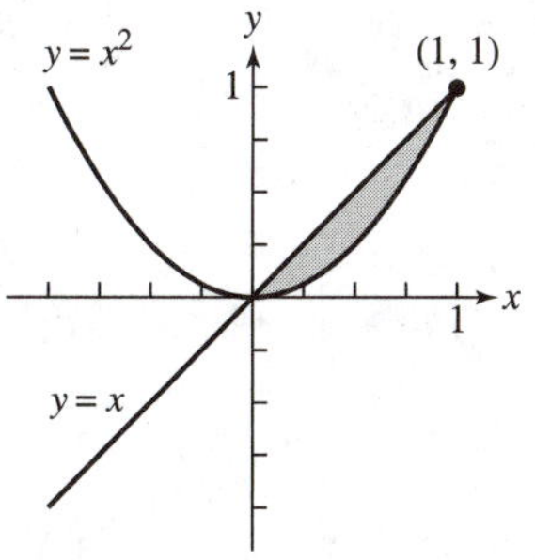

7. $\int_{-3}^{3}\int_0^{(1/2)\sqrt{9-x^2}} y\,dy\,dx = \int_{-3}^3 \left[\frac{y^2}{2}\right]_0^{(1/2)\sqrt{9-x^2}}dx$

$= \int_{-3}^3\frac{1}{8}(9-x^2)\,dx$

$= \left[\frac{9x}{8}-\frac{x^3}{24}\right]_{-3}^3$

$= \left(\frac{27}{8}-\frac{27}{24}\right)-\left(-\frac{27}{8}+\frac{27}{24}\right)$

$= \frac{27}{6}=\frac{9}{2}$

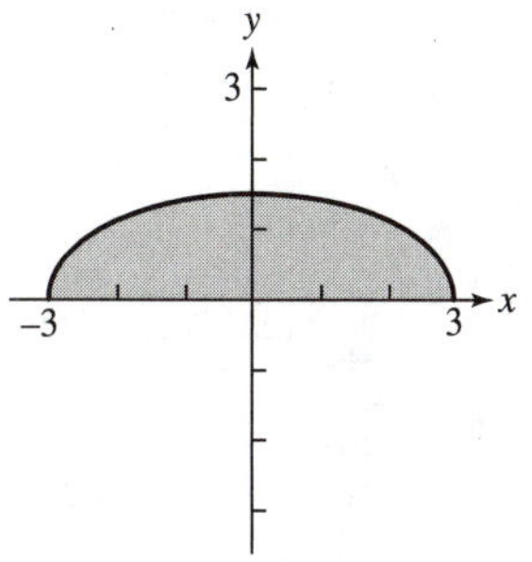

8. $\int_0^4\int_0^{\sqrt{4-y}}2x\,dx\,dy = \int_0^4[x^2]_0^{\sqrt{4-y}}\,dy$

$= \int_0^4(4-y)\,dy = \left[4y-\frac{y^2}{2}\right]_0^4 = 8$

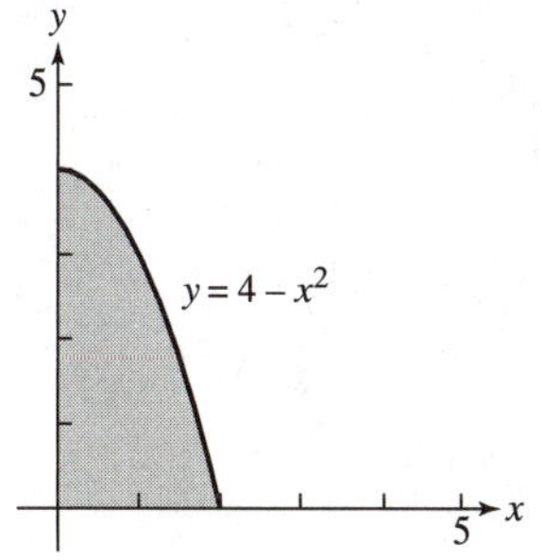

9. $\int_0^1\int_{2y}^2 4\cos(x^2)\,dx\,dy = \int_0^2\int_0^{x/2}4\cos(x^2)\,dy\,dx$

$= \int_0^2 2x\cos(x^2)\,dx = [\sin(x^2)]_0^2$

$= \sin 4$

10. $\int_0^2\int_{y/2}^1 e^{x^2}\,dx\,dy = \int_0^1\int_0^{2x}e^{x^2}\,dy\,dx = \int_0^1 2xe^{x^2}\,dx$

$= [e^{x^2}]_0^1 = e-1$

11. $\int_0^8\int_{\sqrt[3]{x}}^2 \frac{1}{y^4+1}\,dy\,dx = \int_0^2\int_0^{y^3}\frac{1}{y^4+1}\,dx\,dy$

$= \frac{1}{4}\int_0^2\frac{4y^3}{y^4+1}\,dy = \frac{\ln 17}{4}$

12. $\int_0^1\int_{\sqrt[3]{y}}^1 \frac{2\pi\sin(\pi x^2)}{x^2}\,dx\,dy = \int_0^1\int_0^{x^3}\frac{2\pi\sin(\pi x^2)}{x^2}\,dy\,dx$

$= \int_0^1 2\pi x\sin(\pi x^2)\,dx$

$= [-\cos(\pi x^2)]_0^1$

$= -(-1)-(-1) = 2$

13. $A=\int_{-2}^0\int_{2x+4}^{4-x^2}dy\,dx = \int_{-2}^0(-x^2-2x)\,dx=\frac{4}{3}$

14. $A=\int_1^4\int_{2-y}^{\sqrt{y}}dx\,dy = \int_1^4(\sqrt{y}-2+y)\,dy=\frac{37}{6}$

15. $V = \int_0^1 \int_x^{2-x} (x^2 + y^2)\, dy\, dx = \int_0^1 \left[x^2y + \frac{y^3}{3}\right]_x^{2-x} dx$

$= \int_0^1 \left[2x^2 + \frac{(2-x)^3}{3} - \frac{7x^3}{3}\right] dx$

$= \left[\frac{2x^3}{3} - \frac{(2-x)^4}{12} - \frac{7x^4}{12}\right]_0^1 = \left(\frac{2}{3} - \frac{1}{12} - \frac{7}{12}\right) + \frac{2^4}{12} = \frac{4}{3}$

16. $V = \int_{-3}^{2} \int_x^{6-x^2} x^2\, dy\, dx = \int_{-3}^{2} [x^2y]_x^{6-x^2} dx$

$= \int_{-3}^{2} (6x^2 - x^4 - x^3)\, dx = \frac{125}{4}$

17. average value $= \int_0^1 \int_0^1 xy\, dy\, dx = \int_0^1 \left[\frac{xy^2}{2}\right]_0^1 dx$

$= \int_0^1 \frac{x}{2}\, dx = \frac{1}{4}$

18. average value $= \frac{1}{\left(\frac{\pi}{4}\right)} \int_0^1 \int_0^{\sqrt{1-x^2}} xy\, dy\, dx$

$= \frac{4}{\pi} \int_0^1 \left[\frac{xy^2}{2}\right]_0^{\sqrt{1-x^2}} dx$

$= \frac{2}{\pi} \int_0^1 (x - x^3)\, dx = \frac{1}{2\pi}$

19. $M = \int_1^2 \int_{2/x}^2 dy\, dx = \int_1^2 \left(2 - \frac{2}{x}\right) dx = 2 - \ln 4;$

$M_y = \int_1^2 \int_{2/x}^2 x\, dy\, dx = \int_1^2 x\left(2 - \frac{2}{x}\right) dx = 1;$

$M_x = \int_1^2 \int_{2/x}^2 y\, dy\, dx = \int_1^2 \left(2 - \frac{2}{x^2}\right) dx = 1$

$\Rightarrow \bar{x} = \bar{y} = \frac{1}{2 - \ln 4}$

20. $M = \int_0^4 \int_{-2y}^{2y-y^2} dx\, dy = \int_0^4 (4y - y^2)\, dy = \frac{32}{3};$

$M_x = \int_0^4 \int_{-2y}^{2y-y^2} y\, dx\, dy = \int_0^4 (4y^2 - y^3)\, dy$

$= \left[\frac{4y^3}{3} - \frac{y^4}{4}\right]_0^4 = \frac{64}{3};$

$M_y = \int_0^4 \int_{-2y}^{2y-y^2} x\, dx\, dy = \int_0^4 \left[\frac{(2y-y^2)^2}{2} - 2y^2\right] dy$

$= \left[\frac{y^5}{10} - \frac{y^4}{2}\right]_0^4 = -\frac{128}{5} \Rightarrow \bar{x} = \frac{M_y}{M} = -\frac{12}{5}$ and $\bar{y} = \frac{M_x}{M}$

$= 2$

21. $I_o = \int_0^2 \int_{2x}^4 (x^2 + y^2)(3)\, dy\, dx$

$= 3 \int_0^2 \left(4x^2 + \frac{64}{3} - \frac{14x^3}{3}\right) dx = 104$

22. a. $I_o = \int_{-2}^{2} \int_{-1}^{1} (x^2 + y^2)\, dy\, dx = \int_{-2}^{2} \left(2x^2 + \frac{2}{3}\right) dx = \frac{40}{3}$

b. $I_x = \int_{-a}^{a} \int_{-b}^{b} y^2\, dy\, dx = \int_{-a}^{a} \frac{2b^3}{3}\, dx = \frac{4ab^3}{3};$

$I_y = \int_{-b}^{b} \int_{-a}^{a} x^2\, dx\, dy = \int_{-b}^{b} \frac{2a^3}{3}\, dy = \frac{4a^3b}{3}$

$\Rightarrow I_o = I_x + I_y = \frac{4ab^3}{3} + \frac{4a^3b}{3} = \frac{4ab(b^2 + a^2)}{3}$

23. $M = \delta \int_0^3 \int_0^{2x/3} dy\, dx = \delta \int_0^3 \frac{2x}{3}\, dx = 3\delta;$

$I_x = \delta \int_0^3 \int_0^{2x/3} y^2\, dy\, dx = \frac{8\delta}{81} \int_0^3 x^3\, dx = \left(\frac{8\delta}{81}\right)\left(\frac{3^4}{4}\right) = 2\delta$

$\Rightarrow R_x = \sqrt{\frac{2}{3}}$

24. $M = \int_0^1 \int_{x^2}^{x} (x + 1)\, dy\, dx = \int_0^1 (x - x^3)\, dx = \frac{1}{4};$

$M_x = \int_0^1 \int_{x^2}^{x} y(x + 1)\, dy\, dx$

$= \frac{1}{2} \int_0^1 (x^3 - x^5 + x^2 - x^4)\, dx = \frac{13}{120};$

$M_y = \int_0^1 \int_{x^2}^{x} x(x + 1)\, dy\, dx = \int_0^1 (x^2 - x^4)\, dx = \frac{2}{15}$

$\Rightarrow \bar{x} = \frac{8}{15}$ and $\bar{y} = \frac{13}{30};$

$I_x = \int_0^1 \int_{x^2}^{x} y^2(x + 1)\, dy\, dx$

$= \frac{1}{3} \int_0^1 (x^4 - x^7 + x^3 - x^6)\, dx = \frac{17}{280} \Rightarrow R_x = \sqrt{\frac{I_x}{M}}$

$= \sqrt{\frac{17}{70}};$

$I_y = \int_0^1 \int_{x^2}^{x} x^2(x + 1)\, dy\, dx = \int_0^1 (x^3 - x^5)\, dx = \frac{1}{12} \Rightarrow R_y$

$= \sqrt{\frac{I_y}{M}} = \sqrt{\frac{1}{3}}$

25. $M = \int_{-1}^{1} \int_{-1}^{1} \left(x^2 + y^2 + \frac{1}{3}\right) dy\, dx = \int_{-1}^{1} \left(2x^2 + \frac{4}{3}\right) dx = 4;$

$M_x = \int_{-1}^{1} \int_{-1}^{1} y\left(x^2 + y^2 + \frac{1}{3}\right) dy\, dx = \int_{-1}^{1} 0\, dx = 0;$

$M_y = \int_{-1}^{1} \int_{-1}^{1} x\left(x^2 + y^2 + \frac{1}{3}\right) dy\, dx = \int_{-1}^{1} \left(2x^3 + \frac{4}{3}x\right) dx$

$= 0$

26. Place the ΔABC with its vertices at $A(0, 0)$, $B(b, 0)$ and $C(a, h)$. The line through the points A and C is $y = \frac{h}{a}x$; the line through the points C and B is $y = \frac{h}{a - b}(x - b)$.

Thus, $M = \int_0^h \int_{ay/h}^{(a-b)y/h+b} \delta\, dx\, dy = b\delta \int_0^h \left(1 - \frac{y}{h}\right) dy$

$= \frac{\delta bh}{2};$

$I_x = \int_0^h \int_{ay/h}^{(a-b)y/h+b} y^2\delta\, dx\, dy = b\delta \int_0^h \left(y^2 - \frac{y^3}{h}\right) dy$

$= \frac{\delta bh^3}{12};\ R_x = \sqrt{\frac{I_x}{M}} = \frac{h}{\sqrt{6}}$

27. $\displaystyle\int_{-1}^{1}\int_{-\sqrt{1-x^2}}^{\sqrt{1-x^2}} \frac{2}{(1+x^2+y^2)^2}\,dy\,dx = \int_0^{2\pi}\int_0^1 \frac{2r}{(1+r^2)^2}\,dr\,d\theta$

$\displaystyle = \int_0^{2\pi}\left[-\frac{1}{1+r^2}\right]_0^1 d\theta$

$\displaystyle = \frac{1}{2}\int_0^{2\pi} d\theta = \pi$

28. $\displaystyle\int_{-1}^{1}\int_{-\sqrt{1-y^2}}^{\sqrt{1-y^2}} \ln(x^2+y^2+1)\,dx\,dy$

$\displaystyle = \int_0^{2\pi}\int_0^1 r\ln(r^2+1)\,dr\,d\theta = \int_0^{2\pi}\int_1^2 \frac{1}{2}\ln u\,du\,d\theta$

$\displaystyle = \frac{1}{2}\int_0^{2\pi}[u\ln u - u]_1^2\,d\theta = \frac{1}{2}\int_0^{2\pi}(2\ln 2 - 1)\,d\theta$

$= [\ln(4) - 1]\,\pi$

29. $\displaystyle M = \int_{-\pi/3}^{\pi/3}\int_0^3 r\,dr\,d\theta = \frac{9}{2}\int_{-\pi/3}^{\pi/3} d\theta = 3\pi;$

$\displaystyle M_y = \int_{-\pi/3}^{\pi/3}\int_0^3 r^2\cos\theta\,dr\,d\theta = 9\int_{-\pi/3}^{\pi/3}\cos\theta\,d\theta = 9\sqrt{3} \Rightarrow$

$\displaystyle \bar{x} = \frac{3\sqrt{3}}{\pi}$, and $\bar{y} = 0$ by symmetry

30. $\displaystyle M = \int_0^{\pi/2}\int_1^3 r\,dr\,d\theta = 4\int_0^{\pi/2} d\theta = 2\pi;$

$\displaystyle M_y = \int_0^{\pi/2}\int_1^3 r^2\cos\theta\,dr\,d\theta = \frac{26}{3}\int_0^{\pi/2}\cos\theta\,d\theta = \frac{26}{3} \Rightarrow$

$\displaystyle \bar{x} = \frac{13}{3\pi}$, and $\displaystyle \bar{y} = \frac{13}{3\pi}$ by symmetry

31. a. $\displaystyle M = 2\int_0^{\pi/2}\int_1^{1+\cos\theta} r\,dr\,d\theta$

$\displaystyle = \int_0^{\pi/2}\left(2\cos\theta + \frac{1+\cos 2\theta}{2}\right)d\theta = \frac{8+\pi}{4};$

$\displaystyle M_y = \int_{-\pi/2}^{\pi/2}\int_1^{1+\cos\theta}(r\cos\theta)\,r\,dr\,d\theta$

$\displaystyle = \int_{-\pi/2}^{\pi/2}\left(\cos^2\theta + \cos^3\theta + \frac{\cos^4\theta}{3}\right)d\theta$

$\displaystyle = \frac{32+15\pi}{24} \Rightarrow \bar{x} = \frac{15\pi+32}{6\pi+48}$, and $\bar{y} = 0$
by symmetry

b.

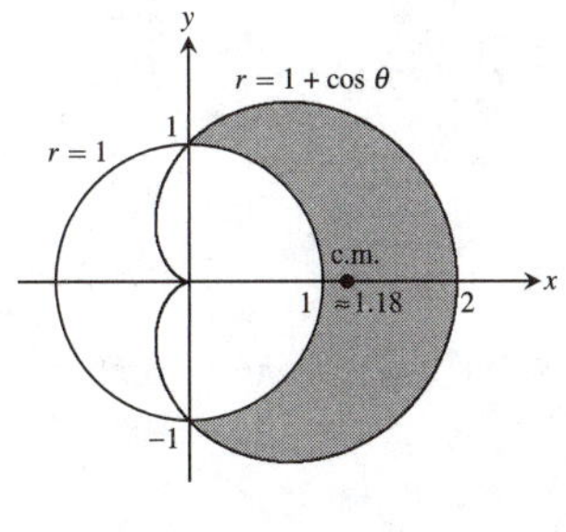

32. a. $\displaystyle M = \int_{-\alpha}^{\alpha}\int_0^{a} r\,dr\,d\theta = \int_{-\alpha}^{\alpha}\frac{a^2}{2}\,d\theta = a^2\alpha;$

$\displaystyle M_y = \int_{-\alpha}^{\alpha}\int_0^{a}(r\cos\theta)\,r\,dr\,d\theta = \int_{-\alpha}^{\alpha}\frac{a^3\cos\theta}{3}\,d\theta$

$\displaystyle = \frac{2a^3\sin\alpha}{3} \Rightarrow \bar{x} = \frac{2a\sin\alpha}{3\alpha}$, and $\bar{y} = 0$ by symmetry;

$\displaystyle = \lim_{\alpha\to\pi^-}\frac{2a\sin\alpha}{3\alpha} = 0$

b. $\displaystyle \bar{x} = \frac{2a}{5\pi}$ and $\bar{y} = 0$

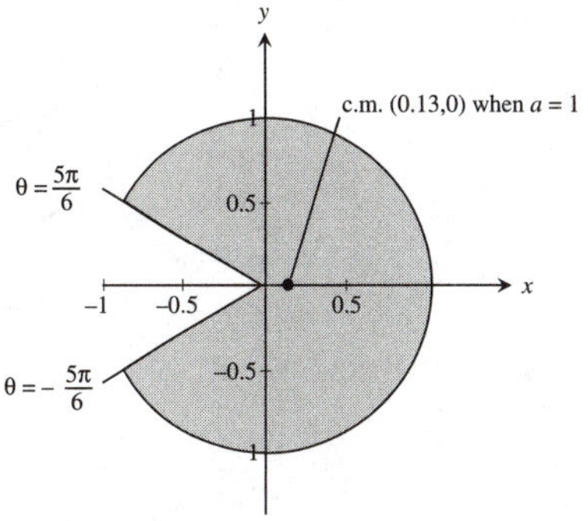

33. $(x^2+y^2)^2 - (x^2-y^2) = 0 \Rightarrow r^4 - r^2\cos 2\theta = 0 \Rightarrow$
$r^2 = \cos 2\theta$ so the integral is

$\displaystyle \int_{-\pi/4}^{\pi/4}\int_0^{\sqrt{\cos 2\theta}} \frac{r}{(1+r^2)^2}\,dr\,d\theta$

$\displaystyle = \int_{-\pi/4}^{\pi/4}\left[-\frac{1}{2(1+r^2)}\right]_0^{\sqrt{\cos 2\theta}} d\theta$

$\displaystyle = \frac{1}{2}\int_{-\pi/4}^{\pi/4}\left(1 - \frac{1}{1+\cos 2\theta}\right)d\theta$

$\displaystyle = \frac{1}{2}\int_{-\pi/4}^{\pi/4}\left(1 - \frac{1}{2\cos^2\theta}\right)d\theta = \frac{1}{2}\int_{-\pi/4}^{\pi/4}\left(1 - \frac{\sec^2\theta}{2}\right)d\theta$

$\displaystyle = \frac{1}{2}\left[\theta - \frac{\tan\theta}{2}\right]_{-\pi/4}^{\pi/4} = \frac{\pi-2}{4}$

34. a. $\displaystyle \iint_R \frac{1}{(1+x^2+y^2)^2}\,dx\,dy$

$\displaystyle = \int_0^{\pi/3}\int_0^{\sec\theta}\frac{r}{(1+r^2)^2}\,dr\,d\theta$

$\displaystyle = \int_0^{\pi/3}\left[-\frac{1}{2(1+r^2)}\right]_0^{\sec\theta} d\theta$

$\displaystyle = \int_0^{\pi/3}\left[\frac{1}{2} - \frac{1}{2(1+\sec^2\theta)}\right]d\theta = \frac{1}{2}\int_0^{\pi/3}\frac{\sec^2\theta}{1+\sec^2\theta}\,d\theta;$

$\displaystyle \begin{bmatrix} u = \tan\theta \\ du = \sec^2\theta\,d\theta \end{bmatrix} \to \frac{1}{2}\int_0^{\sqrt{3}}\frac{du}{2+u^2}$

$\displaystyle = \frac{1}{2}\left[\frac{1}{\sqrt{2}}\tan^{-1}\frac{u}{\sqrt{2}}\right]_0^{\sqrt{3}} = \frac{\sqrt{2}}{4}\tan^{-1}\sqrt{\frac{3}{2}}$

b. $\displaystyle \iint_R \frac{1}{(1+x^2+y^2)^2}\,dx\,dy$

$\displaystyle = \int_0^{\pi/2}\int_0^{\infty}\frac{r}{(1+r^2)^2}\,dr\,d\theta$

$\displaystyle = \int_0^{\pi/2}\lim_{b\to\infty}\left[-\frac{1}{2(1+r^2)}\right]d\theta$

$\displaystyle = \int_0^{\pi/2}\lim_{b\to\infty}\left[\frac{1}{2} - \frac{1}{2(1+b^2)}\right]d\theta = \frac{1}{2}\int_0^{\pi/2}d\theta = \frac{\pi}{4}$

35. $\int_0^{\pi}\int_0^{\pi}\int_0^{\pi} \cos(x+y+z)\,dx\,dy\,dz$
$= \int_0^{\pi}\int_0^{\pi} [\sin(z+y+\pi) - \sin(z+y)]\,dy\,dz$
$= \int_0^{\pi} [-\cos(z+2\pi) + \cos(z+\pi) + \cos z - (z+\pi)]dz$
$= 0$

36. $\int_{\ln 6}^{\ln 7}\int_0^{\ln 2}\int_{\ln 4}^{\ln 5} e^{(x+y+z)}\,dz\,dy\,dx = \int_{\ln 6}^{\ln 7}\int_0^{\ln 2} e^{(x+y)}dy\,dx$
$= \int_{\ln 6}^{\ln 7} e^x\,dx = 1$

37. $\int_0^1\int_0^{x^2}\int_0^{x+y} (2x - y - z)\,dz\,dy\,dx$
$= \int_0^1\int_0^{x^2} \left(\frac{3x^2}{2} - \frac{3y^2}{2}\right) dy\,dx = \int_0^1 \left(\frac{3x^4}{2} - \frac{x^6}{2}\right) dx = \frac{8}{35}$

38. $\int_1^e\int_1^x\int_0^z \frac{2y}{z^3}\,dy\,dz\,dx = \int_1^e\int_1^x \frac{1}{z}\,dz\,dx = \int_1^e \ln x\,dx$
$= [x\ln x - x]_1^e = 1$

39. $V = 2\int_0^{\pi/2}\int_{-\cos y}^{0}\int_0^{-2x} dz\,dx\,dy$
$= 2\int_0^{\pi/2}\int_{-\cos y}^{0} -2x\,dx\,dy = 2\int_0^{\pi/2} \cos^2 y\,dy$
$= 2\left[\frac{y}{2} + \frac{\sin 2y}{4}\right]_0^{\pi/2} = \frac{\pi}{2}$

40. $V = 4\int_0^2\int_0^{\sqrt{4-x^2}}\int_0^{4-x^2} dz\,dy\,dx$
$= 4\int_0^2\int_0^{\sqrt{4-x^2}} (4-x^2)\,dy\,dx = 4\int_0^2 (4-x^2)^{3/2}\,dx$
$= \left[x(4-x^2)^{3/2} + 6x\sqrt{4-x^2} + 24\sin^{-1}\frac{x}{2}\right]_0^2$
$= 24\sin^{-1} 1 = 12\pi$

41. average $= \frac{1}{3}\int_0^1\int_0^3\int_0^1 30xz\sqrt{x^2+y}\,dz\,dy\,dx$
$= \frac{1}{3}\int_0^1\int_0^3 15x\sqrt{x^2+y}\,dy\,dx$
$= \frac{1}{3}\int_0^3\int_0^1 15x\sqrt{x^2+y}\,dx\,dy$
$= \frac{1}{3}\int_0^3 [5(x^2+y)^{3/2}]_0^1\,dy$
$= \frac{1}{3}\int_0^3 [5(1+y)^{3/2} - 5y^{3/2}]\,dy$
$= \frac{1}{3}[2(1+y)^{5/2} - 2y^{5/2}]_0^3$
$= \frac{1}{3}[2(4)^{5/2} - 2(3)^{5/2} - 2] = \frac{1}{3}[2(31 - 3^{5/2})]$

42. average $= \frac{3}{4\pi a^3}\int_0^{2\pi}\int_0^{\pi}\int_0^{a} \rho^3 \sin\phi\,d\rho\,d\phi\,d\theta$
$= \frac{3a}{16\pi}\int_0^{2\pi}\int_0^{\pi} \sin\phi\,d\phi\,d\theta = \frac{3a}{8\pi}\int_0^{2\pi} d\theta = \frac{3a}{4}$

43. $x^2 + y^2 + (z+1)^2 = 1 \Rightarrow$ cylindrical, $r^2 + (z+1)^2 = 1$
$\Rightarrow r^2 + z^2 + 2z + 1 = 1 \Rightarrow r^2 + z^2 = -2z$;
spherical, $x^2 + y^2 + z^2 + 2z = 0 \Rightarrow \rho^2 + 2\rho\cos\phi = 0$
$\Rightarrow \rho(\rho + 2\cos\phi) = 0 \Rightarrow \rho = -2\cos\phi$ (since $\rho \neq 0$),
a sphere of radius 1 centered at $(0, 0, -1)$ (rectangular)

44. $\rho\cos\phi + \rho^2\sin^2\phi = 1 \Rightarrow$ cylindrical, $z + r^2 = 1$
$\Rightarrow z = 1 - r^2$; rectangular, $z = 1 - (x^2 + y^2)$
$\Rightarrow z + x^2 + y^2 = 1$, a circular paraboloid opening downward from the point $(0, 0, 1)$ (rectangular), axis along the z–axis

45. $r = 4\cos\theta \Rightarrow$ rectangular, $r = 4\cos\theta \Rightarrow r = 4\left(\frac{x}{r}\right)$
$\Rightarrow r^2 = 4x \Rightarrow x^2 - 4x + y^2 = 0 \Rightarrow x^2 - 4x + 4 + y^2 = 4$
$\Rightarrow (x-2)^2 + y^2 = 2^2$; spherical, $r = \cos\theta$
$\Rightarrow \rho\sin\phi = 4\cos\theta$, a circular cylinder parallel to the z–axis generated by the circle

46. $\phi = \frac{3\pi}{4} \Rightarrow$ cylindrical, $\tan^{-1}\left(\frac{r}{z}\right) = \frac{3\pi}{4} \Rightarrow \frac{r}{z} = -1$
$\Rightarrow z = -r,\ r \geq 0$; rectangular, $z = -\sqrt{x^2 + y^2}$, the lower nappe of a cone making an angle of $\frac{3\pi}{4}$ with the positive z–axis and having vertex at the origin

47. a. $\int_{-\sqrt{2}}^{\sqrt{2}}\int_{-\sqrt{2-y^2}}^{\sqrt{2-y^2}}\int_{\sqrt{x^2+y^2}}^{\sqrt{4-x^2-y^2}} 3\,dz\,dx\,dy$

b. $\int_0^{2\pi}\int_0^{\pi/4}\int_0^{2} 3\rho^2\sin\phi\,d\rho\,d\phi\,d\theta$

c. $\int_0^{2\pi}\int_0^{\sqrt{2}}\int_r^{\sqrt{4-r^2}} 3\,dz\,dr\,d\theta$
$= 3\int_0^{2\pi}\int_0^{\sqrt{2}} [r(4-r^2)^{1/2} - r^2]\,dr\,d\theta$
$= 3\int_0^{2\pi}\left[-\frac{1}{3}(4-r^2)^{3/2} - \frac{r^3}{3}\right]_0^{\sqrt{2}} d\theta$
$= \int_0^{2\pi} (-2^{3/2} - 2^{3/2} + 4^{3/2})\,d\theta = (8 - 4\sqrt{2})\int_0^{2\pi} d\theta$
$= 2\pi(8 - 4\sqrt{2})$

48. a. $\int_{-\pi/2}^{\pi/2}\int_0^1\int_{-r^2}^{r^2} 21(r\cos\theta)(r\sin\theta)^2\,dz\,r\,dr\,d\theta$
$= \int_{-\pi/2}^{\pi/2}\int_0^1\int_{-r^2}^{r^2} 21r^3\cos\theta\sin^2\theta\,dz\,r\,dr\,d\theta$

b. $\int_{-\pi/2}^{\pi/2}\int_0^1\int_{-r^2}^{r^2} 21r^3\cos\theta\sin^2\theta\,dz\,r\,dr\,d\theta$
$= 84\int_0^{\pi/2}\int_0^1 r^6\sin^2\theta\cos\theta\,dr\,d\theta$
$= 12\int_0^{\pi/2} \sin^2\theta\cos\theta\,d\theta = 4$

49. a. $\int_0^{2\pi}\int_0^{\pi/4}\int_0^{\sec\phi} \rho^2\sin\phi\,d\rho\,d\phi\,d\theta$

b. $\int_0^{2\pi}\int_0^{\pi/4}\int_0^{\sec\phi} \rho^2 \sin\phi \, d\rho\, d\phi\, d\theta$

$= \frac{1}{3}\int_0^{2\pi}\int_0^{\pi/4} (\sec\phi)(\sec\phi\tan\phi)\, d\phi\, d\theta$

$= \frac{1}{3}\int_0^{2\pi}\left[\frac{1}{2}\tan^2\phi\right]_0^{\pi/4} d\theta = \frac{1}{6}\int_0^{2\pi} d\theta = \frac{\pi}{3}$

50. a. $\int_0^1\int_0^{\sqrt{1-x^2}}\int_0^{\sqrt{x^2+y^2}} (6+4y)\, dz\, dy\, dx$

b. $\int_0^{\pi/2}\int_0^1\int_0^r (6+4r\sin\theta)\, dz\, r\, dr\, d\theta$

c. $\int_0^{\pi/2}\int_{\pi/4}^{\pi/2}\int_0^{\csc\phi} (6+4\rho\sin\phi\sin\theta)(\rho^2\sin\phi)\, d\rho\, d\phi\, d\theta$

d. $\int_0^{\pi/2}\int_0^1\int_0^r (6+4r\sin\theta)\, dz\, r\, dr\, d\theta$

$= \int_0^{\pi/2}\int_0^1 (6r^2+4r^3\sin\theta)\, dr\, d\theta$

$= \int_0^{\pi/2} [2r^3+r^4\sin\theta]_0^1\, d\theta = \int_0^{\pi/2} (2+\sin\theta)\, d\theta$

$= [2\theta - \cos\theta]_0^{\pi/2} = \pi + 1$

51. $\int_0^1\int_{\sqrt{1-x^2}}^{\sqrt{3-x^2}}\int_1^{\sqrt{4-x^2-y^2}} z^2yx\, dz\, dy\, dx$

$+ \int_1^{\sqrt{3}}\int_0^{\sqrt{3-x^2}}\int_1^{\sqrt{4-x^2-y^2}} z^2yx\, dz\, dy\, dx$

52. a. Bounded on the top and bottom by the sphere $x^2+y^2+z^2=4$, on the right by the right circular cylinder $(x-1)^2+y^2=1$, on the left by the plane $y=0$

b. $\int_0^{\pi/2}\int_0^{2\cos\theta}\int_{-\sqrt{4-r^2}}^{\sqrt{4-r^2}} dz\, r\, dr\, d\theta$

53. a. $\int_{-\sqrt{3}}^{\sqrt{3}}\int_{-\sqrt{3-x^2}}^{\sqrt{3-x^2}}\int_1^{\sqrt{4-x^2-y^2}} dz\, dy\, dx$

b. $\int_0^{2\pi}\int_0^{\sqrt{3}}\int_1^{\sqrt{4-r^2}} dz\, r\, dr\, d\theta$

c. $\int_0^{2\pi}\int_0^{\pi/3}\int_{\sec\phi}^{2} \rho^2\sin\phi\, d\rho\, d\phi\, d\theta$

54. a. $I_z = \int_{-\sqrt{1-x^2}}^{\sqrt{1-x^2}}\int_0^{\sqrt{1-x^2-y^2}} (x^2+y^2)\, dz\, dy\, dx$

b. $I_z = \int_0^{2\pi}\int_0^1\int_0^{\sqrt{1-r^2}} r^3\, dz\, dr\, d\theta$

c. $I_z = \int_0^{2\pi}\int_0^{\pi/2}\int_0^1 \rho^4\sin^3\phi\, d\rho\, d\phi\, d\theta$

55. a. $V = \int_0^{2\pi}\int_0^2\int_2^{\sqrt{8-r^2}} dz\, r\, dr\, d\theta$

$= \int_0^{2\pi}\int_0^2 (r\sqrt{8-r^2} - 2r)\, dr\, d\theta$

$= \int_0^{2\pi}\left[-\frac{1}{3}(8-r^2)^{3/2} - r^2\right]_0^2 d\theta$

$= \int_0^{2\pi}\left[-\frac{1}{3}(4)^{3/2} - 4 + \frac{1}{3}(8)^{3/2}\right] d\theta$

$= \int_0^{2\pi}\frac{4}{3}(-2-3+2\sqrt{8})\, d\theta$

$= \frac{4}{3}(4\sqrt{2}-5)\int_0^{2\pi} d\theta = \frac{8\pi(4\sqrt{2}-5)}{3}$

b. $V = \int_0^{2\pi}\int_0^{\pi/4}\int_{2\sec\phi}^{\sqrt{8}} \rho^2\sin\phi\, d\rho\, d\phi\, d\theta$

$= \frac{8}{3}\int_0^{2\pi}\int_0^{\pi/4} (2\sqrt{2}\sin\phi - \sec^3\phi\sin\phi)\, d\phi\, d\theta$

$= \frac{8}{3}\int_0^{2\pi}\int_0^{\pi/4} (2\sqrt{2}\sin\phi - \tan\phi\sec^2\phi)\, d\phi\, d\theta$

$= \frac{8}{3}\int_0^{2\pi}\left[-2\sqrt{2}\cos\phi - \frac{1}{2}\tan^2\phi\right]_0^{\pi/4} d\theta$

$= \frac{8}{3}\int_0^{2\pi}\left(-2-\frac{1}{2}+2\sqrt{2}\right) d\theta$

$= \frac{8}{3}\int_0^{2\pi}\left(\frac{-5+4\sqrt{2}}{2}\right) d\theta = \frac{8\pi(4\sqrt{2}-5)}{3}$

56. $I_z = \int_0^{2\pi}\int_0^{\pi/3}\int_0^2 (\rho\sin\phi)^2(\rho^2\sin\phi)\, d\rho\, d\phi\, d\theta$

$= \int_0^{2\pi}\int_0^{\pi/3}\int_0^2 \rho^4\sin^3 d\rho\, d\phi\, d\theta$

$= \frac{32}{5}\int_0^{2\pi}\int_0^{\pi/3} (\sin\phi - \cos^2\phi\sin\phi)\, d\phi\, d\theta$

$= \frac{32}{5}\int_0^{2\pi}\left[-\cos\phi + \frac{\cos^3\phi}{3}\right]_0^{\pi/3} d\theta = \frac{8\pi}{3}$

57. With the centers of the sphere at the origin,

$I_z = \int_0^{2\pi}\int_0^{\pi}\int_a^b \delta(\rho\sin\phi)^2(\rho^2\sin\phi)\, d\rho\, d\phi\, d\theta$

$= \frac{\delta(b^5-a^5)}{5}\int_0^{2\pi}\int_0^{\pi} \sin^3\phi\, d\phi\, d\theta$

$= \frac{\delta(b^5-a^5)}{5}\int_0^{2\pi}\int_0^{\pi} (\sin\phi - \cos^2\phi\sin\phi)\, d\phi\, d\theta$

$= \frac{\delta(b^5-a^5)}{5}\int_0^{2\pi}\left[-\cos\phi + \frac{\cos^3\phi}{3}\right]_0^{\pi} d\theta$

$= \frac{4\delta(b^5-a^5)}{15}\int_0^{2\pi} d\theta = \frac{8\pi\delta(b^5-a^5)}{15}$

58. $I_z = \int_0^{2\pi}\int_0^{\pi}\int_0^{1-\cos\theta} (\rho\sin\phi)^2(\rho^2\sin\phi)\,d\rho\,d\phi\,d\theta$

$= \int_0^{2\pi}\int_0^{\pi}\int_0^{1-\cos\theta} \rho^4\sin^3\phi\,d\rho\,d\phi\,d\theta$

$= \frac{1}{5}\int_0^{2\pi}\int_0^{\pi}(1-\cos\phi)^5\sin^3\phi\,d\phi\,d\theta$

$= \int_0^{2\pi}\int_0^{\pi}(1-\cos\phi)^6(1+\cos\phi)\sin\phi\,d\phi\,d\theta;$

$\begin{bmatrix} u = 1-\cos\phi \\ du = \sin\phi\,d\phi \end{bmatrix} \rightarrow \frac{1}{5}\int_0^{2\pi}\int_0^2 u^6(2-u)\,du\,d\theta$

$= \frac{1}{5}\int_0^{2\pi}\left[\frac{2u^7}{7} - \frac{u^8}{8}\right]_0^2 d\theta = \frac{1}{5}\int_0^{2\pi}\left(\frac{1}{7}-\frac{1}{8}\right)2^8\,d\theta$

$= \frac{1}{5}\int_0^{2\pi}\frac{2^3\cdot 2^5}{56}\,d\theta = \frac{32}{35}\int_0^{2\pi}d\theta = \frac{64\pi}{35}$

59. $x = u + y$ and $y = v \Rightarrow x = u + v$ and $y = v$

$\Rightarrow J(u, v) = \begin{vmatrix} 1 & 1 \\ 0 & 1 \end{vmatrix} = 1$; the boundary of the image G is obtained from the boundary of R as follows:

xy-equations for the boundary of R	Corresponding uv-equations for the boundary of G	Simplified uv-equations
$y = x$	$v = u + v$	$u = 0$
$y = 0$	$v = 0$	$v = 0$

$\Rightarrow \int_0^{\infty}\int_0^{x} e^{-sx} f(x-y, y)\,dy\,dx$

$= \int_0^{\infty}\int_0^{\infty} e^{-s(u+v)} f(u, v)\,du\,dv$

60. If $s = \alpha x + \beta y$ and $t = \gamma x + \delta y$ where $(\alpha\delta - \beta\gamma)^2 = ac - b^2$, then $x = \frac{\delta s - \beta t}{\alpha\delta - \beta\gamma}$, $y = -\frac{\gamma s + \alpha t}{\alpha\delta - \beta\gamma}$, and

$J(s, t) = \frac{1}{(\alpha\delta - \beta\gamma)^2}\begin{vmatrix} \delta & -\beta \\ -\gamma & \alpha \end{vmatrix} = \frac{1}{\alpha\delta - \beta\gamma}$

$\Rightarrow \int_{-\infty}^{\infty}\int_{-\infty}^{\infty} e^{-(s^2+t^2)}\frac{1}{\sqrt{ac-b^2}}\,ds\,dt$

$= \frac{1}{\sqrt{ac-b^2}}\int_0^{2\pi}\int_0^{\infty} re^{-r^2}\,dr\,d\theta = \frac{1}{2\sqrt{ac-b^2}}\int_0^{2\pi}d\theta$

$= \frac{\pi}{\sqrt{ac-b^2}}$. Therefore, $\frac{\pi}{\sqrt{ac-b^2}} = 1 \Rightarrow ac - b^2$

$= \pi^2$

Chapter 15
Integration in Vector Fields

■ Section 15.1 Line Integrals (pp. 829 – 835)

Chapter Opener

Based on the velocity field given, we position the tube so that the base lies in the xy-plane with the positive z-axis as the center of the tube and having the length $z = C$, where C is a positive constant.

The surface $g(x, y, z)$ is a circular region in the plane of $z = C$. We have

$\mathbf{F} = \mathbf{v} = 5(0.25 - r^2)\mathbf{k}$

(Note that $\mathbf{v}$ depends on the radius only)

$g(x, y, z) = z - C$

$\nabla g = \mathbf{k}$

$|\nabla g| = 1$

$\mathbf{n} = \mathbf{k}$

$\mathbf{p} = \mathbf{k}$

$d\sigma = \frac{|\nabla g|}{|\nabla g\cdot\mathbf{k}|} = dA = 1dA$

$\mathbf{F}\cdot\mathbf{n} = 5(0.25 - r^2)\mathbf{k}$

$\iint_S \mathbf{F}\cdot\mathbf{n}\,d\sigma = \iint_S 5(0.25 - r^2)dA$

$= \iint_{R_{xy}} 5(0.25 - r^2)dA$

$= \int_0^{2\pi}\int_0^{1/2} 5(0.25 - r^2)r\,dr\,d\theta$

$= \frac{5\pi}{32}$ cm^3/sec

Quick Review 15.1

1. $\frac{1}{3}\sin 3x + C$
2. $-\frac{1}{2}\cos 2x + C$
3. $-\frac{x^4}{4} + \frac{x^3}{3} + 2x + C$
4. $\frac{21}{4}$
5. 0
6. $x = 2t,\ y = 3t,\ 0 \le t \le 1$
7. $x = at,\ y = bt,\ 0 \le t \le 1$
8. $x = -2 + 5t,\ y = 1 - 3t,\ 0 \le t \le 1$
9. $x = 2t,\ y = 3t,\ z = 4t,\ 0 \le t \le 1$
10. $x = -1 + 4t,\ y = 1 + 2t,\ z = 2 - 4t,\ 0 \le t \le 1$

Section 15.1 Exercises

1. $\mathbf{r} = t\mathbf{i} + (1-t)\mathbf{j}$

$x = t$ and $y = 1 - t$, so $y = 1 - x$

(c)

3. $\mathbf{r} = (2\cos t)\mathbf{i} + (2\sin t)\mathbf{j}$

$x = 2\cos t$ and $y = 2\sin t$, so $x^2 + y^2 = 4$

(g)

5. $\mathbf{r} = t\mathbf{i} + t\mathbf{j} + t\mathbf{k}$

$x = t$, $y = t$, and $z = t$

(d)

7. $\mathbf{r} = (t^2 - 1)\mathbf{j} + 2t\mathbf{k}$

$y = t^2 - 1$ and $z = 2t$, so $y = \frac{z^2}{4} - 1$

(f)

9. $\mathbf{r}(t) = t\mathbf{i} + (1-t)\mathbf{j},\ 0 \le t \le 1$

$\frac{d\mathbf{r}}{dt} = \mathbf{i} - \mathbf{j}$, so $\left|\frac{d\mathbf{r}}{dt}\right| = \sqrt{2}$

$x = t$ and $y = 1 - t$, so $x + y = t + (1 - t) = 1$

$\int_C f(x, y, z)\,ds = \int_0^1 f(t, 1-t, 0)\left|\frac{d\mathbf{r}}{dt}\right|dt$

$= \int_0^1 (1)(\sqrt{2})\,dt$

$= \left[\sqrt{2}t\right]_0^1$

$= \sqrt{2}$

11. $\mathbf{r}(t) = 2t\mathbf{i} + t\mathbf{j} + (2 - 2t)\mathbf{k},\ 0 \le t \le 1$

$$\frac{d\mathbf{r}}{dt} = 2\mathbf{i} + \mathbf{j} - 2\mathbf{k}$$

$$\left|\frac{d\mathbf{r}}{dt}\right| = \sqrt{4 + 1 + 4} = 3$$

$$xy + y + z = (2t)t + t + (2 - 2t)$$

$$\int_C f(x, y, z)\, ds = \int_0^1 (2t^2 - t + 2)3\, dt = 3\left[\frac{2}{3}t^3 - \frac{1}{2}t^2 + 2t\right]_0^1 = 3\left(\frac{2}{3} - \frac{1}{2} + 2\right) = \frac{13}{2}$$

13. $\mathbf{r}(t) = (\mathbf{i} + 2\mathbf{j} + 3\mathbf{k}) + t(-\mathbf{i} - 3\mathbf{j} - 2\mathbf{k})$

$$= (1 - t)\mathbf{i} + (2 - 3t)\mathbf{j} + (3 - 2t)\mathbf{k} \quad 0 \le t \le 1$$

$$\frac{d\mathbf{r}}{dt} = -\mathbf{i} - 3\mathbf{j} - 2\mathbf{k}$$

$$\left|\frac{d\mathbf{r}}{dt}\right| = \sqrt{1 + 9 + 4} = \sqrt{14}$$

$$x + y + z = (1 - t) + (2 - 3t) + (3 - 2t) = 6 - 6t$$

$$\int_C f(x, y, z)\, ds = \int_0^1 (6 - 6t)\sqrt{14}\, dt = 6\sqrt{14}\left[t - \frac{t^2}{2}\right]_0^1 = (6\sqrt{14})\left(\frac{1}{2}\right) = 3\sqrt{14}$$

15. C_1: $\mathbf{r}(t) = t\mathbf{i} + t^2\mathbf{j},\ 0 \le t \le 1$

$$\frac{d\mathbf{r}}{dt} = \mathbf{i} + 2t\mathbf{j}$$

$$\left|\frac{d\mathbf{r}}{dt}\right| = \sqrt{1 + 4t^2}$$

$$x + \sqrt{y} - z^2 = t + \sqrt{t^2} - 0 = t + t = 2t$$

$$\int_{C_1} f(x, y, z)\, ds = \int_0^1 2t\sqrt{1 + 4t^2}\, dt = \left[\frac{1}{6}(1 + 4t^2)^{3/2}\right]_0^1 = \frac{1}{6}(5)^{3/2} - \frac{1}{6} = \frac{1}{6}(5\sqrt{5} - 1)$$

C_2: $\mathbf{r}(t) = \mathbf{i} + \mathbf{j} + t\mathbf{k},\ 0 \le t \le 1$

$$\frac{d\mathbf{r}}{dt} = \mathbf{k}$$

$$\left|\frac{d\mathbf{r}}{dt}\right| = 1$$

$$x + \sqrt{y} - z^2 = 1 + \sqrt{1} - t^2 = 2 - t^2$$

$$\int_{C_2} f(x, y, z)\, ds = \int_0^1 (2 - t^2)(1)\, dt = \left[2t - \frac{1}{3}t^3\right]_0^1 = 2 - \frac{1}{3} = \frac{5}{3};$$

therefore $\int_C f(x, y, z)\, ds = \int_{C_1} f(x, y, z)\, ds + \int_{C_2} f(x, y, z)$

$$= \frac{5}{6}\sqrt{5} + \frac{3}{2}$$

17. $\mathbf{r}(t) = t\mathbf{i} + t\mathbf{j} + t\mathbf{k},\ 0 < a \le t \le b$

$$\frac{d\mathbf{r}}{dt} = \mathbf{i} + \mathbf{j} + \mathbf{k}$$

$$\left|\frac{d\mathbf{r}}{dt}\right| = \sqrt{3}$$

$$\frac{x + y + z}{x^2 + y^2 + z^2} = \frac{t + t + t}{t^2 + t^2 + t^2} = \frac{1}{t}$$

$$\int_C f(x, y, z)\, ds = \int_a^b \left(\frac{1}{t}\right)\sqrt{3}\, dt = \left[\sqrt{3}\ln|t|\right]_a^b = \sqrt{3}\ln\left(\frac{b}{a}\right), \text{ since } 0 < a \le b$$

19. $\mathbf{r}(x) = x\mathbf{i} + y\mathbf{j} = x\mathbf{i} + \frac{x^2}{2}\mathbf{j},\ 0 \le x \le 2$

$$\frac{d\mathbf{r}}{dx} = \mathbf{i} + x\mathbf{j}$$

$$\left|\frac{d\mathbf{r}}{dt}\right| = \sqrt{1 + x^2}$$

$$f(x, y) = f\left(x, \frac{x^2}{2}\right) = \frac{x^3}{x^2/2} = 2x$$

$$\int_C f\, ds = \int_0^2 (2x)\sqrt{1 + x^2}\, dx = \left[\frac{2}{3}(1 + x^2)^{3/2}\right]_0^2 = \frac{2}{3}(5^{3/2} - 1) = \frac{10\sqrt{5} - 2}{3}$$

21. $\mathbf{r}(t) = (2\cos t)\mathbf{i} + (2\sin t)\mathbf{j},\ 0 \le t \le \frac{\pi}{2}$

$$\frac{d\mathbf{r}}{dt} = (-2\sin t)\mathbf{i} + (2\cos t)\mathbf{j}$$

$$\left|\frac{d\mathbf{r}}{dt}\right| = 2$$

$$f(x, y) = f(2\cos t, 2\sin t) = 2\cos t + 2\sin t$$

$$\int_C f\, ds = \int_0^{\pi/2} (2\cos t + 2\sin t)(2)\, dt = \left[4\sin t - 4\cos t\right]_0^{\pi/2} = 4 - (-4) = 8$$

23. $\mathbf{r}(t) = (t^2 - 1)\mathbf{j} + 2t\mathbf{k},\ 0 \le t \le 1$

$$\frac{d\mathbf{r}}{dt} = 2t\mathbf{j} + 2\mathbf{k}$$

$$\left|\frac{d\mathbf{r}}{dt}\right| = 2\sqrt{t^2 + 1}$$

$$M = \int_C \delta(x, y, z)\, ds = \int_0^1 \delta(t)(2\sqrt{t^2 + 1})\, dt$$

$$= \int_0^1 \left(\frac{3}{2}t\right)(2\sqrt{t^2 + 1})\, dt = \left[(t^2 + 1)^{3/2}\right]_0^1 = 2^{3/2} - 1$$

$$= 2\sqrt{2} - 1$$

25. $\mathbf{r}(t) = \sqrt{2}t\mathbf{i} + \sqrt{2}t\mathbf{j} + (4 - t^2)\mathbf{k},\ 0 \le t \le 1$

$$\frac{d\mathbf{r}}{dt} = \sqrt{2}\mathbf{i} + \sqrt{2}\mathbf{j} - 2t\mathbf{k}$$

$$\left|\frac{d\mathbf{r}}{dt}\right| = \sqrt{2 + 2 + 4t^2} = 2\sqrt{1 + t^2}$$

a. $$M = \int_C \delta\, ds = \int_0^1 (3t)(2\sqrt{1 + t^2})\, dt = \left[2(1 + t^2)^{3/2}\right]_0^1 = 2(2^{3/2} - 1) = 4\sqrt{2} - 2$$

b. $$M = \int_C \delta\, ds = \int_0^1 (1)(2\sqrt{1 + t^2})\, dt = \left[t\sqrt{1 + t^2} + \ln\left(t + \sqrt{1 + t^2}\right)\right]_0^1 = [\sqrt{2} + \ln(1 + \sqrt{2})] - (0 + \ln 1) = \sqrt{2} + \ln(1 + \sqrt{2})$$

27. Let $x = a \cos t$ and $y = a \sin t$, $0 \le t \le 2\pi$.

Then $\frac{dx}{dt} = -a \sin t$, $\frac{dy}{dt} = a \cos t$, $\frac{dz}{dt} = 0$

$$\sqrt{\left(\frac{dx}{dt}\right)^2 + \left(\frac{dy}{dt}\right)^2 + \left(\frac{dz}{dt}\right)^2}\, dt = a\, dt;$$

$$I_z = \int_C (x^2 + y^2)\delta\, ds = \int_0^{2\pi} (a^2 \sin^2 t + a^2 \cos^2 t)a\delta\, dt = \int_0^{2\pi} a^3\delta\, dt = 2\pi\delta a^3;$$

$$M = \int_C \delta(x, y, z)\, ds = \int_0^{2\pi} \delta a\, dt = 2\pi\delta a$$

$$R_z = \sqrt{\frac{I_z}{M}} = \sqrt{\frac{2\pi a^3\delta}{2\pi a\delta}} = a.$$

29. $\mathbf{r}(t) = (\cos t)\mathbf{i} + (\sin t)\mathbf{j} + t\mathbf{k},\ 0 \le t \le 2\pi$

$$\frac{d\mathbf{r}}{dt} = (-\sin t)\mathbf{i} + (\cos t)\mathbf{j} + \mathbf{k}$$

$$\left|\frac{d\mathbf{r}}{dt}\right| = \sqrt{\sin^2 t + \cos^2 t + 1} = \sqrt{2}$$

a. $$M = \int_C \delta\, ds = \int_0^{2\pi} \delta\sqrt{2}\, dt = 2\pi\delta\sqrt{2}$$

$$I_z = \int_C (x^2 + y^2)\delta\, ds = \int_0^{2\pi} (\cos^2 t + \sin^2 t)\delta\sqrt{2}\, dt = 2\pi\delta\sqrt{2}$$

$$R_z = \sqrt{\frac{I_z}{M}} = 1$$

b. $$M = \int_C \delta(x, y, z)\, ds = \int_0^{4\pi} \delta\sqrt{2}\, dt = 4\pi\delta\sqrt{2} \text{ and}$$

$$I_z = \int_C (x^2 + y^2)\delta\, ds = \int_0^{4\pi} \delta\sqrt{2}\, dt = 4\pi\delta\sqrt{2}$$

$$R_z = \sqrt{\frac{I_z}{M}} = 1$$

31. $\delta(x, y, z) = 2 - z$ and $\mathbf{r}(t) = (\cos t)\mathbf{j} + (\sin t)\mathbf{k},\ 0 \le t \le \pi$

$M = 2\pi - 2$ as found in Example 4 of the text; also

$$\left|\frac{d\mathbf{r}}{dt}\right| = 1$$

$$I_x = \int_C (y^2 + z^2)\delta\, ds = \int_0^{\pi} (\cos^2 t + \sin^2 t)(2 - \sin t)\, dt = \int_0^{\pi} (2 - \sin t)\, dt = 2\pi - 2$$

$$R_x = \sqrt{\frac{I_x}{M}} = 1$$

33.–35. Example CAS commands for Exercise 36:

Maple:

```
x:= t->cos(2*t); y:=t->sin(2*t);
z:=t->t^(5/2);
f:=(x,y,z)->(1+(9/4)*z^(1/3))^(1/4);
sqrt(D(x)(t)^2 + D(y)(t)^2
  + D(z)(t)^2); absvee:=unapply(%,t);
a:= 0: b:= 2 * Pi:
integrand:=simplify(f(x(t),y(t),z(t))
   *absvee(t));
int(integrand, t=a..b);
evalf(%);
```

Mathematica:

```
Clear[x, y, z, t]
r[t_]:={x[t],y[t],z[t]}
f[x_, y_, z_]=(1+9/4z^(1/3))^(1/4)
x[t_]=Cos[2t]
y[t_]=Sin[2t]
z[t_]=t^(5/2)
{a, b} = {0, 2Pi};
v[t_] = r'[t]
s[t_]=Sqrt[v[t]. v[t]]
integrand = f[x[t], y[t], z[t], s[t]
NIntegrate[integrand, {t,a,b}]
```

Section 15.2 Vector Fields, Work, Circulation and Flux (pp. 835 – 845)

Quick Review 15.2

1. $\frac{\partial f}{\partial x} = \sin y - y \sin x$, $\frac{\partial f}{\partial y} = x \cos y + \cos x$

2. $\frac{\partial f}{\partial x} = \sin(yz) + yz$, $\frac{\partial f}{\partial y} = xz \cos(yz) + xz$, $\frac{\partial f}{\partial z} = xy \cos(yz) + xy$

3. $(3x^2 + 2 \sin x)dx$

4. $(\sin x + x \cos x)\, dx$

5. $dy = 3\cos t\,dt,\ dx = -2\sin t\,dt$

6. $t^2 \sin t$

7. Counterclockwise

8. Clockwise

9. Counterclockwise

10. Clockwise

Section 15.2 Exercises

1. $f(x, y, z) = (x^2 + y^2 + z^2)^{-1/2}$

$\frac{\partial f}{\partial x} = -\frac{1}{2}(x^2 + y^2 + z^2)^{-3/2}(2x) = -x(x^2 + y^2 + z^2)^{-3/2}$;

similarly, $\frac{\partial f}{\partial y} = -y(x^2 + y^2 + z^2)^{-3/2}$

and $\frac{\partial f}{\partial z} = -z(x^2 + y^2 + z^2)^{-3/2}$

$\nabla f = \frac{-x\mathbf{i} - y\mathbf{j} - z\mathbf{k}}{(x^2 + y^2 + z^2)^{3/2}}$

3. $g(x, y, z) = e^z - \ln(x^2 + y^2)$

$\frac{\partial g}{\partial x} = -\frac{2x}{x^2 + y^2},\ \frac{\partial g}{\partial y} = -\frac{2y}{x^2 + y^2}$ and $\frac{\partial g}{\partial z} = e^z$

$\nabla g = \left(\frac{-2x}{x^2 + y^2}\right)\mathbf{i} - \left(\frac{2y}{x^2 + y^2}\right)\mathbf{j} + e^z\mathbf{k}$

5. $|\mathbf{F}|$ inversely proportional to the square of the distance from (x, y) to the origin

$\sqrt{(M(x, y))^2 + (N(x, y))^2} = \frac{k}{x^2 + y^2},\ k > 0$;

$\mathbf{F}$ points toward the origin,

$\mathbf{F}$ is in the direction of $\mathbf{n} = \frac{-x}{\sqrt{x^2 + y^2}}\mathbf{i} - \frac{y}{\sqrt{x^2 + y^2}}\mathbf{j}$

$\mathbf{F} = a\mathbf{n}$, for some constant $a > 0$.

Then $M(x, y) = \frac{-ax}{\sqrt{x^2 + y^2}}$ and $N(x, y) = \frac{-ay}{\sqrt{x^2 + y^2}}$

$\sqrt{(M(x, y))^2 + (N(x, y))^2} = a$

$a = \frac{k}{x^2 + y^2}$

$\mathbf{F} = \frac{-kx}{(x^2 + y^2)^{3/2}}\mathbf{i} - \frac{ky}{(x^2 + y^2)^{3/2}}\mathbf{j}$, for any constant $k > 0$

7. Substitute the parametric representations for $\mathbf{r}(t) = x(t)\mathbf{i} + y(t)\mathbf{j} + z(t)\mathbf{k}$ representing each path into the vector field $\mathbf{F}$, and calculate the work $W = \int_C \mathbf{F} \cdot \frac{d\mathbf{r}}{dt}$.

a. $\mathbf{F} = 3t\mathbf{i} + 2t\mathbf{j} + 4t\mathbf{k}$ and $\frac{d\mathbf{r}}{dt} = \mathbf{i} + \mathbf{j} + \mathbf{k}$

$\mathbf{F} \cdot \frac{d\mathbf{r}}{dt} = 9t$

$W = \int_0^1 9t\,dt = \frac{9}{2}$

b. $\mathbf{F} = 3t^2\mathbf{i} + 2t\mathbf{j} + 4t^4\mathbf{k}$ and $\frac{d\mathbf{r}}{dt} = \mathbf{i} + 2t\mathbf{j} + 4t^3\mathbf{k}$

$\mathbf{F} \cdot \frac{d\mathbf{r}}{dt} = 7t^2 + 16t^7$

$W = \int_0^1 (7t^2 + 16t^7)\,dt = \left[\frac{7}{3}t^3 + 2t^8\right]_0^1 = \frac{7}{3} + 2 = \frac{13}{3}$

c. $\mathbf{r}_1 = t\mathbf{i} + t\mathbf{j}$ and $\mathbf{r}_2 = \mathbf{i} + \mathbf{j} + t\mathbf{k}$; $\mathbf{F}_1 = 3t\mathbf{i} + 2t\mathbf{j}$ and $\frac{d\mathbf{r}_1}{dt} = \mathbf{i} + \mathbf{j}$

$\mathbf{F}_1 \cdot \frac{d\mathbf{r}_1}{dt} = 5t$

$W_1 = \int_0^1 5t\,dt = \frac{5}{2}$

$\mathbf{F}_2 = 3\mathbf{i} + 2\mathbf{j} + 4t\mathbf{k}$ and $\frac{d\mathbf{r}_2}{dt} = \mathbf{k}$ so $\mathbf{F}_2 \cdot \frac{d\mathbf{r}_2}{dt} = 4t$

$W_2 = \int_0^1 4t\,dt = 2$

$W = W_1 + W_2 = \frac{9}{2}$

9. Substitute the parametric representations for $\mathbf{r}(t) = x(t)\mathbf{i} + y(t)\mathbf{j} + z(t)\mathbf{k}$ representing each path into the vector field $\mathbf{F}$, and calculate the work $W = \int_C \mathbf{F} \cdot \frac{d\mathbf{r}}{dt}$.

a. $\mathbf{F} = \sqrt{t}\mathbf{i} - 2t\mathbf{j} + \sqrt{t}\mathbf{k}$ and $\frac{d\mathbf{r}}{dt} = \mathbf{i} + \mathbf{j} + \mathbf{k}$

$\mathbf{F} \cdot \frac{d\mathbf{r}}{dt} = 2\sqrt{t} - 2t$

$W = \int_0^1 (2\sqrt{t} - 2t)\,dt = \left[\frac{4}{3}t^{3/2} - t^2\right]_0^1 = \frac{1}{3}$

b. $\mathbf{F} = t^2\mathbf{i} - 2t\mathbf{j} + t\mathbf{k}$ and $\frac{d\mathbf{r}}{dt} = \mathbf{i} + 2t\mathbf{j} + 4t^3\mathbf{k}$

$\mathbf{F} \cdot \frac{d\mathbf{r}}{dt} = 4t^4 - 3t^2$

$W = \int_0^1 (4t^4 - 3t^2)\,dt = \left[\frac{4}{5}t^5 - t^3\right]_0^1 = -\frac{1}{5}$

c. $\mathbf{r}_1 = t\mathbf{i} + t\mathbf{j}$ and $\mathbf{r}_2 = \mathbf{i} + \mathbf{j} + t\mathbf{k}$; $\mathbf{F}_1 = -2t\mathbf{j} + \sqrt{t}\mathbf{k}$ and $\frac{d\mathbf{r}_1}{dt} = \mathbf{i} + \mathbf{j}$

$\mathbf{F}_1 \cdot \frac{d\mathbf{r}_1}{dt} = -2t$

$W_1 = \int_0^1 -2t\,dt = -1$

$\mathbf{F}_2 = \sqrt{t}\mathbf{i} - 2\mathbf{j} + \mathbf{k}$ and $\frac{d\mathbf{r}_2}{dt} = \mathbf{k}$, so $\mathbf{F}_2 \cdot \frac{d\mathbf{r}_2}{dt} = 1$

$W_2 = \int_0^1 dt = 1$

$W = W_1 + W_2 = 0$

11. Substitute the parametric representations for $\mathbf{r}(t) = x(t)\mathbf{i} + y(t)\mathbf{j} + z(t)\mathbf{k}$ representing each path into the vector field $\mathbf{F}$, and calculate the work $W = \int_C \mathbf{F} \cdot \frac{d\mathbf{r}}{dt}$.

a. $\mathbf{F} = (3t^2 - 3t)\mathbf{i} + 3t\mathbf{j} + \mathbf{k}$ and $\frac{d\mathbf{r}}{dt} = \mathbf{i} + \mathbf{j} + \mathbf{k}$

$\mathbf{F} \cdot \frac{d\mathbf{r}}{dt} = 3t^2 + 1$

$W = \int_0^1 (3t^2 + 1)\,dt = \left[t^3 + t\right]_0^1 = 2$

b. $\mathbf{F} = (3t^2 - 3t)\mathbf{i} + 3t^4\mathbf{j} + \mathbf{k}$ and $\frac{d\mathbf{r}}{dt} = \mathbf{i} + 2t\mathbf{j} + 4t^3\mathbf{k}$

$\mathbf{F} \cdot \frac{d\mathbf{r}}{dt} = 6t^5 + 4t^3 + 3t^2 - 3t$

$W = \int_0^1 (6t^5 + 4t^3 + 3t^2 - 3t)\,dt$

$= \left[t^6 + t^4 + t^3 - \frac{3}{2}t^2\right]_0^1$

$= \frac{3}{2}$

c. $\mathbf{r}_1 = t\mathbf{i} + t\mathbf{j}$ and $\mathbf{r}_2 = \mathbf{i} + \mathbf{j} + t\mathbf{k}$

$$\mathbf{F}_1 = (3t^2 - 3t)\mathbf{i} + \mathbf{k} \text{ and } \frac{d\mathbf{r}_1}{dt} = \mathbf{i} + \mathbf{j}$$
$$\mathbf{F}_1 \cdot \frac{d\mathbf{r}_1}{dt} = 3t^2 - 3t$$
$$W_1 = \int_0^1 (3t^2 - 3t)\,dt = \left[t^3 - \frac{3}{2}t^2\right]_0^1 = -\frac{1}{2}$$
$$\mathbf{F}_2 = 3t\mathbf{j} + \mathbf{k} \text{ and } \frac{d\mathbf{r}_2}{dt} = \mathbf{k}$$
$$\mathbf{F}_2 \cdot \frac{d\mathbf{r}_2}{dt} = 1$$
$$W_2 = \int_0^1 dt = 1$$
$$W = W_1 + W_2 = \frac{1}{2}$$

13. $\mathbf{r} = t\mathbf{i} + t^2\mathrm{j} + t\mathbf{k}$, $0 \le t \le 1$, and $\mathbf{F} = xy\mathbf{i} + y\mathbf{j} - yz\mathbf{k}$

$$\mathbf{F} = t^3\mathbf{i} + t^2\mathbf{j} - t^3\mathbf{k} \text{ and } \frac{d\mathbf{r}}{dt} = \mathbf{i} + 2t\mathbf{j} + \mathbf{k}$$
$$\mathbf{F} \cdot \frac{d\mathbf{r}}{dt} = 2t^3$$
$$\text{work} = \int_0^1 2t^3\,dt = \frac{1}{2}$$

15. $\mathbf{r} = (\sin t)\mathbf{i} + (\cos t)\mathbf{j} + t\mathbf{k}$, $0 \le t \le 2\pi$,

and $\mathbf{F} = z\mathbf{i} + x\mathbf{j} + y\mathbf{k}$

$$\mathbf{F} = t\mathbf{i} + (\sin t)\mathbf{j} + (\cos t)\mathbf{k}$$
$$\text{and } \frac{d\mathbf{r}}{dt} = (\cos t)\mathbf{i} - (\sin t)\mathbf{j} + \mathbf{k}$$
$$\mathbf{F} \cdot \frac{d\mathbf{r}}{dt} = t\cos t - \sin^2 t + \cos t$$
$$\text{work} = \int_0^{2\pi} (t\cos t - \sin^2 t + \cos t)\,dt$$
$$= \left[\cos t + t\sin t - \frac{t}{2} + \frac{\sin 2t}{4} + \sin t\right]_0^{2\pi}$$
$$= -\pi$$

17. $x = t$ and $y = x^2 = t^2$

$$\mathbf{r} = t\mathbf{i} + t^2\mathbf{j},\ -1 \le t \le 2, \text{ and } \mathbf{F} = xy\mathbf{i} + (x + y)\mathbf{j}$$
$$\mathbf{F} = t^3\mathbf{i} + (t + t^2)\mathbf{j} \text{ and } \frac{d\mathbf{r}}{dt} = \mathbf{i} + 2t\mathbf{j}$$
$$\mathbf{F} \cdot \frac{d\mathbf{r}}{dt} = t^3 + (2t^2 + 2t^3) = 3t^3 + 2t^2$$
$$\int_C xy\,dx + (x + y)\,dy = \int_C \mathbf{F} \cdot \frac{d\mathbf{r}}{dt}\,dt$$
$$= \int_{-1}^2 (3t^3 + 2t^2)\,dt$$
$$= \left[\frac{3}{4}t^4 + \frac{2}{3}t^3\right]_{-1}^2$$
$$= \left(12 + \frac{16}{3}\right) - \left(\frac{3}{4} - \frac{2}{3}\right)$$
$$= \frac{45}{4} + \frac{18}{3} = \frac{207}{12}$$

19. $\mathbf{r} = x\mathbf{i} + y\mathbf{j} = y^2\mathbf{i} + y\mathbf{j}$, $2 \ge y \ge -1$, and

$$\mathbf{F} = x^2\mathbf{i} - y\mathbf{j} = y^4\mathbf{i} - y\mathbf{j}$$
$$\frac{d\mathbf{r}}{dy} = 2y\mathbf{i} + \mathbf{j} \text{ and } \mathbf{F} \cdot \frac{d\mathbf{r}}{dy} = 2y^5 - y$$
$$\int_C \mathbf{F} \cdot \mathbf{T}\,ds = \int_2^{-1} \mathbf{F} \cdot \frac{d\mathbf{r}}{dy}\,dy$$
$$= \int_2^{-1} (2y^5 - y)\,dy$$
$$= \left[\frac{1}{3}y^6 - \frac{1}{2}y^2\right]_2^{-1}$$
$$= \left(\frac{1}{3} - \frac{1}{2}\right) - \left(\frac{64}{3} - \frac{4}{2}\right)$$
$$= \frac{3}{2} - \frac{63}{3} = -\frac{39}{2}$$

21. $\mathbf{r} = (\mathbf{i} + \mathbf{j}) + t(\mathbf{i} + 2\mathbf{j}) = (1 + t)\mathbf{i} + (1 + 2t)\mathbf{j}$, $0 \le t \le 1$,

and $\mathbf{F} = xy\mathbf{i} + (y - x)\mathbf{j}$

$$\mathbf{F} = (1 + 3t + 2t^2)\mathbf{i} + t\mathbf{j} \text{ and } \frac{d\mathbf{r}}{dt} = \mathbf{i} + 2\mathbf{j}$$
$$\mathbf{F} \cdot \frac{d\mathbf{r}}{dt} = 1 + 5t + 2t^2$$
$$\text{work} = \int_C \mathbf{F} \cdot \frac{d\mathbf{r}}{dt}\,dt$$
$$= \int_0^1 (1 + 5t + 2t^2)\,dt$$
$$= \left[t + \frac{5}{2}t^2 + \frac{2}{3}t^3\right]_0^1 = \frac{25}{6}$$

23. a. $\mathbf{r} = (\cos t)\mathbf{i} + (\sin t)\mathbf{j}$, $0 \le t \le 2\pi$,

$\mathbf{F}_1 = x\mathbf{i} + y\mathbf{j}$, and $\mathbf{F}_2 = -y\mathbf{i} + x\mathbf{j}$

$\frac{d\mathbf{r}}{dt} = (-\sin t)\mathbf{i} + (\cos t)\mathbf{j}$, $\mathbf{F}_1 = (\cos t)\mathbf{i} + (\sin t)\mathbf{j}$, and

$\mathbf{F}_2 = (-\sin t)\mathbf{i} + (\cos t)\mathbf{j}$

$$\mathbf{F}_1 \cdot \frac{d\mathbf{r}}{dt} = 0 \text{ and } \mathbf{F}_2 \cdot \frac{d\mathbf{r}}{dt} = \sin^2 t + \cos^2 t = 1$$
$$\text{Circ}_1 = \int_0^{2\pi} 0\,dt = 0 \text{ and } \text{Circ}_2 = \int_0^{2\pi} dt = 2\pi$$
$$\mathbf{n} = (\cos t)\mathbf{i} + (\sin t)\mathbf{j}$$
$$\mathbf{F}_1 \cdot \mathbf{n} = \cos^2 t + \sin^2 t = 1 \text{ and } \mathbf{F}_2 \cdot \mathbf{n} = 0$$
$$\text{Flux}_1 = \int_0^{2\pi} dt = 2\pi \text{ and } \text{Flux}_2 = \int_0^{2\pi} 0\,dt = 0$$

b. $\mathbf{r} = (\cos t)\mathbf{i} + (4\sin t)\mathbf{j}$, $0 \le t \le 2\pi$

$\frac{d\mathbf{r}}{dt} = (-\sin t)\mathbf{i} + (4\cos t)\mathbf{j}$, $\mathbf{F}_1 = (\cos t)\mathbf{i} + (4\sin t)\mathbf{j}$,

and $\mathbf{F}_2 = (-4\sin t)\mathbf{i} + (\cos t)\mathbf{j}$

$$\mathbf{F}_1 \cdot \frac{d\mathbf{r}}{dt} = 15\sin t\cos t \text{ and } \mathbf{F}_2 \cdot \frac{d\mathbf{r}}{dt} = 4$$
$$\text{Circ}_1 = \int_0^{2\pi} 15\sin t\cos t\,dt = \left[\frac{15}{2}\sin^2 t\right]_0^{2\pi} = 0 \text{ and}$$
$$\text{Circ}_2 = \int_0^{2\pi} 4\,dt = 8\pi$$
$$\mathbf{v} = -\sin t\mathbf{i} + 4\cos t\mathbf{j}$$
$$\mathbf{n} = \frac{4\cos t}{\sqrt{16\cos^2 t + \sin^2 t}}\mathbf{i} + \frac{\sin t}{\sqrt{16\cos^2 t + \sin^2 t}}\mathbf{j}$$

$\mathbf{F}_1 \cdot \mathbf{n} = \frac{4}{\sqrt{16\cos^2 t + \sin^2 t}}$

$\mathbf{F}_2 \cdot \mathbf{n} = -\frac{15\cos t \sin t}{\sqrt{16\cos^2 t + \sin^2 t}}$

$\text{Flux}_1 = \int_0^{2\pi} (\mathbf{F}_1 \cdot \mathbf{n})|\mathbf{v}|\,dt$

$= \int_0^{2\pi} 4\,dt = 8\pi$

$\text{Flux}_2 = \int_0^{2\pi} (\mathbf{F}_2 \cdot \mathbf{n})|\mathbf{v}|\,dt$

$= \int_0^{2\pi} -15\sin t\cos t\,dt$

$= \left[-\frac{15}{2}\sin^2 t\right]_0^{2\pi} = 0$

25. $\mathbf{F}_1 = (a\cos t)\mathbf{i} + (a\sin t)\mathbf{j}, \frac{d\mathbf{r}_1}{dt} = (-a\sin t)\mathbf{i} + (a\cos t)\mathbf{j}$

$\mathbf{F}_1 \cdot \frac{d\mathbf{r}_1}{dt} = 0$

$\text{Circ}_1 = 0; M_1 = a\cos t, N_1 = a\sin t, dx = -a\sin t\,dt$

$dy = a\cos t\,dt$

$\text{Flux}_1 = \int_C M_1\,dy - N_1\,dx$

$= \int_0^{\pi} (a^2\cos^2 t + a^2\sin^2 t)\,dt$

$= \int_0^{\pi} a^2\,dt$

$= a^2\pi$

$\mathbf{F}_2 = t\mathbf{i}, \frac{d\mathbf{r}_2}{dt} = \mathbf{i}$

$\mathbf{F}_2 \cdot \frac{d\mathbf{r}_2}{dt} = t$

$\text{Circ}_2 = \int_{-a}^{a} t\,dt = 0; M_2 = t, N_2 = 0, dx = dt, dy = 0$

$\text{Flux}_2 = \int_C M_2\,dy - N_2\,dx = \int_{-a}^{a} 0\,dt = 0$

therefore, $\text{Circ} = \text{Circ}_1 + \text{Circ}_2 = 0$ and

$\text{Flux} = \text{Flux}_1 + \text{Flux}_2 = a^2\pi$

27. $\mathbf{F}_1 = (-a\sin t)\mathbf{i} + (a\cos t)\mathbf{j}, \frac{d\mathbf{r}_1}{dt} = (-a\sin t)\mathbf{i} + (a\cos t)\mathbf{j}$

$\mathbf{F}_1 \cdot \frac{d\mathbf{r}_1}{dt} = a^2\sin^2 t + a^2\cos^2 t = a^2$

$\text{Circ}_1 = \int_0^{\pi} a^2\,dt = a^2\pi;$

$M_1 = -a\sin t, N_1 = a\cos t, dx = -a\sin t\,dt,$

$dy = a\cos t\,dt$

$\text{Flux}_1 = \int_C M_1\,dy - N_1\,dx$

$= \int_0^{\pi} (-a^2\sin t\cos t + a^2\sin t\cos t)\,dt = 0$

$\mathbf{F}_2 = t\mathbf{j}, \frac{d\mathbf{r}_2}{dt} = \mathbf{i}$

$\mathbf{F}_2 \cdot \frac{d\mathbf{r}_2}{dt} = 0$

$\text{Circ}_2 = 0; M_2 = 0, N_2 = t, dx = dt, dy = 0$

$\text{Flux}_2 = \int_C M_2\,dy - N_2\,dx = \int_{-a}^{a} -t\,dt = 0$; therefore,

$\text{Circ} = \text{Circ}_1 + \text{Circ}_2 = a^2\pi$ and

$\text{Flux} = \text{Flux}_1 + \text{Flux}_2 = 0$

29. a. $\mathbf{r} = (\cos t)\mathbf{i} + (\sin t)\mathbf{j}, 0 \le t \le \pi$, and

$\mathbf{F} = (x + y)\mathbf{i} - (x^2 + y^2)\mathbf{j}$

$\frac{d\mathbf{r}}{dt} = (-\sin t)\mathbf{i} + (\cos t)\mathbf{j}$ and

$\mathbf{F} = (\cos t + \sin t)\mathbf{i} - (\cos^2 t + \sin^2 t)\mathbf{j}$

$\mathbf{F} \cdot \frac{d\mathbf{r}}{dt} = -\sin t\cos t - \sin^2 t - \cos t$

$\int_C \mathbf{F} \cdot \mathbf{T}\,ds = \int_0^{\pi} (-\sin t\cos t - \sin^2 t - \cos t)\,dt$

$= \left[-\frac{1}{2}\sin^2 t - \frac{t}{2} + \frac{\sin 2t}{4} - \sin t\right]_0^{\pi} = -\frac{\pi}{2}$

b. $\mathbf{r} = (1 - 2t)\mathbf{i}; 0 \le t \le 1$, and

$\mathbf{F} = (x + y)\mathbf{i} - (x^2 + y^2)\mathbf{j}$

$\frac{d\mathbf{r}}{dt} = -\mathbf{i}$ and $\mathbf{F} = (1 - 2t)\mathbf{i} - (1 - 2t)^2\mathbf{j}$

$\mathbf{F} \cdot \frac{d\mathbf{r}}{dt} = 2t - 1$

$\int_C \mathbf{F} \cdot \mathbf{T}\,ds = \int_0^1 (2t - 1)\,dt = \left[t^2 - t\right]_0^1 = 0$

c. $\mathbf{r}_1 = (1 - t)\mathbf{i} - t\mathbf{j}, 0 \le t \le 1$, and

$\mathbf{F} = (x + y)\mathbf{i} - (x^2 + y^2)\mathbf{j}$

$\frac{d\mathbf{r}_1}{dt} = -\mathbf{i} - \mathbf{j}$ and $\mathbf{F} = (1 - 2t)\mathbf{i} - (1 - 2t + 2t^2)\mathbf{j}$

$\mathbf{F} \cdot \frac{d\mathbf{r}_1}{dt} = (2t - 1) + (1 - 2t + 2t^2) = 2t^2$

$\text{Flow}_1 = \int_{C_1} \mathbf{F} \cdot \frac{d\mathbf{r}_1}{dt} = \int_0^1 2t^2\,dt = \frac{2}{3}$

$\mathbf{r}_2 = -t\mathbf{i} + (t - 1)\mathbf{j}, 0 \le t \le 1$, and

$\mathbf{F} = (x + y)\mathbf{i} - (x^2 + y^2)\mathbf{j}$

$\frac{d\mathbf{r}_2}{dt} = -\mathbf{i} + \mathbf{j}$ and

$\mathbf{F} = -\mathbf{i} - (t^2 + t^2 - 2t + 1)\mathbf{j} = -\mathbf{i} - (2t^2 - 2t + 1)\mathbf{j}$

$\mathbf{F} \cdot \frac{d\mathbf{r}_2}{dt} = 1 - (2t^2 - 2t + 1) = 2t - 2t^2$

$\text{Flow}_2 = \int_{C_2} \mathbf{F} \cdot \frac{d\mathbf{r}_2}{dt}$

$= \int_0^1 (2t - 2t^2)\,dt$

$= \left[t^2 - \frac{2}{3}t^3\right]_0^1 = \frac{1}{3}$

$\text{Flow} = \text{Flow}_1 + \text{Flow}_2 = \frac{2}{3} + \frac{1}{3} = 1$

31. $\mathbf{F} = -\frac{y}{\sqrt{x^2+y^2}}\mathbf{i} + \frac{x}{\sqrt{x^2+y^2}}\mathbf{j}$ on $x^2 + y^2 = 4$ at $(2, 0)$, $\mathbf{F} = \mathbf{j}$; at $(0, 2)$, $\mathbf{F} = -\mathbf{i}$; at $(-2, 0)$, $\mathbf{F} = -\mathbf{j}$; at $(0, -2)$, $\mathbf{F} = \mathbf{i}$; at $(1, \sqrt{3})$, $\mathbf{F} = -\frac{\sqrt{3}}{2}\mathbf{i} + \frac{1}{2}\mathbf{j}$ at $(1, -3)$, $\mathbf{F} = \frac{\sqrt{3}}{2}\mathbf{i} + \frac{1}{2}\mathbf{j}$; at $(-1, \sqrt{3})$, $\mathbf{F} = -\frac{\sqrt{3}}{2}\mathbf{i} - \frac{1}{2}\mathbf{j}$; at $(-1, -3)$, $\mathbf{F} = \frac{\sqrt{3}}{2}\mathbf{i} - \frac{1}{2}\mathbf{j}$

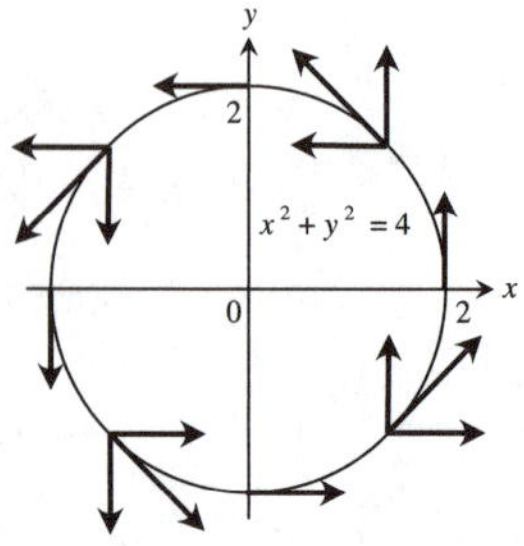

33. a. $\mathbf{G} = P(x, y)\mathbf{i} + Q(x, y)\mathbf{j}$ is to have a magnitude $\sqrt{a^2 + b^2}$ and to be tangent to $x^2 + y^2 = a^2 + b^2$ in a counterclockwise direction. Thus $x^2 + y^2 = a^2 + b^2$

$2x + 2yy' = 0$

$y' = -\frac{x}{y}$ is the slope of the tangent line at any point on the circle

$y' = -\frac{a}{b}$ at (a, b). Let $\mathbf{v} = -b\mathbf{i} + a\mathbf{j}$, $|\mathbf{v}| = \sqrt{a^2 + b^2}$, with $\mathbf{v}$ in a counterclockwise direction and tangent to the circle. Then let $P(x, y) = -y$ and $Q(x, y) = x$

$\mathbf{G} = -y\mathbf{i} + x\mathbf{j}$ for (a, b) on $x^2 + y^2 = a^2 + b^2$

we have $\mathbf{G} = -b\mathbf{i} + a\mathbf{j}$ and $|\mathbf{G}| = \sqrt{a^2 + b^2}$.

b. $\mathbf{G} = (\sqrt{x^2 + y^2})\mathbf{F} = (\sqrt{a^2 + b^2})\mathbf{F}$, since $x^2 + y^2 = a^2 + b^2$

35. The slope of the line through (x, y) and the origin is $\frac{y}{x}$

$\mathbf{v} = x\mathbf{i} + y\mathbf{j}$ is a vector parallel to that line and pointing away from the origin

$\mathbf{F} = -\frac{x\mathbf{i} + y\mathbf{j}}{\sqrt{x^2 + y^2}}$ is the unit vector pointing toward the origin.

37. $\mathbf{F} = -4t^3\mathbf{i} + 8t^2\mathbf{j} + 2\mathbf{k}$ and $\frac{d\mathbf{r}}{dt} = \mathbf{i} + 2t\mathbf{j}$

$\mathbf{F} \cdot \frac{d\mathbf{r}}{dt} = 12t^3$

$\text{Flow} = \int_0^2 12t^3\, dt = \left[3t^4\right]_0^2 = 48$

39. $\mathbf{F} = (\cos t - \sin t)\mathbf{i} + (\cos t)\mathbf{k}$

and $\frac{d\mathbf{r}}{dt} = (-\sin t)\mathbf{i} + (\cos t)\mathbf{k}$

$\mathbf{F} \cdot \frac{d\mathbf{r}}{dt} = -\sin t \cos t + 1$

$\text{Flow} = \int_0^{\pi} (-\sin t \cos t + 1)\, dt$

$= \left[\frac{1}{2}\cos^2 t + t\right]_0^{\pi}$

$= \left(\frac{1}{2} + \pi\right) - \left(\frac{1}{2} + 0\right) = \pi$

41. C_1: $\mathbf{r} = (\cos t)\mathbf{i} + (\sin t)\mathbf{j} + t\mathbf{k}$, $0 \le t \le \frac{\pi}{2}$

$\mathbf{F} = (2\cos t)\mathbf{i} + 2t\mathbf{j} + (2\sin t)\mathbf{k}$

and $\frac{d\mathbf{r}}{dt} = (-\sin t)\mathbf{i} + (\cos t)\mathbf{j} + \mathbf{k}$

$\mathbf{F} \cdot \frac{d\mathbf{r}}{dt} = -2\cos t \sin t + 2t\cos t + 2\sin t$

$= -\sin 2t + 2t\cos t + 2\sin t$

$\text{Flow}_1 = \int_0^{\pi/2} (-\sin 2t + 2t\cos t + 2\sin t)\, dt$

$= \left[\frac{1}{2}\cos 2t + 2t\sin t + 2\cos t - 2\cos t\right]_0^{\pi/2}$

$= -1 + \pi$

C_2: $\mathbf{r} = \mathbf{j} + \frac{\pi}{2}(1 - t)\mathbf{k}$, $0 \le t \le 1$

$\mathbf{F} = \pi(1 - t)\mathbf{j} + 2\mathbf{k}$ and $\frac{d\mathbf{r}}{dt} = -\frac{\pi}{2}\mathbf{k}$

$\mathbf{F} \cdot \frac{d\mathbf{r}}{dt} = -\pi$

$\text{Flow}_2 = \int_0^1 -\pi\, dt = \left[-\pi t\right]_0^1 = -\pi$

C_3: $\mathbf{r} = t\mathbf{i} + (1 - t)\mathbf{j}$, $0 \le t \le 1$

$\mathbf{F} = 2t\mathbf{i} + 2(1 - t)\mathbf{k}$ and $\frac{d\mathbf{r}}{dt} = \mathbf{i} - \mathbf{j}$

$\mathbf{F} \cdot \frac{d\mathbf{r}}{dt} = 2t$

$\text{Flow}_3 = \int_0^1 2t\, dt = \left[t^2\right]_0^1 = 1$

Circulation $= (-1 + \pi) - \pi + 1 = 0$

43. Let $x = t$ be the parameter

$y = x^2 = t^2$ and $z = x = t$

$r = t\mathbf{i} + t^2\mathbf{j} + t\mathbf{k}$, $0 \le t \le 1$ from $(0, 0, 0)$ to $(1, 1, 1)$

$\frac{d\mathbf{r}}{dt} = \mathbf{i} + 2t\mathbf{j} + \mathbf{k}$ and

$\mathbf{F} = xy\mathbf{i} + y\mathbf{j} - yz\mathbf{k} = t^3\mathbf{i} + t^2\mathbf{j} - t^3\mathbf{k}$

$\mathbf{F} \cdot \frac{d\mathbf{r}}{dt} = t^3 + 2t^3 - t^3 = 2t^3$

$\text{Flow} = \int_0^1 2t^3\, dt = \frac{1}{2}$

45. Yes. The work and area have the same numerical value because work $= \int_C \mathbf{F} \cdot d\mathbf{r} = \int_C y\mathbf{i} \cdot d\mathbf{r}$

$= \int_b^a [f(t)\mathbf{i}] \cdot \left[\mathbf{i} + \frac{df}{dt}\mathbf{j}\right] dt$

[On the path, y equals $f(t)$]

$= \int_a^b f(t)\, dt =$ Area under the curve [because $f(t) > 0$]

47.–51. Example CAS commands:

Maple:

```
with(linalg):
x:= t -> cos(t);
y:= t -> sin(t);
z:= t -> 0;
a:=0; b:= Pi;
M:= (x, y, z) -> 3/(1 + x^2);
N:=(x, y, z) -> 2/(1 + y^2);
P:=(x, y, z) -> 0;
F:= t-> vector([M(x(t), y(t), z(t)),
  N(x(t), y(t), z(t)),P(x(t), y(t),
  z(t))]);
dr:= t-> vector([D(x)(t), D(y)(t),
  D(z)(t)]);
integrand:=dotprod(F(t), dr(t),
  orthogonal);
int(integrand, t=a..b);
evalf(%);
```

Mathematica:

```
Clear[x,y,z,t]
r[t_] = {x[t],y[t],z[t]}
f[x_,y_] = {
  3/(1+x^2),
  2/(1+y^2)}
x[t_] = Cos[t]
y[t_] = Sin[t]
z[t_] = 0
{a,b} = {0, Pi};
v[t_] = r'[t]
integrand = f[x[t], y[t], z[t]] .
  v[t]
integrand = Simplify[integrand]
Integrate [integrand, {t, a, b}]
```

Section 15.3 Path Independence, Potential Functions, and Conservative Fields (pp. 845 – 855)

Quick Review 15.3

1. 1

2. 2

3. $-\dfrac{yz}{\sqrt{1-x^2y^2z^2}}$

4. $-\dfrac{xz}{\sqrt{1-x^2y^2z^2}}$

5. $-\dfrac{xy}{\sqrt{1-x^2y^2z^2}}$

6. $-\dfrac{z}{\sqrt{1-x^2y^2z^2}} - \dfrac{x^2y^2z^3}{(1-x^2y^2z^2)^{3/2}}$

7. $-\dfrac{y}{\sqrt{1-x^2y^2z^2}} - \dfrac{x^2y^3z^2}{(1-x^2y^2z^2)^{3/2}}$

8. $g(x, y, z) = h(y, z)$, $h(y, z)$ any function of y and z

9. $g(x, y, z) = y^2 + h(x, z)$, $h(x, z)$ any function of x and z

10. $g(x, y, z) = \dfrac{z^2}{2} + h(x, y)$, $h(x, y)$ any function of x and y

Section 15.3 Exercises

1. $\dfrac{\partial P}{\partial y} = x = \dfrac{\partial N}{\partial z}, \dfrac{\partial M}{\partial z} = y = \dfrac{\partial P}{\partial x}, \dfrac{\partial N}{\partial x} = z = \dfrac{\partial M}{\partial y}$

Conservative

3. $\dfrac{\partial P}{\partial y} = -1 \neq 1 = \dfrac{\partial N}{\partial z}$

Not Conservative

5. $\dfrac{\partial N}{\partial x} = 0 \neq 1 = \dfrac{\partial M}{\partial y}$

Not Conservative

7. $\dfrac{\partial f}{\partial x} = 2x$

$f(x, y, z) = x^2 + g(y, z)$

$\dfrac{\partial f}{\partial y} = \dfrac{\partial g}{\partial y} = 3y$

$g(y, z) = \dfrac{3y^2}{2} + h(z)$

$f(x, y, z) = x^2 + \dfrac{3y^2}{2} + h(z)$

$\dfrac{\partial f}{\partial z} = h'(z) = 4z$

$h(z) = 2z^2 + C$

$f(x, y, z) = x^2 + \dfrac{3y^2}{2} + 2z^2 + C$

9. $\dfrac{\partial f}{\partial x} = e^{y+2z}$

$f(x, y, z) = xe^{y+2z} + g(y, z)$

$\dfrac{\partial f}{\partial y} = xe^{y+2z} + \dfrac{\partial g}{\partial y} = xe^{y+2z}$

$\dfrac{\partial g}{\partial y} = 0$

$f(x, y, z) = xe^{y+2z} + h(z)$

$\dfrac{\partial f}{\partial z} = 2xe^{y+2z} + h'(z) = 2xe^{y+2z}$

$h'(z) = 0$

$h(z) = C$

$f(x, y, z) = xe^{y+2z} + C$

11. $\dfrac{\partial f}{\partial z} = \dfrac{z}{y^2 + z^2}$

$f(x, y, z) = \dfrac{1}{2}\ln(y^2 + z^2) + g(x, y)$

$\dfrac{\partial f}{\partial x} = \dfrac{\partial g}{\partial x} = \ln x + \sec^2(x + y)$

$g(x, y) = (x \ln x - x) + \tan(x + y) + h(y)$

$f(x, y, z) = \dfrac{1}{2}\ln(y^2 + z^2) + (x \ln x - x) + \tan(x + y) + h(y)$

$\dfrac{\partial f}{\partial y} = \dfrac{y}{y^2 + z^2} + \sec^2(x + y) + h'(y)$

$= \sec^2(x + y) + \dfrac{y}{y^2 + z^2}$

$h'(y) = 0$

$h(y) = C$

$f(x, y, z) = \dfrac{1}{2}\ln(y^2 + z^2) + (x \ln x - x) + \tan(x + y) + C$

13. Let $\mathbf{F}(x, y, z) = 2x\mathbf{i} + 2y\mathbf{j} + 2z\mathbf{k}$

$$\frac{\partial P}{\partial y} = 0 = \frac{\partial N}{\partial z}, \frac{\partial M}{\partial z} = 0 = \frac{\partial P}{\partial x}, \frac{\partial N}{\partial x} = 0 = \frac{\partial M}{\partial y}$$

$M\,dx + N\,dy + P\,dz$ is exact;

$$\frac{\partial f}{\partial x} = 2x$$

$$f(x, y, z) = x^2 + g(y, z)$$

$$\frac{\partial f}{\partial y} = \frac{\partial g}{\partial y} = 2y$$

$$g(y, z) = y^2 + h(z)$$

$$f(x, y, z) = x^2 + y^2 = h(z)$$

$$\frac{\partial f}{\partial z} = h'(z) = 2z$$

$$h(z) = z^2 + C$$

$$f(x, y, z) = x^2 + y^2 + z^2 + C$$

$$\int_{(0,0,0)}^{(2,3,-6)} 2x\,dx + 2y\,dy + 2z\,dz$$

$$= f(2, 3, -6) - f(0, 0, 0)$$

$$= 2^2 + 3^2 + (-6)^2 = 49$$

15. Let $\mathbf{F}(x, y, z) = 2xy\mathbf{i} + (x^2 - z^2)\mathbf{j} - 2yz\mathbf{k}$

$$\frac{\partial P}{\partial y} = -2z = \frac{\partial N}{\partial z}, \frac{\partial M}{\partial z} = 0 = \frac{\partial P}{\partial x}, \frac{\partial N}{\partial x} = 2x = \frac{\partial M}{\partial y}$$

$M\,dx + N\,dy + P\,dz$ is exact;

$$\frac{\partial f}{\partial x} = 2xy$$

$$f(x, y, z) = x^2y + g(y, z)$$

$$\frac{\partial f}{\partial y} = x^2 + \frac{\partial g}{\partial y} = x^2 - z^2$$

$$\frac{\partial g}{\partial y} = -z^2$$

$$g(y, z) = -yz^2 + h(z)$$

$$f(x, y, z) = x^2y - yz^2 + h(z)$$

$$\frac{\partial f}{\partial z} = -2yz + h'(z) = -2yz$$

$$h'(z) = 0$$

$$h(z) = C$$

$$f(x, y, z) = x^2y - yz^2 + C$$

$$\int_{(0,0,0)}^{(1,2,3)} 2xy\,dx + (x^2 - z^2)\,dy - 2yz\,dz$$

$$= f(1, 2, 3) - f(0, 0, 0)$$

$$= 2 - 2(3)^2 = -16$$

17. Let $\mathbf{F}(x, y, z) = (\sin y \cos x)\mathbf{i} + (\cos y \sin x)\mathbf{j} + \mathbf{k}$

$$\frac{\partial P}{\partial y} = 0 = \frac{\partial N}{\partial z}, \frac{\partial M}{\partial z} = 0 = \frac{\partial P}{\partial x}, \frac{\partial N}{\partial x} = \cos y \cos x = \frac{\partial M}{\partial y}$$

$M\,dx + N\,dy + P\,dz$ is exact;

$$\frac{\partial f}{\partial x} = \sin y \cos x$$

$$f(x, y, z) = \sin y \sin x + g(y, z)$$

$$\frac{\partial f}{\partial y} = \cos y \sin x + \frac{\partial g}{\partial y} = \cos y \sin x$$

$$\frac{\partial g}{\partial y} = 0$$

$$g(y, z) = h(z)$$

$$f(x, y, z) = \sin y \sin x + h(z)$$

$$\frac{\partial f}{\partial z} = h'(z) = 1$$

$$h(z) = z + C$$

$$f(x, y, z) = \sin y \sin x + z + C$$

$$\int_{(1,0,0)}^{(0,1,1)} \sin y \cos x\,dx + \cos y \sin x\,dy + dz$$

$$= f(0, 1, 1) - f(1, 0, 0)$$

$$= (0 + 1 + C) - (0 + 0 + C) = 1$$

19. Let $\mathbf{F}(x, y, z) = 3x^2\mathbf{i} + \left(\frac{z^2}{y}\right)\mathbf{j} + (2z \ln y)\mathbf{k}$

$$\frac{\partial P}{\partial y} = \frac{2z}{y} = \frac{\partial N}{\partial z}, \frac{\partial M}{\partial z} = 0 = \frac{\partial P}{\partial x}, \frac{\partial N}{\partial x} = 0 = \frac{\partial M}{\partial y}$$

$M\,dx + N\,dy + P\,dz$ is exact;

$$\frac{\partial f}{\partial x} = 3x^2$$

$$f(x, y, z) = x^3 + g(y, z)$$

$$\frac{\partial f}{\partial y} = \frac{\partial g}{\partial y} = \frac{z^2}{y}$$

$$g(y, z) = z^2 \ln y + h(z)$$

$$f(x, y, z) = x^3 + z^2 \ln y + h(z)$$

$$\frac{\partial f}{\partial z} = 2z \ln y + h'(z) = 2z \ln y$$

$$h'(z) = 0$$

$$h(z) = C$$

$$f(x, y, z) = x^3 + z^2 \ln y + C$$

$$\int_{(1,1,1)}^{(1,2,3)} 3x^2\,dx + \frac{z^2}{y}\,dy + 2z \ln y\,dz$$

$$= f(1, 2, 3) - f(1, 1, 1)$$

$$= (1 + 9 \ln 2 + C) - (1 + 0 + C) = 9 \ln 2$$

21. Let $\mathbf{F}(x, y, z) = \left(\frac{1}{y}\right)\mathbf{i} + \left(\frac{1}{z} - \frac{x}{y^2}\right)\mathbf{j} - \left(\frac{y}{z^2}\right)\mathbf{k}$

$$\frac{\partial P}{\partial y} = -\frac{1}{z^2} = \frac{\partial N}{\partial z}, \frac{\partial M}{\partial z} = 0 = \frac{\partial P}{\partial x}, \frac{\partial N}{\partial x} = -\frac{1}{y^2} = \frac{\partial M}{\partial y}$$

$M\,dx + N\,dy + P\,dz$ is exact;

$$\frac{\partial f}{\partial x} = \frac{1}{y}$$

$$f(x, y, z) = \frac{x}{y} + g(y, z)$$

$$\frac{\partial f}{\partial y} = -\frac{x}{y^2} + \frac{\partial g}{\partial y} = \frac{1}{z} - \frac{x}{y^2}$$

$$\frac{\partial g}{\partial y} = \frac{1}{z}$$

$$g(y, z) = \frac{y}{z} + h(z)$$

$$f(x, y, z) = \frac{x}{y} + \frac{y}{z} + h(z)$$

$$\frac{\partial f}{\partial z} = -\frac{y}{z^2} + h'(z) = -\frac{y}{z^2}$$

$$h'(z) = 0$$

$$h(z) = C$$

$$f(x, y, z) = \frac{x}{y} + \frac{y}{z} + C$$

$$\int_{(1,1,1)}^{(2,2,2)} \frac{1}{y}\,dx + \left(\frac{1}{z} - \frac{x}{y^2}\right)dy - \frac{y}{z^2}\,dz$$
$$= f(2,2,2) - f(1,1,1)$$
$$= \left(\frac{2}{2} + \frac{2}{2} + C\right) - \left(\frac{1}{1} + \frac{1}{1} + C\right) = 0$$

23. $\mathbf{r} = (\mathbf{i} + \mathbf{j} + \mathbf{k}) + t(\mathbf{i} + 2\mathbf{j} - 2\mathbf{k})$
$= (1+t)\mathbf{i} + (1+2t)\mathbf{j} + (1-2t)\mathbf{k}$
$dx = dt,\ dy = 2\,dt,\ dz = -2\,dt$

$$\int_{(1,1,1)}^{(2,3,-1)} y\,dx + x\,dy + 4\,dz$$
$$= \int_0^1 (2t+1)\,dt + (t+1)(2\,dt) + 4(-2)\,dt$$
$$= \int_0^1 (4t-5)\,dt$$
$$= \left[2t^2 - 5t\right]_0^1 = -3$$

25. $\dfrac{\partial P}{\partial y} = 0 = \dfrac{\partial N}{\partial z}, \dfrac{\partial M}{\partial z} = 2z = \dfrac{\partial P}{\partial x}, \dfrac{\partial N}{\partial x} = 0 = \dfrac{\partial M}{\partial y}$

$M\,dx + N\,dy + P\,dz$ is exact;
F is conservative
path independence

27. $\dfrac{\partial N}{\partial x} = \dfrac{2x}{y^2} = \dfrac{\partial M}{\partial y}$

F is conservative

there exists an f so that $\mathbf{F} = \nabla f$; $\dfrac{\partial f}{\partial x} = \dfrac{2x}{y}$

$$f(x, y) = \frac{x^2}{y} + g(y)$$
$$\frac{\partial f}{\partial y} = -\frac{x^2}{y^2} + g'(y) = \frac{1 - x^2}{y^2}$$
$$g'(y) = \frac{1}{y^2}$$
$$g(y) = -\frac{1}{y} + C$$
$$f(x, y) = \frac{x^2}{y} - \frac{1}{y} + C$$
$$\mathbf{F} = \nabla\left(\frac{x^2 - 1}{y}\right)$$

29. $\dfrac{\partial P}{\partial y} = 0 = \dfrac{\partial N}{\partial z}, \dfrac{\partial M}{\partial z} = 0 = \dfrac{\partial P}{\partial x}, \dfrac{\partial N}{\partial x} = 1 = \dfrac{\partial M}{\partial y}$

F is conservative
there exists an f so that $\mathbf{F} = \nabla f$;

$$\frac{\partial f}{\partial x} = x^2 + y$$
$$f(x, y, z) = \frac{1}{3}x^3 + xy + g(y, z)$$
$$\frac{\partial f}{\partial y} = x + \frac{\partial g}{\partial y} = y^2 + x$$
$$\frac{\partial g}{\partial y} = y^2$$
$$g(y, z) = \frac{1}{3}y^3 + h(z)$$
$$f(x, y, z) = \frac{1}{3}x^3 + xy + \frac{1}{3}y^3 + h(z)$$
$$\frac{\partial f}{\partial z} = h'(z) = ze^z$$
$$h(z) = ze^z - e^z + C$$
$$f(x, y, z) = \frac{1}{3}x^3 + xy + \frac{1}{3}y^3 + ze^z - e^z + C$$
$$\mathbf{F} = \nabla\left(\frac{1}{3}x^3 + xy + \frac{1}{3}y^3 + ze^z - e^z\right)$$

a. $\text{work} = \displaystyle\int_A^B \mathbf{F}\cdot\frac{d\mathbf{r}}{dt}\,dt = \int_A^B \mathbf{F}\cdot d\mathbf{r}$
$$= \left[\frac{1}{3}x^3 + xy + \frac{1}{3}y^3 + ze^z - e^z\right]_{(1,0,0)}^{(1,0,1)}$$
$$= \left(\frac{1}{3} + 0 + 0 + e - e\right) - \left(\frac{1}{3} + 0 + 0 + 0 - 1\right)$$
$$= 1$$

b. $\text{work} = \displaystyle\int_A^B \mathbf{F}\cdot d\mathbf{r}$
$$= \left[\frac{1}{3}x^3 + xy + \frac{1}{3}y^3 + ze^z - e^z\right]_{(1,0,0)}^{(1,0,1)} = 1$$

c. $\text{work} = \displaystyle\int_A^B \mathbf{F}\cdot d\mathbf{r}$
$$= \left[\frac{1}{3}x^3 + xy + \frac{1}{3}y^3 + ze^z - e^z\right]_{(1,0,0)}^{(1,0,1)} = 1$$

Note: Since F is conservative, $\displaystyle\int_A^B \mathbf{F}\cdot d\mathbf{r}$ is independent of the path from (1, 0, 0) to (1, 0, 1).

31. a. $\mathbf{F} = \nabla(x^3y^2)$
$\mathbf{F} = 3x^2y^2\mathbf{i} + 2x^3y\mathbf{j}$; let C_1 be the path from $(-1, 1)$ to (0 0)
$x = t - 1$ and $y = -t + 1,\ 0 \le t \le 1$
$\mathbf{F} = 3(t-1)^2(-t+1)^2\mathbf{i} + 2(t-1)^3(-t+1)\mathbf{j}$
$= 3(t-1)^4\mathbf{i} - 2(t-1)^4\mathbf{j}$ and
$\mathbf{r}_1 = (t-1)\mathbf{i} + (-t+1)\mathbf{j}$
$d\mathbf{r}_1 = dt\,\mathbf{i} - dt\,\mathbf{j}$
$$\int_{C_1} \mathbf{F}\cdot d\mathbf{r} = \int_0^1 [3(t-1)^4 + 2(t-1)^4]\,dt$$
$= \displaystyle\int_0^1 5(t-1)^4\,dt = [(t-1)^5] = 1$; let C_2 be the path from (0, 0) to (1, 1)

$x = t$ and $y = t\ 0 \le t \le 1$
$\mathbf{F} = 3t^4\mathbf{i} + 2t^4\mathbf{j}$ and $\mathbf{r}_2 = t\mathbf{i} + t\mathbf{j}$
$d\mathbf{r}_2 = dt\mathbf{i} + dt\mathbf{j}$
$$\int_{C_2} \mathbf{F}\cdot d\mathbf{r}_2 = \int_0^1 (3t^4 + 2t^4)\,dt = \int_0^1 5t^4\,dt = 1$$
$$\int_C \mathbf{F}\cdot d\mathbf{r} = \int_{C_1} \mathbf{F}\cdot d\mathbf{r}_1 + \int_{C_2} \mathbf{F}\cdot d\mathbf{r}_2 = 2$$

b. Since $f(x, y) = x^3y^2$ is a potential function for **F**,
$$\int_{(-1,1)}^{(1,1} \mathbf{F}\cdot d\mathbf{r} = f(1, 1) - f(-1, 1) = 2$$

33. Let $-GmM = C$

$$\mathbf{F} = C\left[\frac{x}{(x^2+y^2+z^2)^{3/2}}\mathbf{i} + \frac{y}{(x^2+y^2+z^2)^{3/2}}\mathbf{j} + \frac{z}{(x^2+y^2+z^2)^{3/2}}\mathbf{k}\right]$$
$$\frac{\partial P}{\partial y} = \frac{-3yzC}{(x^2+y^2+z^2)^{5/2}} = \frac{\partial N}{\partial z}, \frac{\partial M}{\partial z} = \frac{-3xzC}{(x^2+y^2+z^2)^{5/2}}$$
$$= \frac{\partial P}{\partial x}, \frac{\partial N}{\partial x} = \frac{-3xyC}{(x^2+y^2+z^2)^{5/2}} = \frac{\partial M}{\partial y}$$

33. (continued)

$\mathbf{F} = \nabla f$ for some f; $\frac{\partial f}{\partial x} = \frac{xC}{(x^2+y^2+z^2)^{3/2}}$

$f(x, y, z) = -\frac{C}{(x^2+y^2+z^2)^{1/2}} + g(y, z)$

$\frac{\partial f}{\partial y} = \frac{yC}{(x^2+y^2+z^2)^{3/2}} + \frac{\partial g}{\partial y} = \frac{yC}{(x^2+y^2+z^2)^{3/2}}$

$\frac{\partial g}{\partial y} = 0$

$g(y, z) = h(z)$

$\frac{\partial f}{\partial z} = \frac{zC}{(x^2+y^2+z^2)^{3/2}} + h'(z) = \frac{zC}{(x^2+y^2+z^2)^{3/2}}$

$h(z) = C_1$

$f(x, y, z) = \frac{C}{(x^2+y^2+z^2)^{1/2}} + C_1$. Let $C_1 = 0$

$f(x, y, z) = \frac{GmM}{(x^2+y^2+z^2)^{1/2}}$ is a potential function for **F**.

35 a. If the differential form is exact, then $\frac{\partial P}{\partial y} = \frac{\partial N}{\partial z}$

$2ay = cy$ for all y

$2a = c$, $\frac{\partial M}{\partial z} = \frac{\partial P}{\partial x}$

$2cx = 2cx$ for all x and $\frac{\partial N}{\partial x} = \frac{\partial M}{\partial y}$

$by = 2ay$ for all y

$b = 2a$ and $c = 2a$

b. $\mathbf{F} = \nabla f$

the differential form with $a = 1$ in part (a) is exact

$b = 2$ and $c = 2$

37. The path will not matter; the work along any path will be the same because the field is conservative.

39. a. Partition the string into small pieces. Let $\Delta_i s$ be the length of the ith piece. Let (x_i, y_i) be a point in the ith piece. The work done by gravity in moving the ith piece to the x-axis approximately $W_i = (gx_iy_i\Delta_i s)y_i$ where $x_iy_i\Delta_i s$ is approximately the mass of the ith piece. The total work done by gravity in moving the string to the x-axis is $\sum_i W_i = \sum_i gx_iy_i^2\Delta_i s$

$\Rightarrow$ Work $\int_C gxy^2\, ds$

b. Work $\int_C gxy^2\, ds$

$= \int_0^{\pi/2} g(2\cos t)(4\sin^2 t)\sqrt{4\sin^2 t + 4\cos^2 t}\, dt$

$= 16\, g\int_0^{\pi/2} \cos t \sin^2 t\, dt = \left[16g\left(\frac{\sin^3 t}{3}\right)\right]_0^{\pi/2} = \frac{16}{3}g$

c. $\bar{x} = \frac{\int_C x(xy)\, ds}{\int_C xy\, ds}$ and y

$= \frac{\int_C y(xy)\, ds}{\int_C xy\, ds}$; the mass of the string is $\int_C xy\, ds$ and the weight of the string is $g\int_C xy\, ds$. Therefore, the work done in moving the point mass at (x, y) to the x-axis is $W - \left(g\int_C xy\, ds\right)\bar{y} = g\int_C xy^2\, ds = \frac{16}{3}g.$

Section 15.4 Green's Theorem in the Plane (pp 855-867)

Quick Review 15.4

1. $x = t, y = 0, 0 \le t \le 1$
2. $x = 1, y = t, 0 \le t \le 1$
3. $x = 1 - t, y = 1, 0 \le t \le 1$
4. $x = 0, y = 1 - t, 0 \le t \le 1$

5. 0

6. $\frac{1}{2}$

7. 1

8. 0

9. $\frac{2xy}{(x^2+y^2)^2}$

10. $\frac{y^2 - x^2}{(x^2+y^2)^2}$

Section 15.4 Exercises

1. $M = -y = -a\sin t$, $N = x = a\cos t$, $dx = -a\sin t\, dt$, $dy = a\cos t\, dt$

$\frac{\partial M}{\partial x} = 0$, $\frac{\partial M}{\partial y} = -1$, $\frac{\partial N}{\partial x} = 1$, and $\frac{\partial N}{\partial y} = 0$

Equation (11): $\oint_C M\, dy - N\, dx$

$= \int_0^{2\pi} [(-a\sin t)(a\cos t) - (a\cos t)(-a\sin t)]\, dt$

$= \int_0^{2\pi} 0\, dt = 0$; $\iint_R \left(\frac{\partial M}{\partial x} + \frac{\partial N}{\partial y}\right) dx\, dy = \iint_R 0\, dx\, dy$

$= 0$, Flux

Equation (12): $\oint_C M\, dx + N\, dy$

$= \int_0^{2\pi} [(-a\sin t)(-a\sin t) + (a\cos t)(a\cos t)]\, dt$

$= \int_0^{2\pi} a^2\, dt = 2\pi a^2$; $\iint_R \left(\frac{\partial N}{\partial x} - \frac{\partial M}{\partial y}\right) dx\, dy$

$= \int_{-a}^{a}\int_{-\sqrt{a^2-x^2}}^{\sqrt{a^2-x^2}} 2\, dy\, dx = \int_{-a}^{a} 4\sqrt{a^2 - x^2}\, dx$

$= 4\left[\frac{x}{2}\sqrt{a^2 - x^2} + \frac{a^2}{2}\sin^{-1}\frac{x}{a}\right]_{-a}^{a} = 2a^2\left(\frac{\pi}{2} + \frac{\pi}{2}\right) = 2a^2\pi,$

Circulation

3. $M = 2x = 2a\cos t$, $N = -3y = -3a\sin t$, $dx = -a\sin t\, dt$, $dy = a\cos t\, dt$

$\frac{\partial M}{\partial x} = 2$, $\frac{\partial M}{\partial y} = 0$, $\frac{\partial N}{\partial x} = 0$, and $\frac{\partial N}{\partial y} = -3$;

Equation (11): $\oint_C M\, dy - N\, dx$

$= \int_0^{2\pi} [(2a\cos t)(a\cos t) + (3a\sin t)(-a\sin t)]\, dt$

$= \int_0^{2\pi} (2a^2\cos^2 t - 3a^2\sin^2 t)\, dt$

$= 2a^2\left[\frac{t}{2} + \frac{\sin 2t}{4}\right]_0^{2\pi} - 3a^2\left[\frac{t}{2} - \frac{\sin 2t}{4}\right]_0^{2\pi}$

$= 2\pi a^2 - 3\pi a^2 = -\pi a^2$; $\iint_R \left(\frac{\partial M}{\partial x} + \frac{\partial N}{\partial y}\right)$

$= \iint_R -1\, dx\, dy = \int_0^{2\pi}\int_0^{a} -r\, dr\, d\theta = \int_0^{2\pi} -\frac{a^2}{2}\, d\theta$

$= -\pi a^2$, Flux

Equation (12): $\oint_C M\,dx + N\,dy$

$= \int_0^{2\pi} [(2a\cos t)(-a\sin t) + (-3a\sin t)(a\cos t)]\,dt$

$= \int_0^{2\pi} (-2a^2 \sin t\cos t - 3a^2 \sin t\cos t)\,dt$

$= -5a^2\left[\frac{1}{2}\sin^2 t\right]_0^{2\pi} = 0;\ \int\int_R 0\,dx\,dy = 0$, Circulation

5. $M = x - y,\ N = y - x$

$\frac{\partial M}{\partial x} = 1,\ \frac{\partial M}{\partial y} = -1,\ \frac{\partial N}{\partial x} = -1,\ \frac{\partial N}{\partial y} = 1$

Flux $= \int\int_R 2\,dx\,dy = \int_0^1\int_0^1 2\,dx\,dy = 2;$

Circ $= \int\int_R [-1 - (-1)]\,dx\,dy = 0$

7. $M = y^2 - x^2,\ N = x^2 + y^2$

$\frac{\partial M}{\partial x} = -2x,\ \frac{\partial M}{\partial y} = 2y,\ \frac{\partial N}{\partial x} = 2x,\ \frac{\partial N}{\partial y} = 2y$

Flux $= \int\int_R (-2x + 2y)\,dx\,dy$

$= \int_0^3\int_0^x (-2x + 2y)\,dy\,dx = \int_0^3 (-2x^2 + x^2)\,dx$

$= \left[-\frac{1}{3}x^3\right]_0^3 = -9;$

Circ $= \int\int_R (2x - 2y)\,dx\,dy$

$= \int_0^3\int_0^x (2x - 2y)\,dy\,dx = \int_0^3 x^2\,dx = 9$

9. $M = x + e^x \sin y,\ N = x + e^x \cos y$

$\frac{\partial M}{\partial x} = 1 + e^x \sin y,\ \frac{\partial M}{\partial y} = e^x \cos y,$

$\frac{\partial N}{\partial x} = 1 + e^x \cos y,\ \frac{\partial N}{\partial y} = -e^x \sin y$

Flux $= \int\int_R dx\,dy = \int_{-\pi/4}^{\pi/4}\int_0^{\sqrt{\cos 2\theta}} r\,dr\,d\theta$

$= \int_{-\pi/4}^{\pi/4}\left(\frac{1}{2}\cos 2\theta\right)d\theta = \left[\frac{1}{4}\sin 2\theta\right]_{-\pi/4}^{\pi/4} = \frac{1}{2};$

Circ $= \int\int_R (1 + e^x\cos y - e^x \cos y)\,dx\,dy$

$= \int\int_R dx\,dy = \int_{-\pi/4}^{\pi/4}\int_0^{\sqrt{\cos 2\theta}} r\,dr\,d\theta$

$= \int_{-\pi/4}^{\pi/4}\left(\frac{1}{2}\cos 2\theta\right)d\theta = \frac{1}{2}$

11. $M = xy;\ N = y^2$

$\frac{\partial M}{\partial x} = y,\ \frac{\partial M}{\partial y} = x,\ \frac{\partial N}{\partial x} = 0,\ \frac{\partial N}{\partial y} = 2y$

Flux $= \int\int_R (y + 2y)\,dy\,dx = \int_0^1\int_{x^2}^x 3y\,dy\,dx$

$= \int_0^1 \left(\frac{3x^2}{2} - \frac{3x^4}{2}\right)dx = \frac{1}{5};$

Circ $= \int\int_R -x\,dy\,dx = \int_0^1\int_{x^2}^x -x\,dy\,dx$

$= \int_0^1 (-x^2 + x^3)\,dx = -\frac{1}{12}$

13. $M = 3xy - \frac{x}{1+y^2},\ N = e^x + \tan^{-1} y$

$\frac{\partial M}{\partial x} = 3y - \frac{1}{1+y^2},\ \frac{\partial N}{\partial y} = \frac{1}{1+y^2}$

Flux $= \int\int_R \left(3y - \frac{1}{1+y^2} + \frac{1}{1+y^2}\right)dx\,dy$

$= \int\int_R 3y\,dx\,dy = \int_0^{2\pi}\int_0^{a(1+\cos\theta)} (3r\sin\theta)\,r\,dr\,d\theta$

$= \int_0^{2\pi} a^3(1 + \cos\theta)^3(\sin\theta)\,d\theta$

$= \left[-\frac{a^3}{4}(1 + \cos\theta)^4\right]_0^{2\pi} = -4a^3 - (-4a^3) = 0$

15. $M = 2xy^3,\ N = 4x^2y^2$

$\frac{\partial M}{\partial y} = 6xy^2,\ \frac{\partial N}{\partial x} = 8xy^2$

work $= \oint_C 2xy^3\,dx + 4x^2y^2\,dy$

$= \int\int_R (8xy^2 - 6xy^2)\,dx\,dy = \int_0^1\int_0^{x^3} 2xy^2\,dy\,dx$

$= \int_0^1 \frac{2}{3}x^{10}\,dx = \frac{2}{33}$

17. $M = y^2,\ N = x^2$

$\frac{\partial M}{\partial y} = 2y,\ \frac{\partial N}{\partial x} = 2x$

$\oint_C y^2 + dx + x^2\,dy = \int\int_R (2x - 2y)\,dy\,dx$

$= \int_0^1\int_0^{1-x} (2x - 2y)\,dy\,dx$

$= \int_0^1 (-3x^2 + 4x - 1)\,dx$

$= \left[-x^3 + 2x^2 - x\right]_0^1$

$= -1 + 2 - 1 = 0$

19. $M = 6y + x,\ N = y + 2x$

$\frac{\partial M}{\partial y} = 6,\ \frac{\partial N}{\partial x} = 2$

$\oint_C (6y + x)\,dx + (y + 2x)\,dy = \int\int_R (2 - 6)\,dy\,dx$

$= -4$(Area of the circle)

$= -16\pi$

21. a. By Eq. (11), run backward,

Area of $R = \int\int_R dy\,dx = \int\int_R \left(\frac{1}{2} + \frac{1}{2}\right)dy\,dx$

$= \oint_C \frac{1}{2}x\,dy - \frac{1}{2}y\,dx$

b. $M = x = a\cos t,\ N = y = a\sin t$

$dx = -a\sin t\,dt,\ dy = \cos t\,dt$

Area $= \frac{1}{2}\oint_C x\,dy - y\,dx$

$= \frac{1}{2}\int_0^{2\pi} (a^2\cos^2 t + a^2\sin^2 t)\,dt = \frac{1}{2}\int_0^{2\pi} a^2\,dt$

$= \pi a^2$

23. $M = x = a\cos^3 t,\ N = y = \sin^3 t$
$dx = -3\cos^2 t \sin t\, dt,\ dy = 3\sin^2 t \cos t\, dt$

$$\text{Area} = \frac{1}{2}\oint_C x\,dy - y\,dx$$
$$= \frac{1}{2}\int_0^{2\pi} (3\sin^2 t\cos^2 t)(\cos^2 t + \sin^2 t)\,dt$$
$$= \frac{1}{2}\int_0^{2\pi} (3\sin^2 t\cos^2 t)\,dt = \frac{3}{8}\int_0^{2\pi} \sin^2 2t\,dt$$
$$= \frac{3}{16}\int_0^{4\pi} \sin^2 u\,du = \frac{3}{16}\left[\frac{u}{2} - \frac{\sin 2u}{4}\right]_0^{4\pi} = \frac{3}{8}\pi$$

25. a. $M = f(x),\ N = g(y)$

$$\frac{\partial M}{\partial y} = 0,\ \frac{\partial N}{\partial x} = 0$$
$$\oint_C f(x)\,dx + g(y)\,dy = \iint_R \left(\frac{\partial N}{\partial x} - \frac{\partial M}{\partial y}\right) dx\,dy$$
$$= \iint_R 0\,dx\,dy = 0$$

b. $M = ky,\ N = hx$

$$\frac{\partial M}{\partial y} = k,\ \frac{\partial M}{\partial y} = h$$
$$\oint_C ky\,dx + hx\,dy = \iint_R \left(\frac{\partial N}{\partial x} - \frac{\partial M}{\partial y}\right) dx\,dy$$
$$= \iint_R (h - k)\,dx\,dy = (h - k)(\text{Area of the region})$$

27. The integral is 0 for any simple closed plane curve C. The reasoning: By the tangential form of Green's Theorem, with

$$M = 4x^3y \text{ and } N = x^4,\ \oint_C 4x^3y\,x + x^4\,dy$$
$$= \iint_R \left[\frac{\partial}{\partial x}(x^4) - \frac{\partial}{\partial y}(4x^3y)\right] dx\,dy$$
$$= \iint_R \underbrace{(4x^3 - 4x^3)}_{0}\,dx\,dy = 0$$

29. Let $M = x$ and $N = 0$

$$\frac{\partial M}{\partial x} = 1 \text{ and } \frac{\partial N}{\partial y} = 0$$
$$\oint_C M\,dy - N\,dx = \iint_R \left(\frac{\partial M}{\partial x} + \frac{\partial N}{\partial y}\right) dx\,dy$$
$$\oint_C x\,dy = \iint_R (1 + 0)\,dx\,dy$$
$$\text{Area of } R = \iint_R dx\,dy = \oint_C x\,dy;\ \text{similarly,}$$

$M = y$ and $N = 0$

$$\frac{\partial M}{\partial y} = 1 \text{ and } \frac{\partial N}{\partial x} = 0$$
$$\oint_C M\,dx + N\,dy = \iint_R \left(\frac{\partial N}{\partial x} + \frac{\partial M}{\partial y}\right) dy\,dx =$$
$$= -\oint_C y\,dx = \iint_R (0 - 1)\,dy\,dx =$$
$$-\oint_C y\,dx = \iint_R dx\,dy = \text{Area of } R$$

31. a. Let $\delta(x, y) = 1$; $\bar{x} = \dfrac{M_y}{M} = \dfrac{\iint_R x\delta(x, y)\,dA}{\iint_R \delta(x, y)\,dA}$

$$= \frac{\iint_R x\,dA}{\iint_R dA} = \frac{\iint_R x\,dA}{A}$$
$$A\bar{x} = \iint_R x\,dA$$
$$A\bar{x} = \iint_R (x + 0)\,dx\,dy = \oint_C \frac{x^2}{2}\,dy$$
$$A\bar{x} = \iint_R (0 + x)\,dx\,dy = -\oint_C xy\,dx, \text{ and}$$
$$A\bar{x} = \iint_R x\,dA = \iint_R \left(\frac{2}{3}x + \frac{1}{3}x\right) dx\,dy$$
$$= \oint_C \frac{1}{3}x^2\,dy - \frac{1}{3}xy\,dx$$
$$\frac{1}{2}\oint_C x^2\,dy = -\oint_C xy\,dx = \frac{1}{3}\oint_C x^2\,dy - xy\,dx = A\bar{x}$$

b. $-\dfrac{1}{2}\oint_C y^2\,dx = \oint_C xy\,dy = \dfrac{1}{3}\oint_C xy\,dy - y^2\,dx = A\bar{y}$

33. $M = \dfrac{\partial f}{\partial y},\ N = -\dfrac{\partial f}{\partial x}$

$$\frac{\partial M}{\partial y} = \frac{\partial^2 f}{\partial y^2},\ \frac{\partial N}{\partial x} = \frac{\partial^2 f}{\partial x^2}$$
$$\oint_C \frac{\partial f}{\partial y}\,dx - \frac{\partial f}{\partial x}\,dy = \iint_R \left(-\frac{\partial^2 f}{\partial x^2} - \frac{\partial^2 f}{\partial y^2}\right) dx\,dy$$
$= 0$ for such curves C

35. a. $\mathbf{F} = \nabla f = \left(\dfrac{2x}{x^2 + y^2}\right)\mathbf{i} + \left(\dfrac{2y}{x^2 + y^2}\right)\mathbf{j}$

$M = \dfrac{2x}{x^2 + y^2},\ N = \dfrac{2y}{x^2 + y^2}$; since M, N are discontinuous at $(0, 0)$, we cannot apply Green's Theorem over C. Thus let C_h be the circle $x = h\cos\theta$, $y = h\sin\theta$, $0 < h \le a$ and let C_1 be the circle $x = a\cos t$, $y = a\sin t$, $a > 0$. Then $\oint_C \mathbf{F}\cdot\mathbf{n}\,ds$

$$= \oint_{C_1} M\,dy - N\,dx + \oint_{C_h} M\,dy - N\,dx$$
$$= \oint_{C_1} \frac{2x}{x^2 + y^2}\,dx - \frac{2y}{x^2 + y^2}\,dx$$
$$+ \oint_{C_h} \frac{2x}{x^2 + y^2}\,dy - \frac{2y}{x^2 + y^2}\,dx.$$ In the first integral,

let $x = a\cos t$, $y = a\sin t$
$dx = -a\sin t\,dt$, $dy = a\cos t\,dt$, $M = 2a\cos t$, $N = 2a\sin t$, $0 \le t \le 2\pi$. In the second integral, let $x = h\cos\theta$, $y = h\sin\theta$
$dx = -h\sin\theta\,d\theta$, $dy = h\cos\theta\,d\theta$, $M = 2h\cos\theta$, $N = 2h\sin\theta$, $0 \le \theta \le 2\pi$. Then $\oint_C \mathbf{F}\cdot\mathbf{n}\,ds$

$$= \oint_{C_1} \frac{2x}{x^2 + y^2}\,dy - \frac{2y}{x^2 + y^2}\,dx$$
$$+ \oint_{C_h} \frac{2x}{x^2 + y^2}\,dy - \frac{2}{x^2 + y^2}\,dx$$

$$= \oint_{C_1} \frac{(2a \cos t)(a \cos t)\, dt}{a^2} - \frac{(2a \sin t)(-a \sin t) dt}{a^2}$$
$$+ \oint_{C_h} \frac{(2h \cos \theta)(h \cos \theta)\, d\theta}{h^2} - \frac{(2h \sin \theta)(-h \sin \theta)\, d\theta}{h^2}$$
$$= \int_0^{2\pi} 2\, dt + \int_{2\pi}^{0} 2\, d\theta = 0 \text{ for every } h$$

b. If K is any simple closed curve surrounding C_h (K contains $(0, 0)$), then $\oint_K \mathbf{F} \cdot \mathbf{n}\, ds$ $= \oint_{C_1} M\, dy - N\, dx + \oint_{C_h} M\, dy - N\, dx$, and in polar coordinates, $\nabla f \cdot \mathbf{n} = M\, dy - N\, dx$

$$= \left(\frac{2r \cos \theta}{r^2}\right)(r \cos \theta\, d\theta + \sin \theta\, dr)$$
$$- \left(\frac{2r \sin \theta}{r^2}\right)(-r \sin \theta\, d\theta + \cos \theta\, dr) = \frac{2r^2}{r^2}\, d\theta$$

$= 2\, d\theta$. Now, 2θ increases by 4π as K is traversed once counterclockwise from $\theta = 0$ to $\theta = 2\pi$

$\oint_C \mathbf{F} \cdot \mathbf{n}\, ds = 0$ (since $\oint_{C_h} M\, dy - N\, dx = -4\pi$) when $(0, 0)$ is in the region, but $\oint_K \mathbf{F} \cdot \mathbf{n}\, ds = 4\pi$ when $(0, 0)$ is not in the region.

37. $\int_{g_1(y)}^{g_2(y)} \frac{\partial N}{\partial x}\, dx\, dy = N(g_2(y), y) - N(g_1(y), y)$

$$\int_c^d \int_{g_1(y)}^{g_2(y)} \left(\frac{\partial N}{\partial x}\, dx\right) dy = \int_c^d [N(g_2(y), y) - N(g_1(y), y)]\, dy$$
$$= \int_c^d N(g_2(y), y)\, dy - \int_c^d N(g_1(y), y)\, dy$$
$$= \int_c^d N(g_2(y), y)\, dy + \int_d^c N(g_1(y), y)\, dy$$
$$= \int_{C_2} N\, dy + \int_{C_1} N\, dy = \oint_C N\, dy$$
$$\oint_C N\, dy = \int\int_R \frac{\partial N}{\partial x}\, dx\, dy$$

39. The curl of a conservative two-dimensional field is zero. The reasoning: A two-dimension field $\mathbf{F} = M\mathbf{i} + N\mathbf{j}$ can be considered to be the restriction to the xy-plane of a three-dimensional field whose $\mathbf{k}$ component is zero, and whose $\mathbf{i}$ and $\mathbf{j}$ components are independent of z. For such a field to be conservative, we must have $\frac{\partial N}{\partial x} = \frac{\partial M}{\partial y}$ by the component test in Section 14.3

$$\text{curl } \mathbf{F} = \frac{\partial N}{\partial x} - \frac{\partial M}{\partial y} = 0$$

41–43. Example CAS commands:

Maple:

```
with (plots):
implicitplot({y=4 - 2*x}, x=0..3,
  y=0..5, scaling=CONSTRAINED);
M:= (x,y) -> x*exp(y);
N:=(x,y) -> 4*x^2*ln(y);
My:= diff(M(x,y),y);
Nx:= diff(N(x,y),x);
int(int(Nx - My, y=0..4 - 2*x), x=0..2);
evalf(%);
```

Mathematica:

```
<< Graphics`ImplicitPlot1`
Clear[x,y, y1, y2]
f[x_,y_] = {x  E^y, 4x^2Log[y]}
y1= y == 0
y2 = y == -2x+4
ImplicitPlot[{y1, y2}, {x, 0, 2},
AspectRatio -> Automatic]
y==0
y==4 - 2x
integrand = D[f[x, y][[2]], x] - D[f[x,
y][[1]], y]
c = Solve[x == 0, x]
d = Solve[y2, x]
x1 = c[[1, 1, 2]]
x2 = d[1, 1, 2]]
Integrate[Integrate[integrand, {x, x1,
  x2}], {y, 0, 4}]
N[%]
```

■ Section 15.5 Surface Area and Surface Integrals (pp 867-877)

Quick Review 15.5

1. $-\frac{\sqrt{2}}{2}$
2. $-\frac{\sqrt{2}}{2}$
3. $\frac{\sqrt{2}}{2}\mathbf{i} + \frac{\sqrt{2}}{2}\mathbf{j} + \mathbf{k}$
4. $x + y + \sqrt{2}z = 4$
5. $x = 1 + \frac{\sqrt{2}}{2}t, y = 1 + \frac{\sqrt{2}}{2}t, z = \sqrt{2} + t$
6. $\frac{(17\sqrt{17} - 1)\pi}{6} \approx 36.18$
7. $2(\sqrt{2} - 1)\pi \approx 2.60$
8. π
9. $(e^2 - 1)^2 \approx 40.82$
10. ≈ 0.014

Section 15.5 Exercises

1. $\mathbf{p} = \mathbf{k}$, $\nabla f = 2x\mathbf{i} + 2y\mathbf{j} - \mathbf{k}$
$|\nabla f| = \sqrt{(2x)^2 + (2y)^2 + (-1)^2} = \sqrt{4x^2 + 4y^2 + 1}$ and $|\nabla f \cdot \mathbf{p}| = 1$; $z = 2$

$x^2 + y^2 = 2$; thus $S = \int\int_R \frac{|\nabla f|}{|\nabla f \cdot \mathbf{p}|}\, dA$

$$= \int\int_R \sqrt{4x^2 + 4y^2 + 1}\, dx\, dy$$
$$= \int\int_R \sqrt{4r^2 \cos^2 \theta + 4r^2 \sin^2 \theta + 1}\, r\, dr\, d\theta$$
$$= \int_0^{2\pi} \int_0^{\sqrt{2}} \sqrt{4r^2 + 1}\, r\, dr\, d\theta$$
$$= \int_0^{2\pi} \left[\frac{1}{12}(4r^2 + 1)^{3/2}\right]_0^{\sqrt{2}} d\theta$$
$$= \int_0^{2\pi} \frac{13}{6}\, d\theta = \frac{13}{3}\pi$$

3. $\mathbf{p} = \mathbf{k}$, $\nabla f = \mathbf{i} + 2\mathbf{j} + 2\mathbf{k}$

$|\nabla f| = 3$ and $|\nabla f \cdot \mathbf{p}| = 2$; $x = y^2$ and $x = 2 - y^2$ intersect at $(1, 1)$ and $(1, -1)$

$$S = \int\int_R \frac{|\nabla f|}{|\nabla f \cdot \mathbf{p}|}\, dA = \int\int_R \frac{3}{2}\, dx\, dy$$

$$= \int_{-1}^{1}\int_{y^2}^{2-y^2} \frac{3}{2}\, dx\, dy = \int_{-1}^{1} (3 - 3y^2)\, dy = 4$$

5. $\mathbf{p} = \mathbf{k}$, $\nabla f = 2x\mathbf{i} - 2\mathbf{j} - 2\mathbf{k}$

$|\nabla f| = \sqrt{(2x)^2 + (-2)^2 + (-2)^2} = \sqrt{4x^2 + 8}$
$= 2\sqrt{x^2 + 2}$ and $|\nabla f \cdot \mathbf{p}| = 2$

$$S = \int\int_R \frac{|\nabla f|}{|\nabla f \cdot \mathbf{p}|}\, dA = \int\int_R \frac{2\sqrt{x^2 + 2}}{2}\, dx\, dy$$

$$= \int_0^2\int_0^{3x} \sqrt{x^2 + 2}\, dy\, dx = \int_0^2 3x\sqrt{x^2 + 2}\, dx$$

$$= \left[(x^2 + 2)^{3/2}\right]_0^2 = 6\sqrt{6} - 2\sqrt{2}$$

7. $\mathbf{p} = \mathbf{k}$, $\nabla f = c\mathbf{i} - \mathbf{k}$

$|\nabla f| = \sqrt{c^2 + 1}$ and $|\nabla f \cdot \mathbf{p}| = 1$

$$S = \int\int_R \frac{|\nabla f|}{|\nabla f \cdot \mathbf{p}|}\, dA = \int\int_R \sqrt{c^2 + 1}\, dx\, dy$$

$$= \int_0^{2\pi}\int_0^1 \sqrt{c^2 + 1}\, r\, dr\, d\theta = \int_0^{2\pi} \frac{\sqrt{c^2 + 1}}{2}\, d\theta$$

$$= \pi\sqrt{c^2 + 1}$$

9. $\mathbf{p} = \mathbf{i}$, $\nabla f = \mathbf{i} + 2y\mathbf{j} + 2z\mathbf{k}$

$|\nabla f| = \sqrt{1^2 + (2y)^2 + (2z)^2} = \sqrt{1 + 4y^2 + 4z^2}$ and $|\nabla f \cdot \mathbf{p}| = 1$; $1 \le y^2 + z^2 \le 4$

$$S = \int\int_R \frac{|\nabla f|}{|\nabla f \cdot \mathbf{p}|}\, dA = \int\int_R \sqrt{1 + 4y^2 + 4z^2}\, dy\, dz$$

$$= \int_0^{2\pi}\int_1^2 \sqrt{1 + 4r^2\cos^2\theta + 4r^2\sin^2\theta}\; r\, dr\, d\theta$$

$$= \int_0^{2\pi}\int_1^2 \sqrt{1 + 4r^2}\; r\, dr\, d\theta$$

$$= \int_0^{2\pi} \left[\frac{1}{12}(1 + 4r^2)^{3/2}\right]_1^2 d\theta$$

$$= \int_0^{2\pi} \frac{1}{12}\left(17\sqrt{17} - 5\sqrt{5}\right) d\theta = \frac{\pi}{6}\left(17\sqrt{17} - 5\sqrt{5}\right)$$

11. $\mathbf{p} = \mathbf{k}$, $\nabla f = \left(2x - \frac{2}{x}\right)\mathbf{i} + \sqrt{15}\mathbf{j} - \mathbf{k}$

$$|\nabla f| = \sqrt{\left(2x - \frac{2}{x}\right)^2 + (\sqrt{15})^2 + (-1)^2}$$

$$= \sqrt{4x^2 + 8 + \frac{4}{x^2}} = \sqrt{\left(2x + \frac{2}{x}\right)^2} = 2x + \frac{2}{x},$$

on $1 \le x \le 2$ and $|\nabla f \cdot \mathbf{p}| = 1$

$$S = \int\int_R \frac{|\nabla f|}{|\nabla f \cdot \mathbf{p}|}\, dA = \int\int_R (2x + 2x^{-1})\, dx\, dy$$

$$= \int_0^1\int_1^2 (2x + 2x^{-1})\, dx\, dy = \int_0^1 \left[x^2 + 2\ln x\right]_1^2 dy$$

$$= \int_0^1 (3 + 2\ln 2)\, dy = 3 + 2\ln 2$$

13. The bottom face S of the cube is in the xy-plane
$z = 0$
$g(x, y, 0) = x + y$ and $f(x, y, z) = z = 0$
$\mathbf{p} = \mathbf{k}$ and $\nabla f = \mathbf{k}$
$|\nabla f| = 1$ and $|\nabla f \cdot \mathbf{p}| = 1$
$d\sigma = dx\, dy$

$$\int\int_S g\, d\sigma = \int\int_R (x + y)\, dx\, dy = \int_0^a\int_0^a (x + y)\, dx\, dy$$

$= \int_0^a \left(\frac{a^2}{2} + ay\right) dy = a^3$. Because of symmetry, we also get a^3 over the face of the cube in the xz-plane and a^3 over the face of the cube in the yz-plane. Next, on the top of the cube, $g(x, y, z) = g(x, y, a) = x + y + a$ and $f(x, y, z) = z = a$

$\mathbf{p} = \mathbf{k}$ and $\nabla f = \mathbf{k}$
$|\nabla f| = 1$ and $|\nabla f \cdot \mathbf{p}| = 1$

$d\sigma = dx\, dy$

$$\int\int_S g\, d\sigma = \int\int_R (x + y + a)\, dx\, dy$$

$$= \int_0^a\int_0^a (x + y + a)\, dx\, dy = \int_0^a\int_0^a (x + y)\, dx\, dy$$

$= \int_0^a\int_0^a a\, dx\, dy = 2a^3$. Because of symmetry, the integral is also $2a^3$ over each of the other two faces. Therefore,

$$\int\int_{\text{cube}} (x + y + z)\, d\sigma = 3(a^3 + 2a^3) = 9a^3.$$

15. On the faces in the coordinate planes, $g(x, y, z) = 0$ the integral over these faces is 0.
On the face $x = a$, we have $f(x, y, z) = x = a$ and $g(x, y, z) = g(a, y, z) = ayz$
$\mathbf{p} = \mathbf{i}$ and $\nabla f = \mathbf{i}$
$|\nabla f| = 1$ and $|\nabla f \cdot \mathbf{p}| = 1$
$d\sigma = dy\, dz$

$$\int\int_S g\, d\sigma = \int\int_S ayz\, d\sigma = \int_0^c\int_0^b ayz\, dy\, dz = \frac{ab^2c^2}{4}.$$

On the face $y = b$, we have $f(x, y, z) = y = b$ and $g(x, y, z) = g(x, b, z) = bxz$
$\mathbf{p} = \mathbf{j}$ and $\nabla f = \mathbf{j}$
$|\nabla f| = 1$ and $|\nabla f \cdot \mathbf{p}| = 1$
$d\sigma = dy\, dz$

$$\int\int_S g\, d\sigma = \int\int_S bxz\, d\sigma = \int_0^c\int_0^a bxz\, dz\, dx = \frac{a^2bc^2}{4}.$$

On the face $z = c$, we have $f(x, y, z) = z = c$ and $g(x, y, z) = g(x, y, c) = cxy$
$\mathbf{p} = \mathbf{k}$ and $\nabla f = \mathbf{k}$
$|\nabla f| = 1$ and $|\nabla f \cdot \mathbf{p}| = 1$
$d\sigma = dy\, dz$

$$\int\int_S g\, d\sigma = \int\int_S cxy\, d\sigma = \int_0^b\int_0^a cxy\, dx\, dy = \frac{a^2b^2c}{4}.$$

Therefore, $\displaystyle\int\int_S g(x, y, z)\, d\sigma = \frac{abc(ab + ac + bc)}{4}.$

17. $f(x, y, z) = 2x + 2y + z = 2$

$\nabla f = 2\mathbf{i} + 2\mathbf{j} + \mathbf{k}$ and $g(x, y, z)$

$= x + y + (2 - 2x - 2y) = 2 - x - y$

$\mathbf{p} = \mathbf{k}$, $|\nabla f| = 3$ and $|\nabla f \cdot \mathbf{p}| = 1$

$d\sigma = 3\, dy\, dx$; $z = 0$

$2x + 2y = 2$

$y = 1 - x$

$$\int\int_S g\, d\sigma = \int\int_S (2 - x - y)\, d\sigma$$

$$= 3\int_0^1 \int_0^{1-x} (2 - x - y)\, dy\, dx$$

$$= 3\int_0^1 \left[(2 - x)(1 - x) - \frac{1}{2}(1 - x)^2\right] dx$$

$$= 3\int_0^1 \left(\frac{3}{2} - 2x + \frac{x^2}{2}\right) dx = 2$$

19. $g(x, y, z) = z$, $\mathbf{p} = \mathbf{k}$

$\nabla g = \mathbf{k}$

$|\nabla g| = 1$ and $|\nabla g \cdot \mathbf{p}| = 1$

$$\text{Flux} = \int\int_S \mathbf{F} \cdot \mathbf{n}\, d\sigma = \int\int_R (\mathbf{F} \cdot \mathbf{k})\, dA$$

$$= \int_0^2 \int_0^3 3\, dy\, dx = 18$$

21. $\nabla g = 2x\mathbf{i} + 2y\mathbf{j} + 2z\mathbf{k}$

$|\nabla g| = \sqrt{4x^2 + 4y^2 + 4z^2} = 2a$; $\mathbf{n} = \dfrac{2x\mathbf{i} + 2y\mathbf{j} + 2z\mathbf{k}}{\sqrt{2x^2 + y^2 + z^2}}$

$= \dfrac{x\mathbf{i} + y\mathbf{j} + z\mathbf{k}}{a}$

$\mathbf{F} \cdot \mathbf{n} = \dfrac{z^2}{a}$; $|\nabla g \cdot \mathbf{k}| = 2z$

$d\sigma = \dfrac{2a}{2z}\, dA$

$$\text{Flux} = \int\int_R \left(\frac{z^2}{a}\right)\left(\frac{a}{z}\right) dA = \int\int_R z\, dA$$

$$= \int\int_R \sqrt{a^2 - (x^2 + y^2)}\, dx\, dy$$

$$= \int_0^{\pi/2} \int_0^a \sqrt{a^2 - r^2}\, r\, dr\, d\theta = \frac{\pi a^3}{6}$$

23. From Exercise 21, $\mathbf{n} = \dfrac{x\mathbf{i} + y\mathbf{j} + z\mathbf{k}}{\text{a}}$ and $d\sigma = \dfrac{a}{z}\, dA$

$\mathbf{F} \cdot \mathbf{n} = \dfrac{xy}{a} - \dfrac{xy}{a} + \dfrac{z}{a} = \dfrac{z}{a}$

$$\text{Flux} = \int\int_R \left(\frac{z}{a}\right)\left(\frac{a}{z}\right) dA = \int\int_R 1\, dA = \frac{\pi a^2}{4}$$

25. From Exercise 21, $\mathbf{n} = \dfrac{x\mathbf{i} + y\mathbf{j} + z\mathbf{k}}{a}$ and

$d\sigma = \dfrac{a}{z}\, dA$

$\mathbf{F} \cdot \mathbf{n} = \dfrac{x^2}{a} + \dfrac{y^2}{a} + \dfrac{z^2}{a} = a$

$$\text{Flux} = \int\int_R a\left(\frac{a}{z}\right) dA = \int\int_R \frac{a^2}{z}\, dA$$

$$= \int\int_R \frac{a^2}{\sqrt{a^2 - (x^2 + y^2)}}\, dA$$

$$= \int_0^{\pi/2} \int_0^a \frac{a^2}{\sqrt{a^2 - r^2}}\, r\, dr\, d\theta$$

$$= \int_0^{\pi/2} a^2\left[-\sqrt{a^2 - r^2}\right]_0^a d\theta$$

$$= \frac{\pi a^3}{2}$$

27. $g(x, y, z) = y^2 + z = 4$

$\nabla g = 2y\mathbf{j} + \mathbf{k}$

$|\nabla g| = \sqrt{4y^2 + 1}$

$\mathbf{n} = \dfrac{2y\mathbf{j} + \mathbf{k}}{\sqrt{4y^2 + 1}}$

$\mathbf{F} \cdot \mathbf{n} = \dfrac{2xy - 3z}{\sqrt{4y^2 + 1}}$; $\mathbf{p} = \mathbf{k}$

$|\nabla g \cdot \mathbf{p}| = 1$ $d\sigma = \sqrt{4y^2 + 1}\, dA$

$$\text{Flux} = \int\int_R \left(\frac{2xy - 3z}{\sqrt{4y^2 + 1}}\right)\sqrt{4y^2 + 1}\, dA$$

$= \displaystyle\int\int_R (2xy - 3z)\, dA$; $z = 0$ and $z = 4 - y^2$

$y^2 - 4 = 0$

$$\text{Flux} = \int\int_R [2xy - 3(4 - y^2)]\, dA$$

$$= \int_0^1 \int_{-2}^2 (2xy - 12 + 3y^2)\, dy\, dx$$

$$= \int_0^1 \left[xy^2 - 12y + y^3\right]_{-2}^2 dx = \int_0^1 -32\, dx = -32$$

29. $g(x, y, z) = y - e^x = 0$

$\nabla g = -e^x\mathbf{i} + \mathbf{j}$

$|\nabla g| = \sqrt{e^{2x} + 1}$

$\mathbf{n} = \dfrac{e^x\mathbf{i} - \mathbf{j}}{\sqrt{e^{2x} + 1}}$

$\mathbf{F} \cdot \mathbf{n} \dfrac{-2e^x - 2y}{\sqrt{e^{2x} + 1}}$; $\mathbf{p} = \mathbf{i}$

$|\nabla g \cdot \mathbf{p}| = e^x$

$d\sigma = \dfrac{\sqrt{e^{2x} + 1}}{e^x}\, dA$

$$\text{Flux} = \int\int_R \left(\frac{-2e^x - 2y}{\sqrt{e^{2x} + 1}}\right)\left(\frac{\sqrt{e^{2x} + 1}}{e^x}\right) dA$$

$$= \int\int_R = \frac{-2e^x - 2e^x}{e^x}\, dA = \int\int_R -4\, dA$$

$$= \int_0^1 \int_1^2 -4\, dy\, dz = -4$$

31. On the face $z = a$; $g(x, y, z) = z$

$\nabla g = \mathbf{k}$

$|\nabla g| = 1$; $\mathbf{n} = \mathbf{k}$

$\mathbf{F} \cdot \mathbf{n} = 2xz = 2ax$ since $z = a$; $d\sigma = dx\, dy$

$$\text{Flux} = \int\int_R 2ax\, dx\, dy = \int_0^a \int_0^a 2ax\, dx\, dy = a^4$$

On the face $z = 0$: $g(x, y, z) = z$

$\nabla g = \mathbf{k}$

$|\nabla g| = 1$; $\mathbf{n} = \mathbf{k}$

$\mathbf{F} \cdot \mathbf{n} = 2xz = 0$ since $z = 0$; $d\sigma = dy\, dz$

$$\text{Flux} = \int\int_R 0\, dx\, dy = 0$$

On the face $x = a$: $g(x, y, z) = x$

$\nabla g = \mathbf{i}$

$|\nabla g| = 1$; $\mathbf{n} = \mathbf{i}$

$\mathbf{F} \cdot \mathbf{n} = -2xy = 2ay$ since $x = a$; $d\sigma = dy\, dz$

$$\text{Flux} = \int_0^a \int_0^a 2ay\, dy\, dz = a^4$$

31. (continued)

On the face $x = 0$: $g(x, y, z) = x\nabla g = i$

$|\nabla g| = 1$; $\mathbf{n} = -\mathbf{i}$

$\mathbf{F} \cdot \mathbf{n} = 2yz = 0$ since $x = 0$

Flux $= 0$

On the face $y = a$: $g(x, y, z) = y$

$\nabla g = \mathbf{j}$

$|\nabla g| = 1$; $\mathbf{n} = \mathbf{j}$

$\mathbf{F} \cdot \mathbf{n} = 2yz = 2az$ since $y = a$; $d\sigma = dz\,dx$

$$\text{Flux} = \int_0^a \int_0^a 2az\,dz\,dx = a^4$$

On the face $y = 0$: $g(x, y, z) = y$

$\nabla g = \mathbf{j}$

$|\nabla g| = 1$; $\mathbf{n} = -\mathbf{j}$

$\mathbf{F} \cdot \mathbf{n} = -2yz = 0$ since $y = 0$

Flux $= 0$. Therefore, Total Flux $= 3a^4$.

33. $\nabla f = 2x\mathbf{i} + 2y\mathbf{j} + 2z\mathbf{k}$

$|\nabla f| = \sqrt{4x^2 + 4y^2 + 4z^2} = 2a$; $\mathbf{p} = \mathbf{k}$

$|\nabla f \cdot \mathbf{p}| = 2z$ since $z \geq 0$

$$d\sigma = \frac{2a}{2z}\,dA = \frac{a}{z}\,dA;\ M = \int\int_S \delta\,d\sigma$$

$$\frac{\delta}{8}(\text{surface area of sphere}) = \frac{\delta\pi a^2}{2}$$

$$M_{xy} = \int\int_S 2\delta\,d\sigma$$

$$= \delta \int\int_R z\left(\frac{a}{z}\right) dA = a\delta = \int\int_R dA$$

$$= a\delta \int_0^{\pi/2} \int_0^a r\,dr\,d\theta = \frac{\delta\pi a^3}{4}$$

$\bar{z} = \dfrac{M_{xy}}{M} = \left(\dfrac{\delta\pi a^3}{4}\right)\left(\dfrac{2}{\delta\pi a^2}\right) = \dfrac{a}{2}$. Because of symmetry,

$\bar{x} = \bar{y} = \dfrac{a}{2}$

the centroid is $\left(\dfrac{a}{2}, \dfrac{a}{2}, \dfrac{a}{2}\right)$.

35. Because of symmetry, $\bar{x} = \bar{y} = 0$; $M = \int\int_S \delta\,d\sigma$

$= \delta \int\int_S d\sigma = (\text{Area of S})\,\delta = 3\pi\sqrt{2}\delta$;

$\nabla f = 2x\mathbf{i} + 2y\mathbf{j} + 2z\mathbf{k}$

$|\nabla f| = \sqrt{4x^2 + 4y^2 + 4z^2} = 2\sqrt{x^2 + y^2 + z^2}$; $\mathbf{p} = \mathbf{k}$

$|\nabla f \cdot \mathbf{p}| = 2z$

$$d\sigma = \frac{2\sqrt{x^2 + y^2 + z^2}}{2z}\,dA = \frac{\sqrt{x^2 + y^2 + (x^2 + y^2)}}{z}\,dA$$

$$= \frac{\sqrt{2}\sqrt{x^2 + y^2}}{z}\,dA$$

$$M_{xy} = \delta \int\int_S z\left(\frac{\sqrt{2}\sqrt{x^2 + y^2}}{z}\right) dA$$

$$= \delta \int\int_S \sqrt{2}\sqrt{x^2 + y^2}\,dA = \delta \int_0^{2\pi} \int_1^2 \sqrt{2}\,r^2\,dr\,d\theta$$

$$= \frac{14\pi\sqrt{2}}{3}\delta$$

$$\bar{z} = \frac{\left(\frac{14\pi\sqrt{2}}{3}\delta\right)}{3\pi\sqrt{2}\delta} = \frac{14}{9}$$

$(\bar{x}, \bar{y}, \bar{z}) = \left(0, 0, \dfrac{14}{9}\right)$. Next, $I_z = \int\int_S (x^2 + y^2)\,\delta\,d\sigma$

$$= \int\int_S (x^2 + y^2)\left(\frac{\sqrt{2}\sqrt{x^2 + y^2}}{z}\right)\delta\,dA$$

$$= \delta\sqrt{2} \int\int_S (x^2 + y^2)\,dA = \delta\sqrt{2} \int_0^{2\pi} \int_1^2 r^3\,dr\,d\theta$$

$$= \frac{15\pi\sqrt{2}}{2}\delta$$

$$R_z = \sqrt{\frac{I_z}{M}} = \frac{\sqrt{10}}{2}$$

37. a. Let the diameter lie on the z-axis and let $f(x, y, z) = x^2 + y^2 + z^2 = a^2$, $z \geq 0$ be the upper hemisphere

$\nabla f = 2x\mathbf{i} + 2y\mathbf{j} + 2z\mathbf{k}$

$|\nabla f| = \sqrt{4x^2 + 4y^2 + 4z^2} = 2a$, $a > 0$; $\mathbf{p} = \mathbf{k}$

$|\nabla f \cdot \mathbf{p}| = 2z$ since $z \geq 0$

$$d\sigma = \frac{a}{z}\,dA$$

$$I_z = \int\int_S \delta(x^2 + y^2)\left(\frac{a}{z}\right) d\sigma$$

$$= a\delta \int\int_R \frac{x^2 + y^2}{\sqrt{a^2 - (x^2 + y^2)}}\,dA$$

$$= a\delta \int_0^{2\pi} \int_0^a \frac{r^2}{\sqrt{a^2 - r^2}}\,r\,dr\,d\theta$$

$$= a\delta \int_0^{2\pi} \left[-r^2\sqrt{a^2 - r^2} - \frac{2}{3}(a^2 - r^2)^{3/2}\right]_0^a d\theta$$

$$= a\delta \int_0^{2\pi} \frac{2}{3}a^3\,d\theta = \frac{4\pi}{3}a^4\delta$$

the moment inertia is $\dfrac{8\pi}{3}a^4\delta$ for the whole sphere

b. $I_L = I_{c.m.} + mh^2$, where m is the mass of the body and h is the distance between the parallel lines;

now, $I_{c.m.} = \dfrac{8\pi}{3}a^4\delta$ (from part a) and

$$\frac{m}{2} = \int\int_S \delta\,d\sigma = \delta \int\int_R \left(\frac{a}{z}\right) dA = a\delta$$

$$= \int\int_R \frac{1}{\sqrt{a^2 - (x^2 + y^2)}}\,dy\,dx$$

$$= a\delta \int_0^{2\pi} \int_0^a \frac{1}{\sqrt{a^2 - r^2}}\,r\,dr\,d\theta$$

$$= a\delta \int_0^{2\pi} \left[-\sqrt{a^2 - r^2}\right]_0^a d\theta = a\delta \int_0^{2\pi} a\,d\theta$$

$= 2\pi a^2\delta$ and $h = a$

$$I_L = \frac{8\pi}{3}a^4\delta + 4\pi a^2\,\delta a^2 = \frac{20\pi}{3}a^4\delta$$

39. $f_x(x, y) = 2x$, $f_y(x, y) = 2y$

$$\sqrt{f_x^2 + f_x^2 + 1} = \sqrt{4x^2 + 4y^2 + 1}$$

$$\text{Area} = \int\int_R \sqrt{4x^2 + 4y^2 + 1}\,dx\,dy$$

$$= \int_0^{2\pi} \int_0^{\sqrt{3}} \sqrt{4r^2 + 1}\,r\,dr\,d\theta = \frac{\pi}{6}(13\sqrt{13} - 1)$$

41. $f_x(x, y) = \frac{x}{\sqrt{x^2 + y^2}}, f_y(x, y) \frac{y}{\sqrt{x^2 + y^2}}$

$\sqrt{f_x^2 + f_y^2 + 1} = \sqrt{\frac{x^2}{x^2 + y^2} + \frac{y^2}{x^2 + y^2} + 1} = \sqrt{2}$

Area $= \int\int_{R_{xy}} \sqrt{2}\, dx\, dy = \sqrt{2}$(Area between the ellipse and the circle) $= \sqrt{2}(6\pi - \pi) = 5\pi\sqrt{2}$

43. $y = \frac{2}{3}z^{3/2}$

$f_x(x, z) = 0, f_z(x, z) = z^{1/2}$

$\sqrt{f_x^2 + f_z^2 + 1} = \sqrt{z + 1}$

Area $= \int_0^4 \int_0^1 \sqrt{z + 1}\, dx\, dz = \int_0^4 \sqrt{z + 1}\, dz$
$= \frac{2}{3}(5\sqrt{5} - 1)$

■ Section 15.6 Parametrized Surfaces
(pp. 878 – 886)

Quick Review 15.6

1. $\left(\sqrt{2}, \frac{\pi}{4}, 1\right)$

2. $\left(\sqrt{3}, \frac{\pi}{4}, \tan^{-1}\sqrt{2}\right)$ and $\tan^{-1}\sqrt{2} \approx 54.74°$

3. Right circular cylinder parallel to the z-axis generated by the circle $x^2 + (y + 2)^2 = 4$ in the xy-plane

4. a. $r = -4 \sin \theta$

b. $\rho \sin \phi + 4 \sin \theta = 0$

5. The sphere with center at the origin and radius 3

6. a. $r^2 + z^2 = 9$

b. $\rho = 3$

7. The plane $y = 2$

8. a. $y = 2$

b. $\rho \sin \phi \sin \theta = 2$

9. $2\mathbf{i} + 2\mathbf{j} + 2\mathbf{k}$

10. $-4\mathbf{i} - 4\mathbf{j} - 4\mathbf{k}$

Section 15.6 Exercises

1. In cylindrical coordinates, let $x = r \cos \theta$, $y = r \sin \theta$, $z = (\sqrt{x^2 + y^2})^2 = r^2$. Then $\mathbf{r}(r, \theta) = (r \cos \theta)\mathbf{i} + (r \sin \theta)\mathbf{j} + r^2\mathbf{k}$, $0 \le r \le 2$, $0 \le \theta \le 2\pi$.

3. In cylindrical coordinates, let $x = r \cos \theta$, $y = r \sin \theta$,

$z = \frac{\sqrt{x^2 + y^2}}{2}$

$z = \frac{r}{2}$. Then $\mathbf{r}(r, \theta) = (r \cos \theta)\mathbf{i} + (r \sin \theta)\mathbf{j} + \left(\frac{r}{2}\right)\mathbf{k}$.

For $0 \le z \le 3$, $0 \le \frac{r}{2} \le 3 \Rightarrow 0 \le r \le 6$; to get only the first octant, let $0 \le \theta \le \frac{\pi}{2}$.

5. In cylindrical coordinates, let $x = r \cos \theta$, $y = r \sin \theta$; since

$x^2 + y^2 + z^2 = 9$

$z^2 = 9 - (x^2 + y^2) = 9 - r^2$

$z = \sqrt{9 - r^2}$, $z \ge 0$. Then $\mathbf{r}(r, \theta) = (r \cos \theta)\mathbf{i} + (r \sin \theta)\mathbf{j} + \sqrt{9 - r^2}\mathbf{k}$. Let $0 \le \theta \le 2\pi$. For the domain of r: $z = \sqrt{x^2 + y^2}$ and $x^2 + y^2 + z^2 = 9$

$x^2 + y^2 + (x^2 + y^2) = 9$

$2(x^2 + y^2) = 9$

$2r^2 = 9 \Rightarrow r = \frac{3}{\sqrt{2}}$

$0 \le r \le \frac{3}{\sqrt{2}}$.

7. In spherical coordinates, $x = \rho \sin \phi \cos \theta$, $y = \rho \sin \phi \sin \theta$, $\rho = \sqrt{x^2 + y^2 + z^2}$

$\rho^2 = 3$

$\rho = \sqrt{3}$

$z = \sqrt{3} \cos \phi$ for the sphere; $z = \frac{\sqrt{3}}{2} = \sqrt{3} \cos \phi$

$\cos \phi = \frac{1}{2} \Rightarrow \phi = \frac{\pi}{3}$; $z = -\frac{\sqrt{3}}{2}$

$-\frac{\sqrt{3}}{2} = \sqrt{3} \cos \phi$

$\cos \phi = -\frac{1}{2} \Rightarrow \phi = \frac{2\pi}{3}$. Then $\mathbf{r}(\phi, \theta)$

$= (\sqrt{3} \sin \phi \cos \theta)\mathbf{i} + (\sqrt{3} \sin \phi \sin \theta)\mathbf{j} + (\sqrt{3} \cos \phi)\mathbf{k}$,

$\frac{\pi}{3} \le \phi \le \frac{2\pi}{3}$ and $0 \le \theta \le 2\pi$.

9. Since $z = 4 - y^2$, we can let $\mathbf{r}$ be a function of x and y

$\mathbf{r}(x, y) = x\mathbf{i} + y\mathbf{j} + (4 - y^2)\mathbf{k}$. Then $z = 0$.

$0 = 4 - y^2$

$y = \pm 2$. Thus, let $-2 \le y \le 2$ and $0 \le x \le 2$.

11. When $x = 0$, let $y^2 + z^2 = 9$ be the circular section in the yz-plane. Use polar coordinates in the yz-plane

$y = 3 \cos \theta$ and $z = 3 \sin \theta$. Thus let $x = u$ and $\theta = v$

$\mathbf{r}(u, v) = u\mathbf{i} + (3 \cos v)\mathbf{j} + (3 \sin v)\mathbf{k}$ where $0 \le u \le 3$, and $0 \le v \le 2\pi$.

13. a. $x + y + z = 1$

$z = 1 - x - y$. In cylindrical coordinates, let $x = r \cos \theta$ and $y = r \sin \theta$

$z = 1 - r \cos \theta - r \sin \theta$

$\mathbf{r}(r, \theta) = (r \cos \theta)\mathbf{i} + (r \sin \theta)\mathbf{j} + (1 - r \cos \theta - r \sin \theta)\mathbf{k}$, $0 \le \theta \le 2\pi$ and $0 \le r \le 3$.

b. In a fashion similar to cylindrical coordinates, but working in the yz-plane instead of the xy-plane,

let $y = u \cos v$, $z = u \sin v$ where $u = \sqrt{y^2 + z^2}$ and v is the angle formed by (x, y, z), $(x, 0, 0)$, and $(x, y, 0)$ with $(x, 0, 0)$ as vertex. Since $x + y + z = 1$

$x = 1 - y - z$

$x = 1 - u \cos v - u \sin v$, then $\mathbf{r}$ is a function of u and v

$\mathbf{r}(u, v) = (1 - u \cos v - u \sin v)\mathbf{i} + (u \cos v)\mathbf{j} + (u \sin v)\mathbf{k}$, $0 \le u \le 3$ and $0 \le v \le 2\pi$.

15. Let $x = w\cos v$ and $z = w\sin v$. Then $(x-2)^2 + z^2 = 4$
$x^2 - 4x + z^2 = 0$
$w^2\cos^2 v - 4w\cos v + w^2\sin^2 v = 0$
$w^2 - 4w\cos v = 0$
$w = 0$ or $w - 4\cos v = 0$
$w = 0$ or $w = 4\cos v$.
Now $w = 0$
$x = 0$ and $y = 0$, which is a line not a cylinder. Therefore, let $w = 4\cos v$
$x = (4\cos v)(\cos v) = 4\cos^2 v$ and $z = 4\cos v\sin v$.
Finally, let $y = u$. Then $\mathbf{r}(u, v)$
$= 4(\cos^2 v)\mathbf{i} + u\mathbf{j} + (4\cos v\sin v)\mathbf{k}, -\frac{\pi}{2} \le v \le \frac{\pi}{2}$ and $0 \le u \le 3$.

17. Let $x = r\cos\theta$ and $y = r\sin\theta$. Then $\mathbf{r}(r, \theta) = (r\cos\theta)\mathbf{i}$
$+ (r\sin\theta)\mathbf{j} + \left(\frac{2 - r\sin\theta}{2}\right)\mathbf{k}$. $0 \le r \le 1$ and $0 \le \theta \le 2\pi$
$\mathbf{r}_r = (\cos\theta)\mathbf{i} + (\sin\theta)\mathbf{j} - \left(\frac{\sin\theta}{2}\right)\mathbf{k}$ and $r_\theta = (-r\sin\theta)\mathbf{i}$
$+ (r\cos\theta)\mathbf{j} - \left(\frac{r\cos\theta}{2}\right)\mathbf{k}$

$$\mathbf{r}_r \times \mathbf{r}_\theta = \begin{vmatrix} \mathbf{i} & \mathbf{j} & \mathbf{k} \\ \cos\theta & \sin\theta & -\frac{\sin\theta}{2} \\ -r\sin\theta & r\cos\theta & -\frac{r\cos\theta}{2} \end{vmatrix}$$

$$= \left(\frac{-r\sin\theta\cos\theta}{2} + \frac{(\sin\theta)(r\cos\theta)}{2}\right)\mathbf{i}$$
$$+ \left(\frac{r\sin^2\theta}{2} + \frac{r\cos^2\theta}{2}\right)\mathbf{j} + (r\cos^2\theta + r\sin^2\theta)\mathbf{k} = \frac{r}{2}\mathbf{j} + r\mathbf{k}$$

$$|\mathbf{r}_r \times \mathbf{r}_\theta| = \sqrt{\frac{r^2}{4} + r^2} = \frac{\sqrt{5}r}{2}$$

$$A = \int_0^{2\pi}\int_0^1 \frac{\sqrt{5}r}{2}\,dr\,d\theta = \int_0^{2\pi}\left[\frac{5r^2}{4}\right]_0^1 d\theta = \int_0^{2\pi}\frac{\sqrt{5}}{4}\,d\theta$$
$$= \frac{\pi\sqrt{5}}{2}$$

19. Let $x = r\cos\theta$ and $y = r\sin\theta$
$z = 2\sqrt{x^2 + y^2} = 2r$, $1 \le r \le 3$ and $0 \le \theta \le 2\pi$.
Then $\mathbf{r}(r, \theta) = (r\cos\theta)\mathbf{i} + (r\sin\theta)\mathbf{j} + 2r\mathbf{k}$
$\mathbf{r}_r = (\cos\theta)\mathbf{i} + (\sin\theta)\mathbf{j} + 2\mathbf{k}$ and
$\mathbf{r}_\theta = (-r\sin\theta)\mathbf{i} + (r\cos\theta)\mathbf{j}$

$$\mathbf{r}_r \times \mathbf{r}_\theta = \begin{vmatrix} \mathbf{i} & \mathbf{j} & \mathbf{k} \\ \cos\theta & \sin\theta & 2 \\ -r\sin\theta & r\cos\theta & 0 \end{vmatrix}$$
$$= (-2r\cos\theta)\mathbf{i} - (2r\sin\theta)\mathbf{j} + (r\cos^2\theta + r\sin^2\theta)\mathbf{k}$$
$$= (-2r\cos\theta)\mathbf{i} - (2r\sin\theta)\mathbf{j} + r\mathbf{k}$$

$$|\mathbf{r}_r \times \mathbf{r}_\theta| = \sqrt{4r^2\cos^2\theta + 4r^2\sin^2\theta + r^2} = \sqrt{5r^2} = r\sqrt{5}$$

$$A = \int_0^{2\pi}\int_1^3 r\sqrt{5}\,dr\,d\theta = \int_0^{2\pi}\left[\frac{r^2\sqrt{5}}{2}\right]_1^3 d\theta$$
$$= \int_0^{2\pi} 4\sqrt{5}\,d\theta = 8\pi\sqrt{5}$$

21. Let $x = r\cos\theta$ and $y = r\sin\theta$
$r^2 = x^2 + y^2 = 1$, $1 \le z \le 4$ and $0 \le \theta \le 2\pi$.
Then $\mathbf{r}(z, \theta) = (\cos\theta)\mathbf{i} + (\sin\theta)\mathbf{j} + z\mathbf{k}$
$\mathbf{r}_z = \mathbf{k}$ and $\mathbf{r}_\theta = (-\sin\theta)\mathbf{i} + (\cos\theta)\mathbf{j}$

$$\mathbf{r}_\theta \times \mathbf{r}_z = \begin{vmatrix} \mathbf{i} & \mathbf{j} & \mathbf{k} \\ -\sin\theta & \cos\theta & 0 \\ 0 & 0 & 1 \end{vmatrix} = (\cos\theta)\mathbf{i} + (\sin\theta)\mathbf{j}$$
$$= |\mathbf{r}_\theta \times \mathbf{r}_z| = \sqrt{\cos^2\theta + \sin^2\theta} = 1$$
$$A = \int_0^{2\pi}\int_1^4 1\,dr\,d\theta = \int_0^{2\pi} 3\,d\theta = 6\pi$$

23. $z = 2 - x^2 - y^2$ and $z = \sqrt{x^2 + y^2}$
$z = 2 - z^2$
$z^2 + z - 2 = 0$
$z = -2$ or $z = 1$. Since $z = \sqrt{x^2 + y^2} \ge 0$, we get $z = 1$ where the cone intersects the paraboloid. When $x = 0$ and $y = 0$, $z = 2$
The vertex of the paraboloid is $(0, 0, 2)$. Therefore, z ranges from 1 to 2 on the "cap"
r ranges from 1 (when $x^2 + y^2 = 1$) to 0 (when $x = 0$ and $y = 0$ at the vertex). Let $x = r\cos\theta$, $y = r\sin\theta$, and $z = 2 - r^2$. Then $\mathbf{r}(r, \theta) = (r\cos\theta)\mathbf{i} + (r\sin\theta)\mathbf{j}$
$+ (2 - r^2)\mathbf{k}$, $0 \le r \le 1$, $0 \le \theta \le 2\pi$
$\mathbf{r}_r = (\cos\theta)\mathbf{i} + (\sin\theta)\mathbf{j} - 2r\mathbf{k}$ and $\mathbf{r}_\theta = (-r\sin\theta)\mathbf{i}$
$+ (r\cos\theta)\mathbf{j}$

$$\mathbf{r}_r \times \mathbf{r}_\theta = \begin{vmatrix} \mathbf{i} & \mathbf{j} & \mathbf{k} \\ \cos\theta & \sin\theta & -2r \\ -r\sin\theta & r\cos\theta & 0 \end{vmatrix}$$

$(2r^2\cos\theta)\mathbf{i} + (2r^2\sin\theta)\mathbf{j} + r\mathbf{k}$

$$|\mathbf{r}_r \times \mathbf{r}_\theta| = \sqrt{4r^4\cos^2\theta + 4r^2\sin^4\theta + r^2} = r\sqrt{4r^2 + 1}$$

$$A = \int_0^{2\pi}\int_0^1 r\sqrt{4r^2 + 1}\,dr\,d\theta = \int_0^{2\pi}\left[\frac{1}{12}(4r^2 + 1)^{3/2}\right]_0^1 d\theta$$
$$= \int_0^{2\pi}\left(\frac{5\sqrt{5} - 1}{12}\right)d\theta = \frac{\pi}{6}(5\sqrt{5} - 1)$$

25. Let $x = \rho\sin\phi\cos\theta$, $y = \rho\sin\phi\sin\theta$, and $z = \rho\cos\phi$
$\rho = \sqrt{x^2 + y^2 + z^2} = \sqrt{2}$ on the sphere. Next,
$x^2 + y^2 + z^2 = 2$ and $z = \sqrt{x^2 + y^2}$
$z^2 + z^2 = 2$
$z^2 = 1$
$z = 1$ since $z \ge 0$
$\phi = \frac{\pi}{4}$. For the lower portion of the sphere cut by the cone, we get $\phi = \pi$. Then $\mathbf{r}(\phi, \theta) = (\sqrt{2}\sin\phi\cos\theta)\mathbf{i}$
$+ (\sqrt{2}\sin\phi\sin\theta)\mathbf{j} + (\sqrt{2}\cos\phi)\mathbf{k}, \frac{\pi}{4}, \le \phi \le \pi,$
$0 \le \theta \le 2\pi$
$\mathbf{r}_\phi = (\sqrt{2}\cos\phi\cos\theta)\mathbf{i} + (\sqrt{2}\cos\phi\sin\theta)\mathbf{j}$
$- (\sqrt{2}\sin\phi)\mathbf{k}$ and $\mathbf{r}_\theta = (-\sqrt{2}\sin\phi\sin\theta)\mathbf{i}$
$+ (\sqrt{2}\sin\phi\cos\theta)\mathbf{j}$

$$\mathbf{r}_r \times \mathbf{r}_\theta = \begin{vmatrix} \mathbf{i} & \mathbf{j} & \mathbf{k} \\ \sqrt{2}\cos\phi\cos\theta & \sqrt{2}\cos\phi\sin\theta & -\sqrt{2}\sin\phi \\ -\sqrt{2}\sin\phi\sin\theta & \sqrt{2}\sin\phi\cos\theta & 0 \end{vmatrix}$$

$= (2\sin^2\phi\cos\theta)\mathbf{i} + (2\sin^2\phi\sin\theta)\mathbf{j} + (2\sin\phi\cos\phi)\mathbf{k}$
$|\mathbf{r}_\phi \times \mathbf{r}_\theta|$

$= \sqrt{4\sin^4\phi\cos^2\theta + 4\sin^4\phi\sin^2\theta + 4\sin^2\phi\cos^2\phi}$

$= \sqrt{4\sin^2\phi} = 2\,|\sin\phi| = 2\sin\phi$

$A = \int_0^{2\pi}\int_{\pi/4}^{\pi} 2\sin\phi\, d\phi\, d\theta = \int_0^{2\pi} (2 + \sqrt{2})\, d\theta$

$= (4 + 2\sqrt{2})\pi$

27. Let the parameterization be $\mathbf{r}(x, z) = x\mathbf{i} + x^2\mathbf{j} + z\mathbf{k}$
$\mathbf{r}_x = \mathbf{i} + 2x\mathbf{j}$ and $\mathbf{r}_z = \mathbf{k}$

$$\mathbf{r}_x \times \mathbf{r}_z = \begin{vmatrix} \mathbf{i} & \mathbf{j} & \mathbf{k} \\ 1 & 2x & 0 \\ 0 & 0 & 1 \end{vmatrix} = 2x\mathbf{i} - \mathbf{j}$$

$|\mathbf{r}_x \times \mathbf{r}_z| = \sqrt{4x^2 + 1}$

$$\iint_S G(x, y, z)\, d\sigma = \int_0^3\int_0^2 x\sqrt{4x^2 + 1}\, dx\, dz$$

$$= \int_0^3 \left[\frac{1}{12}(4x^2 + 1)^{3/2}\right]_0^2 dz$$

$$= \int_0^3 \frac{1}{12}(17\sqrt{17} - 1)\, dz$$

$$= \frac{17\sqrt{17} - 1}{4}$$

29. Let the parametrization be $\mathbf{r}(\phi, \theta) = (\sin\phi\cos\theta)\mathbf{i} + (\sin\phi\sin\theta)\mathbf{j} + (\cos\phi)\mathbf{k}$ (spherical coordinates with $\rho = 1$ on the sphere), $0 \le \phi \le \pi$, $0 \le \theta \le 2\pi$
$\mathbf{r}_\phi = (\cos\phi\cos\theta)\mathbf{i} + (\cos\phi\sin\theta)\mathbf{j} - (\sin\phi)\mathbf{k}$ and
$\mathbf{r}_\theta = (-\sin\phi\sin\theta)\mathbf{i} + (\sin\phi\cos\theta)\mathbf{j}$

$$\mathbf{r}_\phi \times \mathbf{r}_\theta = \begin{vmatrix} \mathbf{i} & \mathbf{j} & \mathbf{k} \\ \cos\phi\cos\theta & \cos\phi\sin\theta & -\sin\phi \\ -\sin\phi\sin\theta & \sin\phi\cos\theta & 0 \end{vmatrix}$$

$= (\sin^2\phi\cos\theta)\mathbf{i} + (\sin^2\phi\sin\theta)\mathbf{j} + (\sin\phi\cos\phi)\mathbf{k}$

$|\mathbf{r}_\phi \times \mathbf{r}_\theta|$

$= \sin^4\phi\cos^2\theta + \sin^4\phi\sin^2\theta + \sin^2\phi\cos^2\phi$

$= \sin\phi$; $x = \sin\phi\cos\theta$

$G(x, y, z) = \cos^2\theta\sin^2\phi$

$$\iint_S G(x, y, z)\, d\sigma$$

$$= \int_0^{2\pi}\int_0^{\pi} (\cos^2\theta\sin^2\phi)(\sin\phi)\, d\phi\, d\theta$$

$$= \int_0^{2\pi}\int_0^{\pi} (\cos^2\theta)(1 - \cos^2\phi)(\sin\phi)\, d\phi\, d\theta;$$

$$\begin{bmatrix} u = \cos\phi \\ du = -\sin\phi\, d\phi \end{bmatrix} \to \int_0^{2\pi}\int_1^{-1} (\cos^2\theta)(u^2 - 1)\, du\, d\theta$$

$$= \int_0^{2\pi} (\cos^2\theta)\left[\frac{u^3}{3} - u\right]_1^{-1} d\theta = \frac{4}{3}\int_0^{2\pi} \cos^2\theta\, d\theta$$

$$= \frac{4}{3}\left[\frac{\theta}{2} + \frac{\sin 2\theta}{4}\right]_0^{2\pi} = \frac{4\pi}{3}$$

31. Let the parametrization be $\mathbf{r}(x, y) = x\mathbf{i} + y\mathbf{j} + (4 - x - y)\mathbf{k}$
$\mathbf{r}_x = \mathbf{i} - \mathbf{k}$ and $\mathbf{r}_y = \mathbf{j} - \mathbf{k}$

$$\mathbf{r}_x \times \mathbf{r}_y = \begin{vmatrix} \mathbf{i} & \mathbf{j} & \mathbf{k} \\ 1 & 0 & -1 \\ 0 & 1 & -1 \end{vmatrix} = \mathbf{i} + \mathbf{j} + \mathbf{k}$$

$|\mathbf{r}_x \times \mathbf{r}_y| = \sqrt{3}$

$$\iint_S F(x, y, z)\, d\sigma = \int_0^1\int_0^1 (4 - x - y)\sqrt{3}\, dy\, dx$$

$$= \int_0^1 \sqrt{3}\left[4y - xy - \frac{y^2}{2}\right]_0^1 dx$$

$$= \int_0^1 \sqrt{3}\left(\frac{7}{2} - x\right) dx$$

$$= \sqrt{3}\left[\frac{7}{2}x - \frac{x^2}{2}\right]_0^1 = 3\sqrt{3}$$

33. Let the parametrization be $\mathbf{r}(r, \theta) = (r\cos\theta)\mathbf{i} + (r\sin\theta)\mathbf{j} + (1 - r^2)\mathbf{k}$, $0 \le r \le 1$ (since $0 \le z \le 1$) and $0 \le \theta \le 2\pi$
$\mathbf{r}_r = (\cos\theta)\mathbf{i} + (\sin\theta)\mathbf{j} - 2r\mathbf{k}$ and
$\mathbf{r}_\theta = (-r\sin\theta)\mathbf{i} + (r\cos\theta)\mathbf{j}$

$$\mathbf{r}_r \times \mathbf{r}_\theta = \begin{vmatrix} \mathbf{i} & \mathbf{j} & \mathbf{k} \\ \cos\theta & \sin\theta & -2r \\ -r\sin\theta & r\cos\theta & 0 \end{vmatrix}$$

$(2r^2\cos\theta)\mathbf{i} + (2r^2\sin\theta)\mathbf{j} + r\mathbf{k}$

$|\mathbf{r}_r \times \mathbf{r}_\theta| = \sqrt{(2r^2\cos\theta)^2 + (2r^2\sin\theta)^2 + r^2}$

$= r\sqrt{1 + 4r^2}$; $z = 1 - r^2$ and $x = r\cos\theta$

$H(x, y, z) = (r^2\cos^2\theta)\sqrt{1 + 4r^2}$

$$\iint_S H(x, y, z)\, d\sigma$$

$$= \int_0^{2\pi}\int_0^1 (r^2\cos^2\theta)(\sqrt{1 + 4r^2})(r\sqrt{1 + 4r^2})\, dr\, d\theta$$

$$= \int_0^{2\pi}\int_0^1 r^3(1 + 4r^2)\cos^2\theta\, dr\, d\theta = \frac{11\pi}{12}$$

35. Let the parametrization be $\mathbf{r}(x, y) = x\mathbf{i} + y\mathbf{j} + (4 - y^2)\mathbf{k}$, $0 \le x \le 1$, $-2 \le y \le 2$; $\mathbf{r}_x = \mathbf{i}$ and $\mathbf{r}_y = \mathbf{j} - 2y\mathbf{k}$

$$\mathbf{r}_x \times \mathbf{r}_y = \begin{vmatrix} \mathbf{i} & \mathbf{j} & \mathbf{k} \\ 1 & 0 & 0 \\ 0 & 1 & -2y \end{vmatrix} = 2y\mathbf{j} + \mathbf{k}$$

$$\mathbf{F} \cdot \mathbf{n}\, d\sigma = \mathbf{F} \cdot \frac{\mathbf{r}_x \times \mathbf{r}_y}{|\mathbf{r}_x \times \mathbf{r}_y|}|\mathbf{r}_x \times \mathbf{r}_y|\, dy\, dx = (2xy - 3z)\, dy\, dx$$

$$= [2xy - 3(4 - y^2)]\, dy\, dx$$

$$\iint_S \mathbf{F} \cdot \mathbf{n}\, d\sigma = \int_0^1\int_{-2}^2 (2xy + 3y^2 - 12)\, dy\, dx$$

$$= \int_0^1 \left[xy^2 + y^3 - 12y\right]_{-2}^2 dx$$

$$= \int_0^1 -32\, dx = -32$$

37. Let the parametrization be $\mathbf{r}(\phi, \theta) = (a \sin\phi \cos\theta)\mathbf{i} + (a \sin\phi \sin\theta)\mathbf{j} + (a \cos\phi)\mathbf{k}$ (spherical coordinates with $\rho = a$, $a \ge 0$, on the sphere), $0 \le \theta \le \frac{\pi}{2}$, $0 \le \phi \le \frac{\pi}{2}$ (for the first octant)

$\mathbf{r}_\phi = (a \cos\phi \cos\theta)\mathbf{i} + (a \cos\phi \sin\theta)\mathbf{j} - (a \sin\phi)\mathbf{k}$ and $\mathbf{r}_\theta = (-a \sin\phi \sin\theta)\mathbf{i} + (a \sin\phi \cos\theta)\mathbf{j}$

$$\mathbf{r}_\phi \times \mathbf{r}_\theta = \begin{vmatrix} \mathbf{i} & \mathbf{j} & \mathbf{k} \\ a\cos\phi\cos\theta & a\cos\phi\sin\theta & -a\sin\phi \\ -a\sin\phi\sin\theta & a\sin\phi\cos\theta & 0 \end{vmatrix}$$

$= (a^2 \sin^2\phi \cos\theta)\mathbf{i} + (a^2 \sin^2\phi \sin\theta)\mathbf{j} + (a^2 \sin\phi \cos\phi)\mathbf{k}$

$$\mathbf{F} \cdot \mathbf{n}\, d\sigma = \mathbf{F} \cdot \frac{\mathbf{r}_\phi \times \mathbf{r}_\theta}{|\mathbf{r}_\phi \times \mathbf{r}_\theta|} |\mathbf{r}_\phi \times \mathbf{r}_\theta|\, d\theta\, d\phi$$

$= a^3 \cos^2\phi \sin\phi\, d\theta\, d\phi$ since

$\mathbf{F} = z\mathbf{k} = (a \cos\phi)\mathbf{k}$

$$\iint_S \mathbf{F} \cdot \mathbf{n}\, d\sigma = \int_0^{\pi/2} \int_0^{\pi/2} a^3 \cos^2\phi \sin\phi\, d\phi\, d\theta = \frac{\pi a^3}{6}$$

39. Let the parametrization be $\mathbf{r}(x, y) = x\mathbf{i} + y\mathbf{j} + (2a - x - y)\mathbf{k}$, $0 \le x \le a$, $0 \le y \le a$

$\mathbf{r}_x = \mathbf{i} - \mathbf{k}$ and $\mathbf{r}_y = \mathbf{j} - \mathbf{k}$

$$\mathbf{r}_x \times \mathbf{r}_y = \begin{vmatrix} \mathbf{i} & \mathbf{j} & \mathbf{k} \\ 1 & 0 & -1 \\ 0 & 1 & -1 \end{vmatrix} = \mathbf{i} + \mathbf{j} + \mathbf{k}$$

$$\mathbf{F} \cdot \mathbf{n}\, d\sigma = \mathbf{F} \cdot \frac{\mathbf{r}_x \times \mathbf{r}_y}{|\mathbf{r}_x \times \mathbf{r}_y|} |\mathbf{r}_x \times \mathbf{r}_y|\, dy\, dx =$$

$= [2xy + 2y(2a - x - y) + 2x(2a - x - y)]\, dy\, dx$

$=$ since $\mathbf{F} = 2xy\mathbf{i} + 2yz\mathbf{j} + 2xz\mathbf{k}$

$= 2xy\mathbf{i} + 2y(2a - x - y)\mathbf{j} + 2x(2a - x - y)\mathbf{k}$

$$\iint_S \mathbf{F} \cdot \mathbf{n}\, d\sigma$$

$$= \int_0^a \int_0^a [2xy + 2y(2a - x - y) + 2x(2a - x - y)]\, dy\, dx$$

$$= \int_0^a \int_0^a (4ay - 2y^2 + 4ax - 2x^2 - 2xy)\, dy\, dx$$

$$= \int_0^a \left(\frac{4}{3}a^3 + 3a^2x - 2ax^2\right) dx = \left(\frac{4}{3} + \frac{3}{2} - \frac{2}{3}\right)a^4 = \frac{13a^4}{6}$$

41. Let the parametrization be $\mathbf{r}(r, \theta) = (r \cos\theta)\mathbf{i} + (r \sin\theta)\mathbf{j} + r\mathbf{k}$, $0 \le r \le 1$ (since $0 \le z \le 1$) and $0 \le \theta \le 2\pi$

$\mathbf{r}_r = (\cos\theta)\mathbf{i} + (\sin\theta)\mathbf{j} + \mathbf{k}$ and $\mathbf{r}_\theta = (-r \sin\theta)\mathbf{i} + (r \cos\theta)\mathbf{j}$

$$\mathbf{r}_\theta \times \mathbf{r}_r = \begin{vmatrix} \mathbf{i} & \mathbf{j} & \mathbf{k} \\ -r\sin\theta & r\cos\theta & 0 \\ \cos\theta & \sin\theta & 1 \end{vmatrix}$$

$= (r \cos\theta)\mathbf{i} + (r \sin\theta)\mathbf{j} - r\mathbf{k}$

$$\mathbf{F} \cdot \mathbf{n}\, d\sigma = \mathbf{F} \cdot \frac{\mathbf{r}_\theta \times \mathbf{r}_r}{|\mathbf{r}_\theta \times \mathbf{r}_r|} |\mathbf{r}_\theta \times \mathbf{r}_r|\, d\theta\, dr$$

$= (r^3 \sin\theta \cos^2\theta + r^2)\, d\theta\, dr$ since $\mathbf{F} = (r^2 \sin\theta \cos\theta)\,\mathbf{i} - r\mathbf{k}$

$$\iint_S \mathbf{F} \cdot \mathbf{n}\, d\sigma = \int_0^{2\pi} \int_0^1 (r^3 \sin\theta \cos^2\theta + r^2)\, dr\, d\theta$$

$$= \int_0^{2\pi} \left(\frac{1}{4} \sin\theta \cos^2\theta + \frac{1}{3}\right) d\theta$$

$$= \left[-\frac{1}{12}\cos^3\theta + \frac{\theta}{3}\right]_0^{2\pi}$$

$$= \frac{2\pi}{3}$$

43. Let the parametrization be $\mathbf{r}(r, \theta) = (r \cos\theta)\mathbf{i} + (r \sin\theta)\mathbf{j} + r\mathbf{k}$, $1 \le r \le 2$ (since $1 \le z \le 2$) and $0 \le \theta \le 2\pi$

$\mathbf{r}_r = (\cos\theta)\mathbf{i} + (\sin\theta)\mathbf{j} + \mathbf{k}$ and $\mathbf{r}_\theta = (-r \sin\theta)\mathbf{i} + (r \cos\theta)\mathbf{j}$

$$\mathbf{r}_\theta \times \mathbf{r}_r = \begin{vmatrix} \mathbf{i} & \mathbf{j} & \mathbf{k} \\ -r\sin\theta & r\cos\theta & 0 \\ \cos\theta & \sin\theta & 1 \end{vmatrix}$$

$= (r \cos\theta)\mathbf{i} + (r \sin\theta)\mathbf{j} - r\mathbf{k}$

$$\mathbf{F} \cdot \mathbf{n}\, d\sigma = \mathbf{F} \cdot \frac{\mathbf{r}_\theta \times \mathbf{r}_r}{|\mathbf{r}_\theta \times \mathbf{r}_r|} |\mathbf{r}_\theta \times \mathbf{r}_r|\, d\theta\, dr$$

$= (-r^2 \cos^2\theta - r^2 \sin^2\theta - r^3)\, d\theta\, dr$

$= (-r^2 - r^3)\, d\theta\, dr$ since

$\mathbf{F} = (-r \cos\theta)\mathbf{i} - (r \sin\theta)\mathbf{j} + r^2\mathbf{k}$,

$$\iint_S \mathbf{F} \cdot \mathbf{n}\, d\sigma = \int_0^{2\pi} \int_1^2 (-r^2 - r^3)\, dr\, d\theta = -\frac{73\pi}{6}$$

45. Let the parametrization be $\mathbf{r}(\phi, \theta) = (a \sin\phi \cos\theta)\mathbf{i} + (a \sin\phi \sin\theta)\mathbf{j} + (a \cos\phi)\mathbf{k}$, $0 \le \phi \le \frac{\pi}{2}$, $0 \le \theta \le \frac{\pi}{2}$

$\mathbf{r}_\phi = (a \cos\phi \cos\theta)\mathbf{i} + (a \cos\phi \sin\theta)\mathbf{j} - (a \sin\phi)\mathbf{k}$ and $\mathbf{r}_\theta = (-a \sin\phi \sin\theta)\mathbf{i} + (a \sin\phi \cos\theta)\mathbf{j}$

$$\mathbf{r}_\phi \times \mathbf{r}_\theta = \begin{vmatrix} \mathbf{i} & \mathbf{j} & \mathbf{k} \\ a\cos\phi\cos\theta & a\cos\phi\sin\theta & -a\sin\phi \\ -a\sin\phi\sin\theta & a\sin\phi\cos\theta & 0 \end{vmatrix}$$

$= (a^2 \sin^2\phi \cos\theta)\mathbf{i} + (a^2 \sin^2\phi \sin\theta)\mathbf{j} + (a^2 \sin\phi \cos\phi)\mathbf{k}$

$|\mathbf{r}_\phi \times \mathbf{r}_\theta|$

$= \sqrt{a^4 \sin^4\phi \cos^2\theta + a^4 \sin^4\phi \sin^2\theta + a^4 \sin^2\phi \cos^2\phi}$

$= \sqrt{a^4 \sin^2\phi} = a^2 \sin\phi.$

The mass is $M = \iint_S d\sigma = \int_0^{\pi/2} \int_0^{\pi/2} (a^2 \sin\phi)\, d\phi\, d\theta$

$= \frac{a^2\pi}{2}$; the first moment is $M_{yz} = \iint_S x\, d\sigma$

$$= \int_0^{\pi/2} \int_0^{\pi/2} (a \sin\phi \cos\theta)(a^2 \sin\phi)\, d\phi\, d\theta = \frac{a^3\pi}{4}$$

$$\bar{x} = \frac{\left(\frac{a^3\pi}{4}\right)}{\left(\frac{a^2\pi}{2}\right)} = \frac{a}{2}$$

The centroid is located at $\left(\frac{a}{2}, \frac{a}{2}, \frac{a}{2}\right)$ by symmetry.

47. Let the parametrization be $\mathbf{r}(\phi, \theta) = (a \sin\phi \cos\theta)\mathbf{i} + (a \sin\phi \sin\theta)\mathbf{j} + (a \cos\phi)\mathbf{k}$, $0 \le \phi \le \pi$, $0 \le \theta \le 2\pi$

$\mathbf{r}_\phi = (a \cos\phi \cos\theta)\mathbf{i} + (a \cos\phi \sin\theta)\mathbf{j} - (a \sin\phi)\mathbf{k}$ and $\mathbf{r}_\theta = (-a \sin\phi \sin\theta)\mathbf{i} + (a \sin\phi \cos\theta)\mathbf{j}$

$$\mathbf{r}_\phi \times \mathbf{r}_\theta = \begin{vmatrix} \mathbf{i} & \mathbf{j} & \mathbf{k} \\ a\cos\phi\cos\theta & a\cos\phi\sin\theta & -a\sin\phi \\ -a\sin\phi\sin\theta & a\sin\phi\cos\theta & 0 \end{vmatrix}$$

$= (a^2 \sin^2\phi \cos\theta)\mathbf{i} + (a^2 \sin^2\phi \sin\theta)\mathbf{j} + (a^2 \sin\phi \cos\phi)\mathbf{k}$

$|\mathbf{r}_\phi \times \mathbf{r}_\theta|$
$= \sqrt{a^4 \sin^4 \phi \cos^2 \theta + a^4 \sin^4 \phi \sin^2 \theta + a^4 \sin^2 \phi \cos^2 \phi}$
$= \sqrt{a^4 \sin^2 \phi} = a^2 \sin \phi$. The moment of inertia is

$$I_z = \iint_S \delta(x^2 + y^2) d\sigma$$
$$= \int_0^{2\pi} \int_0^{\pi} \delta(a^2 \sin^2 \phi)(a^2 \sin \phi)\, d\phi\, d\theta$$
$$= \int_0^{2\pi} \int_0^{\pi} \delta a^4 \sin^3 \phi\, d\phi\, d\theta$$
$$= \int_0^{2\pi} \delta a^4 \left[\left(-\frac{1}{3}\cos \phi\right)(\sin^2 \phi + 2)\right]_0^{\pi} d\theta = \frac{8\delta\pi a^4}{3}$$

49.

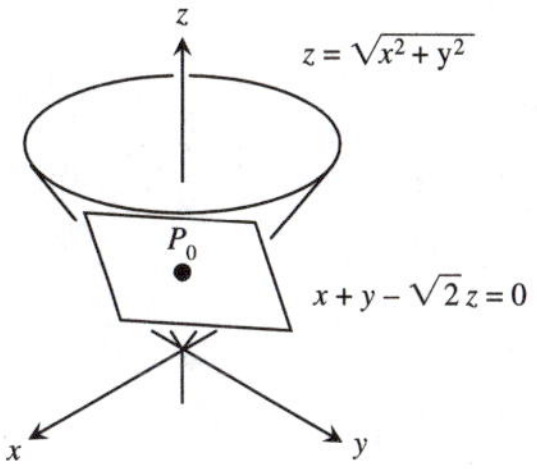

The parametrization $\mathbf{r}(r, \theta) = (r \cos \theta)\mathbf{i} + (r \sin \theta)\mathbf{j} + r\mathbf{k}$ at $P_0 = (\sqrt{2}, \sqrt{2}, 2)$

$\theta = \frac{\pi}{4}$, $r = 2$, $\mathbf{r}_r = (\cos \theta)\mathbf{i} + (\sin \theta)\mathbf{j} + \mathbf{k}$

$= \frac{\sqrt{2}}{2}\mathbf{i} + \frac{\sqrt{2}}{2}\mathbf{j} + \mathbf{k}$ and $\mathbf{r}_\theta = (-r \sin \theta)\mathbf{i} + (r \cos \theta)\mathbf{j}$

$= -\sqrt{2}\mathbf{i} + \sqrt{2}\mathbf{j}$

$$\mathbf{r}_r \times \mathbf{r}_\theta = \begin{vmatrix} \mathbf{i} & \mathbf{j} & \mathbf{k} \\ \frac{\sqrt{2}}{2} & \frac{\sqrt{2}}{2} & 1 \\ -\sqrt{2} & \sqrt{2} & 0 \end{vmatrix} = -\sqrt{2}\mathbf{i} - \sqrt{2}\mathbf{j} + 2\mathbf{k}$$

The tangent plane is $(-\sqrt{2}\mathbf{i} - \sqrt{2}\mathbf{j} + 2\mathbf{k})$
$\cdot [(x - \sqrt{2})\mathbf{i} + (y - \sqrt{2})\mathbf{j} + (z - 2)\mathbf{k}] = 0$
$\sqrt{2}x + \sqrt{2}y - 2z = 0$, or $x + y - \sqrt{2}z = 0$. The parametrization $\mathbf{r}(r, \theta)$
$x = r \cos \theta$, $y = r \sin \theta$ and $z = r$
$x^2 + y^2 = r^2 = z^2$
The surface is $z = \sqrt{x^2 + y^2}$.

51.

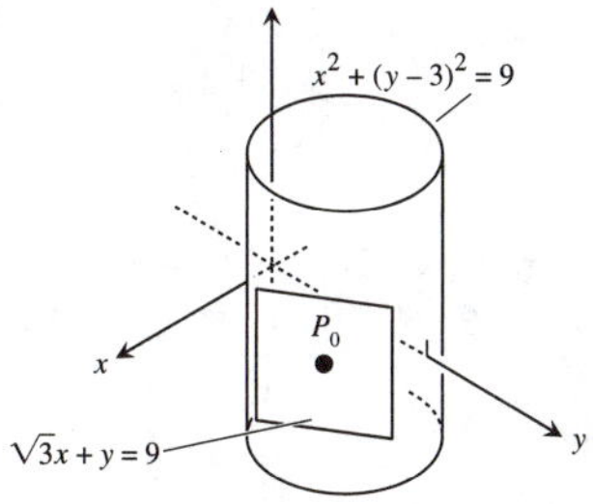

The parametrization $\mathbf{r}(\theta, z) = (3 \sin 2\theta)\mathbf{i} + (6 \sin^2 \theta)\mathbf{j} + z\mathbf{k}$

at $P_0 = \left(\frac{3\sqrt{3}}{2}, \frac{9}{2}, 0\right)$

$\theta = \frac{\pi}{3}$ and $z = 0$. Then

$\mathbf{r}_\theta = (6 \cos 2\theta)\mathbf{i} + (12 \sin \theta \cos \theta)\mathbf{j} = -3\mathbf{i} + 3\sqrt{3}\mathbf{j}$ and $\mathbf{r}_z = \mathbf{k}$ at P_0

$$\mathbf{r}_\theta \times \mathbf{r}_z = \begin{vmatrix} \mathbf{i} & \mathbf{j} & \mathbf{k} \\ -3 & 3\sqrt{3} & 0 \\ 0 & 0 & 1 \end{vmatrix} = 3\sqrt{3}\mathbf{i} + 3\mathbf{j}$$

the tangent plane is $(3\sqrt{3}\mathbf{i} + 3\mathbf{j})$

$$\cdot\left[\left(x - \frac{3\sqrt{3}}{2}\right)\mathbf{i} + \left(y - \frac{9}{2}\right)\mathbf{j} + (z - 0)\mathbf{k}\right] = 0$$

$\sqrt{3}x + y = 9$. The parametrization
$x = 3 \sin 2\theta$ and $y = 6 \sin^2 \theta$
$x^2 + y^2 = 9 \sin^2 2\theta + (6 \sin^2 \theta)^2 = 9(4 \sin^2 \theta \cos^2 \theta)$
$+ 36 \sin^4 \theta = 6(6 \sin^2 \theta) = 6y$
$x^2 + y^2 - 6y + 9 = 9$
$x^2 + (y - 3)^2 = 9$

53. a. An arbitrary point on the circle C is (x, z)
$= (R + r \cos u, r \sin u)$
(x, y, z) is on the torus with $x = (R + r \cos u) \cos v$,
$y = (R + r \cos u) \sin v$, and $z = r \sin u$, $0 \le u \le 2\pi$,
$0 \le v \le 2\pi$

b. $\mathbf{r}_u = (-r \sin u \cos v)\mathbf{i} - (r \sin u \sin v)\mathbf{j} + (r \cos u)\mathbf{k}$
and $\mathbf{r}_v = (-(R + r \cos u) \sin v)\mathbf{i}$
$+ ((R + r \cos u) \cos v)\mathbf{j}$

$\mathbf{r}_u \times \mathbf{r}_v$

$$= \begin{vmatrix} \mathbf{i} & \mathbf{j} & \mathbf{k} \\ -r \sin u \cos v & -r \sin u \sin v & r \cos u \\ -(R + r \cos u) \sin v & (R + r \cos u) \cos v & 0 \end{vmatrix}$$

$= -(R + r \cos u)(r \cos v \cos u)\mathbf{i}$
$- (R + r \cos u)(r \sin v \cos u)\mathbf{j}$
$+ (-r \sin u)(R + r \cos u)\mathbf{k}$

$|\mathbf{r}_u \times \mathbf{r}_v|^2$
$= (R + r \cos u)^2(r^2 \cos^2 v \cos^2 u + r^2 \sin^2 v \cos^2 u$
$+ r^2 \sin^2 u)$

$|\mathbf{r}_u \times \mathbf{r}_v| = r(R + r \cos u)$

$$A = \int_0^{2\pi} \int_0^{2\pi} (rR + r^2 \cos u)\, du\, dv = \int_0^{2\pi} 2\pi rR\, dv$$
$= 4\pi^2 rR$

■ Section 15.7 Stokes's Theorem
(pp. 886-896)

Quick Review 15.7

1. $\dfrac{e^{-xyz}}{x + y + z} - yze^{-xyz} \ln(x + y + z)$

2. $\dfrac{e^{-xyz}}{x + y + z} - xye^{-xyz} \ln(x + y + z)$

3. $(xyz^2 - z)e^{-xyz} \ln(x + y + z)$
$- \dfrac{(y^2z + 2xyz + yz^2 + x^2z + xz^2 + 1)e^{-xyz}}{(x + y + z)^2}$

4. $(xyz^2 - z)e^{-xyz}\ln(x+y+z) - \dfrac{(y^2z + 2xyz + yz^2 + x^2z + xz^2 + 1)e^{-xyz}}{(x+y+z)^2}$

5. $(x^2yz - x)e^{-xyz}\ln(x+y+z) - \dfrac{(xz^2 + x^2z + 2xyz + x^2y + xy^2 + 1)e^{-xyz}}{(x+y+z)^2}$

6. $(x^2yz - x)e^{-xyz}\ln(x+y+z) - \dfrac{(xz^2 + x^2z + 2xyz + x^2y + xy^2 + 1)e^{-xyz}}{(x+y+z)^2}$

7. $(cy + bz)\mathbf{i} - (cx + az)\mathbf{j} + (bx - ay)\mathbf{k}$

8. $-(cy + bz)\mathbf{i} + (cx + az)\mathbf{j} - (bx - ay)\mathbf{k}$

9. $\nabla f = (\sin(y+z))\mathbf{i} + (x\cos(y+z))\mathbf{j} + (x\cos(y+z))\mathbf{k}$

10. $\dfrac{-1 \pm \sqrt{5}}{2}$

Section 15.7 Exercises

1. $\text{curl}\,\mathbf{F} = \nabla \times \mathbf{F} = \begin{vmatrix} \mathbf{i} & \mathbf{j} & \mathbf{k} \\ \frac{\partial}{\partial x} & \frac{\partial}{\partial y} & \frac{\partial}{\partial z} \\ x^2 & 2x & z^2 \end{vmatrix} = 0\mathbf{i} + 0\mathbf{j} + (2-0)\mathbf{k}$

$= 2\mathbf{k}$ and $\mathbf{n} = \mathbf{k}$

$\text{curl}\,\mathbf{F}\cdot\mathbf{n} = 2$

$d\sigma = dx\,dy$

$\oint_C \mathbf{F}\cdot d\mathbf{r} = \iint_R 2\,dA = 2(\text{Area of the ellipse}) = 4\pi$

3. $\text{curl}\,\mathbf{F} = \nabla \times \mathbf{F} = \begin{vmatrix} \mathbf{i} & \mathbf{j} & \mathbf{k} \\ \frac{\partial}{\partial x} & \frac{\partial}{\partial y} & \frac{\partial}{\partial z} \\ y & xz & x^2 \end{vmatrix} = -x\mathbf{i} - 2x\mathbf{j}$

$+ (z-1)\mathbf{k}$ and $\mathbf{n} = \dfrac{\mathbf{i}+\mathbf{j}+\mathbf{k}}{\sqrt{3}}$

$\text{curl}\,\mathbf{F}\cdot\mathbf{n} = \dfrac{1}{\sqrt{3}}(-3x + z - 1)$

$d\sigma = \dfrac{\sqrt{3}}{1}\,dA$

$\oint_C \mathbf{F}\cdot d\mathbf{r} = \iint_R \dfrac{1}{\sqrt{3}}(-3x + z - 1)\sqrt{3}\,dA$

$= \int_0^1\int_0^{1-x} [-3x + (1 - x - y) - 1]\,dy\,dx$

$= \int_0^1\int_0^{1-x} (-4x - y)\,dy\,dx$

$= \int_0^1 -\left[4x(1-x) + \frac{1}{2}(1-x)^2\right]dx$

$= -\int_0^1 \left(\frac{1}{2} + 3x - \frac{7}{2}x^2\right)dx = -\frac{5}{6}$

5. $\text{curl}\,\mathbf{F} = \nabla \times \mathbf{F} = \begin{vmatrix} \mathbf{i} & \mathbf{j} & \mathbf{k} \\ \frac{\partial}{\partial x} & \frac{\partial}{\partial y} & \frac{\partial}{\partial z} \\ y^2+z^2 & x^2+y^2 & x^2+y^2 \end{vmatrix}$

$= 2y\mathbf{i} + (2z - 2x)\mathbf{j} + (2x - 2y)\mathbf{k}$ and $\mathbf{n} = \mathbf{k}$

$\text{curl}\,\mathbf{F}\cdot\mathbf{n} = 2x - 2y$

$d\sigma = dx\,dy$

$\oint_C \mathbf{F}\cdot d\mathbf{r} = \int_{-1}^1\int_{-1}^1 (2x - 2y)\,dx\,dy$

$= \int_{-1}^1 \left[x^2 - 2xy\right]_{-1}^1 dy = \int_{-1}^1 -4y\,dy = 0$

7. $x = 3\cos t$ and $y = 2\sin t$

$\mathbf{F} = (2\sin t)\mathbf{i} + (9\cos^2 t)\mathbf{j}$

$+ (9\cos^2 t + 16\sin^4 t)^{3/2}\sin e^{\sqrt{(6\sin t\cos t)(0)}}\,\mathbf{k}$ at the base of the shell; $\mathbf{r} = (3\cos t)\mathbf{i} + (2\sin t)\mathbf{j}$

$\dfrac{d\mathbf{r}}{dt} = (-3\sin t)\mathbf{i} + (2\cos t)\mathbf{j}$

$\mathbf{F}\cdot\dfrac{d\mathbf{r}}{dt} = -6\sin^2 t + 18\cos^3 t$

$\iint_S \nabla\times\mathbf{F}\cdot\mathbf{n}\,d\sigma = \int_0^{2\pi}(-6\sin^2 t + 18\cos^3 t)\,dt$

$= \left[-3t + \frac{3}{2}\sin 2t + 6(\sin t)(\cos^2 t + 2)\right]_0^{2\pi} = -6\pi$

9. Flux of $\nabla\times\mathbf{F} = \iint_S \nabla\times\mathbf{F}\cdot\mathbf{n}\,d\sigma = \oint_C \mathbf{F}\cdot d\mathbf{r}$, so let C be parametrized by $\mathbf{r} = (a\cos t)\mathbf{i} + (a\sin t)\mathbf{j}$, $0 \le t \le 2\pi$

$\dfrac{d\mathbf{r}}{dt} = (-a\sin t)\mathbf{i} + (a\cos t)\mathbf{j}$

$\mathbf{F}\cdot\dfrac{d\mathbf{r}}{dt} = ay\sin t + ax\cos t = a^2\sin^2 t + a^2\cos^2 t = a^2$

Flux of $\nabla\times\mathbf{F} = \oint_C \mathbf{F}\cdot d\mathbf{r} = \int_0^{2\pi} a^2\,dt = 2\pi a^2$

11. Let S_1 and S_2 be oriented surfaces that span C and that induce the same positive direction on C. Then

$\iint_{S_1} \nabla\times\mathbf{F}\cdot\mathbf{n}_1\,d\sigma_1 = \oint_C \mathbf{F}\cdot d\mathbf{r} = \iint_{S_2} \nabla\times\mathbf{F}\cdot\mathbf{n}_2\,d\sigma_2$

13. $\nabla\times\mathbf{F} = \begin{vmatrix} \mathbf{i} & \mathbf{j} & \mathbf{k} \\ \frac{\partial}{\partial x} & \frac{\partial}{\partial y} & \frac{\partial}{\partial z} \\ 2z & 3x & 5y \end{vmatrix} = 5\mathbf{i} + 2\mathbf{j} + 3\mathbf{k}$;

$\mathbf{r}_r = (\cos\theta)\mathbf{i} + (\sin\theta)\mathbf{j} - 2r\mathbf{k}$ and

$\mathbf{r}_\theta = (-r\sin\theta)\mathbf{i} + (r\cos\theta)\mathbf{j}$

$\mathbf{r}_r\times\mathbf{r}_\theta = \begin{vmatrix} \mathbf{i} & \mathbf{j} & \mathbf{k} \\ \cos\theta & \sin\theta & -2r \\ -r\sin\theta & r\cos\theta & 0 \end{vmatrix}$

$= (2r^2\cos\theta)\mathbf{i} + (2r^2\sin\theta)\mathbf{j} + r\mathbf{k}$; $\mathbf{n} = \dfrac{\mathbf{r}_r\times\mathbf{r}_\theta}{|\mathbf{r}_r\times\mathbf{r}_\theta|}$ and $d\sigma$

$= |\mathbf{r}_r\times\mathbf{r}_\theta|\,dr\,d\theta$

$\nabla\times\mathbf{F}\cdot\mathbf{n}\,d\sigma = (\nabla\times\mathbf{F})\cdot(\mathbf{r}_r\times\mathbf{r}_\theta)\,dr\,d\theta$

$= (10r^2\cos\theta + 4r^2\sin\theta + 3r)\,dr\,d\theta$

$$\iint_S \nabla \times \mathbf{F} \cdot \mathbf{n}\, d\sigma$$

$$= \int_0^{2\pi} \int_0^2 (10r^2 \cos\theta + 4r^2 \sin\theta + 3r)\, dr\, d\theta$$

$$= \int_0^{2\pi} \left[\frac{10}{3} r^3 \cos\theta + \frac{4}{3} r^3 \sin\theta + \frac{3}{2} r^2\right]_0^2 d\theta$$

$$= \int_0^{2\pi} \left(\frac{80}{3}\cos\theta + \frac{32}{3}\sin\theta + 6\right) d\theta = 6(2\pi) = 12\pi$$

15. $$\nabla \times \mathbf{F} = \begin{vmatrix} \mathbf{i} & \mathbf{j} & \mathbf{k} \\ \frac{\partial}{\partial x} & \frac{\partial}{\partial y} & \frac{\partial}{\partial z} \\ x^2 y & 2y^3 z & 3z \end{vmatrix} = -2y^3\mathbf{i} + 0\mathbf{j} - x^2\mathbf{k}$$

$$\mathbf{r}_r \times \mathbf{r}_\theta = \begin{vmatrix} \mathbf{i} & \mathbf{j} & \mathbf{k} \\ \cos\theta & \sin\theta & 1 \\ -r\sin\theta & r\cos\theta & 0 \end{vmatrix}$$

$= (-r\cos\theta)\mathbf{i} - (r\sin\theta)\mathbf{j} + r\mathbf{k}$ and $\nabla \times \mathbf{F} \cdot \mathbf{n}\, d\sigma$
$= (\nabla \times \mathbf{F}) \cdot (\mathbf{r}_r \times \mathbf{r}_\theta)\, dr\, d\theta$ (see Exercise 13 above)

$$\iint_S \nabla \times \mathbf{F} \cdot \mathbf{n}\, d\sigma$$

$$= \iint_R (2ry^3 \cos\theta - rx^2)\, dr\, d\theta$$

$$= \int_0^{2\pi} \int_0^1 (2r^4 \sin^3\theta \cos\theta - r^3 \cos^2\theta)\, dr\, d\theta$$

$$= \int_0^{2\pi} \left(\frac{2}{5}\sin^3\theta\cos\theta - \frac{1}{4}\cos^2\theta\right) d\theta$$

$$= \left[\frac{1}{10}\sin^4\theta - \frac{1}{4}\left(\frac{\theta}{2} + \frac{\sin 2\theta}{4}\right)\right]_0^{2\pi} = -\frac{\pi}{4}$$

17. $$\nabla \times \mathbf{F} = \begin{vmatrix} \mathbf{i} & \mathbf{j} & \mathbf{k} \\ \frac{\partial}{\partial x} & \frac{\partial}{\partial y} & \frac{\partial}{\partial z} \\ 3y & 5 - 2x & z^2 - 2 \end{vmatrix} = 0\mathbf{i} + 0\mathbf{j} - 5\mathbf{k};$$

$$\mathbf{r}_\phi \times \mathbf{r}_\theta = \begin{vmatrix} \mathbf{i} & \mathbf{j} & \mathbf{k} \\ \sqrt{3}\cos\phi\cos\theta & \sqrt{3}\cos\phi\sin\theta & -\sqrt{3}\sin\phi \\ -\sqrt{3}\sin\phi\sin\theta & \sqrt{3}\sin\phi\cos\theta & 0 \end{vmatrix}$$

$= (3\sin^2\phi\cos\theta)\mathbf{i} + (3\sin^2\phi\sin\theta)\mathbf{j} + (3\sin\phi\cos\phi)\mathbf{k};$
$\nabla \times \mathbf{F} \cdot \mathbf{n} = d\sigma = (\nabla \times \mathbf{F}) \cdot (\mathbf{r}_\phi \times \mathbf{r}_\theta)\, d\phi\, d\theta$ (see Exercise 13 above)

$$\iint_S \nabla \times \mathbf{F} \cdot \mathbf{n}\, d\sigma = \int_0^{2\pi} \int_0^{\pi/2} -15\cos\phi\sin\phi\, d\phi\, d\theta$$

$$= \int_0^{2\pi} \left[\frac{15}{2}\cos^2\phi\right]_0^{\pi/2} d\theta$$

$$= \int_0^{2\pi} -\frac{15}{2}\, d\theta = -15\pi$$

19. a. $\mathbf{F} = 2x\mathbf{i} + 2y\mathbf{j} + 2z\mathbf{k}$
curl $\mathbf{F} = 0$

$$\oint_C \mathbf{F} \cdot d\mathbf{r} = \iint_S \nabla \times \mathbf{F} \cdot \mathbf{n}\, d\sigma = \iint_S 0\, d\sigma = 0$$

b. Let $f(x, y, z) = xy^2z^3$
$\nabla \times \mathbf{F} = \nabla \times \nabla f = 0$
curl $\mathbf{F} = 0$

$$\oint_C \mathbf{F} \cdot d\mathbf{r} = \iint_S \nabla \times \mathbf{F} \cdot \mathbf{n}\, d\sigma = \iint_S 0\, d\sigma = 0$$

c. $\mathbf{F} = \nabla \times (x\mathbf{i} + y\mathbf{j} + z\mathbf{k}) = \mathbf{0}$
$\nabla \times \mathbf{F} = 0$

$$\oint_C \mathbf{F} \cdot d\mathbf{r} = \iint_S \nabla \times \mathbf{F} \cdot \mathbf{n}\, d\sigma = \iint_S 0\, d\sigma = 0$$

d. $\mathbf{F} = \nabla f$
$\nabla \times \mathbf{F} = \nabla \times \nabla f = 0$

$$\oint_C \mathbf{F} \cdot d\mathbf{r} = \iint_S \nabla \times \mathbf{F} \cdot \mathbf{n}\, d\sigma = \iint_S 0\, d\sigma = 0$$

21. Let $\mathbf{F} = 2y\mathbf{i} + 3z\mathbf{j} - x\mathbf{k}$

$$\nabla \times \mathbf{F} = \begin{vmatrix} \mathbf{i} & \mathbf{j} & \mathbf{k} \\ \frac{\partial}{\partial x} & \frac{\partial}{\partial y} & \frac{\partial}{\partial z} \\ 2y & 3z & -x \end{vmatrix} = -3\mathbf{i} + \mathbf{j} - 2\mathbf{k};$$

$$\mathbf{n} = \frac{2\mathbf{i} + 2\mathbf{j} + \mathbf{k}}{3}$$

$\nabla \times \mathbf{F} \cdot \mathbf{n} = -2$

$$\oint_C 2y\, dx + 3z\, dy - x\, dz = \oint_C \mathbf{F} \cdot d\mathbf{r}$$

$$= \iint_S \nabla \times \mathbf{F} \cdot \mathbf{n}\, d\sigma = \iint_S -2\, d\sigma = -2\iint_S d\sigma,$$

where $\iint_S d\sigma =$ is the area of the region enclosed by C on the plane S: $2x + 2y + z = 2$.

23. Suppose $\mathbf{F} = M\mathbf{i} + N\mathbf{j} + P\mathbf{k}$ exists such that $\nabla \times \mathbf{F}$
$= \left(\frac{\partial P}{\partial y} - \frac{\partial N}{\partial z}\right)\mathbf{i} + \left(\frac{\partial M}{\partial z} - \frac{\partial P}{\partial x}\right)\mathbf{j} + \left(\frac{\partial N}{\partial x} - \frac{\partial M}{\partial y}\right)\mathbf{k}$
$= x\mathbf{i} + y\mathbf{j} + z\mathbf{k}$. Then $\frac{\partial}{\partial x}\left(\frac{\partial P}{\partial y} - \frac{\partial N}{\partial z}\right) = \frac{\partial}{\partial x}(x)$
$= \frac{\partial^2 P}{\partial x \partial y} - \frac{\partial^2 N}{\partial x \partial z} = 1$. Likewise, $\frac{\partial}{\partial y}\left(\frac{\partial M}{\partial z} - \frac{\partial P}{\partial x}\right) = \frac{\partial}{\partial y}(y)$
$= \frac{\partial^2 M}{\partial y \partial z} - \frac{\partial^2 P}{\partial y \partial x} = 1$ and $\frac{\partial}{\partial z}\left(\frac{\partial N}{\partial x} - \frac{\partial M}{\partial y}\right)$
$= \frac{\partial}{\partial z}(z) = \frac{\partial^2 N}{\partial z \partial x} - \frac{\partial^2 M}{\partial z \partial y} = 1$. Summing the calculated equations $\left(\frac{\partial^2 P}{\partial x \partial y} - \frac{\partial^2 P}{\partial y \partial x}\right) + \left(\frac{\partial^2 N}{\partial z \partial x} - \frac{\partial^2 N}{\partial x \partial z}\right) + \left(\frac{\partial^2 M}{\partial y \partial z} - \frac{\partial^2 M}{\partial z \partial y}\right)$
$= 3$ or $0 = 3$ (assuming the second mixed partials are equal). This result is a contradiction, so there is no field F such that curl $\mathbf{F} = x\mathbf{i} + y\mathbf{j} + z\mathbf{k}$.

25. $r = \sqrt{x^2 + y^2}$

$r^4 = (x^2 + y^2)^2$

$\mathbf{F} = \nabla(r^4) = 4x(x^2 + y^2)\mathbf{i} + 4y(x^2 + y^2)\mathbf{j} = M\mathbf{i} + N\mathbf{j}$

$$\oint_C \nabla(r^4) \cdot \mathbf{n}\, ds = \oint_C \mathbf{F} \cdot \mathbf{n}\, ds = \oint_C M\, dy - N\, dx$$

$$= \oint_C (-N\mathbf{i} + M\mathbf{j}) \cdot d\mathbf{r}$$

$$= \iint_R \left(\frac{\partial M}{\partial x} + \frac{\partial N}{\partial y}\right) dx\, dy$$

$$= \iint_R [4(x^2 + y^2) + 8x^2 + 4(x^2 + y^2) + 8y^2]\, dA$$

$$= \iint_R 16(x^2 + y^2)\, dA$$

$$= 16 \iint_R x^2\, dA + 16 \iint_R y^2\, dA = 16\, I_y + 16\, I_x$$

■ Section 15.8 The Divergence Theorem and a Unified Theory (pp. 896-907)

Quick Review 15.8

1. $\frac{4}{3}$

2. $\frac{2}{3}$

3. $\frac{2}{3}$

4. $\frac{2}{3}$

5. $\int_0^1 \int_0^1 \int_{\sqrt{z}}^1 dy\, dz\, dx$

6. $\int_0^1 \int_0^1 \int_{\sqrt{z}}^1 dy\, dx\, dz$

7. $\int_0^1 \int_{\sqrt{z}}^1 \int_0^1 dx\, dy\, dz$

8. $\int_0^1 \int_0^{y^2} \int_0^1 dx\, dz\, dy$

9. $\cos^{-1}\left(\frac{1}{\sqrt{3}}\right) \approx 54.74°$

10. $\cos^{-1}\left(\frac{1}{\sqrt{3}}\right) \approx 54.74°$

Section 15.8 Exercises

1. $\mathbf{F} = \dfrac{-y\mathbf{i} + x\mathbf{j}}{\sqrt{x^2 + y^2}}$

$$\text{div F} = \frac{xy - xy}{(x^2 + y^2)^{3/2}} = 0$$

3. $\mathbf{F} = -\dfrac{GM(x\mathbf{i} + y\mathbf{j} + z\mathbf{k})}{(x^2 + y^2 + z^2)^{3/2}}$

$$\text{div } \mathbf{F} = -GM\left[\frac{(x^2 + y^2 + z^2)^{3/2} - 3x^2(x^2 + y^2 + z^2)^{1/2}}{(x^2 + y^2 + z^2)^3}\right]$$

$$-GM\left[\frac{(x^2 + y^2 + z^2)^{3/2} - 3y^2(x^2 + y^2 + z^2)^{1/2}}{(x^2 + y^2 + z^2)^3}\right]$$

$$-GM\left[\frac{(x^2 + y^2 + z^2)^{3/2} - 3z^2(x^2 + y^2 + z^2)^{1/2}}{(x^2 + y^2 + z^2)^3}\right]$$

$$= -GM\left[\frac{3(x^2 + y^2 + z^2)^2 - 3(x^2 + y^2 + z^2)(x^2 + y^2 + z^2)}{(x^2 + y^2 + z^2)^{7/2}}\right] = 0$$

5. $\frac{\partial}{\partial x}(y - x) = -1, \frac{\partial}{\partial y}(z - y) = -1, \frac{\partial}{\partial z}(y - x) = 0$

$\nabla \cdot \mathbf{F} = -2$

$$\text{Flux} = \int_{-1}^1 \int_{-1}^1 \int_{-1}^1 -2\, dx\, dy\, dz = -2(2^3) = -16$$

7. $\frac{\partial}{\partial x}(y) = 0, \frac{\partial}{\partial y}(xy) = x, \frac{\partial}{\partial z}(-z) = -1$

$\nabla \cdot \mathbf{F} = x - 1;\ z = x^2 + y^2$

$z = r^2$ in cylindrical coordinates

$$\text{Flux} = \iiint_D (x - 1)\, dz\, dy\, dx$$

$$= \int_0^{2\pi} \int_0^2 \int_0^{r^2} (r\cos\theta - 1)\, dz\, r\, dr\, d\theta$$

$$= \int_0^{2\pi} \int_0^2 (r^3 \cos\theta - r^2)\, r\, dr\, d\theta$$

$$= \int_0^{2\pi} \left[\frac{r^5}{5}\cos\theta - \frac{r^4}{4}\right]_0^2 d\theta$$

$$= \int_0^{2\pi} \left(\frac{32}{5}\cos\theta - 4\right) d\theta$$

$$= \left[\frac{32}{5}\sin\theta - 4\theta\right]_0^{2\pi} = -8\pi$$

9. $\frac{\partial}{\partial x}(x^2) = 2x, \frac{\partial}{\partial x}(-2xy) = -2x, \frac{\partial}{\partial z}(3xz) = 3x$

$$\text{Flux} = \iiint_D 3x\, dx\, dy\, dz$$

$$= \int_0^{\pi/2} \int_0^{\pi/2} \int_0^2 (3\rho \sin\phi \cos\theta)(\rho^2 \sin\phi)\, d\rho\, d\phi\, d\theta$$

$$= \int_0^{\pi/2} \int_0^{\pi/2} 12 \sin^2\phi \cos\theta\, d\phi\, d\theta$$

$$= \int_0^{\pi/2} 3\pi \cos\theta\, d\theta = 3\pi$$

11. $\frac{\partial}{\partial x}(2xz) = 2z, \frac{\partial}{\partial y}(-xy) = -x, \frac{\partial}{\partial z}(-z^2) = -2z$

$\nabla \cdot \mathbf{F} = -x$

$$\text{Flux} = \iiint_D -x\, dV$$

$$= \int_0^2 \int_0^{2\sqrt{16-4x^2}} \int_0^{4-y} -x\, dz\, dy\, dx$$

$$= \int_0^2 \int_0^{2\sqrt{16-4x^2}} (xy - 4x)\, dy\, dx$$

$$= \int_0^2 \left[\frac{1}{2}x(16 - 4x^2) - 4x\sqrt{16 - 4x^2}\right] dx$$

$$= \left[4x^2 - \frac{1}{2}x^4 + \frac{1}{3}(16 - 4x^2)^{3/2}\right]_0^2 = -\frac{40}{3}$$

13. Let $\rho = \sqrt{x^2 + y^2 + z^2}$. Then $\frac{\partial \rho}{\partial x} = \frac{x}{\rho}, \frac{\partial \rho}{\partial y} = \frac{y}{\rho}, \frac{\partial \rho}{\partial z} = \frac{z}{\rho}$

$\frac{\partial}{\partial x}(\rho x) = \left(\frac{\partial \rho}{\partial x}\right)x + \rho = \frac{x^2}{\rho} + \rho, \frac{\partial}{\partial y}(\rho y) = \left(\frac{\partial \rho}{\partial y}\right)y + \rho$

$= \frac{y^2}{\rho} + \rho, \frac{\partial}{\partial z}(\rho z) = \left(\frac{\partial \rho}{\partial z}\right)z + \rho = \frac{z^2}{\rho} + \rho$

$\nabla \cdot \mathbf{F} = \frac{x^2 + y^2 + z^2}{\rho} + 3\rho = 4\rho$, since $\rho = \sqrt{x^2 + y^2 + z^2}$

$\text{Flux} = \iiint_D 4\rho \, dV$

$= \int_0^{2\pi}\int_0^{\pi}\int_1^{\sqrt{2}} (4\rho)(\rho^2 \sin\phi)\, d\rho\, d\phi\, d\theta$

$= \int_0^{2\pi}\int_0^{\pi} 3\sin\phi\, d\phi\, d\theta = \int_0^{2\pi} 6\, d\theta = 12\pi$

15. $\frac{\partial}{\partial x}(5x^3 + 12xy^2) = 15x^2 + 12y^2$,

$\frac{\partial}{\partial y}(y^3 + e^y \sin z) = 3y^2 + e^y \sin z$,

$\frac{\partial}{\partial z}(5z^3 + e^y \cos z) = 15z^2 - e^y \sin z$

$\nabla \cdot \mathbf{F} = 15x^2 + 15y^2 + 15z^2 = 15\rho^2$

$\text{Flux} = \iiint_D 15\rho^2\, dV$

$= \int_0^{2\pi}\int_0^{\pi}\int_1^{\sqrt{2}} (15\rho^2)(\rho^2 \sin\phi)\, d\rho\, d\phi\, d\theta$

$\int_0^{2\pi}\int_0^{\pi} (12\sqrt{2} - 3)\sin\phi\, d\phi\, d\theta$

$= \int_0^{2\pi} (24\sqrt{2} - 6)\, d\theta = (48\sqrt{2} - 12)\pi$

17. a. $\mathbf{G} = M\mathbf{i} + N\mathbf{j} + P\mathbf{k}$

$\nabla \times \mathbf{G} = \text{curl } \mathbf{G}$

$= \left(\frac{\partial P}{\partial y} - \frac{\partial N}{\partial z}\right)\mathbf{i} + \left(\frac{\partial M}{\partial z} - \frac{\partial P}{\partial x}\right)\mathbf{k} + \left(\frac{\partial N}{\partial x} - \frac{\partial M}{\partial y}\right)\mathbf{k}$

$\nabla \cdot \nabla \times \mathbf{G} = \text{div}(\text{curl } \mathbf{G})$

$= \frac{\partial}{\partial x}\left(\frac{\partial P}{\partial y} - \frac{\partial N}{\partial z}\right) + \frac{\partial}{\partial y}\left(\frac{\partial M}{\partial z} - \frac{\partial P}{\partial x}\right)$

$+ \frac{\partial}{\partial z}\left(\frac{\partial N}{\partial x} - \frac{\partial M}{\partial y}\right)$

$= \frac{\partial^2 P}{\partial x \partial y} - \frac{\partial^2 N}{\partial x \partial z} + \frac{\partial^2 M}{\partial y \partial z} - \frac{\partial^2 P}{\partial y \partial x} + \frac{\partial^2 N}{\partial z \partial x} - \frac{\partial^2 M}{\partial z \partial y} = 0$

if all first and second partical derivatives are continuous

b. By the Divergence Theorem, the outward flux of $\nabla \times \mathbf{G}$ across a closed surface is zero because outward

flux of $\nabla \times \mathbf{G} = \iint_S (\nabla \times \mathbf{G}) \cdot \mathbf{n}\, d\sigma$

$= \iiint_D \nabla \cdot \nabla \times \mathbf{G}\, dV$ [Divergence Theorem with $\mathbf{F} = \nabla \times \mathbf{G}$]

$= \iiint_D (0)\, dV = 0$ [by part (a)]

19. a. $\text{div}(g\mathbf{F}) = \nabla \cdot g\mathbf{F} = \frac{\partial}{\partial x}(gM) + \frac{\partial}{\partial y}(gN) + \frac{\partial}{\partial z}(gP)$

$= \left(g\frac{\partial M}{\partial x} + M\frac{\partial g}{\partial x}\right) + \left(g\frac{\partial N}{\partial y} + N\frac{\partial g}{\partial y}\right) + \left(g\frac{\partial P}{\partial z} + P\frac{\partial g}{\partial x}\right)$

$= \left(M\frac{\partial g}{\partial x} + N\frac{\partial g}{\partial y} + P\frac{\partial g}{\partial z}\right) + g\left(\frac{\partial M}{\partial x} + \frac{\partial N}{\partial y} + \frac{\partial P}{\partial z}\right)$

$= g\nabla \cdot \mathbf{F} + \nabla g \cdot \mathbf{F}$

b. $\nabla \times (g\mathbf{F}) = \left[\frac{\partial}{\partial y}(gP) - \frac{\partial}{\partial z}(gN)\right]\mathbf{i}$

$+ \left[\frac{\partial}{\partial z}(gM) - \frac{\partial}{\partial x}(gP)\right]\mathbf{j}$

$+ \left[\frac{\partial}{\partial x}(gN) - \frac{\partial}{\partial y}(gM)\right]\mathbf{k}$

$= \left(P\frac{\partial g}{\partial y} + g\frac{\partial P}{\partial y} - N\frac{\partial g}{\partial z} - g\frac{\partial N}{\partial z}\right)\mathbf{i}$

$+ \left(M\frac{\partial g}{\partial z} + g\frac{\partial M}{\partial z} - P\frac{\partial g}{\partial x} - g\frac{\partial P}{\partial x}\right)\mathbf{j}$

$+ \left(N\frac{\partial g}{\partial x} + g\frac{\partial N}{\partial x} - M\frac{\partial g}{\partial y} - g\frac{\partial M}{\partial y}\right)\mathbf{k}$

$= \left(P\frac{\partial g}{\partial y} - N\frac{\partial g}{\partial z}\right)\mathbf{i} + \left(g\frac{\partial P}{\partial y} - g\frac{\partial N}{\partial z}\right)\mathbf{i} + \left(M\frac{\partial g}{\partial z} - P\frac{\partial g}{\partial x}\right)\mathbf{j}$

$+ \left(g\frac{\partial M}{\partial z} - g\frac{\partial P}{\partial x}\right)\mathbf{j} + \left(N\frac{\partial g}{\partial x} - M\frac{\partial g}{\partial y}\right)\mathbf{k}$

$+ \left(g\frac{\partial N}{\partial x} - g\frac{\partial M}{\partial y}\right)\mathbf{k} = g\nabla \times \mathbf{F} + \nabla g \times \mathbf{F}$

21. The integral's value never exceeds the surface area of S. Since $|\mathbf{F}| \le 1$, we have $|\mathbf{F} \cdot \mathbf{n}| \le |\mathbf{F}|\,|\mathbf{n}| \le (1)(1) = 1$ and

$\iiint_D \nabla \cdot \mathbf{F}\, dV = \iint_S \mathbf{F} \cdot \mathbf{n}\, d\sigma$ [Divergence Theorem]

$\le \iint_S |\mathbf{F} \cdot \mathbf{n}|\, d\sigma$ [A property of integrals]

$\le \iint_S (1)\, d\sigma$ [$|\mathbf{F} \cdot \mathbf{n}| \le 1$]

= Area of S

23 a. $\frac{\partial}{\partial x}(x) = 1, \frac{\partial}{\partial y}(y) = 1, \frac{\partial}{\partial z}(z) = 1$

$\nabla \cdot \mathbf{F} = 3$

$\text{Flux} = \iiint_D 3\, dV = 3\iiint_D dV = 3$ (Volume of the solid)

b. If $\mathbf{F}$ is orthogonal to $\mathbf{n}$ at every point of S, then $\mathbf{F} \cdot \mathbf{n} = 0$ everywhere

$\text{Flux} = \iint_S \mathbf{F} \cdot \mathbf{n}\, d\sigma = 0$. But the flux is 3(Volume of the solid) $\ne 0$, so $\mathbf{F}$ is not orthogonal to $\mathbf{n}$ at every point.

25. $\iint_S \mathbf{F} \cdot \mathbf{n}\, d\sigma = \iiint_D \nabla \cdot \mathbf{F}\, dV = \iiint_D 3\, dV$

$\frac{1}{3}\iint_S \mathbf{F} \cdot \mathbf{n}\, d\sigma = \iiint_D dV = \text{Volume of } D$

27. a. From the Divergence Theorem,

$$\iint_S \nabla f \cdot \mathbf{n}\, d\sigma = \iiint_D \nabla \cdot \nabla f\, dV$$

$$= \iiint_D \nabla^2 f\, dV = \iiint_D 0\, dV$$

$$= 0$$

b. From the Divergence Theorem,

$\iint_S f\nabla f \cdot \mathbf{n}\, d\sigma = \iiint_D \nabla \cdot f\nabla f\, dV$. Now,

$$f\nabla f = \left(f\frac{\partial f}{\partial x}\right)\mathbf{i} + \left(f\frac{\partial f}{\partial y}\right)\mathbf{j} + \left(f\frac{\partial f}{\partial z}\right)\mathbf{k}$$

$$\nabla \cdot f\nabla f = \left[f\frac{\partial^2 f}{\partial x^2} + \left(\frac{\partial f}{\partial x}\right)^2\right] + \left[f\frac{\partial^2 f}{\partial y^2} + \left(\frac{\partial f}{\partial y}\right)^2\right]$$

$$+ \left[f\frac{\partial^2 f}{\partial z^2} + \left(\frac{\partial f}{\partial z}\right)^2\right]$$

$= f\nabla^2 f + |\nabla f|^2 = 0 + |\nabla f|^2$ since f is harmonic

$\iint_S f\nabla f \cdot \mathbf{n}\, d\sigma = \iiint_D |\nabla f|^2\, dV$, as claimed.

29. $$\iint_S f\nabla g \cdot \mathbf{n}\, d\sigma = \iiint_D \nabla \cdot f\nabla g\, dV$$

$$= \iiint_D \nabla \cdot \left(f\frac{\partial g}{\partial x}\mathbf{i} + f\frac{\partial g}{\partial y}\mathbf{j} + f\frac{\partial g}{\partial z}\mathbf{k}\right) dV$$

$$= \iiint_D \left(f\frac{\partial^2 g}{\partial x^2} + \frac{\partial f}{\partial x}\frac{\partial g}{\partial x} + f\frac{\partial^2 g}{\partial y^2} + \frac{\partial f}{\partial y}\frac{\partial g}{\partial y} + f\frac{\partial^2 g}{\partial z^2}\right.$$

$$\left. + \frac{\partial f}{\partial z}\frac{\partial g}{\partial z}\right) dV$$

$$= \iiint_D \left[f\left(\frac{\partial^2 g}{\partial x^2} + \frac{\partial^2 g}{\partial y^2} + \frac{\partial^2 g}{\partial z^2}\right)\right.$$

$$\left. + \left(\frac{\partial f}{\partial x}\frac{\partial g}{\partial x} + \frac{\partial f}{\partial y}\frac{\partial g}{\partial y} + \frac{\partial f}{\partial x}\frac{\partial g}{\partial z}\right)\right] dV$$

$$= \iiint_D (f\nabla^2 g + \nabla f \cdot \nabla g)\, dV$$

31. a. The integral $\iiint_D p(t, x, y, z)\, dV$ represents the mass of the fluid at any time, t. The equation says that the instantaneous rate of change of mass equals the flux of the fluid through the surface S enclosing the region D: the mass decreases if the flux is outward (so the fluid flows out of D), and increases if the flow is inward (interpreting $\mathbf{n}$ as the outward pointing unit normal to the surface).

b. $$\iiint_D \frac{\partial p}{\partial t}\, dV = \frac{d}{dt}\iiint_D p\, dV$$

$$= -\iint_S p\mathbf{v} \cdot \mathbf{n}\, d\sigma = -\iiint_D \nabla \cdot p\mathbf{v}\, dV$$

$$\frac{\partial p}{\partial t} = -\nabla \cdot p\mathbf{v}$$

$\nabla \cdot p\mathbf{v} + \dfrac{\partial p}{\partial t} = 0$ as claimed

Chapter 15 Review Exercises
(pp 908-911)

1. Path 1: $\mathbf{r} = t\mathbf{i} + t\mathbf{j} + t\mathbf{k}$

$x = t, y = t, z = t, 0 \le t \le 1$

$f(g(t), h(t), k(t)) = 3 - 3t^2$ and $\frac{dx}{dt} = 1, \frac{dy}{dt} = 1, \frac{dz}{dt} = 1$

$$\sqrt{\left(\frac{dx}{dt}\right)^2 + \left(\frac{dy}{dt}\right)^2 + \left(\frac{dz}{dt}\right)^2}\, dt = \sqrt{3}\, dt$$

$$\int_C f(x, y, z)\, ds = \int_0^1 \sqrt{3}(3 - 3t^2)\, dt = 2\sqrt{3}$$

Path 2: $\mathbf{r}_1 = t\mathbf{i} + t\mathbf{j}, 0 \le t \le 1$

$x = t, y = t, z = 0$

$f(g(t), h(t), k(t)) = 2t - 3t^2 + 3$ and $\frac{dx}{dt} = 1, \frac{dy}{dt} = 1,$

$\frac{dz}{dt} = 0$

$$\sqrt{\left(\frac{dx}{dt}\right)^2 + \left(\frac{dy}{dt}\right)^2 + \left(\frac{dz}{dt}\right)^2}\, dt = \sqrt{2}\, dt$$

$\int_{C_1} f(x, y, z)\, ds = \int_0^1 \sqrt{2}(2t - 3t^2 + 3)\, dt = 3\sqrt{2};$

$\mathbf{r}_2 = \mathbf{i} + \mathbf{j} + t\mathbf{k}$

$x = 1, y = 1, z = t$

$f(g(t), h(t), k(t)) = 2 - 2t$ and $\frac{dx}{dt} = 0, \frac{dy}{dt} = 0, \frac{dz}{dt} = 1$

$$\sqrt{\left(\frac{dx}{dt}\right)^2 + \left(\frac{dy}{dt}\right)^2 + \left(\frac{dz}{dt}\right)^2}\, dt = dt$$

$$\int_{C_2} f(x, y, z)\, ds = \int_0^1 (2 - 2t)\, dt = 1$$

$$f(x, y, z)\, ds = \int_{C_1} f(x, y, z)\, ds + \int_{C_2} f(x, y, z) = 3\sqrt{2} + 1$$

2. Path 1: $\mathbf{r}_1 = t\mathbf{i}$

$x = t, y = 0, z = 0$

$f(g(t), h(t), k(t)) = t^2$ and $\frac{dx}{dt} = 1, \frac{dy}{dt} = 0, \frac{dz}{dt} = 0$

$$\sqrt{\left(\frac{dx}{dt}\right)^2 + \left(\frac{dy}{dt}\right)^2 + \left(\frac{dz}{dt}\right)^2}\, dt = dt$$

$\int_{C_1} f(x, y, z)\, ds = \int_0^1 t^2\, dt = \frac{1}{3}; \mathbf{r}_2 = \mathbf{i} + t\mathbf{j}$

$x = 1, y = t, z = 0$

$f(g(t), h(t), k(t)) = 1 + t$ and $\frac{dx}{dt} = 0, \frac{dy}{dt} = 1, \frac{dz}{dt} = 0$

$$\sqrt{\left(\frac{dx}{dt}\right)^2 + \left(\frac{dy}{dt}\right)^2 + \left(\frac{dz}{dt}\right)^2}\, dt = dt$$

$\int_{C_2} f(x, y, z)\, ds = \int_0^1 (t + 1)\, dt = \frac{3}{2}; \mathbf{r}_3 = \mathbf{i} + \mathbf{j} + t\mathbf{k}$

$x = 1, y = 1, z = t$

$f(g(t), h(t), k(t)) = 2 - t$ and $\frac{dx}{dt} = 0, \frac{dy}{dt} = 0, \frac{dz}{dt} = 1$

$$\sqrt{\left(\frac{dx}{dt}\right)^2 + \left(\frac{dy}{dt}\right)^2 + \left(\frac{dz}{dt}\right)^2}\, dt = dt$$

$$\int_{C_3} f(x, y, z)\, ds = \int_0^1 (2 - t)\, dt = \frac{3}{2}$$

$$\int_{\text{Path 1}} f(x, y, z)\, ds = \int_{C_1} f(x, y, z)\, ds + \int_{C_2} f(x, y, z)\, ds + \int_{C_3} f(x, y, z)\, ds = \frac{10}{3}$$

Path 2: $\mathbf{r}_4 = t\mathbf{i} + t\mathbf{j}$

$x = t, y = t, z = 0$

$f(g(t), h(t), k(t)) = t^2 + t$ and $\frac{dx}{dt} = 1, \frac{dy}{dt} = 1, \frac{dz}{dt} = 0$

$$\sqrt{\left(\frac{dx}{dt}\right)^2 + \left(\frac{dy}{dt}\right)^2 + \left(\frac{dz}{dt}\right)^2}\, dt = \sqrt{2}\, dt$$

$\int_{C_4} f(x, y, z)\, ds = \int_0^1 \sqrt{2}(t^2 + t)dt = \frac{5}{6}\sqrt{2}$; $\mathbf{r}_3 = \mathbf{i} + \mathbf{j} + t\mathbf{k}$ (see above)

$$\int_{C_3} f(x, y, z)\, ds = \frac{3}{2}$$

$$\int_{\text{Path 2}} f(x, y, z)\, ds = \int_{C_3} f(x, y, z)\, ds + \int_{C_4} f(x, y, z)\, ds = \frac{5}{6}\sqrt{2} + \frac{3}{2} = \frac{5\sqrt{2} + 9}{6}$$

Path 3: $\mathbf{r}_5 = t\mathbf{k}$

$x = 0, y = 0, z = t, 0 \le t \le 1$

$f(g(t), h(t), k(t)) = -t$ and $\frac{dx}{dt} = 0, \frac{dy}{dt} = 1, \frac{dz}{dt} = 0$

$$\sqrt{\left(\frac{dx}{dt}\right)^2 + \left(\frac{dy}{dt}\right)^2 + \left(\frac{dz}{dt}\right)^2}\, dt = dt$$

$$\int_{C_5} f(x, y, z)\, ds = \int_0^1 -t\, dt = -\frac{1}{2};$$

$\mathbf{r}_6 = t\mathbf{j} + \mathbf{k}$

$x = 0, y = t, z = 1, 0 \le t \le 1$

$f(g(t), h(t), k(t) = t - 1$ and $\frac{dx}{dt} = 0, \frac{dy}{dt} = 1, \frac{dz}{dt} = 0$

$$\sqrt{\left(\frac{dx}{dt}\right)^2 + \left(\frac{dy}{dt}\right)^2 + \left(\frac{dz}{dt}\right)^2}\, dt = dt$$

$\int_{C_6} f(x, y, z)\, ds = \int_0^1 (t - 1)\, dt = -\frac{1}{2}$; $\mathbf{r}_7 = t\mathbf{i} + \mathbf{j} + \mathbf{k}$

$x = t, y = 1, z = 1, 0 \le t \le 1$

$f(g(t), h(t), k(t)) = t^2$ and $\frac{dx}{dt} = 1, \frac{dy}{dt} = 0, \frac{dz}{dt} = 0$

$$\sqrt{\left(\frac{dx}{dt}\right)^2 + \left(\frac{dy}{dt}\right)^2 + \left(\frac{dz}{dt}\right)^2}\, dt = dt$$

$$\int_{C_7} f(x, y, z)\, ds = \int_0^1 t^2\, dt = \frac{1}{3}$$

$$\int_{\text{Path 3}} f(x, y, z)\, ds = \int_{C_5} f(x, y, z)\, ds + \int_{C_6} f(x, y, z)\, ds + \int_{C_7} f(x, y, z)\, ds = -\frac{1}{2} - \frac{1}{2} + \frac{1}{3} = -\frac{2}{3}$$

3. $\mathbf{r} = (a \cos t)\mathbf{j} + (a \sin t)\mathbf{k}$

$x = 0, y = a \cos t, z = a \sin t$

$f(g(t), h(t), k(t)) = \sqrt{a^2 \sin^2 t} = a\,|\sin t|$ and $\frac{dx}{dt} = 0$, $\frac{dy}{dt} = -a \sin t, \frac{dz}{dt} = a \cos t$

$$\sqrt{\left(\frac{dx}{dt}\right)^2 + \left(\frac{dy}{dt}\right)^2 + \left(\frac{dz}{dt}\right)^2}\, dt = a\, dt$$

$$\int_C f(x, y, z)\, ds = \int_0^{2\pi} a^2\,|\sin t|\, dt = \int_0^{\pi} a^2 \sin t\, dt + \int_{\pi}^{2\pi} -a^2 \sin t\, dt = 4a^2$$

4. $\mathbf{r} = (\cos t + t \sin t)\mathbf{i} + (\sin t - t \cos t)\mathbf{j}$

$x = \cos t + t \sin t, y = \sin t - t \cos t, z = 0$

$f(g(t), h(t), k(t)) = \sqrt{(\cos t + t \sin t)^2 + (\sin t - \cos t)^2}$ $= \sqrt{1 + t^2}$ and $\frac{dx}{dt} = -\sin t + \sin t + t \cos t = t \cos t$, $\frac{dy}{dt} = \cos t - \cos t + t \sin t = t \sin t, \frac{dz}{dt} = 0$

$$\sqrt{\left(\frac{dx}{dt}\right)^2 + \left(\frac{dy}{dt}\right)^2 + \left(\frac{dz}{dt}\right)^2}\, dt = \sqrt{t^2 \cos^2 t + t^2 \sin^2 t}\, dt$$

$= |t|\, dt = t\, dt$ since $0 \le t \le \sqrt{3}$

$$\int_C f(x, y, z)\, ds = \int_0^{\sqrt{3}} t\sqrt{1 + t^2}\, dt = \frac{7}{3}$$

5. $\frac{\partial P}{\partial y} = -\frac{1}{2}(x + y + z)^{-3/2} = \frac{\partial N}{\partial z}$, $\frac{\partial M}{\partial z} = -\frac{1}{2}(x + y + z)^{-3/2} = \frac{\partial P}{\partial x}$, $\frac{\partial N}{\partial x} = -\frac{1}{2}(x + y + z)^{-3/2} = \frac{\partial M}{\partial y}$

$M\, dx + N\, dy + P\, dz$ is exact; $\frac{\partial f}{\partial x} = \frac{1}{\sqrt{x + y + z}}$

$f(x, y, z) = 2\sqrt{x + y + z} + g(y, z)$

$\frac{\partial f}{\partial y} = \frac{1}{\sqrt{x + y + z}} + \frac{\partial g}{\partial y} = \frac{1}{\sqrt{x + y + z}}$

$\frac{\partial g}{\partial y} = 0$

$g(y, z) = h(z)$

$f(x, y, z) = 2\sqrt{x + y + z} + h(z)$

$\frac{\partial f}{\partial z} = \frac{1}{\sqrt{x + y + z}} + h'(z) = \frac{1}{\sqrt{x + y + z}}$

$h'(z) = 0$

$h(z) = C$

$f(x, y, z) = 2\sqrt{x + y + z} + C$

$$\int_{(-1, 1, 1)}^{(4, -3, 0)} \frac{dx + dy + dz}{\sqrt{x + y + z}} = f(4, -3, 0) - f(-1, 1, 1) = 2\sqrt{1} - 2\sqrt{1} = 0$$

6. $\frac{\partial P}{\partial y} = -\frac{1}{2\sqrt{yz}} = \frac{\partial N}{\partial z}, \frac{\partial M}{\partial z} = 0 = \frac{\partial P}{\partial x}, \frac{\partial N}{\partial x} = 0 = \frac{\partial M}{\partial y}$

$M\, dx + N\, dy + P\, dz$ is exact; $\frac{\partial f}{\partial x} = 1$

$f(x, y, z) = x + g(y, z)$

$\frac{\partial f}{\partial y} = \frac{\partial g}{\partial y} = -\sqrt{\frac{z}{y}}$

$g(y, z) = -2\sqrt{yz} + h(z)$

$f(x, y, z) = x - 2\sqrt{yz} + h(z)$

$\frac{\partial f}{\partial z} = -\sqrt{\frac{y}{z}} + h'(z) = -\sqrt{\frac{y}{z}}$

$h'(z) = 0$

$h(z) = C$

$f(x, y, z) = x - 2\sqrt{yz} + C$

$$\int_{(1,1,1)}^{(10,3,3)} dx - \sqrt{\frac{z}{y}}\,dy - \sqrt{\frac{y}{z}}\,dz = f(10,3,3) - f(1,1,1)$$
$$= (10 - 2\cdot 3) - (1 - 2\cdot 1) = 4 + 1 = 5$$

7. $\frac{\partial P}{\partial y} = x\cos z = \frac{\partial N}{\partial z}, \frac{\partial M}{\partial z} = y\cos z = \frac{\partial P}{\partial x}, \frac{\partial N}{\partial x} = \sin z = \frac{\partial M}{\partial y}$

F is conservative

$$\int_C \mathbf{F}\cdot d\mathbf{r} = 0$$

8. $\frac{\partial P}{\partial y} = 0 = \frac{\partial N}{\partial z}, \frac{\partial M}{\partial z} = 0 = \frac{\partial P}{\partial x}, \frac{\partial N}{\partial x} = 3x^2 = \frac{\partial M}{\partial y}$

F is conservative

$$\int_C \mathbf{F}\cdot d\mathbf{r} = 0$$

9. Let $M = 8x\sin y$ and $N = -8y\cos x$

$\frac{\partial M}{\partial y} = 8x\cos y$ and $\frac{\partial N}{\partial x} = 8y\sin x$

$$\int_C 8x\sin y\,dx - 8y\cos x\,dy$$
$$= \int\int_R (8y\sin x - 8x\cos y)\,dy\,dx$$
$$= \int_0^{\pi/2}\int_0^{\pi/2} (8y\sin x - 8x\cos y)\,dy\,dx$$
$$= \int_0^{\pi/2} (\pi^2\sin x - 8x)\,dx = -\pi^2 + \pi^2 = 0$$

10. Let $M = y^2$ and $N = x^2$

$\frac{\partial M}{\partial y} = 2y$ and $\frac{\partial N}{\partial x} = 2x$

$$\int_C y^2\,dx + x^2\,dy = \int\int_R (2x - 2y)\,dx\,dy$$
$$= \int_0^{2\pi}\int_0^2 (2r\cos\theta - 2r\sin\theta)\,r\,dr\,d\theta$$
$$= \int_0^{2\pi} \frac{16}{3}(\cos\theta - \sin\theta)\,d\theta = 0$$

11. Let $z = 1 - x - y$

$f_x(x, y) = -1, f_y(x, y) = -1$

$\sqrt{f_x^2 + f_y^2 + 1} = \sqrt{3}$

Surface Area $= \int\int_R \sqrt{3}\,dx\,dy = \sqrt{3}$(Area of the circular region in the xy-plane) $= \pi\sqrt{3}$

12. $\nabla f = -3\mathbf{i} + 2y\mathbf{j} + 2z\mathbf{k}, \mathbf{p} = \mathbf{i}$

$|\nabla f| = \sqrt{9 + 4y^2 + 4z^2}$ and $|\nabla f\cdot\mathbf{p}| = 3$

$$\text{Surface Area} = \int\int_R \frac{\sqrt{9 + 4y^2 + 4z^2}}{3}\,dy\,dz$$
$$= \int_0^{2\pi}\int_0^{\sqrt{3}} \frac{\sqrt{9 + 4r^2}}{3}\,r\,dr\,d\theta$$
$$= \frac{1}{3}\int_0^{2\pi}\left(\frac{7}{4}\sqrt{21} - \frac{9}{4}\right)d\theta$$
$$= \frac{\pi}{6}(7\sqrt{21} - 9)$$

13. $\nabla f = 2x\mathbf{i} + 2y\mathbf{j} + 2z\mathbf{k}, \mathbf{p} = \mathbf{k}$

$|\nabla f| = \sqrt{4x^2 + 4y^2 + 4z^2} = 2\sqrt{x^2 + y^2 + z^2} = 2$ and

$|\nabla f\cdot\mathbf{p}| = |2z| = 2z$ since $z \ge 0$

$$\text{Surface Area} = \int\int_R \frac{2}{2z}\,dA = \int\int_R \frac{1}{z}\,dA$$
$$= \int\int_R \frac{1}{\sqrt{1 - x^2 - y^2}}\,dx\,dy$$
$$= \int_0^{2\pi}\int_0^{1/\sqrt{2}} \frac{1}{\sqrt{1 - r^2}}\,r\,dr\,d\theta = \int_0^{2\pi}\left[-\sqrt{1 - r^2}\right]_0^{1/\sqrt{2}} d\theta$$
$$= \int_0^{2\pi}\left(1 - \frac{1}{\sqrt{2}}\right)d\theta = 2\pi\left(1 - \frac{1}{\sqrt{2}}\right)$$

14. a. $\nabla f = 2x\mathbf{i} + 2y\mathbf{j} + 2z\mathbf{k}, \mathbf{p} = \mathbf{k}$

$|\nabla f| = \sqrt{4x^2 + 4y^2 + 4z^2} = 2\sqrt{x^2 + y^2 + z^2} = 4$ and

$|\nabla f\cdot\mathbf{p}| = 2z$ since $z \ge 0$

$$\text{Surface Area} = \int\int_R \frac{4}{2z}\,dA = \int\int_R \frac{2}{z}\,dA$$
$$= 2\int_0^{\pi/2}\int_0^{2\cos\theta} \frac{2}{\sqrt{4 - r^2}}\,r\,dr\,d\theta$$
$$= 4\pi - 8$$

b. $\mathbf{r} = 2\cos\theta$

$d\mathbf{r} = -2\sin\theta\,d\theta$; $ds^2 = r^2\,d\theta^2 + dr^2$ (Arc length in polar coordinates)

$ds^2 = (2\cos\theta)^2\,d\theta^2 + dr^2 = 4\cos^2\theta\,d\theta^2 + 4\sin^2\theta\,d\theta^2 = 4\,d\theta^2$

$ds = 2\,d\theta$; the height of the cylinder is $z = \sqrt{4 - r^2} = \sqrt{4 - 4\cos^2\theta} = 2|\sin\theta| = 2\sin\theta$ if $0 \le \theta \le \frac{\pi}{2}$

$$\text{Surface Area} = \int_{-\pi/2}^{\pi/2} h\,ds = 2\int_0^{\pi/2}(2\sin\theta)(2\,d\theta) = 8$$

15. $f(x, y, z) = \frac{x}{a} + \frac{y}{b} + \frac{z}{c} = 1$

$\nabla f = \left(\frac{1}{a}\right)\mathbf{i} + \left(\frac{1}{b}\right)\mathbf{j} + \left(\frac{1}{c}\right)\mathbf{k}$

$|\nabla f| = \sqrt{\frac{1}{a^2} + \frac{1}{b^2} + \frac{1}{c^2}}$ and $\mathbf{p} = \mathbf{k}$

$|\nabla f\cdot\mathbf{p}| = \frac{1}{c}$ since $c > 0$

$$\text{Surface Area} = \int\int_R \frac{\sqrt{\frac{1}{a^2} + \frac{1}{b^2} + \frac{1}{c^2}}}{\left(\frac{1}{c}\right)}\,dA$$
$$= c\sqrt{\frac{1}{a^2} + \frac{1}{b^2} + \frac{1}{c^2}}\int\int_R dA = \frac{1}{2}abc\sqrt{\frac{1}{a^2} + \frac{1}{b^2} + \frac{1}{c^2}},$$

since the area of the triangular region R is $\frac{1}{2}ab$

16. a. $\nabla f = 2y\mathbf{j} - \mathbf{k}, \mathbf{p} = \mathbf{k}$

$|\nabla f| = \sqrt{4y^2 + 1}$ and $|\nabla f\cdot\mathbf{p}| = 1$

$d\sigma = \sqrt{4y^2 + 1}\,dx\,dy$

$$\int\int_S g(x, y, z)\,d\sigma$$
$$= \int\int_R \frac{yz}{\sqrt{4y^2 + 1}}\sqrt{4y^2 + 1}\,dx\,dy$$

$$= \int\int_R y(y^2 - 1)\,dx\,dy = \int_{-1}^{1}\int_0^3 (y^3 - y)\,dx\,dy$$

$$= \int_{-1}^{1} 3(y^3 - y)\,dy = 3\left[\frac{y^4}{4} - \frac{y^2}{2}\right]_{-1}^{1} = 0$$

b. $\int\int_S g(x, y, z)\,d\sigma$

$$= \int\int_R \frac{z}{\sqrt{4y^2+1}}\sqrt{4y^2+1}\,dx\,dy$$

$$= \int_{-1}^{1}\int_0^3 (y^2 - 1)\,dx\,dy = \int_{-1}^{1} 3(y^2 - 1)\,dy$$

$$= 3\left[\frac{y^3}{3} - y\right]_{-1}^{1} = -4$$

17. $\nabla f = 2y\mathbf{j} + 2z\mathbf{k}$, $\mathbf{p} = \mathbf{k}$

$|\nabla f| = \sqrt{4y^2 + 4z^2} = 2\sqrt{y^2 + z^2} = 10$ and $|\nabla f \cdot \mathbf{p}| = 2z$ since $z \ge 0$

$$d\sigma = \frac{10}{2z}\,dx\,dy = \frac{5}{z}\,dx\,dy = \int\int_S g(x, y, z)\,d\sigma$$

$$= \int\int_R (x^4 y)(y^2 + z^2)\left(\frac{5}{z}\right)dx\,dy$$

$$= \int\int_R (x^4 y)(25)\left(\frac{5}{\sqrt{25 - y^2}}\right)dx\,dy$$

$$= \int_0^4\int_0^1 \frac{125y}{\sqrt{25 - y^2}}x^4\,dx\,dy = \int_0^4 \frac{25y}{\sqrt{25 - y^2}}\,dy = 50$$

18. Define the coordinate system so that the origin is at the center of the earth, the z-axis is the earth's axis (north is the positive z direction), and the xz-plane contains the earth's prime meridian. Let S denote the surface which is Wyoming so then S is part of the surface $z = (R^2 - x^2 - y^2)^{1/2}$. Let R_{xy} be the projection of S onto the xy-plane. The surface area of Wyoming is $\int\int_S 1\,d\sigma$

$$= \int\int_{R_{xy}} \sqrt{1 + \left(\frac{\partial z}{\partial x}\right)^2 + \left(\frac{\partial z}{\partial y}\right)^2}\,dA$$

$$\int\int_{R_{xy}} \sqrt{\frac{x^2}{R^2 - x^2 - y^2} + \frac{y^2}{R^2 - x^2 - y^2} + 1}\,dA$$

$$= \int\int_{R_{xy}} \frac{R}{(R^2 - x^2 - y^2)^{1/2}}\,dA$$

$= \int_{\theta_1}^{\theta_2}\int_{R\sin 45°}^{R\sin 49°} R(R^2 - r^2)^{-1/2}\,r\,dr\,d\theta$ (where θ_1 and θ_2 are the radian equivalent to 104°3′ and 111°3′, respectively)

$$= \int_{\theta_1}^{\theta_2} -R(R^2 - r^2)^{1/2}\Big|_{R\sin 45°}^{R\sin 49°}$$

$$= \int_{\theta_1}^{\theta_2} R(R^2 - R^2\sin^2 45°)^{1/2} - R(R^2 - R^2\sin^2 49°)^{1/2}\,d\theta$$

$$= (\theta_2 - \theta_1)R^2(\cos 45° - \cos 49°)$$

$$= \frac{7\pi}{180}R^2(\cos 45° - \cos 49°)$$

$$= \frac{7\pi}{180}(3959)^2(\cos 45° - \cos 49°) \approx 97{,}751 \text{ sq mi}$$

19. A possible parametrization is $\mathbf{r}(\phi, \theta) = (6\sin\phi\cos\theta)\mathbf{i} + (6\sin\phi\sin\theta)\mathbf{j} + (6\cos\phi)\mathbf{k}$ (spherical coordinates). For $z = -3$

$-3 = 6\cos\phi$

$\cos\phi = -\frac{1}{2}$

$\phi = \frac{2\pi}{3}$ and for $z = 3\sqrt{3}$

$3\sqrt{3} = 6\cos\phi$

$\cos\phi = \frac{\sqrt{3}}{2}$

$\phi = \frac{\pi}{6}$

$\frac{\pi}{6} \le \phi \le \frac{2\pi}{3}$; also $0 \le \theta \le 2\pi$

20. A possible parametrization is $\mathbf{r}(r, \theta) = (r\cos\theta)\mathbf{i} + (r\sin\theta)\mathbf{j} - \left(\frac{r^2}{2}\right)\mathbf{k}$ (cylindrical coordinates);

now $r = \sqrt{x^2 + y^2}$

$z = -\frac{r^2}{2}$ and $-2 \le z \le 0$

$-2 \le -\frac{r^2}{2} \le 0$

$4 \ge r^2 \ge 0$

$0 \le r \le 2$ since $r \ge 0$; also $0 \le \theta \le 2\pi$

21. A possible parametrization is $\mathbf{r}(r, \theta) = (r\cos\theta)\mathbf{i} + (r\sin\theta)\mathbf{j} + (1 + r)\mathbf{k}$ (cylindrical coordinates);

now $r = \sqrt{x^2 + y^2}$

$z = 1 + r$ and $1 \le z \le 3$

$1 \le 1 + r \le 3$

$0 \le r \le 2$; also $0 \le \theta \le 2\pi$

22. A possible parametrization is $\mathbf{r}(x, y) = x\mathbf{i} + y\mathbf{j} + \left(3 - x - \frac{y}{2}\right)\mathbf{k}$ for $0 \le x \le 2$ and $0 \le y \le 2$

23. Let $x = u\cos v$ and $z = u\sin v$, where $u = \sqrt{x^2 + z^2}$ and v is the angle in the xz-plane with the x-axis

$\mathbf{r}(u, v) = (u\cos v)\mathbf{i} + 2u^2\mathbf{j} + (u\sin v)\mathbf{k}$ is a possible parametrization; $0 \le y \le 2$

$2u^2 \le 2$

$u^2 \le 1$

$0 \le u \le 1$ since $u \ge 0$; also, for just the upper half of the paraboloid, $0 \le v \le \pi$

24. A possible parametrization is $(\sqrt{10}\sin\phi\cos\theta)\mathbf{i} + (\sqrt{10}\sin\phi\sin\theta)\mathbf{j} + (\sqrt{10}\cos\phi)\mathbf{k}$, $0 \le \phi \le \frac{\pi}{2}$ and $0 \le \theta \le \frac{\pi}{2}$

25. $\mathbf{r}_u = \mathbf{i} + \mathbf{j}$, $\mathbf{r}_v = \mathbf{i} - \mathbf{j} + \mathbf{k}$

$$\mathbf{r}_u \times \mathbf{r}_v = \begin{vmatrix} \mathbf{i} & \mathbf{j} & \mathbf{k} \\ 1 & 1 & 0 \\ 1 & -1 & 1 \end{vmatrix} = \mathbf{i} - \mathbf{j} - 2\mathbf{k}$$

$$\text{Surface Area} = \int\int_{R_{uv}} |\mathbf{r}_u \times \mathbf{r}_v|\,du\,dv = \int_0^1\int_0^1 \sqrt{6}\,du\,dv$$

$$= \sqrt{6}$$

26. $\iint_S (xy - z^2)\, d\sigma$

$= \int_0^1 \int_0^1 [(u + v)(u - v) - v^2]\sqrt{6}\, du\, dv$

$= \sqrt{6} \int_0^1 \int_0^1 (u^2 - 2v^2)\, du\, dv = \sqrt{6} \int_0^1 \left[\frac{u^3}{3} - 2uv^2\right]_0^1 dv$

$= \sqrt{6} \int_0^1 \left(\frac{1}{3} - 2v^2\right) dv = \sqrt{6}\left[\frac{1}{3}v - \frac{2}{3}v^3\right]_0^1 = -\frac{\sqrt{6}}{3}$

$= -\sqrt{\frac{2}{3}}$

27. $\mathbf{r}_r = (\cos\theta)\mathbf{i} + (\sin\theta)\mathbf{j}$, $\mathbf{r}_\theta = (-r\sin\theta)\mathbf{i} + (r\cos\theta)\mathbf{j} + \mathbf{k}$

$$\mathbf{r}_r \times \mathbf{r}_\theta = \begin{vmatrix} \mathbf{i} & \mathbf{j} & \mathbf{k} \\ \cos\theta & \sin\theta & 0 \\ -r\sin\theta & r\cos\theta & 1 \end{vmatrix}$$

$= (\sin\theta)\mathbf{i} - (\cos\theta)\mathbf{j} + r\mathbf{k}$

$|\mathbf{r}_r \times \mathbf{r}_\theta|^2 = \sin^2\theta + \cos^2\theta + r^2 = 1 + r^2$

Surface Area $= \iint_{R_{r\theta}} |\mathbf{r}_r \times \mathbf{r}_\theta|\, dr\, d\theta$

$= \int_0^{2\pi} \int_0^1 \sqrt{1 + r^2}\, dr\, d\theta$

$= \int_0^{2\pi} \left[\frac{r}{2}\sqrt{1 + r^2} + \frac{1}{2}\ln(r + \sqrt{1 + r^2})\right]_0^1 d\theta$

$= \int_0^{2\pi} \left[\frac{1}{2}\sqrt{2} + \frac{1}{2}\ln(1 + \sqrt{2})\right] d\theta = \pi[\sqrt{2} + \ln(1 + \sqrt{2})]$

28. $\iint_S \sqrt{x^2 + y^2 + 1}\, d\sigma$

$= \int_0^{2\pi} \int_0^1 \sqrt{r^2\cos^2\theta + r^2\sin^2\theta + 1}\, \sqrt{1 + r^2}\, dr\, d\theta$

$= \int_0^{2\pi} \int_0^1 (1 + r^2)\, dr\, d\theta = \int_0^{2\pi} \left[r + \frac{r^3}{3}\right]_0^1 d\theta$

$= \int_0^{2\pi} \frac{4}{3}\, d\theta = \frac{8}{3}\pi$

29. $\frac{\partial P}{\partial y} = 0 = \frac{\partial N}{\partial z}, \frac{\partial M}{\partial z} = 0 = \frac{\partial P}{\partial x}, \frac{\partial N}{\partial x} = 0 = \frac{\partial M}{\partial y}$

Conservative

30. $\frac{\partial P}{\partial y} = \frac{-3zy}{(x^2 + y^2 + z^2)^{5/2}} = \frac{\partial N}{\partial z},$

$\frac{\partial M}{\partial z} = \frac{-3xz}{(x^2 + y^2 + z^2)^{5/2}} = \frac{\partial P}{\partial x},$

$\frac{\partial N}{\partial x} = \frac{-3xy}{(x^2 + y^2 + z^2)^{5/2}} = \frac{\partial M}{\partial y}$

Conservative

31. $\frac{\partial P}{\partial y} = 0 \neq ye^z = \frac{\partial N}{\partial z}$

Not conservative

32. $\frac{\partial P}{\partial y} = \frac{x}{(x + yz)^2} = \frac{\partial N}{\partial z},$

$\frac{\partial M}{\partial z} = \frac{-y}{(x + yz)^2} = \frac{\partial P}{\partial x},$

$\frac{\partial N}{\partial x} = \frac{-z}{(x + yz)^2} = \frac{\partial M}{\partial y}$

Conservative

33. $\frac{\partial f}{\partial x} = 2$

$f(x, y, z) = 2x + g(y, z)$

$\frac{\partial f}{\partial y} = \frac{\partial g}{\partial y} = 2y + z$

$g(y, z) = y^2 + zy + h(z)$

$f(x, y, z) = 2x + y^2 + zy + h(z)$

$\frac{\partial f}{\partial z} = y + h'(z) = y + 1$

$h'(z) = 1$

$h(z) = z + C$

$f(x, y, z) = 2x + y^2 + zy + z$

34. $\frac{\partial f}{\partial x} = z\cos xz$

$f(x, y, z) = \sin xz + g(y, z)$

$\frac{\partial f}{\partial y} = \frac{\partial g}{\partial y} = e^y$

$g(y, z) = e^y + h(z)$

$f(x, y, z) = \sin xz + e^y + h(z)$

$\frac{\partial f}{\partial z} = x\cos xz + h'(z) = x\cos xz$

$h'(z) = 0$

$h(z) = C$

$f(x, y, z) = \sin xz + e^y$

35. Over Path 1: $\mathbf{r} = t\mathbf{i} + t\mathbf{j} + t\mathbf{k}$, $0 \le t \le 1$

$x = t, y = t, z = t$ and $d\mathbf{r} = (\mathbf{i} + \mathbf{j} + \mathbf{k})\, dt$

$\mathbf{F} = 2t^2\mathbf{i} + \mathbf{j} + t^2\mathbf{k}$

$\mathbf{F} \cdot d\mathbf{r} = (3t^2 + 1)\, dt$

Work $= \int_0^1 (3t^2 + 1)\, dt = 2;$

Over Path 2: $\mathbf{r}_1 = t\mathbf{i} + t\mathbf{j}$, $0 \le t \le 1$

$x = t, y = t, z = 0$ and $d\mathbf{r}_1 = (\mathbf{i} + \mathbf{j})\, dt$

$\mathbf{F}_1 = 2t^2\mathbf{i} + \mathbf{j} + t^2\mathbf{k}$

$\mathbf{F}_1 \cdot d\mathbf{r}_1 = (2t^2 + 1)\, dt$

$\text{Work}_1 = \int_0^1 (2t^2 + 1)\, dt = \frac{5}{3}$; $\mathbf{r}_2 = \mathbf{i} + \mathbf{j} + t\mathbf{k}$, $0 \le t \le 1$

$x = 1, y = 1, z = t$ and $d\mathbf{r}_2 = \mathbf{k}\, dt$

$\mathbf{F}_2 = 2\mathbf{i} + \mathbf{j} + \mathbf{k}$

$\mathbf{F}_2 \cdot d\mathbf{r}_2 = dt$

$\text{Work}_2 = \int_0^1 dt = 1$

$\text{Work} = \text{Work}_1 + \text{Work}_2 = \frac{5}{3} + 1 = \frac{8}{3}$

36. Over Path 1: $\mathbf{r} = t\mathbf{i} + t\mathbf{j} + t\mathbf{k}$, $0 \le t \le 1$

$x = t$, $y = t$, $z = t$ and $d\mathbf{r} = (\mathbf{i} + \mathbf{j} + \mathbf{k})\,dt$

$\mathbf{F} = 2t^2\mathbf{i} + t^2\mathbf{j} + \mathbf{k}$

$\mathbf{F} \cdot d\mathbf{r} = (3t^2 + 1)\,dt$

$\text{Work} = \int_0^1 (3t^2 + 1)\,dt = 2;$

Over Path 2: Since $\mathbf{F}$ is conservative, $\oint_C \mathbf{F} \cdot d\mathbf{r} = 0$ around any simple closed curve C. Thus consider $\int_{\text{curve}} \mathbf{F} \cdot d\mathbf{r}$

$= \int_{C_1} \mathbf{F} \cdot d\mathbf{r} + \int_{C_2} \mathbf{F} \cdot d\mathbf{r}$, where C_1 is the path from $(0, 0, 0)$ to $(1, 1, 0)$ to $(1, 1, 1)$ and C_2 is the path from $(1, 1, 1)$ to $(0, 0, 0)$. Now, from Path 1 above,

$\int_{C_2} \mathbf{F} \cdot d\mathbf{r} = -2$

$\int_{\text{curve}} \mathbf{F} \cdot d\mathbf{r} = \int_{C_1} \mathbf{F} \cdot d\mathbf{r} + (-2) = 0$

$\int_{C_1} \mathbf{F} \cdot d\mathbf{r} = 2$

37. a. $\mathbf{r} = (e^t \cos t)\mathbf{i} + (e^t \sin t)\mathbf{j}$

$x = e^t \cos t$, $y = e^t \sin t$ from $(1, 0)$ to $(e^{2\pi}, 0)$

$0 \le t \le 2\pi$

$\frac{d\mathbf{r}}{dt} = (e^t \cos t - e^t \sin t)\mathbf{i} + (e^t \sin t + e^t \cos t)\mathbf{j}$ and

$\mathbf{F} = \frac{x\mathbf{i} + y\mathbf{j}}{(x^2 + y^2)^{3/2}} = \frac{(e^t \cos t)\mathbf{i} + (e^t \sin t)\mathbf{j}}{(e^{2t} \cos^2 t + e^{2t} \sin^2 t)^{3/2}}$

$= \left(\frac{\cos t}{e^{2t}}\right)\mathbf{i} + \left(\frac{\sin t}{e^{2t}}\right)\mathbf{j}$

$\mathbf{F} \cdot \frac{d\mathbf{r}}{dt} = \left(\frac{\cos^2 t}{e^t} - \frac{\sin t \cos t}{e^t} + \frac{\sin^2 t}{e^t} + \frac{\sin t \cos t}{e^t}\right) = e^{-t}$

$\text{Work} = \int_0^{2\pi} e^{-t}\,dt = 1 - e^{-2\pi}$

b. $\mathbf{F} = \frac{x\mathbf{i} + y\mathbf{j}}{(x^2 + y^2)^{3/2}}$

$\frac{\partial f}{\partial x} = \frac{x}{(x^2 + y^2)^{3/2}}$

$f(x, y, z) = -(x^2 + y^2)^{-1/2} + g(y, z)$

$\frac{\partial f}{\partial y} = \frac{y}{(x^2 + y^2)^{3/2}} + \frac{\partial g}{\partial y} = \frac{y}{(x^2 + y^2)^{3/2}}$

$g(y, z) = C$

$f(x, y, z) = -(x^2 + y^2)^{-1/2}$ is a potential function for $\mathbf{F}$

$\int_C = \mathbf{F} \cdot d\mathbf{r} = f(e^{2\pi}, 0) - f(1, 0) = 1 - e^{-2\pi}$

38. a. $\mathbf{F} = \nabla(x^2 z e^y)$

$\mathbf{F}$ is conservative

$\oint_C \mathbf{F} \cdot d\mathbf{r} = 0$ for <u>any</u> closed path C

b. $\int_C \mathbf{F} \cdot d\mathbf{r} = \int_{(1,0,0)}^{(1,0,2\pi)} \nabla(x^2 z e^y) \cdot d\mathbf{r}$

$= (x^2 z e^y)\Big|_{(1,0,2\pi)} - (x^2 z e^y)\Big|_{(1,0,0)}$

$= 2\pi - 0 = 2\pi$

39. a. $x^2 + y^2 = 1$

$\mathbf{r} = (\cos t)\mathbf{i} + (\sin t)\mathbf{j}$, $0 \le t \le \pi$

$x = \cos t$ and $y = \sin t$

$\mathbf{F} = (\cos t + \sin t)\mathbf{i} - \mathbf{j}$ and $\frac{d\mathbf{r}}{dt} = (-\sin t)\mathbf{i} + (\cos t)\mathbf{j}$

$\mathbf{F} \cdot \frac{d\mathbf{r}}{dt} = -\sin t \cos t - \sin^2 t - \cos t$

$\text{Flow} = \int_0^{\pi} (-\sin t \cos t - \sin^2 t - \cos t)\,dt$

$= \left[\frac{1}{2}\cos^2 t - \frac{t}{2} + \frac{\sin 2t}{4} - \sin t\right]_0^{\pi} = -\frac{\pi}{2}$

b. $\mathbf{r} = -t\mathbf{i}$, $-1 \le t \le 1$

$x = -t$ and $y = 0$

$\mathbf{F} = -t\mathbf{i} - t^2\mathbf{j}$ and $\frac{d\mathbf{r}}{dt} = -\mathbf{i}$

$\mathbf{F} \cdot \frac{d\mathbf{r}}{dt} = t$

$\text{Flow} = \int_{-1}^{1} t\,dt = 0$

c. $\mathbf{r}_1 = (1 - t)\mathbf{i} - t\mathbf{j}$, $0 \le t \le 1$

$x = 1 - t$, $y = -t$

$\mathbf{F}_1 = (1 - 2t)\mathbf{i} - (1 - 2t + 2t^2)\mathbf{j}$ and $\frac{d\mathbf{r}_1}{dt} = -\mathbf{i} - \mathbf{j}$

$\mathbf{F}_1 \cdot \frac{d\mathbf{r}_1}{dt} = 2t^2$

$\text{Flow}_1 = \int_0^1 2t^2\,dt = \frac{2}{3}$; $\mathbf{r}_2 = -t\mathbf{i} + (t - 1)\mathbf{j}$,

$0 \le t \le 1$

$x = -t$, $y = t - 1$

$\mathbf{F}_2 = -\mathbf{i} - (2t^2 - 2t + 1)\mathbf{j}$ and $\frac{d\mathbf{r}_2}{dt} = -\mathbf{i} + \mathbf{j}$

$\mathbf{F}_2 \cdot \frac{d\mathbf{r}}{dt} = -2t^2 + 2t$

$\text{Flow}_2 = \int_0^1 (-2t^2 + 2t)\,dt = \frac{1}{3}$

$\text{Flow} = \text{Flow}_1 + \text{Flow}_2 = \frac{2}{3} + \frac{1}{3} = 1$

40. $\mathbf{r}_1 = (\cos t)\mathbf{i} + (\sin t)\mathbf{j} + t\mathbf{k},\ 0 \le t \le \frac{\pi}{2}$

$\mathbf{F}_1 = (2\cos t)\mathbf{i} + 2t\mathbf{j} + (2\sin t)\mathbf{k}$ and $\frac{d\mathbf{r}_1}{dt}$

$= (-\sin t)\mathbf{i} + (\cos t)\mathbf{j} + \mathbf{k}$

$\mathbf{F}_1 \cdot \frac{d\mathbf{r}}{dt} = -2\sin t\cos t + 2t\cos t + 2\sin t$

$$\text{Circ}_1 = \int_0^{\pi/2} (-2\sin t\cos t + 2t\cos t + 2\sin t)\,dt$$

$$= \left[\cos^2 t + 2t\sin t\right]_0^{\pi/2} = \pi - 1;$$

$\mathbf{r}_2 = \mathbf{j} + \left(\frac{\pi}{2}\right)(1-t)\mathbf{k},\ 0 \le t \le 1$

$\mathbf{F}_2 = \pi(1-t)\mathbf{j} + 2\mathbf{k}$ and $\frac{d\mathbf{r}_2}{dt} = -\frac{\pi}{2}\mathbf{k}$

$\mathbf{F}_2 \cdot \frac{d\mathbf{r}_2}{dt} = -\pi$

$\text{Circ}_2 = \int_0^1 -\pi\,dt = -\pi$; $\mathbf{r}_3 = t\mathbf{i} + (1-t)\mathbf{j},\ 0 \le t \le 1$

$\mathbf{F}_3 = 2t\mathbf{i} + 2(1-t)\mathbf{k}$ and $\frac{d\mathbf{r}_3}{dt} = \mathbf{i} - \mathbf{j}$

$\mathbf{F}_3 \cdot \frac{d\mathbf{r}_3}{dt} = 2t$

$\text{Circ}_3 = \int_0^1 2t\,dt = 1$

$\text{Circ} = \text{Circ}_1 + \text{Circ}_2 + \text{Circ}_3 = 0$

Or: Circ = 0 because $\nabla \times \mathbf{F} = \mathbf{0}$ and the path is closed.

41. $\nabla \times \mathbf{F} = \begin{vmatrix} \mathbf{i} & \mathbf{j} & \mathbf{k} \\ \frac{\partial}{\partial x} & \frac{\partial}{\partial y} & \frac{\partial}{\partial z} \\ y^2 & -y & 3z^2 \end{vmatrix} = -2y\mathbf{k}$; unit normal to the

plane is $\mathbf{n} = \frac{2\mathbf{i} + 6\mathbf{j} - 3\mathbf{k}}{\sqrt{4+36+9}} = \frac{2}{7}\mathbf{i} + \frac{6}{7}\mathbf{j} - \frac{3}{7}\mathbf{k}$

$\nabla \times \mathbf{F} \cdot \mathbf{n} = \frac{6}{7}y$; $\mathbf{p} = \mathbf{k}$ and $f(x, y, z) = 2x + 6y - 3z$

$|\nabla f \cdot \mathbf{p}| = 3$

$d\sigma = \frac{|\nabla f|}{|\nabla f \cdot \mathbf{p}|}\,dA = \frac{7}{3}\,dA$

$$\oint_C \mathbf{F} \cdot d\mathbf{r} = \iint_R \frac{6}{7}y\,d\sigma = \iint_R \left(\frac{6}{7}y\right)\left(\frac{7}{3}\,dA\right)$$

$$= \iint_R 2y\,dA = \int_0^{2\pi}\int_0^1 2r\sin\theta\, r\,dr\,d\theta$$

$$= \int_0^{2\pi} \frac{2}{3}\sin\theta\,d\theta = 0$$

42. $\nabla \times \mathbf{F} = \begin{vmatrix} \mathbf{i} & \mathbf{j} & \mathbf{k} \\ \frac{\partial}{\partial x} & \frac{\partial}{\partial y} & \frac{\partial}{\partial z} \\ x^2 + y & x + y & 4y^2 - z \end{vmatrix} = 8y\mathbf{i}$; the circle

lies in the plane $f(x, y, z) = y + z = 0$ with unit normal

$\mathbf{n} = \frac{1}{\sqrt{2}}\mathbf{j} + \frac{1}{\sqrt{2}}\mathbf{k}$

$\nabla \times \mathbf{F} \cdot \mathbf{n} = 0$

$$\oint_C \mathbf{F} \cdot d\mathbf{r} = \iint_R \nabla \times \mathbf{F} \cdot \mathbf{n}\,d\sigma = \iint_R 0\,d\sigma = 0$$

43. a. $\mathbf{r} = \sqrt{2}t\mathbf{i} + \sqrt{2}t\mathbf{j} + (4 - t^2)\mathbf{k},\ 0 \le t \le 1$

$x = \sqrt{2}t,\ y = \sqrt{2}t,\ z = 4 - t^2$

$\frac{dx}{dt} = \sqrt{2},\ \frac{dy}{dt} = \sqrt{2},\ \frac{dz}{dt} = -2t$

$$\sqrt{\left(\frac{dx}{dt}\right)^2 + \left(\frac{dy}{dt}\right)^2 + \left(\frac{dz}{dt}\right)^2}\,dt = \sqrt{4 + 4t^2}\,dt$$

$$M = \int_C \delta(x, y, z)\,ds = \int_0^1 3t\sqrt{4 + 4t^2}\,dt$$

$$= \left[\frac{1}{4}(4 + 4t)^{3/2}\right]_0^1 = 4\sqrt{2} - 2$$

b. $$M = \int_C \delta(x, y, z)\,ds = \int_0^1 \sqrt{4 + 4t^2}\,dt$$

$$= \left[t\sqrt{1 + t^2} + \ln(t + \sqrt{1 + t^2})\right]_0^1$$

$$= \sqrt{2} + \ln(1 + \sqrt{2})$$

44. $\mathbf{r} = t\mathbf{i} + 2t\mathbf{j} + \frac{2}{3}t^{3/2}\mathbf{k},\ 0 \le t \le 2$

$x = t,\ y = 2t,\ z = \frac{2}{3}t^{3/2}$

$\frac{dx}{dt} = 1,\ \frac{dy}{dt} = 2,\ \frac{dz}{dt} = t^{1/2}$

$$\sqrt{\left(\frac{dx}{dt}\right)^2 + \left(\frac{dy}{dt}\right)^2 + \left(\frac{dz}{dt}\right)^2}\,dt = \sqrt{t + 5}\,dt$$

$$M = \int_C \delta(x, y, z)\,ds = \int_0^2 3\sqrt{5 + t}\sqrt{t + 5}\,dt$$

$$= \int_0^2 3(t + 5)\,dt = 36;$$

$$M_{yz} = \int_C x\delta\,ds = \int_0^2 3t(t + 5)\,dt = 38;$$

$$M_{xz} = \int_C y\delta\,ds = \int_0^2 6t(t + 5)\,dt = 76;$$

$$M_{xy} = \int_C = z\delta\,ds = \int_0^2 2t^{3/2}(t + 5)\,dt = \frac{144}{7}\sqrt{2}$$

$$\bar{x} = \frac{M_{yz}}{M} = \frac{38}{36} = \frac{19}{18},\ \bar{y} = \frac{M_{xz}}{M} = \frac{76}{36} = \frac{19}{9},$$

$$\bar{z} = \frac{M_{xy}}{M} = \frac{\left(\frac{144}{7}\sqrt{2}\right)}{36} = \frac{4}{7}\sqrt{2}$$

45. $\mathbf{r} = t\mathbf{i} + \left(\frac{2\sqrt{2}}{3}t^{3/2}\right)\mathbf{j} + \left(\frac{t^2}{2}\right)\mathbf{k},\ 0 \le t \le 2$

$x = t,\ y = \frac{2\sqrt{2}}{3}t^{3/2},\ z = \frac{t^2}{2}$

$\frac{dx}{dt} = 1,\ \frac{dy}{dt} = \sqrt{2}t^{1/2},\ \frac{dz}{dt} = t$

$$\sqrt{\left(\frac{dx}{dt}\right)^2 + \left(\frac{dy}{dt}\right)^2 + \left(\frac{dz}{dt}\right)^2}\,dt = \sqrt{1 + 2t + t^2}\,dt$$

$= \sqrt{(t + 1)^2}\,dt = |t + 1|\,dt = (t + 1)\,dt$ on the domain

given. Then $M = \int_C \delta\,ds = \int_0^2 \left(\frac{1}{t + 1}\right)(t + 1)\,dt$

$$= \int_0^2 dt = 2;$$

$$M_{yz} = \int_C x\delta\,ds = \int_0^2 t\left(\frac{1}{t + 1}\right)(t + 1)\,dt$$

$$= \int_0^2 t\, dt = 2;$$

$$M_{xz} = \int_C y\delta\, ds = \int_0^2 \left(\frac{2\sqrt{2}}{3}t^{3/2}\right)\left(\frac{1}{t+1}\right)(t+1)\, dt$$

$$= \int_0^2 \frac{2\sqrt{2}}{3}t^{3/2}\, dt = \frac{32}{15};$$

$$M_{xy} = \int_C z\delta\, ds = \int_0^2 \left(\frac{t^2}{2}\right)\left(\frac{1}{t+1}\right)(t+1)\, dt = \int_0^2 \frac{t^2}{2}\, dt = \frac{4}{3}$$

$$\bar{x} = \frac{M_{yz}}{M} = \frac{2}{2} = 1;\ \bar{y} = \frac{M_{xz}}{M} = \frac{\left(\frac{32}{15}\right)}{2} = \frac{16}{15};$$

$$\bar{z} = \frac{M_{xy}}{M} = \frac{\left(\frac{4}{3}\right)}{2} = \frac{2}{3};$$

$$I_x = \int_C (y^2 + z^2)\,\delta\, ds = \int_0^2 \left(\frac{8}{9}t^3 + \frac{t^4}{4}\right) dt = \frac{232}{45};$$

$$I_y = \int_C (x^2 + z^2)\,\delta\, ds = \int_0^2 \left(t^2 + \frac{t^4}{4}\right) dt = \frac{64}{15};$$

$$I_z = \int_C (y^2 + x^2)\,\delta\, ds = \int_0^2 \left(t^2 + \frac{8}{9}t^3\right) dt = \frac{56}{9};$$

$$R_x = \sqrt{\frac{I_x}{M}} = \sqrt{\frac{\left(\frac{232}{45}\right)}{2}} = \sqrt{\frac{116}{45}};$$

$$R_y = \sqrt{\frac{I_y}{M}} = \sqrt{\frac{\left(\frac{64}{15}\right)}{2}} = \sqrt{\frac{32}{15}};$$

$$R_z = \sqrt{\frac{I_z}{M}} = \sqrt{\frac{\left(\frac{56}{9}\right)}{2}} = \sqrt{\frac{28}{9}}$$

46. $\bar{z} = 0$ because the arch is in the xy-plane, and $\bar{x} = 0$ because the mass is distributed symmetrically with respect to the y-axis; $\mathbf{r}(t) = (a\cos t)\mathbf{i} + (a\sin t)\mathbf{j},\ 0 \le t \le \pi$

$$ds = \sqrt{\left(\frac{dx}{dt}\right)^2 + \left(\frac{dy}{dt}\right)^2 + \left(\frac{dz}{dt}\right)^2}\, dt$$

$$= \sqrt{(-a\sin t)^2 + (a\cos t)^2}\, dt = a\, dt, \text{ since } a \ge 0;$$

$$M = \int_C \delta\, ds = \int_C (2a - y)\, ds = \int_0^\pi (2a - a\sin t)\, a\, dt$$

$$= 2a^2\pi - 2a^2;$$

$$M_{xz} = \int_C y\delta\, dt = \int_C = y(2a - y)\, ds$$

$$= \int_0^\pi (a\sin t)(2a - a\sin t)\, dt$$

$$= \int_0^\pi (2a^2 \sin t - a^2 \sin^2 t)\, dt$$

$$= \left[-2a^2\cos t - a^2\left(\frac{t}{2} - \frac{\sin 2t}{4}\right)\right]_0^\pi = 4a^2 - \frac{a^2\pi}{2}$$

$$\bar{y} = \frac{\left(4a^2 - \frac{a^2\pi}{2}\right)}{2a^2\pi - 2a^2} = \frac{8 - \pi}{4\pi - 4}$$

$$(\bar{x}, \bar{y}, \bar{z}) = \left(0, \frac{8 - \pi}{4\pi - 4}, 0\right)$$

47. $\mathbf{r}(t) = (e^t\cos t)\mathbf{i} + (e^t \sin t)\mathbf{j} + e^t\mathbf{k},\ 0 \le t \le \ln 2$

$x = e^t\cos t,\ y = e^t\sin t,\ z = e^t$

$$\frac{dx}{dt} = (e^t\cos t - e^t\sin t),\ \frac{dy}{dt} = (e^t\sin t + e^t\cos t),\ \frac{dz}{dt} = e^t$$

$$\sqrt{\left(\frac{dx}{dt}\right)^2 + \left(\frac{dy}{dt}\right)^2 + \left(\frac{dz}{dt}\right)^2}\, dt$$

$$= \sqrt{(e^t\cos t - e^t\sin t)^2 + (e^t\sin t + e^t\cos t)^2 + (e^t)^2}\, dt$$

$$= \sqrt{3e^{2t}}\, dt = \sqrt{3}e^t\, dt;$$

$$M = \int_C \delta\, ds = \int_0^{\ln 2} \sqrt{3}e^t\, dt = \sqrt{3};$$

$$M_{xy} = \int_C z\delta\, ds = \int_0^{\ln 2} (\sqrt{3}e^t)(e^t)\, dt = \int_0^{\ln 2} \sqrt{3}e^{2t}\, dt$$

$$= \frac{3\sqrt{3}}{2}$$

$$\bar{z} = \frac{M_{xy}}{M} = \frac{\frac{3\sqrt{3}}{2}}{\sqrt{3}} = \frac{3}{2};$$

$$I_z = \int_C (x^2 + y^2)\delta\, ds$$

$$= \int_0^{\ln 2} (e^{2t}\cos^2 t + e^{2t}\sin^2 t)(\sqrt{3}e^t)\, dt = \int_0^{\ln 2} \sqrt{3}\, e^{3t}\, dt$$

$$= \frac{7\sqrt{3}}{3}$$

$$R_z = \sqrt{\frac{I_z}{M}} = \sqrt{\frac{7\sqrt{3}}{3\sqrt{3}}} = \sqrt{\frac{7}{3}}$$

48. $\mathbf{r}(t) = (2\sin t)\mathbf{i} + (2\cos t)\mathbf{j} + 3t\mathbf{k},\ 0 \le t \le 2\pi$

$x = 2\sin t,\ y = 2\cos t,\ z = 3t$

$$\frac{dx}{dt} = 2\cos t,\ \frac{dy}{dt} = -2\sin t,\ \frac{dz}{dt} = 3$$

$$\sqrt{\left(\frac{dx}{dt}\right)^2 + \left(\frac{dy}{dt}\right)^2 + \left(\frac{dz}{dt}\right)^2}\, dt = \sqrt{4 + 9}\, dt = \sqrt{13}\, dt;$$

$$M = \int_C \delta\, ds = \int_0^{2\pi} \delta\sqrt{13}\, dt = 2\pi\delta\sqrt{13};$$

$$M_{xy} = \int_C z\delta\, ds = \int_0^{2\pi} (3t)(\delta\sqrt{13})\, dt = 6\delta\pi^2\sqrt{13};$$

$$M_{yz} = \int_C x\delta\, ds = \int_0^{2\pi} (2\sin t)(\delta\sqrt{13})\, dt = 0;$$

$$M_{xz} = \int_C y\delta\, ds = \int_0^{2\pi} (2\cos t)(\delta\sqrt{13})\, dt = 0$$

$$\bar{x} = \bar{y} = 0 \text{ and } \bar{z} = \frac{M_{xy}}{M} = \frac{6\delta\pi^2\sqrt{13}}{2\delta\pi\sqrt{13}} = 3\pi$$

$(0, 0, 3\pi)$ is the center of mass

49. Because of symmetry $\bar{x} = \bar{y} = 0$.

Let $f(x, y, z) = x^2 + y^2 + z^2 = 25$

$\nabla f = 2x\mathbf{i} + 2y\mathbf{j} + 2z\mathbf{k}$

$|\nabla f| = \sqrt{4x^2 + 4y^2 + 4z^2} = 10$ and $\mathbf{p} = \mathbf{k}$

$|\nabla f \cdot \mathbf{p}| = 2z$, since $z \geq 0$

$$M = \iint_R \delta(x, y, z)\, d\sigma = \iint_R z\left(\frac{10}{2z}\right) dA = \iint_R 5\, dA$$

$= 5$(Area of the circular region) $= 80\pi$;

$$M_{xy} = \iint_R z\delta\, d\sigma = \iint_R 5z\, dA$$

$$= \iint_R 5\sqrt{25 - x^2 - y^2}\, dx\, dy$$

$$= \int_0^{2\pi}\int_0^4 (5\sqrt{25 - r^2})\, r\, dr\, d\theta = \int_0^{2\pi} \frac{490}{3}\, d\theta$$

$$= \frac{980}{3}\pi$$

$$\bar{z} = \frac{\left(\frac{980}{3}\pi\right)}{80\pi} = \frac{49}{12}$$

$$(\bar{x}, \bar{y}, \bar{z}) = \left(0, 0, \frac{49}{12}\right);\ I_z = \iint_R (x^2 + y^2)\, \delta\, d\sigma$$

$$= \iint_R 5(x^2 + y^2)\, dx\, dy = \int_0^{2\pi}\int_0^4 5r^3\, dr\, d\theta$$

$$= \int_0^{2\pi} 320\, d\theta = 640\pi;\ R_z = \sqrt{\frac{I_z}{M}} = \sqrt{\frac{640\pi}{80\pi}} = 2\sqrt{2}$$

50. On the face $z = 1$: $g(x, y, z) = z = 1$ and $\mathbf{p} = \mathbf{k}$

$\nabla g = \mathbf{k}$

$|\nabla g| = 1$ and $|\nabla g \cdot \mathbf{p}| = 1$

$d\sigma = dA$

$$I = \iint_R (x^2 + y^2)\, dA = 2\int_0^{\pi/4}\int_0^{\sec\theta} r^3\, dr\, d\theta = \frac{2}{3};$$

On the face $z = 0$: $g(x, y, z) = z = 0$

$\nabla g = \mathbf{k}$ and $\mathbf{p} = \mathbf{k}$

$|\nabla g| = 1$

$|\nabla g \cdot \mathbf{p}| = 1$

$d\sigma = dA$

$$I = \iint_R (x^2 + y^2)\, dA = \frac{2}{3}$$

On the face $y = 0$: $g(x, y, z) = y = 0$

$\nabla g = \mathbf{j}$ and $\mathbf{p} = \mathbf{j}$

$|\nabla g| = 1$

$|\nabla g \cdot \mathbf{p}| = 1$

$d\sigma = dA$

$$I = \iint_R (x^2 + 0)\, dA = \int_0^1\int_0^1 x^2\, dx\, dz = \frac{1}{3};$$

On the face $y = 1$: $g(x, y, z) = y = 1$

$\nabla g = \mathbf{j}$ and $\mathbf{p} = \mathbf{j}$

$|\nabla g| = 1$

$|\nabla g \cdot \mathbf{p}| = 1$

$d\sigma = dA$

$$I = \iint_R (x^2 + 1)\, dA = \int_0^1\int_0^1 (x^2 + 1)\, dx\, dz = \frac{4}{3};$$

On the face $x = 1$: $g(x, y, z) = x = 1$

$\nabla g = \mathbf{i}$ and $\mathbf{p} = \mathbf{i}$

$|\nabla g| = 1$

$|\nabla g \cdot \mathbf{p}| = 1$

$d\sigma = dA$

$$I = \iint_R (1^2 + y^2)\, dA = \int_0^1\int_0^1 (1 + y^2)\, dy\, dz = \frac{4}{3};$$

On the face $x = 0$: $g(x, y, z) = x = 0$

$\nabla g = \mathbf{i}$ and $\mathbf{p} = \mathbf{i}$

$|\nabla g| = 1$

$|\nabla g \cdot \mathbf{p}| = 1$

$d\sigma = dA$

$$I = \iint_R (0^2 + y^2)\, dA = \int_0^1\int_0^1 y^2\, dy\, dz = \frac{1}{3}$$

$$I_z = \frac{2}{3} + \frac{2}{3} + \frac{1}{3} + \frac{4}{3} + \frac{4}{3} + \frac{1}{3} = \frac{14}{3}$$

51. $M = 2xy + x$ and $N = xy - y$

$$\frac{\partial M}{\partial x} = 2y + 1,\ \frac{\partial M}{\partial y} = 2x,\ \frac{\partial N}{\partial x} = y,\ \frac{\partial N}{\partial y} = x - 1$$

$$\text{Flux} = \iint_R \left(\frac{\partial M}{\partial x} + \frac{\partial N}{\partial y}\right) dx\, dy$$

$$= \iint_R (2y + 1 + x - 1)\, dy\, dx$$

$$= \int_0^1\int_0^1 (2y + x)\, dy\, dx = \frac{3}{2};$$

$$\text{Circ} = \iint_R \left(\frac{\partial N}{\partial x} - \frac{\partial M}{\partial y}\right) dx\, dy$$

$$= \iint_R (y - 2x)\, dy\, dx = \int_0^1\int_0^1 (y - 2x)\, dy\, dx = -\frac{1}{2}$$

52. $M = y - 6x^2$ and $N = x + y^2$

$$\frac{\partial M}{\partial x} = -12x,\ \frac{\partial M}{\partial y} = 1,\ \frac{\partial N}{\partial x} = 1,\ \frac{\partial N}{\partial y} = 2y$$

$$\text{Flux} = \iint_R \left(\frac{\partial M}{\partial x} + \frac{\partial N}{\partial y}\right) dx\, dy$$

$$= \iint_R (-12x + 2y)\, dx\, dy$$

$$= \int_0^1\int_y^1 (-12x + 2y)\, dx\, dy$$

$$= \int_0^1 (4y^2 + 2y - 6)\, dy = -\frac{11}{3};$$

$$\text{Circ} = \iint_R \left(\frac{\partial N}{\partial x} - \frac{\partial M}{\partial y}\right) dx\, dy = \iint_R (1 - 1)\, dx\, dy = 0$$

53. $M = -\dfrac{\cos y}{x}$ and $N = \ln x \sin y$

$$\frac{\partial M}{\partial y} = \frac{\sin y}{x} \text{ and } \frac{\partial N}{\partial x} = \frac{\sin y}{x}$$

$$\oint_C \ln x \sin y\, dy - \frac{\cos y}{x}\, dx = \iint_R \left(\frac{\partial N}{\partial x} - \frac{\partial M}{\partial y}\right) dx\, dy$$

$$= \iint_R \left(\frac{\sin y}{x} - \frac{\sin y}{x}\right) dx\, dy = 0$$

54. a. Let $M = x$ and $N = y$

$$\frac{\partial M}{\partial x} = 1,\ \frac{\partial M}{\partial y} = 0,\ \frac{\partial N}{\partial x} = 0,\ \frac{\partial N}{\partial y} = 1$$

$$\text{Flux} = \int\int_R \left(\frac{\partial M}{\partial x} + \frac{\partial N}{\partial y}\right) dx\, dy$$

$$= \int\int_R (1 + 1)\, dx\, dy = 2\int\int_R dx\, dy$$

$$= 2(\text{Area of the region})$$

b. Let C be a closed curve to which Green's Theorem applies and let **n** be the unit normal vector to C. Let $\mathbf{F} = x\mathbf{i} + y\mathbf{j}$ and assume **F** is orthogonal to **n** at every point of C. Then the flux density of **F** at every point of C is 0 since $\mathbf{F} \cdot \mathbf{n} = 0$ at every point C

$$\frac{\partial M}{\partial x} + \frac{\partial N}{\partial y} = 0 \text{ at every point of } C$$

$$\text{Flux} = \int\int_R \left(\frac{\partial M}{\partial x} + \frac{\partial N}{\partial y}\right) dx\, dy = \int\int_R 0\, dx\, dy = 0.$$

But part (a) above states that the flux is 2(Area of the region) the area of the region would be 0, a contradiction. Therefore, **F** cannot be orthogonal to **n** at every point of C.

55. $\frac{\partial}{\partial x}(2xy) = 2y,\ \frac{\partial}{\partial y}(2yz) = 2z,\ \frac{\partial}{\partial z}(2xz) = 2x$

$\nabla \cdot \mathbf{F} = 2y + 2z + 2x$

$$\text{Flux} = \int\int\int_D (2x + 2y + 2z)\, dV$$

$$= \int_0^1\int_0^1\int_0^1 (2x + 2y + 2z)\, dx\, dy\, dz$$

$$= \int_0^1\int_0^1 (1 + 2y + 2z)\, dy\, dz = \int_0^1 (2 + 2z)\, dz = 3$$

56. $\frac{\partial}{\partial x}(xz) = z,\ \frac{\partial}{\partial y}(yz) = z,\ \frac{\partial}{\partial z}(1) = 0$

$\nabla \cdot \mathbf{F} = 2z$

$$\text{Flux} = \int\int\int_D 2z\, r\, dr\, d\theta\, dz$$

$$= \int_0^{2\pi}\int_0^4\int_3^{\sqrt{25 - r^2}} 2z\, dz\, r\, dr\, d\theta$$

$$= \int_0^{2\pi}\int_0^4 (16 - r^2) r\, dr\, d\theta = \int_0^{2\pi} 64\, d\theta = 128\pi$$

57. $\frac{\partial}{\partial x}(-2x) = -2,\ \frac{\partial}{\partial y}(-3y) = -3,\ \frac{\partial}{\partial z}(z) = 1$

$\nabla \cdot \mathbf{F} = -4;\ x^2 + y^2 + z^2 = 2$ and $x^2 + y^2 = z$

$z = 1$

$x^2 + y^2 = 1$

$$\text{Flux} = \int\int\int_D -4\, dV$$

$$= -4\int_0^{2\pi}\int_0^1\int_{r^2}^{\sqrt{2 - r^2}} dz\, r\, dr\, d\theta$$

$$= -4\int_0^{2\pi}\int_0^1 (r\sqrt{2 - r^2} - r^3)\, dr\, d\theta$$

$$= 4\int_0^{2\pi}\left(-\frac{7}{12} + \frac{2}{3}\sqrt{2}\right) d\theta = -\frac{2}{3}\pi(7 - 8\sqrt{2})$$

58. $\frac{\partial}{\partial x}(6x + y) = 6,\ \frac{\partial}{\partial y}(-x - z) = 0,\ \frac{\partial}{\partial z}(4yz) = 4y$

$\nabla \cdot \mathbf{F} = 6 + 4y;\ z = \sqrt{x^2 + y^2} = r$

$$\text{Flux} = \int\int\int_D (6 + 4y)\, dV$$

$$= \int_0^{\pi/2}\int_0^1\int_0^r (6 + 4r\sin\theta)\, dz\, r\, dr\, d\theta$$

$$= \int_0^{\pi/2}\int_0^1 (6r^2 + 4r^3\sin\theta)\, dr\, d\theta$$

$$= \int_0^{\pi/2} (2 + \sin\theta)\, d\theta = \pi + 1$$

59. $\mathbf{F} = y\mathbf{i} + z\mathbf{j} + x\mathbf{k}$

$\nabla \cdot \mathbf{F} = 0$

$$\text{Flux} = \int\int_S \mathbf{F} \cdot \mathbf{n}\, d\sigma = \int\int\int_D \nabla \cdot \mathbf{F}\, dV = 0$$

60. $\mathbf{F} = 3xz^2\mathbf{i} + y\mathbf{j} - z^3\mathbf{k}$

$\nabla \cdot \mathbf{F} = 3z^2 + 1 - 3z^2 = 1$

$$\text{Flux} = \int\int_S \mathbf{F} \cdot \mathbf{n}\, d\sigma = \int\int\int_D \nabla \cdot \mathbf{F}\, dV$$

$$= \int_0^4\int_0^{\sqrt{16 - x^2}}\int_0^{y/2} 1\, dz\, dy\, dx = \int_0^4\left(\frac{16 - x^2}{16}\right) dx$$

$$= \left[x - \frac{x^3}{48}\right]_0^4 = \frac{8}{3}$$

61. $\mathbf{F} = xy^2\mathbf{i} + x^2y\mathbf{j} + y\mathbf{k}$

$\nabla \cdot \mathbf{F} = y^2 + x^2 + 0$

$$\text{Flux} = \int\int_S \mathbf{F} \cdot \mathbf{n}\, d\sigma = \int\int\int_D \nabla \cdot \mathbf{F}\, dV$$

$$= \int\int\int_D (x^2 + y^2)\, dV = \int_0^{2\pi}\int_0^1\int_{-1}^1 r^2\, dz\, r\, dr\, d\theta$$

$$= \int_0^{2\pi}\int_0^1 2r^3\, dr\, d\theta = \int_0^{2\pi} \frac{1}{2}\, d\theta = \pi$$

62. a. $\mathbf{F} = (3z + 1)\mathbf{k}$

$\nabla \cdot \mathbf{F} = 3$

$$\text{Flux across the hemisphere} = \int\int_S \mathbf{F} \cdot \mathbf{n}\, d\sigma$$

$$= \int\int\int_D \nabla \cdot \mathbf{F}\, dV = \int\int\int_D 3\, dV$$

$$= 3\left(\frac{1}{2}\right)\left(\frac{4}{3}\pi a^3\right) = 2\pi a^3$$

b. $f(x, y, z) = x^2 + y^2 + z^2 - a^2 = 0$

$\nabla f = 2x\mathbf{i} + 2y\mathbf{j} + 2z\mathbf{k}$

$|\nabla f| = \sqrt{4x^2 + 4y^2 + 4z^2} = \sqrt{4a^2} = 2a$ since $a \geq 0$

$\mathbf{n} = \dfrac{2x\mathbf{i} + 2y\mathbf{j} + 2z\mathbf{k}}{2a} = \dfrac{x\mathbf{i} + y\mathbf{j} + z\mathbf{k}}{a}$

$\mathbf{F} \cdot \mathbf{n} = (3z + 1)\left(\dfrac{z}{a}\right)$; $\mathbf{p} = \mathbf{k}$

$\nabla f \cdot \mathbf{p} = \nabla f \cdot \mathbf{k} = 2z$

$|\nabla f \cdot \mathbf{p}| = 2z$ since $z \geq 0$

$d\sigma = \dfrac{|\nabla f|}{|\nabla f \cdot \mathbf{p}|} = \dfrac{2a}{2z}\,dA = \dfrac{a}{z}\,dA = \displaystyle\int_S\!\!\int \mathbf{F} \cdot \mathbf{n}\,d\sigma$

$= \displaystyle\int_{R_{xy}}\!\!\int (3z + 1)\left(\frac{z}{a}\right)\left(\frac{a}{z}\right) dA = \int_{R_{xy}}\!\!\int (3z + 1)\,dx\,dy$

$= \displaystyle\int_{R_{xy}}\!\!\int (3\sqrt{a^2 - x^2 - y^2} + 1)\,dx\,dy$

$= \displaystyle\int_0^{2\pi}\!\int_0^a (3\sqrt{a^2 - r^2} + 1)\,r\,dr\,d\theta = \int_0^{2\pi}\left(\frac{a^2}{2} + a^3\right) d\theta$

$= \pi a^2 + 2\pi a^3$, which is the flux across the hemisphere. Across the base we find

$\mathbf{F} = [3(0) + 1]\mathbf{k} = \mathbf{k}$ since $z = 0$ in the xy-plane

$\mathbf{n} = -\mathbf{k}$(outward normal)

$\mathbf{F} \cdot \mathbf{n} = -1$

Flux across the base $= \displaystyle\int_S\!\!\int \mathbf{F} \cdot \mathbf{n}\,d\sigma$

$= \displaystyle\int_{R_{xy}}\!\!\int -1\,dx\,dy = -\pi a^2$. Therefore, the total flux across the closed surface is $(\pi a^2 + 2\pi a^3) - \pi a^2$

$= 2\pi a^3$.

Cumulative Review Exercises

Chapters 1-10 (pp. 912–915)

1. Since the function has no discontinuity at $x = 1$, the limit is $\dfrac{2(1)^2 - 1 - 1}{1^2 + 1 - 12} = 0$.

2. By l'Hôpital's Rule, $\displaystyle\lim_{x\to 0} \frac{\sin 3x}{4x} = \lim_{x\to 0} \frac{3\cos 3x}{4} = \frac{3}{4}$.

3. By l'Hôpital's Rule, $\displaystyle\lim_{x\to 0} \frac{\frac{1}{x+1} - 1}{x} = \lim_{x\to 0} \frac{-\frac{1}{(x+1)^2}}{1} = -1$.

4. By l'Hôpital's Rule, $\displaystyle\lim_{x\to\infty} \frac{x + e^x}{x - e^x} = \lim_{x\to\infty} \frac{1 + e^x}{1 - e^x}$

$\displaystyle= \lim_{x\to\infty} \frac{e^x}{-e^x} = -1$.

5. By l'Hôpital's Rule, $\displaystyle\lim_{t\to 0} \frac{t(1 - \cos t)}{t - \sin t}$

$\displaystyle= \lim_{t\to 0} \frac{t\sin t + (1 - \cos t)}{1 - \cos t} = \lim_{t\to 0} \frac{t\cos t + 2\sin t}{\sin t}$

$\displaystyle= \lim_{t\to 0} \frac{-t\sin t + 3\cos t}{\cos t} = 3$.

6. By l'Hôpital's Rule, $\displaystyle\lim_{x\to 0^+} \frac{\ln(e^x - 1)}{\ln x} = \lim_{x\to 0^+} \frac{\frac{e^x}{(e^x - 1)}}{\frac{1}{x}}$

$\displaystyle= \lim_{x\to 0^+} \frac{xe^x}{e^x - 1} = \lim_{x\to 0^+} \frac{xe^x + e^x}{e^x} = 1$.

7. Use $f(x) = (e^x + x)^{1/x}$. Then $\ln f(x) = \dfrac{\ln(e^x + x)}{x}$, and

$\displaystyle\lim_{x\to 0} \frac{\ln(e^x + x)}{x} = \lim_{x\to 0} \frac{(e^x + 1)/(e^x + x)}{1} = 2$.

So $\displaystyle\lim_{x\to 0} (e^x + x)^{1/x} = \lim_{x\to 0} e^{\ln f(x)} = e^2$.

8. $\displaystyle\lim_{x\to 0}\left(\frac{3x + 1}{x} - \frac{1}{\sin x}\right) = \lim_{x\to 0} \frac{(3x + 1)\sin x - x}{x\sin x}$

$\displaystyle= \lim_{x\to 0} \frac{(3x + 1)\cos x + 3\sin x - 1}{x\cos x + \sin x}$

$\displaystyle= \lim_{x\to 0} \frac{-(3x + 1)\sin x + 6\cos x}{-x\sin x + 2\cos x} = 3$

9. **(a)** $2(1) - 1^2 = 1$

(b) $2 - 1 = 1$

(c) 1 [from (a) and (b)]

(d) Yes, since $\displaystyle\lim_{x\to 1} f(x) = f(1) = 1$

(e) No.

Left-hand derivative:

$\displaystyle\lim_{h\to 0^-} \frac{f(1 + h) - f(1)}{h} = \lim_{h\to 0^-} \frac{2(1 + h) - (1 + h)^2 - 1}{h}$

$\displaystyle= \lim_{h\to 0^-} \frac{2 + 2h - 1 - 2h - h^2 - 1}{h}$

$\displaystyle= \lim_{h\to 0^-} -\frac{h^2}{h}$

$\displaystyle= \lim_{h\to 0^-} -h = 0$

Right-hand derivative:

$\displaystyle\lim_{h\to 0^+} \frac{f(1 + h) - f(1)}{h} = \lim_{h\to 0^+} \frac{2 - (1 + h) - 1}{h}$

$\displaystyle= \lim_{h\to 0^+} \frac{-h}{h} = -1$

Since the left- and right-hand derivatives are not equal, f is not differentiable at $x = 1$.

10. Solve $4 - x^2 \leq 0$: all $x \leq -2$ and $x \geq 2$.

11. Horizontal: since as $x \to \pm\infty$, $2x^2 - x \to +\infty$ while $-1 \leq \cos x \leq 1$, the end behavior at both ends is $y = 0$.

Vertical: solve $2x^2 - x = 0$ to find $x = 0$, $x = \dfrac{1}{2}$.

12. One possible function is $y = \begin{cases} -3 + \dfrac{1}{(2 - x)}, & x < 2 \\ 3 - \dfrac{8}{x}, & x \geq 2 \end{cases}$

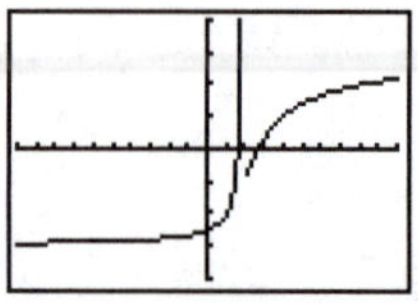

[−10, 10] by [−4, 4]

13. $\dfrac{f(5) - f(0)}{5 - 0} = \dfrac{\sqrt{9} - \sqrt{4}}{5} = \dfrac{1}{5}$

14. $y' = \dfrac{(x - 2)(1) - (x + 1)(1)}{(x - 2)^2} = -\dfrac{3}{(x - 2)^2}$

15. $y' = -\sin(\sqrt{1-3x})\left[\frac{1}{2}(1-3x)^{-1/2}\right](-3)$

$= \frac{3\sin\sqrt{1-3x}}{2\sqrt{1-3x}}$

16. $y' = \sin x \sec^2 x + \tan x \cos x = \frac{\sin x}{\cos^2 x} + \sin x$

$= \frac{(\sin x)(1+\cos^2 x)}{\cos^2 x}$

17. $y' = \left(\frac{1}{x^2+1}\right)(2x) = \frac{2x}{x^2+1}$

18. $y' = (e^{x^2-x})(2x-1) = (2x-1)e^{x^2-x}$

19. $y' = 2x\tan^{-1}x + \frac{x^2}{1+x^2}$

20. $y' = -3x^{-4}e^x + e^x x^{-3} = (x^{-3} - 3x^{-4})e^x$

21. $y' = 3\left(\frac{\csc x}{1+\cos x}\right)^2\left(\frac{(1+\cos x)(-\csc x\cot x) + \csc x\sin x}{(1+\cos x)^2}\right)$

$= \frac{3\csc^2 x}{(1+\cos x)^4}(1 - \csc x\cot x - \cos x\csc x\cot x)$

$= \frac{3\csc^2 x}{(1+\cos x)^4}(1 - \cot x\csc x - \cot^2 x)$

$= \frac{3\csc^2 x}{(1+\cos x)^4}(1 - \csc^2 x + \csc^2 x - \cot x\csc x - \cot^2 x)$

$= \frac{3\csc^2 x}{(1+\cos x)^4}(\csc^2 x - \cot x\csc x - 2\cot^2 x)$

$= \left(\frac{3}{(\sin^2 x)(1+\cos x)^4}\right)\left(\frac{1 - \cos x - 2\cos^2 x}{\sin^2 x}\right)$

$= \left(\frac{3}{(\sin^2 x)(1+\cos x)^4}\right)\left(\frac{(1+\cos x)(1-2\cos x)}{\sin^2 x}\right)$

$= \frac{3(1-2\cos x)}{(\sin^4 x)(1+\cos x)^3}$

22. $y' = \frac{d}{dx}\left(\frac{\pi}{2} - \sin^{-1}x\right) - \frac{d}{dx}\left(\frac{\pi}{2} - \tan^{-1}x\right)$

$= -\frac{1}{\sqrt{1-x^2}} + \frac{1}{1+x^2}$

23. $\frac{d}{dx}[\cos(xy) + y^2 - \ln x] = \frac{d}{dx}(0)$

$-\sin(xy)(xy' + y) + 2yy' - \frac{1}{x} = 0$

$y' = \frac{\frac{1}{x} + y\sin(xy)}{-x\sin(xy) + 2y} = \frac{1 + xy\sin(xy)}{2xy - x^2\sin(xy)}$

24. $y' = \frac{1}{2}|x|^{-1/2}\frac{d}{dx}|x| = \frac{1}{2\sqrt{|x|}}\left(\frac{|x|}{x}\right) = \frac{|x|}{2x\sqrt{|x|}}$

25. $\frac{dy}{dx} = \frac{dy/dt}{dx/dt} = \frac{-\cos t}{-\sin t} = \cot t = \frac{x-1}{1-y}$

26. $\ln y = \ln[(\cos x)^x]$

$\ln y = x\ln(\cos x)$

$\frac{1}{y}\frac{dy}{dx} = x\left(\frac{1}{\cos x}\right)(-\sin x) + \ln\cos x$

$\frac{dy}{dx} = y\cdot\left(\ln(\cos x) - \frac{x\sin x}{\cos x}\right)$

$= (\cos x)^x\left(\ln(\cos x) - \frac{x\sin x}{\cos x}\right)$

$= (\cos x)^{x-1}[\cos x\ln(\cos x) - x\sin x]$

27. By the Fundamental Theorem of Calculus, $y' = \sqrt{1+x^3}$.

28. $y = \left[-\cos t\right]_{2x}^{x^2} = -\cos(x^2) + \cos(2x);$

$y' = 2x\sin(x^2) - 2\sin(2x)$

29. $\frac{d}{dx}(y^2 + 2y) = \frac{d}{dx}(\sec x)$

$2yy' + 2y' = \sec x\tan x,$

$y' = \frac{\sec x\tan x}{2y+2},$

$y'' = \frac{(2y+2)(\sec^3 x + \sec x\tan^2 x) - 2y'\sec x\tan x}{(2y+2)^2}$

$= \frac{(2y+2)^2(\sec^3 x + \sec x\tan^2 x) - 2\sec^2 x\tan^2 x}{(2y+2)^3}$

30. $\left.\frac{(1+v)u' - uv'}{(1+v)^2}\right|_{x=0} = \frac{(1-3)(-1) - (2)(3)}{(1-3)^2} = -1$

31. **(a)** $v = \frac{dx}{dt} = 3t^2 - 12t + 9$

$a = \frac{dv}{dt} = 6t - 12$

(b) Solve $v = 0$ for t: $3(t-1)(t-3) = 0$; $t = 1$ or $t = 3$.

(c) Right: $v > 0$ for $0 \le t < 1,\ 3 < t \le 5$
left: $v < 0$ for $1 < t < 3$

(d) $a = 0$ at $t = 2$, and at that instant $v = 3(2)^2 - 12(2) + 9 = -3$ m/sec

32. For $x = 1$, $y = -1$ and

$\frac{dy}{dx} = 6(1)^2 - 12(1) + 4 = -2$

(a) $y + 1 = -2(x-1)$ or $y = -2x + 1$

(b) $y + 1 = \frac{1}{2}(x-1)$ or $y = \frac{1}{2}x - \frac{3}{2}$

33. For $x = \frac{\pi}{3}$, $y = \frac{\pi}{3}\cos\frac{\pi}{3} = \frac{\pi}{6}$ and

$\frac{dy}{dx} = -\frac{\pi}{3}\sin\frac{\pi}{3} + \cos\frac{\pi}{3} = -\frac{\pi\sqrt{3}}{6} + \frac{1}{2} = \frac{3-\pi\sqrt{3}}{6}.$

(a) $y - \frac{\pi}{6} = \left(\frac{3-\pi\sqrt{3}}{6}\right)\left(x - \frac{\pi}{3}\right)$ or

$y = \left(\frac{3-\pi\sqrt{3}}{6}\right)\left(x - \frac{\pi}{3}\right) + \frac{\pi}{6} \approx -0.407x + 0.950$

(b) $y - \frac{\pi}{6} = \left(\frac{6}{\pi\sqrt{3} - 3}\right)\left(x - \frac{\pi}{3}\right)$ or

$y = \left(\frac{6}{\pi\sqrt{3} - 3}\right)\left(x - \frac{\pi}{3}\right) + \frac{\pi}{6} \approx 2.458x - 2.050$

34. $\frac{1}{4}(2x) + \frac{1}{9}(2yy') = 0$; $y' = -\frac{\frac{x}{2}}{\frac{2y}{9}} = -\frac{9x}{4y}$.

At $x = 1$, $y = \frac{3\sqrt{3}}{2}$, the slope is $y' = -\frac{\sqrt{3}}{2}$.

(a) $y - \frac{3\sqrt{3}}{2} = -\frac{\sqrt{3}}{2}(x - 1)$ or

$y = -\frac{\sqrt{3}}{2}x + 2\sqrt{3} \approx -0.866x + 3.464$

(b) $y - \frac{3\sqrt{3}}{2} = \frac{2}{\sqrt{3}}(x - 1)$

or $y = \frac{2}{\sqrt{3}}x + \frac{5}{2\sqrt{3}} \approx 1.155x + 1.443$

35. At $t = \frac{\pi}{3}$: $x = 1$, $y = \frac{3\sqrt{3}}{2}$, and

$\frac{dy}{dx} = \frac{dy/dt}{dx/dt} = \frac{3\cos(\pi/3)}{-2\sin(\pi/3)} = -\frac{\sqrt{3}}{2}$.

(a) $y - \frac{3\sqrt{3}}{2} = -\frac{\sqrt{3}}{2}(x - 1)$ or

$y = -\frac{\sqrt{3}}{2}x + 2\sqrt{3} \approx -0.866x + 3.464$

(b) $y - \frac{3\sqrt{3}}{2} = \frac{2}{\sqrt{3}}(x - 1)$ or

$y = \frac{2}{\sqrt{3}}x + \frac{5}{2\sqrt{3}} \approx 1.155x + 1.443$

36. At $t = \frac{\pi}{4}$: $\mathbf{r} = \sec\left(\frac{\pi}{4}\right)\mathbf{i} + \tan\left(\frac{\pi}{4}\right)\mathbf{j} = \sqrt{2}\mathbf{i} + \mathbf{j}$ and

$\mathbf{r}' = \sec\left(\frac{\pi}{4}\right)\tan\left(\frac{\pi}{4}\right)\mathbf{i} + \sec^2\left(\frac{\pi}{4}\right)\mathbf{j} = \sqrt{2}\mathbf{i} + 2\mathbf{j}$, so that

$\frac{dy}{dx} = \frac{2}{\sqrt{2}} = \sqrt{2}$.

(a) $y - 1 = \sqrt{2}(x - \sqrt{2})$ or $y = \sqrt{2}x - 1 \approx 1.414x - 1$

(b) $y - 1 = -\frac{1}{\sqrt{2}}(x - \sqrt{2})$ or

$y = -\frac{1}{\sqrt{2}}x + 2 \approx -0.707x + 2$

37. With $f(x) = \begin{cases} -x + C_1, & x < 3 \\ 2x + C_2, & x > 3 \end{cases}$, choose C_1, C_2 so that

$-3 + C_1 = 2(3) + C_2 = 1$.

$f(x) = \begin{cases} -x + 4, & x \le 3 \\ 2x - 5, & x > 3 \end{cases}$.

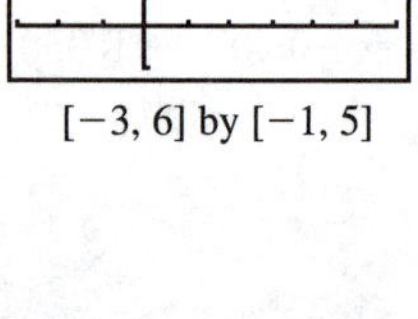

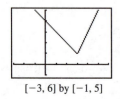

[−3, 6] by [−1, 5]

38. **(a)** $x \ne 0, 2$

(b) $x = 0$

(c) $x = 2$

(d) Absolute maximum of 2 at $x = 0$:
absolute minimum of 0 at $x = -2, 2, 3$

39. According to the Mean Value Theorem the driver's speed at some time was $\frac{111}{1.5} = 74$ mph.

40. **(a)** Increasing in $[-0.7, 2]$ (where $f' \ge 0$), decreasing in $[-2, -0.7]$ (where $f' \le 0$), and has a local minimum at $x \approx -0.7$.

(b) $y \approx -2x^2 + 3x + 3$

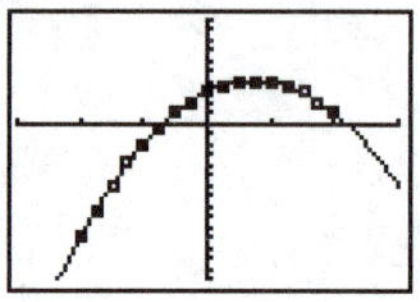

[−3, 3] by [−15, 10]

(c) $f(x) = -\frac{2}{3}x^3 + \frac{3}{2}x^2 + 3x + C$; choose C so that
$f(0) = 1$: $f(x) = -\frac{2}{3}x^3 + \frac{3}{2}x^2 + 3x + 1$.

41. $f(x) = x^2 - 3x - \cos x + C$; choose C so that $f(0) = -2$:
$f(x) = x^2 - 3x - \cos x - 1$.

42.

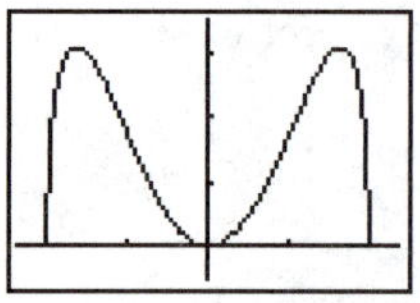

[−2.35, 2.35] by [−0.5, 3.5]

$f(x)$ is defined on $[-2, 2]$.

$f'(x) = 2x\sqrt{4 - x^2} - \frac{x^3}{\sqrt{4 - x^2}} = \frac{8x - 3x^3}{\sqrt{4 - x^2}}$; solve
$f'(x) = 0$ for x to find $x = 0$, $x = \pm\frac{2\sqrt{6}}{3}$.

The graph of $y = f'(x)$ is shown.

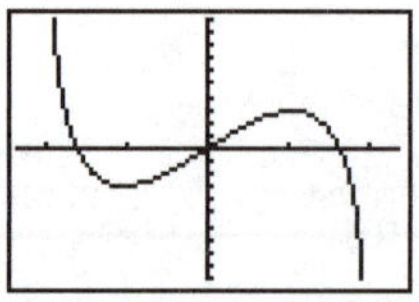

[−2.35, 2.35] by [−10, 10]

(a) $\left[-2, -\frac{2\sqrt{6}}{3}\right]$, $\left[0, \frac{2\sqrt{6}}{3}\right]$

(b) $\left[-\frac{2\sqrt{6}}{3}, 0\right], \left[\frac{2\sqrt{6}}{3}, 2\right]$

Use NDER to plot $f''(x)$ and find that $f''(x) = 0$ for $x \approx \pm 1.042$.

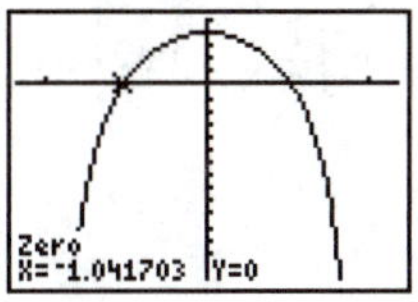

$[-2.35, 2.35]$ by $[-15, 5]$

(c) Approximately $(-1.042, 1.042)$

(d) Approximately $(-2, -1.042)$, $(1.042, 2)$

(e) Local (and absolute) maximum of approximately 3.079 at

$x = -\frac{2\sqrt{6}}{3}$ and $x = \frac{2\sqrt{6}}{3}$;

local (and absolute) minimum of 0 at $x = 0$ and at $x = \pm 2$

(f) $\approx(\pm 1.042, 1.853)$

43. **(a)** f has an absolute maximum at $x = 1$ and an absolute minimum at $x = 3$.

(b) f has a point of inflection at $x = 2$.

(c) The function $f(x) = \begin{cases} -\frac{1}{2}(x-1)^2 + 3, & -1 \le x \le 2 \\ -\frac{7}{2}\sqrt{x-2} + \frac{3}{2}, & 2 < x \le 3 \end{cases}$

is one example of a function with the given properties.

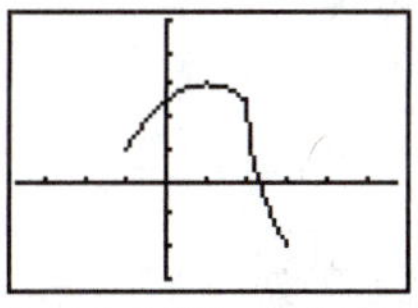

$[-3.7, 5.7]$ by $[-3, 5]$

44. $y = 2\sqrt{1 - \frac{x^2}{16}}$, and the area of the rectangle for $x > 0$ is

$A(x) = 4x\sqrt{1 - \frac{x^2}{16}} = x\sqrt{16 - x^2}$.

$A'(x) = \sqrt{16 - x^2} - \frac{x^2}{\sqrt{16 - x^2}} = \frac{2(8 - x^2)}{\sqrt{16 - x^2}}$, and so

$A'(x) = 0$ when $x = \pm 2\sqrt{2}$ and $y = \sqrt{2}$. The maximum possible area is $A(2\sqrt{2}) = 8$, with dimensions $4\sqrt{2}$ by $\sqrt{2}$.

45. $f\left(\frac{\pi}{4}\right) = \sqrt{2}$ and $f'\left(\frac{\pi}{4}\right) = \sec\left(\frac{\pi}{4}\right)\tan\left(\frac{\pi}{4}\right) = \sqrt{2}$. The equation is $y - \sqrt{2} = \sqrt{2}\left(x - \frac{\pi}{4}\right)$ or

$y = \sqrt{2}\left(x - \frac{\pi}{4}\right) + \sqrt{2} \approx 1.414x + 0.303$

46. $V = s^3$

$dV = 3s^2\,ds$

Since $|ds| = 0.01s$, the error of the volume calculation is approximately $|dV| = 3s^2(0.01s) = 0.03s^3 = 0.03V$, or 3%.

47. Let s be the rope length remaining and x be the horizontal distance from the dock.

(a) $x = \sqrt{s^2 - 5^2}$, $\frac{ds}{dt} = -1.5$, and $\frac{dx}{dt} = \frac{s}{\sqrt{s^2 - 25}}\frac{ds}{dt}$,

which means that for $s = 8$ ft,

$$\text{speed} = -\frac{dx}{dt} = -\frac{8}{\sqrt{64 - 25}}(-1.5) \approx 1.9 \text{ ft/sec}$$

(b) $\theta = \sec^{-1}\left(\frac{s}{5}\right)$, so $\frac{d\theta}{dt} = \frac{5}{|s|\sqrt{s^2 - 25}}\frac{ds}{dt}$, which for $s = 8$ ft becomes $\frac{5}{8\sqrt{64 - 25}}(-1.5) = -0.15$ rad/sec.

48. **(a)** Let h be the level of the coffee in the pot, and let V be the volume of the coffee in the pot.

$$h = \frac{V}{16\pi}, \text{ so } \frac{dh}{dt} = \frac{dV/dt}{16\pi} = \frac{9}{16\pi} \approx 0.179 \text{ in./min.}$$

(b) Now let h be the level of the coffee in the cone, and let V be the volume of the coffee in the cone.

$$V = \frac{1}{3}\pi\left(\frac{h}{2}\right)^2 h = \frac{\pi}{12}h^3, \text{ so } \frac{dV}{dt} = \left(\frac{\pi}{4}h^2\right)\left(\frac{dh}{dt}\right) \text{ and } \frac{dh}{dt}$$
$$= \left(\frac{4}{\pi h^2}\right)\left(\frac{dV}{dt}\right) = \left(\frac{4}{25\pi}\right)(-9) = -\frac{36}{25\pi}$$
$$\approx -0.458 \text{ in./min.}$$

Since $\frac{dh}{dt}$ is negative, the level in the cone is falling at the rate of about 0.458 in./min.

49. **(a)** $(1)(0 + 1.8 + 6.4 + \cdots + 16.2) = 165$ in.

(b) $(1)(1.8 + 6.4 + 12.6 + \cdots + 0) = 165$ in.

50. $\int_{-2}^{1} |x|\,dx = \int_{-2}^{0} -x\,dx + \int_{0}^{1} x\,dx = \left[-\frac{1}{2}x^2\right]_{-2}^{0} + \left[\frac{1}{2}x^2\right]_{0}^{1}$

$= 2 + \frac{1}{2} = 2.5$

51. Using Number 29 in the Table of Integrals, with $a = 2$,

$\int_{-2}^{2} \sqrt{4 - x^2}\,dx = \left[\frac{x}{2}\sqrt{4 - x^2} + 2\sin^{-1}\left(\frac{x}{2}\right)\right]_{-2}^{2}$

$= \pi - (-\pi) = 2\pi$.

Alternately, observe that the region under the curve and above the x-axis is a semicircle of radius 2, so the area is $\frac{1}{2}\pi(2)^2 = 2\pi$.

52. $\int_{1}^{3}\left(x^2 + \frac{1}{x}\right)dx = \left[\frac{1}{3}x^3 + \ln x\right]_{1}^{3} = 9 + \ln 3 - \frac{1}{3}$

$= \ln 3 + \frac{26}{3} \approx 9.765$

53. $\int_0^{\pi/4} \sec^2 x\,dx = \Big[\tan x\Big]_0^{\pi/4} = 1$

54. $\int_1^4 \frac{2+\sqrt{x}}{\sqrt{x}}\,dx = \int_1^4 \left(\frac{2}{\sqrt{x}} + 1\right)dx = \Big[4\sqrt{x} + x\Big]_1^4 = 12 - 5$
$= 7$

55. Let $u = \ln x$, so $du = \frac{1}{x}\,dx$.

Then $\int \frac{dx}{x(\ln x)^2} = \int u^{-2}\,du = -\frac{1}{u} + C = -\frac{1}{\ln x} + C.$

Therefore, $\int_e^{2e} \frac{dx}{x(\ln x)^2} =$

$\left[-\frac{1}{\ln x}\right]_e^{2e} = \left[-\frac{1}{1+\ln 2} + \frac{1}{1}\right] = \frac{\ln 2}{1+\ln 2} \approx 0.409$

56. $\int\left[(3-2t)\mathbf{i} + \left(\frac{1}{t}\right)\mathbf{j}\right]dt =$

$\Big[(3t - t^2)\mathbf{i} + (\ln t)\mathbf{j}\Big]_1^3 = (\ln 3)\mathbf{j} - 2\mathbf{i} = -2\mathbf{i} + (\ln 3)\mathbf{j}$

57. Let $u = e^x + 1$, so $du = e^x\,dx$.

Use the identity $\cot^2 u = \csc^2 u - 1$.

$$\begin{aligned}\int e^x \cot^2(e^x+1)\,dx &= \int \cot^2 u\,du \\ &= \int(\csc^2 u - 1)\,du \\ &= -\cot u - u + C \\ &= -\cot(e^x+1) - (e^x+1) + C\end{aligned}$$

Since $-1 + C$ is an arbitrary constant, we may redefine C and write the solution as $-\cot(e^x+1) - e^x + C$.

58. Let $u = \frac{s}{2}$, so $du = \frac{ds}{2}$.

$\int \frac{ds}{s^2+4} = \int \frac{ds}{4(s/2)^2+4} = \frac{1}{2}\int \frac{ds}{2[(s/2)^2+1]} = \frac{1}{2}\int \frac{du}{u^2+1}$
$= \frac{1}{2}\tan^{-1} u + C = \frac{1}{2}\tan^{-1}\left(\frac{s}{2}\right) + C$

59. Let $u = \cos(x-3)$, so $du = -\sin(x-3)\,dx$.

$\int \frac{\sin(x-3)}{\cos^3(x-3)}\,dx = \int(-u^{-3})\,du = \frac{1}{2}u^{-2} + C$
$= \frac{1}{2\cos^2(x-3)} + C$

60. Use integration by parts.

$u = e^{-x}$ $\quad dv = \cos 2x\,dx$

$du = -e^{-x}\,dx$ $\quad v = \frac{1}{2}\sin 2x$

$\int e^{-x}\cos 2x\,dx = \frac{1}{2}e^{-x}\sin 2x + \int \frac{1}{2}e^{-x}\sin 2x\,dx$

Now let

$u = e^{-x}$ $\quad dv = \frac{1}{2}\sin 2x\,dx$

$du = -e^{-x}\,dx$ $\quad v = -\frac{1}{4}\cos 2x$

Then

$\int e^{-x}\cos 2x\,dx$
$= \frac{1}{2}e^{-x}\sin 2x - \frac{1}{4}e^{-x}\cos 2x - \frac{1}{4}\int e^{-x}\cos 2x\,dx$

so

$\int e^{-x}\cos 2x\,dx = \frac{e^{-x}}{5}(2\sin 2x - \cos 2x) + C$

61. $\frac{x+2}{x^2-5x-6} = \frac{x+2}{(x+1)(x-6)} = \frac{A}{x+1} + \frac{B}{x-6}$

$x + 2 = A(x-6) + B(x+1) = (A+B)x + (B-6A)$

Solving $A + B = 1$, $B - 6A = 2$ yields $A = -\frac{1}{7}$, $B = \frac{8}{7}$ so

$\frac{x+2}{x^2-5x-6} = \frac{8}{7(x-6)} - \frac{1}{7(x+1)}$. Then

$\int \frac{x+2}{x^2-5x-6}\,dx = \int\left(\frac{8}{7(x-6)} - \frac{1}{7(x+1)}\right)dx$
$= \frac{8}{7}\ln|x-6| - \frac{1}{7}\ln|x+1| + C = \frac{1}{7}\ln\frac{(x-6)^8}{|x+1|} + C$

62. Area $\approx \frac{5}{2}[3 + 2(8.3) + 2(9.9) + \cdots + 2(8.3) + 3] = 359;$

Volume $\approx 25 \times 359 = 8975\ \text{ft}^3$

63. $y = -(t+1)^{-1} - \frac{1}{2}e^{-2t} + C$; $y(0) = -1 - \frac{1}{2} + C = 2$, so $C = \frac{7}{2}$ and $y = -\frac{1}{t+1} - \frac{1}{2}e^{-2t} + \frac{7}{2}$.

64. $y' = -\frac{1}{2}\cos 2\theta - \sin\theta + C_1$, and $y'\left(\frac{\pi}{2}\right) = 0 \Rightarrow$

$y' = -\frac{1}{2}\cos 2\theta - \sin\theta + \frac{1}{2}$.

$y = -\frac{1}{4}\sin 2\theta + \cos\theta + \frac{1}{2}\theta + C_2$, and $y\left(\frac{\pi}{2}\right) = 0$

$\Rightarrow y = -\frac{1}{4}\sin 2\theta + \cos\theta + \frac{1}{2}\theta - \frac{\pi}{4}$

65. Use integration by parts.

$u = x^2 \qquad dv = \sin x\,dx$

$du = 2x\,dx \qquad v = -\cos x$

$\int x^2 \sin x\,dx = -x^2 \cos x + \int 2x \cos x\,dx$

Now let

$u = x \qquad dv = 2\cos x\,dx$

$du = dx \qquad v = 2\sin x$

$$\int x^2 \sin x\,dx = -x^2\cos x + 2x\sin x - \int 2\sin x\,dx$$
$$= -x^2\cos x + 2x\sin x + 2\cos x + C$$
$$= (2 - x^2)\cos x + 2x\sin x + C$$

The graph of the slope field of the differential equation $\frac{dy}{dx} = x^2 \sin x$ and the antiderivative $y = (2 - x^2)\cos x + 2x\sin x$ is shown below.

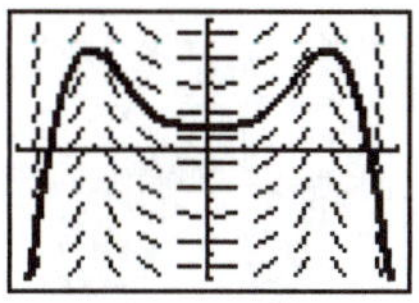

[−5, 5] by [−10, 10]

66. Use integration by parts.

$u = x \qquad dv = e^x\,dx$

$du = dx \qquad v = e^x$

$\int x e^x\,dx = xe^x - \int e^x\,dx = xe^x - e^x + C = e^x(x - 1) + C$

Confirm by differentiation:

$\frac{d}{dx}[e^x(x-1) + C] = e^x + (x-1)e^x = xe^x$

67. **(a)** $y = Ce^{kt}$, with $6{,}000 = Ce^{k(2)}$ and $10{,}000 = Ce^{k(5)}$.

Then $\frac{10{,}000}{6{,}000} = e^{k(5-2)}$, so $\frac{5}{3} = e^{3k}$ and therefore $k = \frac{\ln\left(\frac{5}{3}\right)}{3} \approx 0.170.$

Furthermore, $C = \frac{6{,}000}{e^{2k}} \approx 4268$. The approximate number of bacteria is given by $y = 4268e^{0.170t}$.

(b) About 4268

68. Let t be the time in minutes where $t = 0$ represents right now, and let $T(t)$ be the number of degrees above room temperature. Then we may write $T(t) = T_0e^{-kt}$ where $T(0) = 50$ and $T(-15) = 65$, giving $T_0 = 50$ and $k = \frac{1}{15}\ln\frac{13}{10} \approx 0.0175.$

(a) $50e^{-k(120)} \approx 6.13$°C above room temperature.

(b) Solving $5 = 50e^{-kt}$ gives $t = \frac{\ln 0.1}{-k} \approx 131.6$ minutes, or about 2 hours and 12 minutes from now.

69. $\frac{dy}{dx} = 0.08y\left(1 - \frac{y}{500}\right)$

$$\frac{500\,dy}{y(500-y)} = 0.08\,dx$$
$$\frac{(500-y)+y}{y(500-y)}\,dy = 0.08\,dx$$
$$\left(\frac{1}{y} + \frac{1}{500-y}\right)dy = 0.08\,dx$$

Integrate both sides.

$$\ln|y| - \ln|500-y| = 0.08x + C_1$$
$$\frac{y}{500-y} = C_2e^{0.08x}$$
$$y \cdot (1 + C_2e^{0.08x}) = 500C_2e^{0.08x}$$
$$y = \frac{500}{1 + Ce^{-0.08x}}$$

70.
$$\frac{dy}{dx} = (y-4)(x+3)$$
$$\frac{dy}{y-4} = (x+3)\,dx$$
$$\int\frac{dy}{y-4} = \int(x+3)\,dx$$
$$\ln|y-4| = \frac{x^2}{2} + 3x + C_1$$
$$y - 4 = e^{C_1}e^{(x^2/2)+3x} + 4$$
$$y = Ce^{(x^2/2)+3x} + 4$$

71. Use EULERT.

x	y
0	0
0.1	0.1
0.2	0.2095
0.3	0.3285
0.4	0.4568
0.5	0.5946
0.6	0.7418
0.7	0.8986
0.8	1.0649
0.9	1.2411
1.0	1.4273

72. The region has four congruent portions, so

$$\text{Area} = 4\int_0^{\pi/2} \sin 2x\, dx = 4\left[-\frac{1}{2}\cos 2x\right]_0^{\pi/2} = 4$$

73. Solve $5 - x^2 = x^2 - 3$ to find the integration limits:

$2x^2 = 8 \Rightarrow x = \pm 2$. Then

$$\text{Area} = \int_{-2}^{2} [(5 - x^2) - (x^2 - 3)]\, dx = \int_{-2}^{2} (8 - 2x^2)\, dx$$
$$= \left[8x - \frac{2}{3}x^3\right]_{-2}^{2} = \frac{64}{3}$$

74. Solve $y^2 - 3 = y + 2$ to find the integration limits:

$y^2 - y - 5 = 0 \Rightarrow y = \dfrac{1 \pm \sqrt{21}}{2}$. Then

$$\text{Area} = \int_{(1-\sqrt{21})/2}^{(1+\sqrt{21})/2} [(y + 2) - (y^2 - 3)\, dy \approx 16.039.$$

75. $$\text{Area} = \frac{1}{2}\int_0^{2\pi} r^2\, d\theta = \frac{1}{2}\int_0^{2\pi} 9(1 + \cos\theta)^2\, d\theta$$
$$= \frac{9}{2}\int_0^{2\pi} (1 + 2\cos\theta + \cos^2\theta)\, d\theta$$
$$= \frac{9}{2}\int_0^{2\pi} \left(1 + 2\cos\theta + \frac{1 + \cos 2\theta}{2}\right) d\theta$$
$$= \frac{9}{2}\int_0^{2\pi} \left(\frac{3}{2} + \cos\theta + \frac{1}{2}\cos 2\theta\right) d\theta$$
$$= \frac{9}{2}\left[\frac{3}{2}\theta + 2\sin\theta - \frac{1}{4}\sin 2\theta\right]_0^{2\pi}$$
$$= \frac{9}{2}(3\pi - 0) = \frac{27\pi}{2} \approx 42.412$$

76. $$\text{Volume} = \int_{-1}^{1} \pi\left(\frac{x^3}{2}\right)^2 dx = \frac{\pi}{4}\left[\frac{1}{7}x^7\right]_{-1}^{1} = \frac{\pi}{14} \approx 0.224$$

77. Solve $4x - x^2 = 0$ to find the limit of integration:

$x = 0$ or $x = 4$. By the cylindrical shell method,

$$\text{Volume} = \int_0^4 2\pi x(4x - x^2)\, dx = 2\pi\int_0^4 (4x^2 - x^3)\, dx$$
$$= 2\pi\left[\frac{4}{3}x^3 - \frac{1}{4}x^4\right]_0^4 = \frac{128\pi}{3}$$
$$\approx 134.041.$$

78. The average value is the integral divided by the interval length. Using NINT,

$$\text{average value} = \frac{1}{\pi}\int_0^{\pi} \sqrt{\sin x}\, dx \approx 0.763$$

79. $y' = \sec^2 x$, so we may use NINT to obtain

$$\text{Length} = \int_{-\pi/4}^{\pi/4} \sqrt{1 + (\sec^2 x)^2}\, dx \approx 2.556.$$

80. $\dfrac{dx}{dt} = \cos t$ and $\dfrac{dy}{dt} = 1 - \sin t$, so we may use NINT to obtain

$$\text{Length} = \int_{-\pi/2}^{\pi/2} \sqrt{\cos^2 t + (1 - \sin t)^2}\, dt$$
$$= \int_{-\pi/2}^{\pi/2} \sqrt{2 - 2\sin t}\, dt = 4.$$

81. Using NINT,

$$\text{Length} = \int_0^{\pi} \sqrt{\theta^2 + \left(\frac{dr}{d\theta}\right)^2}\, d\theta = \int_0^{\pi} \sqrt{1 + \theta^2}\, d\theta \approx 6.110.$$

82. $\dfrac{dy}{dx} = -\dfrac{1}{2}e^{-x/2}$, so we may use NINT to obtain

$$\text{Area} = \int_0^2 2\pi e^{-x/2}\sqrt{1 + \left(\frac{-e^{-x/2}}{2}\right)^2}\, dx$$
$$= \int_0^2 \pi e^{-x/2}\sqrt{4 + e^{-x}}\, dx \approx 8.423.$$

83. $\dfrac{dx}{dt} = \cos t$ and $\dfrac{dy}{dt} = 1 - \sin t$, so

$$\text{Area} = \int_0^{\pi/2} 2\pi \sin t\sqrt{\cos^2 t + (1 - \sin t)^2}\, dt$$
$$= \int_0^{\pi/2} 2\pi\sin t\sqrt{2 - 2\sin t}\, dt \approx 3.470.$$

84. $\dfrac{dr}{d\theta} = 1$, so we may use NINT to obtain

$$\text{Area} = \int_{\pi/2}^{\pi} 2\pi\theta\sin\theta\sqrt{\theta^2 + 1}\, d\theta \approx 32.683.$$

85. $$\text{Volume} = \int_0^1 \pi\left(\frac{\sqrt{x} - x^2}{2}\right)^2 dx$$
$$= \int_0^1 \frac{\pi}{4}(x - 2x^{5/2} + x^4)\, dx$$
$$= \frac{\pi}{4}\left[\frac{1}{2}x^2 - \frac{4}{7}x^{7/2} + \frac{1}{5}x^5\right]_0^1$$
$$= \frac{9\pi}{280} \approx 0.101.$$

86. Use the region's symmetry:

$$\text{Volume} = 2\int_0^{\pi/4} \pi(2\tan x)^2\, dx$$
$$= 8\pi\int_0^{\pi/4} \tan^2 x\, dx$$
$$= 8\pi\int_0^{\pi/4} (\sec^2 x - 1)\, dx$$
$$= 8\pi\Big[\tan x - x\Big]_0^{\pi/4}$$
$$= 8\pi\left(1 - \frac{\pi}{4}\right)$$
$$= 8\pi - 2\pi^2 \approx 5.394.$$

87. **(a)** $F = kx \Rightarrow 200 = k(0.8) \Rightarrow k = 250$ N/m, so for $F = 300$ N, $x = \dfrac{300}{250} = 1.2$ m.

(b) $$\text{Work} = \int_0^{1.2} 250x\, dx = \Big[125x^2\Big]_0^{1.2} = 180 \text{ J}$$

88. **(a)** The work required to raise a thin disk at height y from the bottom is

$$(\text{weight})(\text{distance}) = \left[60\pi\left(\frac{y}{2}\right)^2 dy\right](12 - y).$$

$$\text{Total work} = \int_0^{10} 15\pi y^2(12 - y)\, dy$$

$$= 15\pi\int_0^{10} (-y^3 + 12y^2)\, dy = 15\pi\left[-\frac{1}{4}y^4 + 4y^3\right]_0^{10}$$

$$= 22{,}500\pi \approx 70{,}686 \text{ ft-lb}.$$

(b) $\dfrac{22{,}500\pi}{275} \approx 257$ sec $= 4$ min, 17 sec

89. The sideways force exerted by a thin disk at depth y is its edge area times the pressure, or

$(2\pi\, dy)(849y) = 1698\pi y\, dy.$

Total force $= \displaystyle\int_0^H 1698\pi y\, dy = 849\pi H^2$, where H is depth.

Solve: $40{,}000 = 849\pi H^2$

$\Rightarrow H = \sqrt{\dfrac{40{,}000}{849\pi}}$ and $V = \pi H \approx 12.166$ ft^3.

90. Use l'Hôpital's Rule: $\displaystyle\lim_{x\to\infty} \frac{\ln x}{\sqrt{x}} = \lim_{x\to\infty} \frac{\frac{1}{x}}{\frac{x^{-1/2}}{2}} = \lim_{x\to\infty} \frac{2}{\sqrt{x}} = 0.$

$f(x) = \ln x$ grows slower than $g(x) = \sqrt{x}$.

91. Use the limit comparison test with $f(t) = \dfrac{1}{t^2 - 4}$ and $g(t) = \dfrac{1}{t^2}$. Since f and g are both continuous on $[3, \infty)$, $\displaystyle\lim_{t\to\infty} \frac{f(t)}{g(t)} = 1$, and $\displaystyle\int_3^\infty g(t)\, dt$ converges, we conclude that $\displaystyle\int_3^\infty f(t)\, dt = \int_3^\infty \frac{dt}{t^2 - 4}$ converges.

92. Use the comparison test: for $x \ge 2$, $\dfrac{1}{\ln x} > \dfrac{1}{x}$, and $\displaystyle\lim_{b\to\infty} \int_2^b \frac{dx}{x} = \lim_{b\to\infty} (\ln b - \ln 2) = \infty$. Both integrals diverge.

93. $\displaystyle\int_{-\infty}^{\infty} e^{-|x|}\, dx = \int_{-\infty}^0 e^x\, dx + \int_0^\infty e^{-x}\, dx = 2\int_0^\infty e^{-x}\, dx$

Since $\displaystyle\int_0^\infty e^{-x}\, dx = \lim_{b\to\infty} \int_0^b e^{-x}\, dx = \lim_{b\to\infty} \left[-e^{-x}\right]_0^b = 1$, the original integral converges.

94. $\displaystyle\int_0^1 \frac{4r\, dr}{\sqrt{1 - r^2}} = \lim_{b\to 1^-} \int_0^b \frac{4r\, dr}{\sqrt{1 - r^2}} = \lim_{b\to 1^-} \left[-4\sqrt{1 - r^2}\right]_0^b$

$= \displaystyle\lim_{b\to 1^-} (-4\sqrt{1 - b^2} + 4) = 4$. The integral converges.

95. $\displaystyle\int_0^{10} \frac{dx}{1 - x} = \int_0^1 \frac{dx}{1 - x} + \int_1^{10} \frac{dx}{1 - x}$

Since $\displaystyle\int_0^1 \frac{dx}{1 - x} = \lim_{b\to 1^-} \int_0^b \frac{dx}{1 - x}$

$= \displaystyle\lim_{b\to 1^-} \left[-\ln(1 - x)\right]_0^b = \infty$, the original integral diverges.

96. $\displaystyle\int_0^2 \frac{dx}{\sqrt[3]{x - 1}} = \int_0^1 \frac{dx}{\sqrt[3]{x - 1}} + \int_1^2 \frac{dx}{\sqrt[3]{x - 1}}$

$= \displaystyle\lim_{b\to 1^-} \int_0^b \frac{dx}{\sqrt[3]{x - 1}} + \lim_{a\to 1^+} \int_a^2 \frac{dx}{\sqrt[3]{x - 1}}$

$= \displaystyle\lim_{b\to 1^-} \left[\frac{3}{2}(x - 1)^{2/3}\right]_0^b + \lim_{a\to 1^+} \left[\frac{3}{2}(x - 1)^{2/3}\right]_a^2 = 0.$

The whole integral converges.

97. We know that

$\dfrac{1}{1 + x} = 1 - x + x^2 - x^3 + \cdots + (-1)^n x^n + \cdots$

for $-1 < x < 1$. Substituting $2x$ for x yields

$\dfrac{1}{1 + 2x} = 1 - 2x + 4x^2 - 8x^3 + \cdots + (-1)^n 2^n x^n + \cdots$

for $-1 < 2x < 1$, so the interval of convergence is $-\dfrac{1}{2} < x < \dfrac{1}{2}$.

98. **(a)**

$$\cos t^2 = 1 - \frac{(t^2)^2}{2!} + \frac{(t^2)^4}{4!} - \frac{(t^2)^6}{6!} + \cdots + (-1)^n \frac{(t^2)^{2n}}{(2n)!} + \cdots$$

$$= 1 - \frac{t^4}{2!} + \frac{t^8}{4!} - \frac{t^{12}}{6!} + \cdots + (-1)^n \frac{t^{4n}}{(2n)!}$$

Integrating each term with respect to t from 0 to x yields

$$x - \frac{x^5}{5(2!)} + \frac{x^9}{9(4!)} - \frac{x^{13}}{13(6!)} + \cdots + (-1)^n \frac{x^{4n+1}}{(4n + 1)(2n)!} + \cdots.$$

(b) $-\infty < x < \infty$; Since the cosine series converges for all real numbers, so does the integrated series, by, the term-by-term integration theorem (Section 9.1, Theorem 2).

99. $\ln(2 + 2x) = \ln[2(x + 1)] = \ln 2 + \ln(x + 1)$

$= \ln 2 + x - \dfrac{x^2}{2} + \dfrac{x^3}{3} - \dfrac{x^4}{4} + \cdots + (-1)^{n-1}\dfrac{x^n}{n} + \cdots$

Since by the Ratio Test $\displaystyle\lim_{n\to\infty} \left|\frac{a_{n+1}}{a_n}\right| = \lim_{n\to\infty} \left|\frac{n}{n + 1} x\right| = |x|$, the series converges for $-1 < x \le 1$.

$\left(\displaystyle\sum_{n=1}^{\infty} \frac{(-1)^n}{n}\right.$ converges, but $\left.\displaystyle\sum_{n=1}^{\infty} -\frac{1}{n}\right.$ does not.$\Big)$

100. Let $f(x) = \sin x$. Then $f'(x) = \cos x$, $f''(x) = -\sin x$, $f'''(x) = -\cos x$, $f^{(4)}(x) = \sin x$, and so on. At $x = 2\pi$ the sine terms are zero and the cosine terms alternate between 1 and -1, so the Taylor series is

$$(x - 2\pi) - \frac{(x - 2\pi)^3}{3!} + \frac{(x - 2\pi)^5}{5!} - \cdots$$
$$+ (-1)^n \frac{(x - 2\pi)^{2n+1}}{(2n + 1)!} + \cdots.$$

101. The first six terms of the Maclaurin series are

$1 - x + \frac{x^2}{2!} - \frac{x^3}{3!} + \frac{x^4}{4!} - \frac{x^5}{5!} + \frac{x^6}{6!}$. By the Alternating Series Estimation Theorem, $|\text{error}| \le \left|\frac{x^7}{7!}\right| \le \frac{1}{7!} < 0.001$.

102. $f(0) = 1, f'(0) = \frac{1}{3(0+1)^{2/3}} = \frac{1}{3}$,

$f''(0) = -\frac{2}{9(0+1)^{2/3}} = -\frac{2}{9}$,

$\cdots f^{(n)}(0) = (-1)^{n-1}\frac{2 \cdot 5 \cdot \cdots \cdot (3n-4)}{3^n}$, so the Taylor series is

$$1 + \frac{1}{3}x - \frac{2}{2! \cdot 3^2}x^2 + \frac{2 \cdot 5}{3! \cdot 3^3}x^3 - \frac{2 \cdot 5 \cdot 8}{4! \cdot 3^4}x^4 + \cdots$$
$$+ (-1)^{n-1}\frac{2 \cdot 5 \cdot \cdots \cdot (3n-4)}{n! \cdot 3^n}x^n + \cdots.$$

Since by the Ratio Test

$$\lim_{n\to\infty}\left|\frac{a_{n+1}}{a_n}\right|$$
$$= \lim_{n\to\infty}\left|\frac{2 \cdot 5 \cdot \cdots \cdot (3n-4)(3n-1)x^{n+1}}{(n+1)!3^{n+1}} \cdot \frac{n!3^n}{2 \cdot 5 \cdot \cdots \cdot (3n-4)x^n}\right|$$
$$= \lim_{n\to\infty}\left|\frac{(3n-1)x}{(n+1)(3)}\right| = |x|, \text{ the radius of convergence is 1.}$$

103. Using the Ratio Test, $\lim_{n\to\infty}\left|\frac{a_{n+1}}{a_n}\right| = \lim_{n\to\infty}\left|\frac{2}{3^{n+1}} \cdot \frac{3^n}{2}\right| = \frac{1}{3}$, so the series converges.

104. Note that $a_n > \frac{1}{n}$ for every n. By the Direct Comparison Test, since $\sum_{n=1}^{\infty}\frac{1}{n}$ diverges, so does $\sum_{n=1}^{\infty}\frac{2}{\sqrt{n}}$.

105. Use the alternating series test.

Note that $\sum_{n=0}^{\infty}\frac{(-1)^n}{n+1} = \sum_{n=0}^{\infty}(-1)^n u_n$, where $u_n = \frac{1}{n+1}$.

Since each u_n is positive, $u_n > u_{n+1}$ for all n, and $\lim_{n\to\infty} u_n = 0$, the original series converges.

106. Using the Ratio Test,

$\lim_{n\to\infty}\left|\frac{a_{n+1}}{a_n}\right| = \lim_{n\to\infty}\left|\frac{3^{n+1}}{(n+1)!} \cdot \frac{n!}{3^n}\right| = \lim_{n\to\infty}\frac{3}{n+1} = 0$, and the series converges.

107. (a) Using the Ratio Test,

$\lim_{n\to\infty}\left|\frac{a_{n+1}}{a_n}\right| = \lim_{n\to\infty}\left|\frac{(x+2)^{n+1}}{n+1} \cdot \frac{n}{(x+2)^n}\right| = |x+2|$, which means that the series converges for $-1 < x+2 < 1$, or $-3 < x < -1$. Furthermore, at $x = -3$, the series is $\sum_{n=1}^{\infty}\frac{1}{n}$, which diverges, and at $x = -1$, the series is $\sum_{n=1}^{\infty}\frac{(-1)^n}{n}$, which converges. The interval of convergence is $-3 < x \le -1$ and the radius of convergence is 1.

(b) $-3 < x < -1$

(c) At $x = -1$

108. (a) Using the Ratio Test, $\lim_{n\to\infty}\left|\frac{a_{n+1}}{a_n}\right|$

$$= \lim_{n\to\infty}\left|\frac{x^{n+1}}{(n+1)\ln^2(n+1)} \cdot \frac{n\ln^2 n}{x^n}\right|$$
$$= \lim_{n\to\infty}\left|\frac{nx\ln^2(n)}{(n+1)\ln^2(n+1)}\right|$$
$$= |x|\left(\lim_{n\to\infty}\frac{n}{n+1}\right)\left(\lim_{n\to\infty}\frac{\ln n}{\ln(n+1)}\right)^2 = |x|, \text{ which means}$$

that the series converges for $-1 < x < 1$. At $x = \pm 1$, the series converges by the Integral Test:

$$\int_2^{\infty}\frac{1}{x(\ln x)^2}\,dx = \lim_{b\to\infty}\int_2^b\frac{1}{x(\ln x)^2}\,dx = \left[-\frac{1}{\ln x}\right]_2^b$$
$$= \lim_{b\to\infty}\left(-\frac{1}{\ln b} + \frac{1}{\ln 2}\right) = \frac{1}{\ln 2}. \text{ So the}$$

convergence interval is $-1 \le x \le 1$ and the radius of convergence is 1.

(b) $-1 \le x \le 1$

(c) Nowhere

109. $\frac{1}{\sqrt{2^2 + (-3)^2}}\langle 2, -3\rangle = \left\langle\frac{2}{\sqrt{13}}, -\frac{3}{\sqrt{13}}\right\rangle$

110. $\left\langle 1\cos\frac{\pi}{3}, 1\sin\frac{\pi}{3}\right\rangle = \left\langle\frac{1}{2}, \frac{\sqrt{3}}{2}\right\rangle$

111. $\left.\frac{dy}{dx}\right|_{t=3\pi/4} = \left.\frac{dy/dt}{dx/dt}\right|_{t=3\pi/4} = \left.\frac{-3\sin t}{4\cos t}\right|_{t=3\pi/4} = \frac{3}{4}$. The tangent vectors are $\frac{1}{\sqrt{3^2+4^2}}\langle 4, 3\rangle = \left\langle\frac{4}{5}, \frac{3}{5}\right\rangle$ and $\left\langle -\frac{4}{5}, -\frac{3}{5}\right\rangle$. The normal vectors are $\left\langle\frac{3}{5}, -\frac{4}{5}\right\rangle$ and $\left\langle -\frac{3}{5}, \frac{4}{5}\right\rangle$.

112. **(a)** $\mathbf{v}(t) = \dfrac{d\mathbf{r}}{dt} = (-\cos t)\mathbf{i} + (1 + \sin t)\mathbf{j}$

$\mathbf{a}(t) = \dfrac{d\mathbf{v}}{dt} = (\sin t)\mathbf{i} + (\cos t)\mathbf{j}$

(b) Using NINT, the distance traveled is

$$\int_{\pi/2}^{3\pi/2} |\mathbf{v}(t)|dt = \int_{\pi/2}^{3\pi/2} \sqrt{(-\cos t)^2 + (1 + \sin t)^2}\, dt$$

$$= \int_{\pi/2}^{3\pi/2} \sqrt{2 + 2\sin t}\, dt = 4.$$

113. Yes. The path of the ball is given by

$x = 100(\cos 45°)t = 50\sqrt{2}t$ and

$y = -16t^2 + 100(\sin 45°)t = -16t^2 + 50\sqrt{2}t.$

When $x = 130$, we have $t = \dfrac{13}{5\sqrt{2}}$ and so

$y = -16\left(\dfrac{13}{5\sqrt{2}}\right)^2 + 50\sqrt{2}\left(\dfrac{13}{5\sqrt{2}}\right) = 75.92$ ft, high enough

to easily clear the 35-ft tree.

114. Since $r\cos\theta = x$, $r\sin\theta = y$, the Cartesian equation is $x - y = 2$. The graph is a line with slope 1 and y-intercept -2.

115.

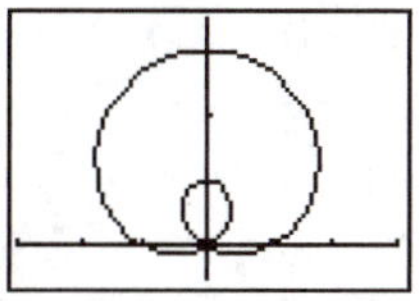

$[-3, 3]$ by $[-0.5, 3.5]$

The shortest possible θ-interval has length 2π.

116. $x = r\cos\theta = \cos\theta - \cos^2\theta,$

$y = r\sin\theta = \sin\theta - \sin\theta\cos\theta$

$$\frac{dy}{dx} = \frac{dy/d\theta}{dx/d\theta} = \frac{\cos\theta + \sin^2\theta - \cos^2\theta}{-\sin\theta + 2\sin\theta\cos\theta}$$

Zeros of $\dfrac{dy}{d\theta}$:

$\cos\theta + \sin^2\theta - \cos^2\theta = 0$

$\cos\theta + 1 - 2\cos^2\theta = 0$

$(2\cos\theta + 1)(\cos\theta - 1) = 0$

$\theta = 0, \theta = \dfrac{2\pi}{3}, \theta = \dfrac{4\pi}{3}$, or $\theta = 2\pi$

Zeros of $\dfrac{dx}{d\theta}$:

$-\sin\theta + 2\sin\theta\cos\theta = 0$

$\sin\theta = 0$ or $\cos\theta = \dfrac{1}{2}$

$\theta = 0, \theta = \pi, \theta = \dfrac{\pi}{3}$ or $\theta = \dfrac{5\pi}{3}, \theta = 2\pi$

There are horizontal tangents $\left(\dfrac{dy}{d\theta} = 0, \dfrac{dx}{d\theta} \neq 0\right)$ at $\theta = \dfrac{2\pi}{3}$ and at $\theta = \dfrac{4\pi}{3}$, and vertical tangents $\left(\dfrac{dx}{d\theta} = 0, \dfrac{dy}{d\theta} \neq 0\right)$ at $\theta = \dfrac{\pi}{3}$, at $\theta = \pi$, and at $\theta = \dfrac{5\pi}{3}$.

For $\theta = 0$ (or 2π), $\dfrac{dy}{dx}$ becomes $\dfrac{0}{0}$ and l'Hôpital's Rule leads to

$\left.\dfrac{dy}{dx}\right|_{\theta=0} = \dfrac{-\sin(0) + 4\sin(0)\cos(0)}{-\cos(0) + 2\cos^2(0) - 2\sin^2(0)} = 0$, so this is another horizontal tangent line.

Horizontal tangents:

At $\theta = 0$ or $\theta = 2\pi$, we have $r = 0$ and the Cartesian coordinates are $(0, 0)$, so the tangent is $y = 0$.

At $\theta = \dfrac{2\pi}{3}$, we have $r = \dfrac{3}{2}$ and the Cartesian coordinates are $\left(\dfrac{3}{2}\cos\dfrac{2\pi}{3}, \dfrac{3}{2}\sin\dfrac{2\pi}{3}\right) = \left(-\dfrac{3}{4}, \dfrac{3\sqrt{3}}{4}\right)$, so the tangent is $y = \dfrac{3\sqrt{3}}{4}$.

At $\theta = \dfrac{4\pi}{3}$, we again have $r = \dfrac{3}{2}$ and the Cartesian coordinates are $\left(\dfrac{3}{2}\cos\dfrac{4\pi}{3}, \dfrac{3}{2}\sin\dfrac{4\pi}{3}\right) = \left(-\dfrac{3}{4}, -\dfrac{3\sqrt{3}}{4}\right)$, so the tangent is $y = -\dfrac{3\sqrt{3}}{4}$.

Vertical tangents:

At $\theta = \dfrac{\pi}{3}$, we have $r = \dfrac{1}{2}$ and the Cartesian coordinates are

$\left(\dfrac{1}{2}\cos\dfrac{\pi}{3}, \dfrac{1}{2}\sin\dfrac{\pi}{3}\right) = \left(\dfrac{1}{4}, \dfrac{\sqrt{3}}{4}\right)$, so the tangent is $x = \dfrac{1}{4}$.

At $\theta = \pi$, we have $r = 2$ and the Cartesian coordinates are $(2\cos\pi, 2\sin\pi) = (-2, 0)$, so the tangent is $x = -2$.

At $\theta = \dfrac{5\pi}{3}$, we have $r = \dfrac{1}{2}$ and the Cartesian coordinates are $\left(\dfrac{1}{2}\cos\dfrac{5\pi}{3}, \dfrac{1}{2}\sin\dfrac{5\pi}{3}\right) = \left(\dfrac{1}{4}, -\dfrac{\sqrt{3}}{4}\right)$, so the tangent is $x = \dfrac{1}{4}$. In summary, the horizontal tangents are $y = 0$, $y = -\dfrac{3\sqrt{3}}{4}$, and $y = \dfrac{3\sqrt{3}}{4}$, and the vertical tangents are $x = -2$ and $x = \dfrac{1}{4}$.

Cumulative Review Exercises
Chapters 11-15 (pp. 916 - 920)

1. a. $|\mathbf{A}| = \sqrt{2^2 + (-2)^2 + 1^2} = \sqrt{9} = 3$

b. $|\mathbf{B}| = \sqrt{1^2 + 1^2 + (-1)^2} = \sqrt{3}$

c. $\mathbf{A} \cdot \mathbf{B} = 2 \cdot 1 + (-2) \cdot 1 + 1 \cdot (-1) = -1$

d. $\mathbf{B} \cdot \mathbf{A} = 1 \cdot 2 + 1 \cdot (-2) + (-1) \cdot 1 = -1$

e. $\mathbf{A} \times \mathbf{B} = \begin{vmatrix} \mathbf{i} & \mathbf{j} & \mathbf{k} \\ 2 & -2 & 1 \\ 1 & 1 & -1 \end{vmatrix} = \begin{vmatrix} -2 & 1 \\ 1 & -1 \end{vmatrix}\mathbf{i} - \begin{vmatrix} 2 & 1 \\ 1 & -1 \end{vmatrix}\mathbf{j} + \begin{vmatrix} 2 & -2 \\ 1 & 1 \end{vmatrix}\mathbf{k}$

$= \mathbf{i} + 3\mathbf{j} + 4\mathbf{k}$

f. $\mathbf{B} \times \mathbf{A} = \begin{vmatrix} \mathbf{i} & \mathbf{j} & \mathbf{k} \\ 1 & 1 & -1 \\ 2 & -2 & 1 \end{vmatrix} = -\mathbf{i} - 3\mathbf{j} - 4\mathbf{k}$

g. $\theta = \cos^{-1}\left(\frac{\mathbf{A} \cdot \mathbf{B}}{|\mathbf{A}|\,|\mathbf{B}|}\right) = \cos^{-1}\left(\frac{-1}{3\sqrt{3}}\right) \approx 1.76$ rad

h. $\text{proj}_{\mathbf{A}}\mathbf{B} = \left(\frac{\mathbf{B} \cdot \mathbf{A}}{\mathbf{A} \cdot \mathbf{A}}\right)\mathbf{A} = \frac{-1}{9}(2\mathbf{i} - 2\mathbf{j} + \mathbf{k})$

$= -\frac{2}{9}\mathbf{i} + \frac{2}{9}\mathbf{j} - \frac{1}{9}\mathbf{k}$

2. $\mathbf{u}$ and $\mathbf{v}$ are orthogonal if and only if $\mathbf{u} \cdot \mathbf{v} = 0$

$\mathbf{u} \cdot \mathbf{v} = a \cdot a + 1 \cdot (-1) + 3 \cdot (-1) = a^2 - 4 = 0$

$a = \pm 2$

3. a. $|\mathbf{A} \times \mathbf{B}| = \left\| \begin{vmatrix} \mathbf{i} & \mathbf{j} & \mathbf{k} \\ 2 & 1 & 0 \\ 1 & -2 & 0 \end{vmatrix} \right\| = |0\mathbf{i} + 0\mathbf{j} - 5\mathbf{k}|$

$= \sqrt{0^2 + 0^2 + (-5)^2} = 5$

b. $|(\mathbf{A} \times \mathbf{B}) \cdot \mathbf{C}| = |(-5\mathbf{k}) \cdot (\mathbf{i} + \mathbf{j} + \mathbf{k})|$

$= |0 \cdot 1 + 0 \cdot 1 + (-5) \cdot 1| = |-5| = 5$

4. $\mathbf{B} \cdot \mathbf{A} = 2 \cdot 1 + 3 \cdot 1 + (-1) \cdot 1 = 4$;

$\mathbf{A} \cdot \mathbf{A} = 1 \cdot 1 + 1 \cdot 1 + 1 \cdot 1 = 3$

$\mathbf{B} = \left(\frac{\mathbf{B} \cdot \mathbf{A}}{\mathbf{A} \cdot \mathbf{A}}\right)\mathbf{A} + \left(\mathbf{B} - \left(\frac{\mathbf{B} \cdot \mathbf{A}}{\mathbf{A} \cdot \mathbf{A}}\right)\mathbf{A}\right)$

$= \frac{4}{3}(\mathbf{i} + \mathbf{j} + \mathbf{k}) + \left((2\mathbf{i} + 3\mathbf{j} - \mathbf{k}) - \frac{4}{3}(\mathbf{i} + \mathbf{j} + \mathbf{k})\right)$

$= \left(\frac{4}{3}\mathbf{i} + \frac{4}{3}\mathbf{j} + \frac{4}{3}\mathbf{k}\right) + \left(\frac{2}{3}\mathbf{i} + \frac{5}{3}\mathbf{j} - \frac{7}{3}\mathbf{k}\right)$

5. $x = x_0 + tA = -2 + t$, $y = y_0 + tB = 1 - 2t$, $z = z_0 + tC = -1 + 3t$

6. $\overrightarrow{PQ} = (1 - 2)\mathbf{i} + (2 - 1)\mathbf{j} + (4 - (-1))\mathbf{k} = -\mathbf{i} + \mathbf{j} + 5\mathbf{k}$, and using $(x_0, y_0, z_0) = (2, 1, -1)$, we have $x = x_0 + tA = 2 - t$, $y = y_0 + tB = 1 + t$, $z = z_0 + tC = -1 + 5t$, $0 \le t \le 1$

7. Let $(2, 0, 1)$ be the point $\mathbf{S}$. The line L given parametrically passes through $P(1, 0, -2)$ and is parallel to $\mathbf{v} = \mathbf{i} + 2\mathbf{j} + \mathbf{k}$

$\overrightarrow{PS} = (2 - 1)\mathbf{i} + (0 - 0)\mathbf{j} + (1 - (-2))\mathbf{k} = \mathbf{i} + 3\mathbf{k}$;

$\overrightarrow{PS} \times \mathbf{v} = \begin{vmatrix} \mathbf{i} & \mathbf{j} & \mathbf{k} \\ 1 & 0 & 3 \\ 1 & 2 & 1 \end{vmatrix} = -6\mathbf{i} + 2\mathbf{j} + 2\mathbf{k}$;

$d = \frac{|\overrightarrow{PS} \times \mathbf{v}|}{|\mathbf{v}|} = \frac{\sqrt{36 + 4 + 4}}{\sqrt{1 + 4 + 1}} = \sqrt{\frac{22}{3}}$

8. Let $(-1, 2, 3)$ be the point S. The point $P(0, 0, 2)$ is on the plane $2x + 3y + z = 2$ and

$\overrightarrow{PS} = (-1 - 0)\mathbf{i} + (2 - 0)\mathbf{j} + (3 - 2)\mathbf{k} = -\mathbf{i} + 2\mathbf{j} + \mathbf{k}$

with $\mathbf{n} = 2\mathbf{i} + 3\mathbf{j} + \mathbf{k}$

$\text{distance} = \frac{|\mathbf{n} \cdot \overrightarrow{PS}|}{|\mathbf{n}|} = \sqrt{\frac{41}{14}}$

9. Using the component form:

$$A(x - x_0) + B(y - y_0) + C(z - z_0) = 0$$
$$3(x - (-1)) + (-1)(y - (-2)) + 2(z - 3) = 0$$
$$(3x + 3) + (-y - 2) + (2z - 6) = 0$$
$$3x - y + 2z = 5$$

10. $\mathbf{n} = 3\mathbf{i} - 2\mathbf{j} + 2\mathbf{k}$

$$A = (x - x_0) + B(y - y_0) + C(z - z_0) = 0$$
$$3(x - (-2)) + (-2)(y - 3) + 2(z - 4) = 0$$
$$3x + 6 - 2y + 6 + 2z - 8 = 0$$
$$3x - 2y + 2z = -4$$

11. $\overrightarrow{PQ} \times \overrightarrow{PR} = \begin{vmatrix} \mathbf{i} & \mathbf{j} & \mathbf{k} \\ -1 & 0 & 1 \\ 1 & 1 & 0 \end{vmatrix} = -\mathbf{i} + \mathbf{j} - \mathbf{k}$

Use the component form of the equation for a plane normal to the vector we just found; use the point $P(1, 1, 0)$.

$$A(x - x_0) + B(y - y_0) + C(z - z_0) = 0$$
$$-1(x - 1) + 1(y - 1) - 1(z - 0) = 0$$
$$-x + y - z = 0$$

12. First, find a vector $\mathbf{v}$ parallel to the line:

$\mathbf{v} = \begin{vmatrix} \mathbf{i} & \mathbf{j} & \mathbf{k} \\ 2 & -1 & 1 \\ 1 & -2 & 1 \end{vmatrix} = \begin{vmatrix} -1 & 1 \\ -2 & 1 \end{vmatrix}\mathbf{i} - \begin{vmatrix} 2 & 1 \\ 1 & 1 \end{vmatrix}\mathbf{j} + \begin{vmatrix} 2 & -1 \\ 1 & -2 \end{vmatrix}\mathbf{k}$

$= \mathbf{i} - \mathbf{j} - 3\mathbf{k}$

To find a point on the line, let $z = 0$ in the plane equations and solve for x and y: $(0, -1, 0)$;

$x = 0 + 1 \cdot t = t$, $y = -1 - 1 \cdot t = -1 - t$, $z = 0 - 3t = -3t$

13. $\mathbf{n}_1 = \mathbf{i} + \mathbf{j} - \mathbf{z}$ and $\mathbf{n}_2 = 2\mathbf{i} - \mathbf{j} + 3\mathbf{z}$

$|\mathbf{n}_1| = \sqrt{1^2 + 1^2 + (-1)^2} = \sqrt{3}$ and

$|\mathbf{n}_2| = \sqrt{2^2 + (-1)^2 + 3^2} = \sqrt{14}$

$\theta_R = \cos^{-1}\left(\frac{\mathbf{n}_1 \cdot \mathbf{n}_2}{|\mathbf{n}_1|\,|\mathbf{n}_2|}\right) = \cos^{-1}\frac{1 \cdot 2 + 1 \cdot (-1) + (-1) \cdot 3}{\sqrt{3} \cdot \sqrt{14}}$

$= \cos^{-1}\frac{-2}{\sqrt{42}}$. Since this angle is in Quadrant II, the acute angle is $\theta = \pi - \theta_R$: $\theta = \pi = \cos^{-1} -\frac{2}{\sqrt{42}}$

≈ 1.26 rad.

14. $x - 2y = 1$ is a line in the xy-plane.
$x - 2y = 1$ is a plane in three-dimensional space

15. $r = 4 \sin \theta$
$r^2 = 4r \sin \theta$
$x^2 + y^2 = 4y$
$x^2 + y^2 = 4y = 0$
$x^2 + (y - 2)^2 = 4.$
The set is a circular cylinder parallel to the z-axis.

16.

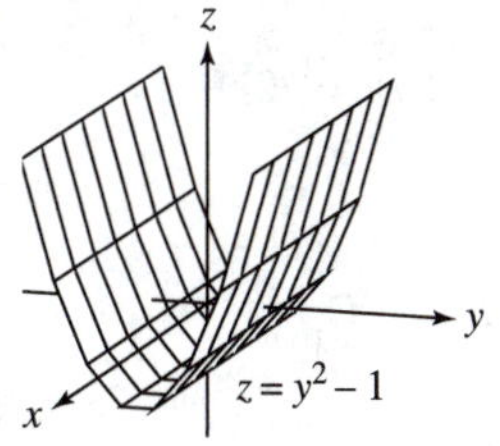

17.

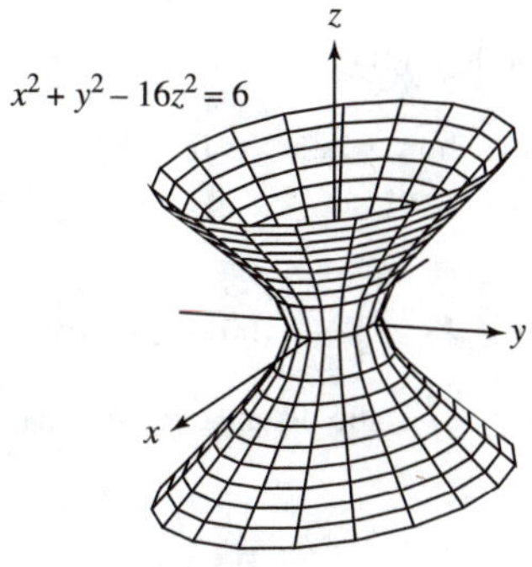

18.

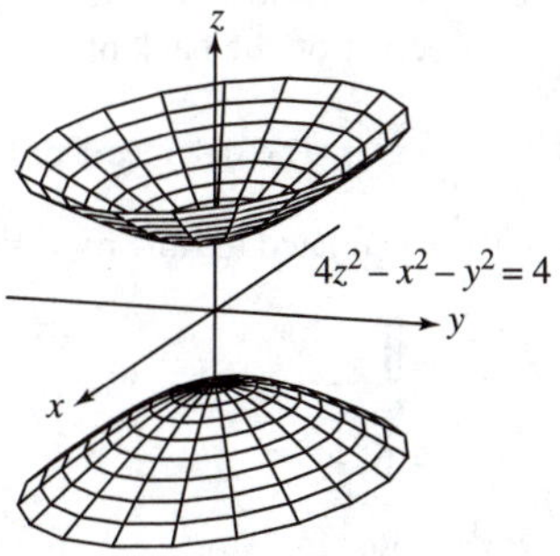

19. $\dfrac{d\mathbf{r}}{dt} = e^t\mathbf{i} + (\cos t)\mathbf{j} - 3t^2\mathbf{k}$

$\mathbf{r}(t) = \int (e^t\mathbf{i} + \cos t\mathbf{j} - 3t^2\mathbf{k})dt = e^t\mathbf{i} + \sin t\mathbf{j} - t^3\mathbf{k} + \mathbf{C}$

$\mathbf{r}(0) = \mathbf{i} + \mathbf{C} = 2\mathbf{i} + 3\mathbf{j}$ so $\mathbf{C} = \mathbf{i} + 3\mathbf{j}$
$\mathbf{r}(t) = (e^t + 1)\mathbf{i} + (\sin t + 3)\mathbf{j} - t^3\mathbf{k}$

20. $L = \displaystyle\int_0^{\pi} \sqrt{(-\sin t)^2 + (\cos t)^2 + (2t)^2}\, dt$

$= \displaystyle\int_0^{\pi} \sqrt{1 + 4t^2}\, dt$

$= \left[\dfrac{t}{2}\sqrt{1 + 4t^2} + \dfrac{1}{4}\ln(2t + \sqrt{1 + 4t^2})\right]_0^{\pi}$

$= \dfrac{\pi}{2}\sqrt{1 + 4\pi^2} + \dfrac{1}{4}\ln(2\pi + \sqrt{1 + 4\pi^2}) \approx 10.63$

21. $\mathbf{r}(t) = (e^t \sin t)\mathbf{i} + (e^t \cos t)\mathbf{j} + e^t\mathbf{k}$
$\mathbf{v} = (e^t \sin t + e^t \cos t)\mathbf{i} + (e^t \cos t - e^t \sin t)\mathbf{j} + e^t\mathbf{k}$
$|\mathbf{v}| = \sqrt{(e^t \sin t + e^t \cos t)^2 + (e^t \cos t - e^t \sin t)^2 + (e^t)^2}$
$= e^t\sqrt{3}$

$\mathbf{T} = \dfrac{\mathbf{v}}{|\mathbf{v}|} = \dfrac{1}{\sqrt{3}}(\sin t + \cos t)\mathbf{i} + \dfrac{1}{\sqrt{3}}(\cos t - \sin t)\mathbf{j} + \dfrac{1}{\sqrt{3}}\mathbf{k}$

$\mathbf{T}(0) = \dfrac{1}{\sqrt{3}}\mathbf{i} + \dfrac{1}{\sqrt{3}}\mathbf{j} + \dfrac{1}{\sqrt{3}}\mathbf{k}$

$\dfrac{d\mathbf{T}}{dt} = \dfrac{1}{\sqrt{3}}(\cos t - \sin t)\mathbf{i} + \dfrac{1}{\sqrt{3}}(-\sin t - \cos t)\mathbf{j}$

$\dfrac{d\mathbf{T}}{dt}(0) = \dfrac{1}{\sqrt{3}}\mathbf{i} - \dfrac{1}{\sqrt{3}}\mathbf{j}$

$\left|\dfrac{d\mathbf{T}}{dt}(0)\right| = \dfrac{\sqrt{2}}{\sqrt{3}}$

$\mathbf{N}(0) = \dfrac{\dfrac{d\mathbf{T}}{dt}(0)}{\left|\dfrac{d\mathbf{T}}{dt}(0)\right|} = \dfrac{\dfrac{1}{\sqrt{3}}\mathbf{i} - \dfrac{1}{\sqrt{3}}\mathbf{j}}{\dfrac{\sqrt{2}}{\sqrt{3}}} = \dfrac{1}{\sqrt{2}}\mathbf{i} - \dfrac{1}{\sqrt{2}}\mathbf{j};$

$\mathbf{B}(0) = \mathbf{T}(0) \times \mathbf{N}(0) = \begin{vmatrix} \mathbf{i} & \mathbf{j} & \mathbf{k} \\ \frac{1}{\sqrt{3}} & \frac{1}{\sqrt{3}} & \frac{1}{\sqrt{3}} \\ \frac{1}{\sqrt{2}} & -\frac{1}{\sqrt{2}} & 0 \end{vmatrix}$

$= \dfrac{1}{\sqrt{6}}\mathbf{i} + \dfrac{1}{\sqrt{6}}\mathbf{j} - \dfrac{2}{\sqrt{6}}\mathbf{k};$

$\mathbf{a} = (2e^t \cos t)\mathbf{i} - (2e^t \sin t)\mathbf{j} + e^t\mathbf{k}$
$\mathbf{a}(0) = 2\mathbf{i} + \mathbf{k}$

$\mathbf{v}(0) \times \mathbf{a}(0) = \begin{vmatrix} \mathbf{i} & \mathbf{j} & \mathbf{k} \\ 1 & 1 & 1 \\ 2 & 0 & 1 \end{vmatrix} = \mathbf{i} + \mathbf{j} - 2\mathbf{k}$

$|\mathbf{v} \times \mathbf{a}| = \sqrt{1 + 1 + 4} = \sqrt{6}$ and $|\mathbf{v}(0)| = \sqrt{3}$

$\kappa(0) = \dfrac{|\mathbf{v} \times \mathbf{a}|}{|\mathbf{v}(0)|^3} = \dfrac{\sqrt{6}}{(\sqrt{3})^3} = \dfrac{\sqrt{2}}{3};$

$\mathring{\mathbf{a}} = (2e^t \cos t - 2e^t \sin t)\mathbf{i} - (2e^t \cos t + 2e^t \sin t)\mathbf{j} + e^t\mathbf{k}$
$\mathring{\mathbf{a}}(0) = 2\mathbf{i} - 2\mathbf{j} + \mathbf{k}$

$\tau(0) = \begin{vmatrix} 1 & 1 & 1 \\ 2 & 0 & 1 \\ 2 & -2 & 1 \end{vmatrix} = -2$

22. $\mathbf{r}(t) = (1 + 2t^3)\mathbf{i} + (t + t^2)\mathbf{j} - (2 \sin t)\mathbf{k};$
$\mathbf{v} = 6t^2\mathbf{i} + (1 + 2t)\mathbf{j} - (2 \cos t)\mathbf{k}$
$|\mathbf{v}| = \sqrt{(36t^4) + (1 + 4t + 4t^2) + (4 \cos^2 t)}$
$= \sqrt{36t^4 + 4t^2 + 4t + 1 + 4 \cos^2 t};$

$a_T = \dfrac{d}{dt}|\mathbf{v}|$

$a_T(0) = \dfrac{2}{\sqrt{5}}$

$\mathbf{a} = \dfrac{d\mathbf{v}}{dt} = 12t\mathbf{i} + 2\mathbf{j} + 2 \sin t\, \mathbf{k}$

$|\mathbf{a}|^2 = 2\sqrt{36t^2 + 1 + \sin^2 t}$
$a_N = \sqrt{|\mathbf{a}|^2 - a_T^2}$

$a_N(0) = \dfrac{\sqrt{6}}{\sqrt{5}}$

$\mathbf{a}(0) = \dfrac{2}{\sqrt{5}}\mathbf{T} + \dfrac{\sqrt{6}}{\sqrt{5}}\mathbf{N}$

23. $\mathbf{r}(t) = t\mathbf{i} + (2\sin t)\mathbf{j} + (3\cos t)\mathbf{k}$;
$\mathbf{v} = \mathbf{i} + (2\cos t)\mathbf{j} - (3\sin t)\mathbf{k}$;
$\mathbf{a} = -(2\sin t)\mathbf{j} - (3\cos t)\mathbf{k}$;
$\mathbf{v}\cdot\mathbf{a} = 0\cdot 1 + (2\cos t)(-2\sin t) + (-3\sin t)(-3\cos t)$
$= 5\sin t\cos t$
$5\sin t\cos t = 0$ when $\sin t = 0$ or $\cos t = 0$
$\mathbf{v}$ and $\mathbf{a}$ are orthogonal when $t = 0, \frac{\pi}{2}, \pi, \frac{3\pi}{2}, 2\pi$

24. $\mathbf{r}(t) = t\mathbf{i} - t^2\mathbf{j} + 2t^3\mathbf{k}$, $(1, -1, 2)$
$t = 1$; $\mathbf{v} = \mathbf{i} - 2t\mathbf{j} + 6t^2\mathbf{k}$
$|\mathbf{v}| = \sqrt{1^2 + (-2t)^2 + (6t^2)^2} = \sqrt{1 + 4t^2 + 36t^4}$
$|\mathbf{v}(1)| = \sqrt{41}$

$\mathbf{T}(1) = \frac{1}{\sqrt{41}}\mathbf{i} - \frac{2}{\sqrt{41}}\mathbf{j} + \frac{6}{\sqrt{41}}\mathbf{k}$, which is normal to the normal plane

$$\frac{1}{\sqrt{41}}(x-1) - \frac{2}{\sqrt{41}}(y+1) + \frac{6}{\sqrt{41}}(z-2) = 0$$

or $x - 2y + 6z = 15$ is an equation of the normal plane. Next we calculate $\mathbf{N}(1)$ which is normal to the rectifying plane. Now, $\mathbf{a} = -2\mathbf{j} + 12t\mathbf{k}$
$\mathbf{a}(1) = -2\mathbf{j} + 12\mathbf{k}$

$$\mathbf{v}(1)\times\mathbf{a}(1) = \begin{vmatrix} \mathbf{i} & \mathbf{j} & \mathbf{k} \\ 1 & -2 & 6 \\ 0 & -2 & 12 \end{vmatrix} = -12\mathbf{i} - 12\mathbf{j} - 2\mathbf{k}$$

$|\mathbf{v}(1)\times\mathbf{a}(1)| = 2\sqrt{73}$

$$\kappa(1) = \frac{|\mathbf{v}(1)\times\mathbf{a}(1)|}{|\mathbf{v}(1)|^3} = \frac{2\sqrt{73}}{\sqrt{41}^3};\ \frac{ds}{dt} = |\mathbf{v}(t)|$$

$$\left.\frac{d^2s}{dt^2}\right|_{t=1} = \frac{1}{2}(1 + 4t^2 + 36t^4)^{-1/2}(8t + 144t^3)\Big|_{t=1} = \frac{76}{\sqrt{41}},$$

so $\mathbf{a} = \frac{d^2s}{dt^2}\mathbf{T} + \mathbf{k}\left(\frac{ds}{dt}\right)^2\mathbf{N}$

$$-2\mathbf{j} + 12\mathbf{k} = \frac{76}{\sqrt{41}}\left(\frac{\mathbf{i} - 2\mathbf{j} + 6\mathbf{k}}{\sqrt{41}}\right) + \frac{2\sqrt{73}}{\sqrt{41}^3}(\sqrt{41})^2\mathbf{N}$$

$$\mathbf{N} = -\frac{1}{\sqrt{2993}}(38\mathbf{i} - 35\mathbf{j} - 18\mathbf{k})$$

$38(x-1) - 35(y+1) - 18(z-2) = 0$ or
$38x - 35y - 18z = 37$ is an equation of the rectifying plane. Finally, $\mathbf{B}(1) = \mathbf{T}(1)\times\mathbf{N}(1)\left(\frac{1}{\sqrt{41}}\right)\left(-\frac{1}{\sqrt{2993}}\right)$

$$\begin{vmatrix} \mathbf{i} & \mathbf{j} & \mathbf{k} \\ 1 & -2 & 6 \\ 38 & -35 & -18 \end{vmatrix} = -\frac{1}{\sqrt{73}}(6\mathbf{i} + 6\mathbf{j} + \mathbf{k})$$

$6(x-1) + 6(y-1) + (z-2) = 0$ or $6x + 6y + z = 6$ is an equation of the osculating plane.

25. a. The velocity of the boat at (x, y) relative to land is the sum of the velocity due to the rower and the velocity of the river, or $\mathbf{v} = \left[-\frac{1}{250}(y-50)^2 + 10\right]\mathbf{i} - 20\mathbf{j}$.
Now, $\frac{dy}{dt} = -20$
$y = -20t + c$; $y(0) = 100$
$c = 100$
$y = -20t + 100$

$$\mathbf{v} = \left[-\frac{1}{250}(-20t + 50)^2 + 10\right]\mathbf{i} - 20\mathbf{j}$$
$$= \left(-\frac{8}{5}t^2 + 8t\right)\mathbf{i} - 20\mathbf{j}$$
$$\mathbf{r}(t) = \left(-\frac{8}{15}t^3 + 4t^2\right)\mathbf{i} - 20t\mathbf{j} + \mathbf{C}_1;$$

$\mathbf{r}(0) = 0\mathbf{i} + 100\mathbf{j}$
$100\mathbf{j} = \mathbf{C}_1$

$$\mathbf{r}(t) = \left(-\frac{8}{15}t^3 + 4t^2\right)\mathbf{i} + (100 - 20t)\mathbf{j}$$

b. The boat reaches the shore when $y = 0$
$0 = -20t + 100$ from part (a)
$t = 5$

$$\mathbf{r}(5) = \left(-\frac{8}{15}\cdot 125 + 4\cdot 25\right)\mathbf{i} + (100 - 20\cdot 5)\mathbf{j}$$

$= \left(-\frac{200}{3} + 100\right)\mathbf{i} = \frac{100}{3}\mathbf{i}$; the distance downstream is therefore $\frac{100}{3}$ m.

26. a. Let $a\mathbf{i} + b\mathbf{j}$ be the velocity of the boat. The velocity of the boat relative to an observer on the bank of the river is $\mathbf{v} = a\mathbf{i} + \left[b - \frac{30x(20-x)}{100}\right]\mathbf{j}$. The distance x of the boat as it crosses the river is related to time by $x = at$

$$\mathbf{v} = a\mathbf{i} + \left[b - \frac{30x(20-x)}{100}\right]\mathbf{j}$$
$$= a\mathbf{i} + \left(b + \frac{3a^24t^2 - 60at}{100}\right)\mathbf{j};$$
$$\mathbf{r}(t) = at\,\mathbf{i} + \left(bt + \frac{a^2t^3}{100} - \frac{30at^2}{100}\right)\mathbf{j} + C;$$

$\mathbf{r}(0) = 0\mathbf{i} + 0\mathbf{j} = \mathbf{c}$
$\mathbf{c} = 0$

$\mathbf{r}(t) = at\,\mathbf{i} + \left(bt + \frac{a^2t3}{100} - \frac{30at^2}{100}\right)\mathbf{j}$. The boat reaches the shore when $x = 20$
$20 = at$

$t = \frac{20}{a}$ and $y = 0$

$$0 = b\left(\frac{20}{a}\right) + \frac{a^2\left(\frac{20}{a}\right)^3 - 30a\left(\frac{20}{a}\right)^2}{100}$$
$$= \frac{20b}{a} + \frac{(20)^3 - 30(20)^2}{100a}$$
$$= \frac{2000b - 8000 - 12{,}000}{100a}$$

$b = 2$; the speed of the boat is $\sqrt{20} = |\mathbf{v}|$
$= \sqrt{a^2 + b^2} = \sqrt{a^2 + 4}$
$a^2 = 16$
$a = 4$; thus, $\mathbf{v} = 4\mathbf{i} + 2\mathbf{j}$ is the velocity of the boat.

b. $\mathbf{r}(t) = at\,\mathbf{i} + \left(bt + \frac{a^2t^3 - 30at^2}{100}\right)\mathbf{j}$
$= 4t\mathbf{i} + \left(2t + \frac{16t^3 - 120t^2}{100}\right)\mathbf{j}$
$= 4t\mathbf{i} + \left(2t + \frac{4}{25}t^3 - \frac{6}{5}t^2\right)\mathbf{j}$ by part (a), where $0 \le t \le 5$.

c. $x = 4t$ and $y = 2t + \frac{4}{25}t^3 - \frac{6}{5}t^2$

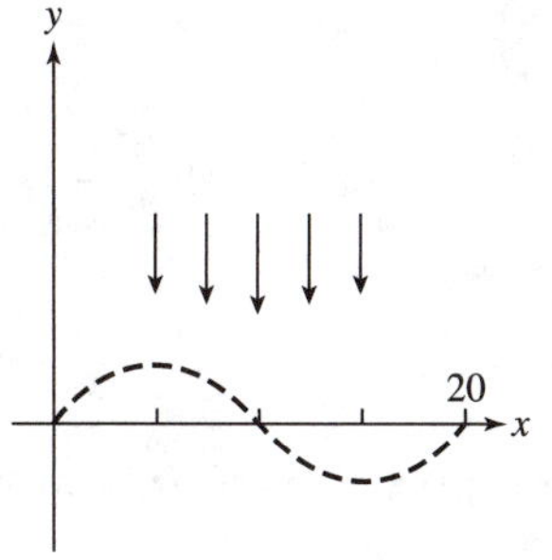

27. Let $\mathbf{a} = \mathbf{i} + \mathbf{j} + \mathbf{k}$ be the vector from 0 to A and $\mathbf{b} = \mathbf{i} + 2\mathbf{j} + 2\mathbf{k}$ be the vector from 0 to B. The vector $\mathbf{v}$ is orthogonal to $\mathbf{a}$ and $\mathbf{b}$
$\mathbf{v}$ is parallel to $\mathbf{b} \times \mathbf{a}$ (since the rotation is clockwise).

Now, $\mathbf{b} \times \mathbf{a} = \mathbf{i} + \mathbf{j} - 2\mathbf{k}$; $\text{proj}_a\,\mathbf{b} = \left(\frac{\mathbf{a}\cdot\mathbf{b}}{\mathbf{a}\cdot\mathbf{a}}\right)\mathbf{a}$

$= 2\mathbf{i} + 2\mathbf{j} + 2\mathbf{k}$
(2, 2, 2) is the center of the circular path (1, 3, 2) takes
radius $= \sqrt{1^2 + (-1)^2 + 0^2} = \sqrt{2}$
arc length per second covered by the point is $\frac{3}{2}\sqrt{2}$ units/sec $= |\mathbf{v}|$ (velocity is constant). A unit vector in the direction of

$\mathbf{v}$ is $\frac{\mathbf{b}\times\mathbf{a}}{|\mathbf{b}\times\mathbf{a}|} = \frac{1}{\sqrt{6}}\mathbf{i} + \frac{1}{\sqrt{6}}\mathbf{j} - \frac{2}{\sqrt{6}}\mathbf{k}$

$\mathbf{v} = |\mathbf{v}| = \left(\frac{\mathbf{b}\times\mathbf{a}}{|\mathbf{b}\times\mathbf{a}|}\right) = \frac{3}{2}\sqrt{2}\left(\frac{1}{\sqrt{6}}\mathbf{i} + \frac{1}{\sqrt{6}}\mathbf{j} - \frac{2}{\sqrt{6}}\mathbf{k}\right)$
$= \frac{\sqrt{3}}{2}\mathbf{i} + \frac{\sqrt{3}}{2}\mathbf{j} - \sqrt{3}\mathbf{k}.$

28. $\mathbf{r} = (2\sqrt{t}\cos t)\mathbf{i} + (3\sqrt{t}\sin t)\mathbf{j} + \sqrt{1-t}\mathbf{k},\ 0 \le t \le 1$
$x = 2\sqrt{t}\cos t$, $y = 3\sqrt{t}\sin t$, and $z = \sqrt{1-t}$
$x^2 = 4t\cos^2 t$ and $y^2 = 9t\sin^2 t$

$\frac{x^2}{4} + \frac{y^2}{9} = t\cos^2 t + t\sin^2 t = t;\ z^2 = 1 - t$

$t = 1 - z^2$

$\frac{x^2}{4} + \frac{y^2}{9} = 1 - z^2$

$\frac{x^2}{4} + \frac{y^2}{9} + z^2 = 1$, an ellipsoid

29 a. $\mathbf{r}(\theta) = (a\cos\theta)\mathbf{i} + (a\sin\theta)\mathbf{j} + b\theta\mathbf{k}$

$\frac{d\mathbf{r}}{dt} = [(-a\sin\theta)\mathbf{i} + (a\cos\theta)\mathbf{j} + b\mathbf{k}]\frac{d\theta}{dt};$

$|\mathbf{v}| = \sqrt{2gz} = \left|\frac{d\mathbf{r}}{dt}\right| = \sqrt{a^2+b^2}\,\frac{d\theta}{dt}$

$\frac{d\theta}{dt} = \sqrt{\frac{2gz}{a^2+b^2}}\quad \left.\frac{d\theta}{dt}\right|_{z=b\theta} = \sqrt{\frac{2gb\theta}{a^2+b^2}}$

$\left.\frac{d\theta}{dt}\right|_{\theta=2\pi} = \sqrt{\frac{4\pi gb}{a^2+b^2}} = 2\sqrt{\frac{\pi gb}{a^2+b^2}}$

b. $\frac{d\theta}{dt} = \sqrt{\frac{2gb\theta}{a^2+b^2}}$

$\frac{d\theta}{\sqrt{\theta}} = \sqrt{\frac{2bg}{a^2+b^2}}\,dt$

$2\theta^{1/2} = \sqrt{\frac{2gb}{a^2+b^2}}\,t + c;$

$t = 0$
$c = 0$

$2\theta^{1/2} = \sqrt{\frac{2gb}{a^2+b^2}}\,t$

$\theta = \frac{gbt^2}{2(a^2+b^2)};\ z = b\theta$

$z = \frac{gb^2t^2}{2(a^2+b^2)}$

c. $\mathbf{v}(t) = \frac{d\mathbf{r}}{dt}[(-a\sin\theta)\mathbf{i} + (a\cos\theta)\mathbf{j} + b\mathbf{k}]\frac{d\theta}{dt}$
$= [(-a\sin\theta)\mathbf{i} + (a\cos\theta)\mathbf{j} + b\mathbf{k}]\left(\frac{gbt}{a^2+b^2}\right),$

from part (b)

$\mathbf{v}(t) = \left[\frac{(-a\sin\theta)\mathbf{i} + (a\cos\theta)\mathbf{j} + b\mathbf{k}}{\sqrt{a^2+b^2}}\right]\left(\frac{gbt}{\sqrt{a^2+b^2}}\right)$

$= \frac{gbt}{\sqrt{a^2+b^2}}\mathbf{T};$

$\frac{d^2\mathbf{r}}{dt^2} = [(-a\cos\theta)\mathbf{i} - (a\sin\theta)\mathbf{j}]\left(\frac{d\theta}{dt}\right)^2$
$+ [(-a\sin\theta)\mathbf{i} + (a\cos\theta)\mathbf{j} + b\mathbf{k}]\frac{d^2\theta}{dt^2}$

$= \left(\frac{gbt}{a^2+b^2}\right)^2[(-a\cos\theta)\mathbf{i} - (a\sin\theta)\mathbf{j}]$
$+ [(-a\sin\theta)\mathbf{i} + (a\cos\theta)\mathbf{j} + b\mathbf{k}]\left(\frac{gb}{a^2+b^2}\right)$

$= \left[\frac{-a\sin\theta\,\mathbf{i} + (a\cos\theta)\mathbf{j} + b\mathbf{k}}{\sqrt{a^2+b^2}}\right]\left(\frac{gb}{\sqrt{a^2+b^2}}\right)$
$+ a\left(\frac{gbt}{a^2+b^2}\right)^2[(-\cos\theta)\mathbf{i} - (\sin\theta)\mathbf{j}]$

$= \frac{gb}{\sqrt{a^2+b^2}}\mathbf{T} + a\left(\frac{gbt}{a^2+b^2}\right)^2\mathbf{N}$ (there is no component in the direction of $\mathbf{B}$).

30. $\mathbf{L}(t) = \mathbf{r}(t) \times m\mathbf{v}(t)$

$\frac{d\mathbf{L}}{dt} = \left(\frac{d\mathbf{r}}{dt} \times m\mathbf{v}\right) + \left(\mathbf{r} + m\frac{d^2\mathbf{r}}{dt^2}\right)$

$\frac{d\mathbf{L}}{dt} = (\mathbf{v} \times m\mathbf{v}) + (\mathbf{r} \times m\mathbf{a}) = \mathbf{r} \times m\mathbf{a};\ \mathbf{F} = m\mathbf{a}$

$-\frac{c}{|\mathbf{r}|^3}\mathbf{r} = m\mathbf{a}$

$\frac{d\mathbf{L}}{dt} = \mathbf{r} \times m\mathbf{a} = \mathbf{r} \times \left(-\frac{c}{|\mathbf{r}|^3}\mathbf{r}\right) = -\frac{c}{|\mathbf{r}|^3}(\mathbf{r} \times \mathbf{r}) = \mathbf{0}$

$\mathbf{L}$ = constant vector

31. $\lim_{(x,y)\to(1,2)} \frac{2x - y}{4x^2 - y^2} = \lim_{(x,y)\to(1,2)} \frac{2x - y}{(2x+y)(2x-y)}$

$= \lim_{(x,y)\to(1,2)} \frac{1}{2x+y} = \frac{1}{4}$

32. $\lim_{(x,y)\to(0,0)} \frac{y^2}{x^2+y^2}$, along the line $y = 0$, the limit is 0; along the line $x = 0$, the limit is 1 therefore the limit does not exist.

33. Yes, $\lim_{(x,y)\to(0,0)} \frac{2\sin(x^2+y^2)}{x^2+y^2} = 2$, so define $f(0, 0) = 2$

34. $f_x = x\left[e^{-(x^2+y^2+z^2)} \cdot (-2x)\right] + e^{-(x^2+y^2+z^2)}$

$= (1 - 2x^2)e^{-(x^2+y^2+z^2)}$

$f_y = x\,e^{-(x^2+y^2+z^2)}(-2y) = -2xye^{-(x^2+y^2+z^2)}$

$f_x = xe^{-(x^2+y^2+z^2)}(-2z) = -2xze^{-(x^2+y^2+z^2)}$

35. $f\left(0, \frac{\pi}{6}\right) = \frac{1}{2}, f_x\left(0, \frac{\pi}{6}\right) = e^x \sin y\Big|_{(0,\pi/6)} = \frac{1}{2},$

$f_y\left(0, \frac{\pi}{6}\right) = e^x \cos y\Big|_{(0,\pi/6)} = \frac{\sqrt{3}}{2}$

$L(x, y) = \frac{1}{2} + \frac{1}{2}(x - 0) + \frac{\sqrt{3}}{2}\left(y - \frac{\pi}{6}\right)$

$= \frac{1}{2}x + \frac{\sqrt{3}}{2}y + \frac{6 - \sqrt{3}\pi}{12}$

36. $f\left(1, \sqrt{2}, \frac{1}{2}\right) = \sin^{-1}\left(\frac{1}{\sqrt{2}}\right) = \frac{\pi}{4},$

$f_x\left(1, \sqrt{2}, \frac{1}{2}\right) = \frac{yz}{\sqrt{1 - x^2y^2z^2}}\Big|_{(1,\sqrt{2},1/2)} = 1,$

$f_y\left(1, \sqrt{2}, \frac{1}{2}\right) = \frac{xz}{\sqrt{1 - x^2y^2z^2}}\Big|_{(1,\sqrt{2},1/2)} = \frac{\sqrt{2}}{2},$

$f_z\left(1, \sqrt{2}, \frac{1}{2}\right) = \frac{xy}{\sqrt{1 - x^2y^2z^2}} = 2$

$L(x, y, z) = \frac{\pi}{4} + 1(x - 1) + \frac{\sqrt{2}}{2}(y - \sqrt{2}) + 2\left(z - \frac{1}{2}\right)$

$= x + \frac{\sqrt{2}}{2}y + 2z + \frac{\pi - 12}{4}$

37. $f(1, 2) = 6, f_x(1, 2) = y^2 - y(\sin(x - 1)\Big|_{(1,2)}$

$= 4, f_y(1, 2) = 2xy + \cos(x - 1)\Big|_{(1,2)} = 5$

$L(x, y) = 6 + 4(x - 1) + 5(y - 2) = 4x + 5y - 8;$

$f_{xx}(x, y) = -y\cos(x - 1), f_{yy}(x, y) = 2x,$

$f_{xy}(x, y) = 2y - \sin(x - 1);\ |x - 1| \le 0.1$

$0.9 \le x \le 1.1$ and $|y - 2| \le 0.1$

$1.9 \le y \le 2.1$, thus, the max of $|f_{xx}(x, y)|$ on R is

$\frac{1}{(0.98)^2} \le 1.04$

$M = 1.04$; thus $|E(x, y)| \le \left(\frac{1}{2}\right)(1.04)(|x - 1| + |y - 1|)^2$

$\le (0.52)(0.2 + 0.2)^2 = 0.0832$

38. $\frac{\partial w}{\partial x} = e^y, \frac{\partial w}{\partial y} = xe^y + \sin z, \frac{\partial w}{\partial z} = y\cos z + \sin z,$

$\frac{dx}{dt} = \frac{1}{\sqrt{t}}, \frac{dy}{dt} = 1 + \frac{1}{t}, \frac{dz}{dt} = \pi$

$\frac{dw}{dt} = \frac{dw}{dx}\frac{dx}{dt} + \frac{dw}{dy}\frac{dy}{dt} + \frac{dw}{dz}\frac{dz}{dt}$

$= e^y\frac{1}{\sqrt{t}} + (xe^y + \sin z)\left(1 + \frac{1}{t}\right) + (y\cos z + \sin z)\,\pi;$

$t = 1$

$x = 2, y = 0, z = \pi; \frac{dw}{dt} = 1 + (2)(2) + (0) = 5$

39. $\frac{\partial w}{\partial x} = 2\cos(2x - y), \frac{\partial w}{\partial y} = -\cos(2x - y), \frac{\partial x}{\partial r} = 1,$

$\frac{\partial x}{\partial s} = \cos s, \frac{\partial y}{\partial r} = s, \frac{\partial y}{\partial s} = r$

$\frac{\partial w}{\partial r} = \frac{\partial w}{\partial x}\frac{\partial x}{\partial r} + \frac{\partial w}{\partial y}\frac{\partial y}{\partial r}$

$= 2\cos(2x - y) \cdot 1 - \cos(2x - y) \cdot s,\ r = \pi$ and $s = 0$

$x = \pi, y = 0; \frac{\partial w}{\partial r}\Big|_{r=\pi, s=0} = 2;$

$\frac{\partial w}{\partial s} = \frac{\partial w}{\partial x}\frac{\partial x}{\partial s} + \frac{\partial w}{\partial y}\frac{\partial y}{\partial s}$

$= 2\cos(2x - y)\cos s - \cos(2x - y)r$

$\frac{\partial w}{\partial s}\Big|_{r=\pi, s=0} = 2 - \pi$

40. $w = \int \frac{\partial w}{\partial x}dx = \int (1 + e^x\cos y) = x + e^x\cos y + g(y)$

$\frac{\partial w}{\partial y} = -e^x\sin y + g'(y) = 2y - e^x\sin y$

$g'(y) = 2y$

$g(y) = y^2 + c$

$w = x + e^x\cos y + y^2 + c$

$w(\ln 2, 0) = \ln 2 + 2 + c = \ln 2$

$c = -2$. thus, $w = x + e^x\cos y + y^2 - 2$

41. $\frac{\partial f}{\partial x} = 0$

$f(x, y) = h(y)$ is a function of y on y. Also, $\frac{\partial f}{\partial y} = 0$

$g(x, y) = k(x)$ is a function of x only. Moreover, $\frac{\partial f}{\partial x} = \frac{\partial g}{\partial x}$

$h'(y) = k'(x)$ for all x and y. This can happen only if $h'(y) = k'(x) = c$ is a constant. Integration gives $h(y) = cy + c$, and $k(x) = cx + c_2$. Where c_1 and c_2 are constants. Therefore $f(x, y) = cy + c$, and $g(x, y) = cx + c_2$.Then $f(1, 2) = g(1, 2) = 5$

$5 = 2c + c_1 = c + c_2$, and $f(0, 0) = 4$

$c_1 = 4$

$c = \frac{1}{2}$

$c_2 = \frac{9}{2}$. Thus, $f(x, y) = \frac{1}{2}y + 4$ and $g(x, y) = \frac{1}{2}x + \frac{9}{2}$.

42. $\mathbf{r} = (\cos 2t)\mathbf{i} + (\sin 2t)\mathbf{j} + 2t\mathbf{k}$

$\mathbf{v}(t) = (-2\sin 2t)\mathbf{i} + (2\cos 2t)\mathbf{j} + 2\mathbf{k}$

$\mathbf{v}\left(\frac{\pi}{2}\right) = -2\mathbf{j} + 2\mathbf{k}$

$\mathbf{u} = \frac{1}{\sqrt{2}}\mathbf{j} + \frac{1}{\sqrt{2}}\mathbf{k}; f(x, y, z) = xyz$

$\nabla f = yz\mathbf{i} + xz\mathbf{j} + xy\mathbf{k}$; $t = \frac{\pi}{2}$ yields the point on the helix $(-1, 0, \pi)$

$\nabla f\Big|_{(1, 0, \pi)} = -\pi\mathbf{j}$

$\nabla f \cdot \mathbf{u} = (-\pi\mathbf{j}); \cdot \left(-\frac{1}{\sqrt{2}}\mathbf{j} + \frac{1}{\sqrt{2}}\mathbf{k}\right) = \frac{\pi}{\sqrt{2}}$

43. a. $\mathbf{k}$ is a vector normal to $z = 10 - x^2 - y^2$ at the point $(0, 0, 10)$. So directions tangential to S at $(0, 0, 10)$ will be unit vectors $\mathbf{u} = a\mathbf{i} + b\mathbf{j}$. Also, $\nabla T(x, y, z)$ $= (2xy + 4)\mathbf{i} + (x^2 + 2yz + 14)\mathbf{j} + (y^2 + 1)\mathbf{k}$
$\nabla T(0, 0, 10) = 4\mathbf{i} + 14\mathbf{j} + \mathbf{k}$. We seek the unit vector $\mathbf{u} = a\mathbf{i} + b\mathbf{j}$ such that $D_{\mathbf{u}}T(0, 0, 10)$ $= (4\mathbf{i} + 14\mathbf{j} + \mathbf{k}) \cdot (a\mathbf{i} + b\mathbf{j}) = (4\mathbf{i} + 14\mathbf{j}) \cdot (a\mathbf{i} + b\mathbf{j})$ is a maximum. The maximum will occur when $a\mathbf{i} + b\mathbf{j}$ has the same direction as $4\mathbf{i} + 14\mathbf{j}$ or

$\mathbf{u} = \frac{1}{\sqrt{53}}(2\mathbf{i} + 7\mathbf{j})$.

b. A vector normal to S at $(1, 1, 8)$ is $\mathbf{n} = 2\mathbf{i} + 2\mathbf{j} + \mathbf{k}$. Now, $\nabla T(1, 1, 8) = 6\mathbf{i} + 31\mathbf{j} + 2\mathbf{k}$ and we seek the unit vector $\mathbf{u}$ such that $D_{\mathbf{u}}T(1, 1, 8) = \nabla T \cdot \mathbf{u}$ has its largest value. Now, write $\nabla T = \mathbf{v} + \mathbf{w}$, where v is parallel to ∇T and $\mathbf{w}$ is orthogonal to ∇T. Then $D_{\mathbf{u}}T = \nabla T \cdot \mathbf{u} = (\mathbf{v} + \mathbf{w}) \cdot \mathbf{u} = \mathbf{v} \cdot \mathbf{u} + \mathbf{w} \cdot \mathbf{u} = \mathbf{w} \cdot \mathbf{u}$.
Thus, $D_{\mathbf{u}}T(1, 1, 8)$ is a maximum when $\mathbf{u}$ has the same direction as $\mathbf{w}$. Now, $\mathbf{w} = \nabla T - \left(\frac{\nabla T \cdot \mathbf{n}}{|\mathbf{n}|^2}\right)\mathbf{n}$

$= (6\mathbf{i} + 31\mathbf{j} + 2\mathbf{k}) - \left(\frac{12 + 62 + 2}{4 + 4 + 1}\right)(2\mathbf{i} + 2\mathbf{j} + \mathbf{k})$

$= \left(6 - \frac{152}{9}\right)\mathbf{i} + \left(31 - \frac{152}{9}\right)\mathbf{j} + \left(2 - \frac{76}{9}\right)\mathbf{k}$

$= -\frac{98}{9}\mathbf{i} + \frac{127}{9}\mathbf{j} - \frac{58}{9}\mathbf{k}$

$\mathbf{u} = \frac{\mathbf{w}}{|\mathbf{w}|} = -\frac{1}{\sqrt{29{,}097}}(98\mathbf{i} - 127\mathbf{j} + 58\mathbf{k})$.

44. $\frac{\partial z}{\partial x} = \cos(x + y)$

$\frac{\partial z}{\partial x}\Big|_{(1, -1, 0)} = 1$ and $\frac{\partial z}{\partial y} = \cos(x + y)$

$\frac{\partial z}{\partial y}\Big|_{(1, -1, 0)} = 1$; thus the tangent plane is $1(x - 1)$ $+ 1(y + 1) - (z - 0) = 0$ or $x + y - z = 0$

45. $f_x = 6x^2 - 18x = 0$

$6x(x - 3) = 0$

$x = 0$ or $x = 3$; $f_y = 6y^2 + 6y - 12 = 0$

$6(y + 2)(y - 1) = 0$

$y = -2$ or $y = 1$

the critical points are $(0, -2)$, $(0, 1)$, $(3, -2)$, and $(3, 1)$;

$f_{xx} = 12x - 18$, $f_{yy} = 12y + 6$, and $f_{xy} = 0$; for $(0, -2)$:

$f_{xx}(0, -2) = -18$, $f_{yy}(0, -2) = -18$, $f_{xy}(0, -2) = 0$

$f_{xx}f_{yy} - f^2_{xy} = 324 > 0$ and $f_{xx} < 0$

local maximum of $f(0, -2) = 20$; for $(0, 1)$:

$f_{xx}(0, 1) = -18$, $f_{yy}(0, 1) = 18$, $f_{xy}(0, 1) = 0$

$f_{xx}f_{yy} - f_{xy}{}^2 = -324 < 0$

saddle point; for $(3, -2)$: $f_{xx}(3, -2) = 18$,

$f_{yy}(3, -2) = -18$, $f_{xy}(3, -2) = 0$

$f_{xx}f_{yy} - f_{xy}{}^2 = -324 < 0$

saddle point; for $(3, 1)$: $f_{xx}(3, 1)$: $f_{xx}(3, 1) = 18$,

$f_{yy}(3, 1) = 18$,

$f_{xy}(3, 1) = 0$

$f_{xx}f_{yy} - f_{xy}{}^2 = 324 > 0$ and $f_{xx} > 0$

local minimum of $f(3, 1) = -34$

46.

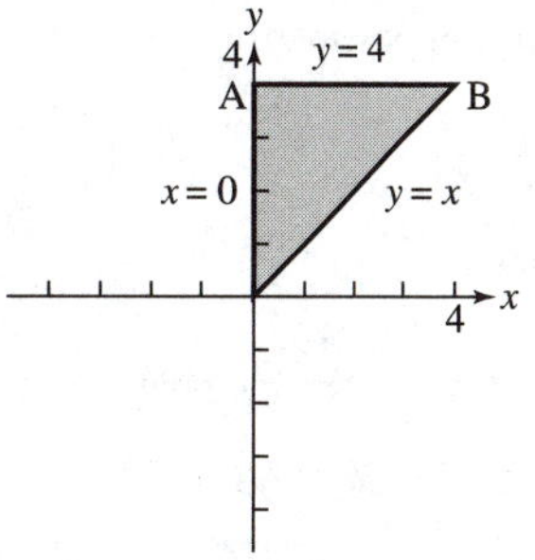

a. On OA, $f(x, y) = f(0, y) = y^2 + 1$ on $0 \le y \le 4$;
$f'(0, y) = 2y = 0$
$y = 0$; $f(0, 0) = 1$ and $f(0, 4) = 17$

b. On AB, $f(x, y) = f(x, 4) = x^2 - 4x + 17$ on $0 \le x \le 4$; $f'(x, y) = 2x - 4 = 0$
$x = 2$ and $(2, 4)$ is an interior point of OA; $f(2, 4) = 13$ and $f(4, 4) = f(0, 4) = 17$

c. On OB, $f(x, y) = f(x, x) = x^2 + 1$ on $0 \le x \le 4$;
$f'(x, x) = 2x = 0$
$x = 0$ and $y = 0$, which is not an interior point of OB; endpoint values have been found above

d. For interior points of the triangular region, $f_x(x, y) = 2x - y = 0$ and $f_y(x, y) = -x + 2y = 0$
$x = 0$ and $y = 0$, which is not an interior point of the region. Therefore, the absolute maximum is 17 at $(0, 4)$ and $(4, 4)$, and the absolute minimum is 1 at $(0, 0)$.

47. $\nabla f = 2x\mathbf{i} + 3y^2\mathbf{j}$ and $\nabla g = 2x\mathbf{i} + 2y\mathbf{j}$

$\nabla f = \lambda \nabla g$

$2x\mathbf{i} + 3y^2\mathbf{j} = \lambda(2x\mathbf{i} + 2y\mathbf{j})$

$2x = \lambda 2x$; $3y^2 = 2y$

$\lambda = 1$ or $x = 0$.

CASE 1: $\lambda = 1$

$3y^2 = 2y$

$y = 0$ or $y = \frac{2}{3}$

If $y = 0$, $x = \pm 1$ yielding the points $(1, 0)$ and $(-1, 0)$

If $y = \frac{2}{3}$, $x = \pm\frac{\sqrt{5}}{3}$ yielding the points $\left(\frac{\sqrt{5}}{3}, \frac{2}{3}\right)$ and $\left(-\frac{\sqrt{5}}{3}, \frac{2}{3}\right)$.

CASE 2: $x = 0$
$y^2 - 1 = 0$
$y = \pm 1$ yielding the points (0, 1) and (0, −1,). Evaluations give $f(\pm 1, 0,) = 1, f\left(\pm\frac{\sqrt{5}}{3}, \frac{2}{3}\right) = \frac{23}{27}, f(0, 1) = 1$, and $f(0, -1,) = -1$. Therefore the absolute maximum is 1 at (±1, 0,) and (0, 1), and the absolute minimum is −1 at (0, −1,).

48. $\nabla f = y\mathbf{i} + (x + z)\mathbf{j} + y\mathbf{k}$, $\nabla g = 2x\mathbf{i} + 2y\mathbf{j}$, and $\nabla h = z\mathbf{j} + y\mathbf{k}$ so that $\nabla f = \lambda\nabla g + \mu\nabla h$
$y\mathbf{i} + (x + z)\mathbf{j} + y\mathbf{k} = \lambda(2x\mathbf{i} + 2y\mathbf{j}) + \mu(z\mathbf{j} + y\mathbf{k})$
$y = \lambda x, x + z = 2\lambda y + \mu z, y = \mu y$
$y = 0$ or $\mu = 1$

CASE 1: $y = 0$ which is impossible since $yz = 1$

CASE 2: $\mu = 1$
$x + z = 2\lambda y + z$
$x = 2\lambda y$ and $y = 2\lambda x$
$x = 2\lambda(2\lambda x) = 4\lambda^2 x$
$x = 0$ or $4\lambda^2 = 1$. If $x = 0$, then $y^2 = 1$
$y = \pm 1$ so with $yz = 1$, we obtain the points (0, 1, 1) and (0, −1, −1). If $4\lambda^2 = 1$ then $\lambda = \pm\frac{1}{2}$. For $\lambda = -\frac{1}{2}$,
$x = -y$ so $x^2 + y^2 = 1$
$y^2 = \frac{1}{2}$
$y = \pm\frac{1}{\sqrt{2}}$ with $yz = 1$
$z = \pm\sqrt{2}$, and we obtain the points $\left(-\frac{1}{\sqrt{2}}, \frac{1}{\sqrt{2}}, \sqrt{2}\right)$ and $\left(\frac{1}{\sqrt{2}}, -\frac{1}{\sqrt{2}}, -\sqrt{2}\right)$. For $\lambda = \frac{1}{2}$, $x = y$
$x^2 = \frac{1}{2}$
$x = \pm\frac{1}{\sqrt{2}}$ with $yz = 1$
$z = \pm\sqrt{2}$, and we obtain the points $\left(\frac{1}{\sqrt{2}}, \frac{1}{\sqrt{2}}, \sqrt{2}\right)$ and $\left(-\frac{1}{\sqrt{2}}, -\frac{1}{\sqrt{2}}, -\sqrt{2}\right)$.

Evaluations give $f(0, 1, 1) = f(0, -1, -1) = 1$,
$f\left(-\frac{1}{\sqrt{2}}, \frac{1}{\sqrt{2}}, \sqrt{2}\right) = f\left(\frac{1}{\sqrt{2}}, -\frac{1}{\sqrt{2}}, -\sqrt{2}\right)$
$= \frac{1}{2}, f\left(\frac{1}{\sqrt{2}}, \frac{1}{\sqrt{2}}, \sqrt{2}\right) = f\left(-\frac{1}{\sqrt{2}}, -\frac{1}{\sqrt{2}}, -\sqrt{2}\right) = \frac{3}{2}$.
Therefore, the absolute maximum is $\frac{3}{2}$ at $\left(\frac{1}{\sqrt{2}}, \frac{1}{\sqrt{2}}, \sqrt{2}\right)$ and $\left(-\frac{1}{\sqrt{2}}, -\frac{1}{\sqrt{2}}, -\sqrt{2}\right)$, and the absolute minimum is $\frac{1}{2}$ at $\left(-\frac{1}{\sqrt{2}}, \frac{1}{\sqrt{2}}, \sqrt{2}\right)$ and $\left(\frac{1}{\sqrt{2}}, -\frac{1}{\sqrt{2}}, -\sqrt{2}\right)$.

49. $f(x, y) = 6xye^{-(2x+3y)}$
$f_x(x, y) = 6y(1 - 2x)e^{-(2x+3y)} = 0$ and
$f_y(x, y) = 6x(1 - 3y)e^{-(2x+3y)} = 0$
$x = 0$ and $y = 0$ or $x = \frac{1}{2}$ and $y = \frac{1}{3}$. The value of $f(0, 0) = 0$ is on the boundary, and $f\left(\frac{1}{2}, \frac{1}{3}\right) = \frac{1}{e^2}$. On the positive y-axis, $f(0, y) = 0$, and on the positive x-axis, $f(x, 0) = 0$. As $x \to \infty$ we see that $f(x, y) \to 0$. Thus the absolute maximum of f in the closed first qudrant is $\frac{1}{e^2}$ at the point $\left(\frac{1}{2}, \frac{1}{3}\right)$.

50. Let $f(x, y, z) = \frac{x^2}{a^2} + \frac{y^2}{b^2} + \frac{z^2}{c^2} - 1 =$
$\nabla f = \frac{2x}{a^2}\mathbf{i} + \frac{2y}{b^2}\mathbf{j} + \frac{2z}{c^2}\mathbf{k}$ and equation of the plane tangent at the point $P_0(x_0, y_0, y_0)$ is $\left(\frac{2x_0}{a^2}\right)x + \left(\frac{2y_0}{b^2}\right)y + \left(\frac{2x_0}{c^2}\right)z$
$= \frac{2x_0^2}{a^2} + \frac{2y_0^2}{b^2} + \frac{2x_0^2}{c^2} = 2$ or $\left(\frac{x_0}{a^2}\right)x + \left(\frac{y_0}{b^2}\right)y + \left(\frac{z_0}{c^2}\right)z = 1$.

The intercepts of the plane are $\left(\frac{a^2}{x_0}, 0, 0\right), \left(0, \frac{b^2}{y_0}, 0\right)$ and $\left(0, 0, \frac{c^2}{z_0}\right)$. The volume of the tetrahedron formed by the plane and the coordinate planes is $V = \left(\frac{1}{3}\right)\left(\frac{1}{2}\right)\left(\frac{a^2}{x_0}\right)\left(\frac{b^2}{y_0}\right)\left(\frac{c^2}{z_0}\right)$ we need to maximize $V(x, y, z) = \frac{(abc)^2}{6xyz}$ subject to the constraint $f(x, y, z) = \frac{x^2}{a^2} + \frac{y^2}{b^2} + \frac{z^2}{c^2} = 1$. Thus,
$\left[-\frac{(abc)^2}{6}\right]\left(\frac{1}{x^2yz}\right) = \frac{2x}{a^2}\lambda, \left[-\frac{(abc)^2}{6}\right]\left(\frac{1}{xy^2z}\right) = \frac{2y}{b^2}\lambda$ and
$\left[-\frac{(abc)^2}{6}\right]\left(\frac{1}{xyz^2}\right) = \frac{2z}{c^2}\lambda$. Multiply the first equation by a^2yz, the second by b^2xz, and the third by c^2xy, Then equate the first and second
$a^2y^2 = b^2x^2$
$y = \frac{b}{a}x, x > 0$; equate the first and third
$a^2z^2 = c^2x^2$
$z = \frac{c}{a}x, x > 0$; substitute into $f(x, y, z) = 0$
$x = \sqrt{\frac{a}{3}}$
$y = \sqrt{\frac{b}{3}}$
$z = \sqrt{\frac{c}{3}}$
$V = \frac{\sqrt{3}}{2}abc$.

51. $\int_0^1\int_{x^2}^{x} f(x, y)\,dy\,dx = \int_0^1\int_y^{\sqrt{y}} f(x, y)\,dx\,dy$

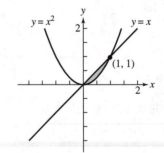

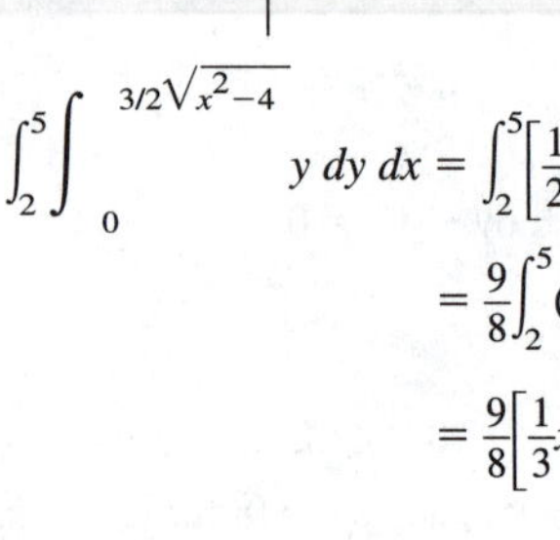

52. $\int_2^5\int_0^{3/2\sqrt{x^2-4}} y\,dy\,dx = \int_2^5\left[\frac{1}{2}y^2\right]_0^{3/2\sqrt{x^2-4}} dx$
$= \frac{9}{8}\int_2^5 (x^2 - 4)\,dx$
$= \frac{9}{8}\left[\frac{1}{3}x^3 - 4x\right]_2^5 = \frac{243}{8}$

53. $\int_0^2 \int_0^{x^2} \int_0^{x+y} (x + 2y - z)\, dz\, dy\, dx$

$= \int_0^2 \int_0^{x^2} \left[xz + 2yz - \frac{1}{2}z^2\right]_0^{x+y} dy\, dx$

$= \int_0^2 \int_0^{x^2} \left(\frac{1}{2}x^2 + 2xy + \frac{3}{2}y^2\right) dy\, dx$

$= \int_0^2 \left[\frac{1}{2}x^2y + xy^2 + \frac{1}{2}y^3\right]_0^{x^2} dx$

$= \int_0^2 \left(\frac{1}{2}x^6 + x^5 + \frac{1}{2}x^4\right)dx = \left[\frac{1}{14}x^7 + \frac{1}{6}x^6 + \frac{1}{10}x^5\right]_0^2$

$= \frac{2416}{105}$

54. $\int_0^{2\pi} \int_0^3 \int_{r^2/3}^{\sqrt{18-r^2}} dz\, r\, dr\, d\theta$

$= \int_0^{2\pi} \int_0^3 \left[z\right]_{r^2/3}^{\sqrt{18-r^2}} r\, dr\, d\theta$

$= \int_0^{2\pi} \int_0^3 \left(r\sqrt{18 - r^2} - \frac{r^3}{3}\right) dr\, d\theta$

$= \int_0^{2\pi} \left[-\frac{1}{3}(18 - r^2)^{3/2} - \frac{r^4}{12}\right]_0^3 d\theta$

$= \int_0^{2\pi} \left(18\sqrt{2} - \frac{63}{4}\right) d\theta = \left(18\sqrt{2} - \frac{63}{4}\right)\left[\theta\right]_0^{2\pi}$

$= \frac{9\pi(8\sqrt{2} - 7)}{2}$

55. $\int_0^{2\pi} \int_0^{\pi} \int_0^{(1-\cos\phi)/2} \rho^2 \sin\phi\, d\rho\, d\phi\, d\theta$

$= \int_0^{2\pi} \int_0^{\pi} \left[\frac{1}{3}\rho^3\right]_0^{(1-\cos\phi)/2} \sin\phi\, d\phi\, d\theta$

$= \frac{1}{24} \int_0^{2\pi} \int_0^{\pi} (1 - \cos\phi)^3 \sin\phi\, d\phi\, d\theta$

$= \frac{1}{24} \int_0^{2\pi} \frac{1}{4}\left[1 - \cos\phi\right]_0^{\pi} d\theta = \frac{1}{96} \int_0^{2\pi} 16\, d\theta$

$= \frac{1}{6}\left[\theta\right]_0^{2\pi} = \frac{\pi}{3}$

56. The lines intersect where $x = y^2$ and $x = 2 - y$

$y^2 + y = 2$

$y^2 + y - 2 = 0$

$(y + 2)(y - 1) = 0$

$y = -2$ or $y = 1$.

$\int_{-2}^{1} [(2 - y) - y^2]\, dy = \int_{-2}^{1} (2 - y - y^2)\, dy = \frac{9}{2}$

57. a. $V = \int_{-3}^{2} \int_x^{6-x^2} x^2\, dy\, dx$

b. $V = \int_{-3}^{2} \int_x^{6-x^2} \int_0^{x^2} dz\, dy\, dx$

c. $V = \int_{-3}^{2} \left[x^2y\right]_x^{6-x^2} dx = \int_{-3}^{2} x^2(6 - x^2 - x)\, dx$

$= \int_{-3}^{2} (-x^4 - x^3 + 6x^2)\, dx$

$= \left[-\frac{1}{5}x^5 - \frac{1}{4}x^4 + 2x^3\right]_{-3}^{2} = \frac{125}{4}$

58. The surfaces intersect where $3 - x^2 - y^2 = 2x^2 + 2y^2$
$x^2 + y^2 = 1$. Thus, the volume is

$V = 4\int_0^1 \int_0^{\sqrt{1-x^2}} \int_{2x^2+2y^2}^{3-x^2-y^2} dz\, dy\, dx$

$= 4\int_0^{\pi/2} \int_0^1 \int_{2r^2}^{3-r^2} dz\, r\, dr\, d\theta = 4\int_0^{\pi/2} \int_0^1 \left[z\right]_{2r^2}^{3-r^2} r\, dr\, d\theta$

$= 4\int_0^{\pi/2} \int_0^1 (3r - 3r^3)\, dr\, d\theta = 4\int_0^{\pi/2} \left[\frac{3}{2}r^2 - \frac{3}{4}r^4\right]_0^1 d\theta$

$= 3\int_0^{\pi/2} d\theta = 3\left[\theta\right]_0^{\pi/2} = \frac{3\pi}{2}$

59. $V = 8\int_0^{\pi/2} \int_0^{\pi/2} \int_0^{2\sin\phi} \rho^2 \sin\phi\, d\rho\, d\phi\, d\theta$

$= 8\int_0^{\pi/2} \int_0^{\pi/2} \left[\frac{1}{3}\rho^3\right]_0^{2\sin\phi} \sin\phi\, d\phi\, d\theta$

$= \frac{64}{3}\int_0^{\pi/2} \int_0^{\pi/2} \sin^4\phi\, d\phi\, d\theta$

$= \frac{64}{3}\int_0^{\pi/2} [-\sin^3\phi\cos\phi\Big|_0^{\pi/2}$

$+ 3\int_0^{\pi/2} \cos^2\phi\sin^2\phi\, d\phi\,]\, d\theta$

$= 64\int_0^{\pi/2} \left[\frac{\phi}{8} - \frac{\sin 4\phi}{32}\right]_0^{\pi/2} d\theta$

$= 4\pi\int_0^{\pi/2} d\theta = 2\pi^2$ (We used integration by parts to evaluate $\int \sin^4\phi\, d\phi = -\sin^3\phi\cos\phi + 3\int \sin^2\phi\cos^2\phi\, d\phi = etc.$)

60.

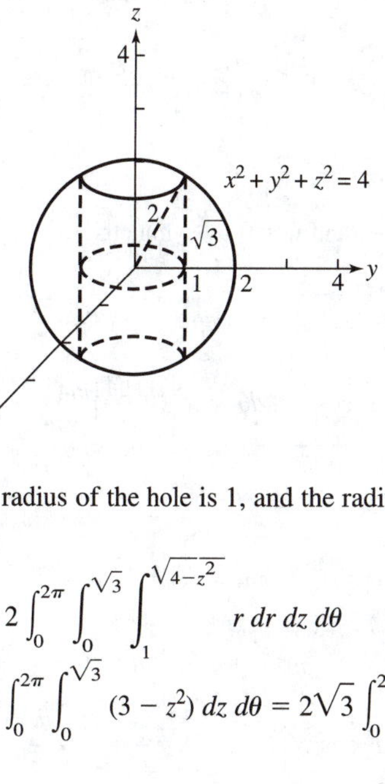

a. The radius of the hole is 1, and the radius of the sphere is 2.

b. $V = 2\int_0^{2\pi} \int_0^{\sqrt{3}} \int_1^{\sqrt{4-z^2}} r\, dr\, dz\, d\theta$

$= \int_0^{2\pi} \int_0^{\sqrt{3}} (3 - z^2)\, dz\, d\theta = 2\sqrt{3}\int_0^{2\pi} d\theta = 4\sqrt{3}\pi$

61. $I_o(a) = \int_0^a \int_0^{x/a^2} (x^2 + y^2)\, dy\, dx$

$= \int_0^a \left[x^2 y + \frac{y^3}{3}\right]_0^{x/a^2} dx = \int_0^a \left(\frac{x^3}{a^2} + \frac{x^3}{3a^6}\right) dx$

$= \left[\frac{x^4}{4a^2} + \frac{x^4}{12a^6}\right]_0^a = \frac{a^2}{4} + \frac{1}{12}a^{-2};$

$I'_o(a) = \frac{1}{2}a - \frac{1}{6}a^{-3} = 0$

$a^4 = \frac{1}{3}$

$a = \sqrt[4]{\frac{1}{3}} = \frac{1}{\sqrt[4]{3}}$. Since $I''_o(a) = \frac{1}{2} + \frac{1}{2}a^{-4} > 0$, the value of a does provide a minimum for the polar moment of inertia $I_o(a)$.

62. $I_o = \int_0^2 \int_{2x}^4 (x^2 + y^2)(3)\, dy\, dx$

$= 3\int_0^2 \left(4x^2 - \frac{14x^2}{3} + \frac{64}{3}\right) dx = 104$

63. $M = 2\int_0^{\pi/2} \int_1^{1+\cos\theta} r\, dr\, d\theta$

$= \int_0^{\pi/2} \left(2\cos\theta + \frac{1 + \cos 2\theta}{2}\right) d\theta = \frac{8 + \pi}{4};$

$M_y = \int_{-\pi/2}^{\pi/2} \int_1^{1+\cos\theta} r^2 \cos\theta\, dr\, d\theta$

$= \int_{-\pi/2}^{\pi/2} \left(\cos^2\theta + \cos^3\theta + \frac{\cos^4\theta}{3}\right) d\theta$

$= \left[\frac{\theta}{2} + \frac{\sin 2\theta}{4} + \frac{\cos^2\theta \sin\theta}{3} + \frac{\cos^3\theta\sin\theta}{12}\right]_{-\pi/2}^{\pi/2}$

$+ \int_{-\pi/2}^{\pi/2} \left(\frac{2}{3}\cos\theta + \frac{1}{4}\cos^2\theta\right) d\theta$

$= \frac{\pi}{2} + \left[\frac{2}{3}\sin\theta + \frac{1}{4}\left(\frac{\theta}{2} + \frac{\sin 2\theta}{4}\right)\right]_{-\pi/2}^{\pi/2} = \frac{\pi}{2} + \frac{4}{3} + \frac{\pi}{8}$

$= \frac{5\pi}{8} + \frac{4}{3} = \frac{32 + 15\pi}{24}$

$\bar{x} = \frac{15\pi + 32}{6\pi + 48}$, and $\bar{y} = 0$ by symmetry

64. $M = \int_{-\pi/4}^{\pi/4} \int_0^{a\sqrt{2\cos 2\theta}} r\, dr\, d\theta$

$= \int_{-\pi/4}^{\pi/4} (a^2 \cos 2\theta)\, d\theta = a^2\left[\frac{\sin 2\theta}{2}\right]_{-\pi/4}^{\pi/4} = a^2;$

$I_x = \int_{-\pi/4}^{\pi/4} \int_0^{a\sqrt{2\cos 2\theta}} (r \sin\theta)^2\, r\, dr\, d\theta$

$= \int_{-\pi/4}^{\pi/4} (a^4 \cos^2 2\theta \sin^2\theta) d\theta = \frac{(3\pi - 8)a^4}{24};$

$I_y = \int_{-\pi/4}^{\pi/4} \int_0^{a\sqrt{2\cos 2\theta}} (r\cos\theta)^2\, r\, dr\, d\theta$

$= \int_{-\pi/4}^{\pi/4} (a^4 \cos^2 2\theta \cos^2\theta) = \frac{(8 + 3\pi)a^4}{24}$

$R_x = \sqrt{\frac{I_x}{M}} = \frac{a\sqrt{3\pi - 8}}{2\sqrt{6}}$ and

$R_y = \sqrt{\frac{I_y}{M}} = \frac{a\sqrt{3\pi + 8}}{2\sqrt{6}}$

65. a. $M = \int_0^{2\pi} \int_0^1 \int_r^1 z\, dz\, r\, dr\, d\theta$

$= \frac{1}{2}\int_0^{2\pi} \int_0^1 (r - r^3)\, dr\, d\theta = \frac{1}{8}\int_0^{2\pi} d\theta = \frac{\pi}{4};$

$M_{xy} = \int_0^{2\pi} \int_0^1 \int_r^1 z^2\, dz\, r\, dr\, d\theta$

$= \frac{1}{3}\int_0^{2\pi} \int_0^1 (r - r^4)\, dr\, d\theta = \frac{1}{10}\int_0^{2\pi} d\theta = \frac{\pi}{5}$

$\bar{z} = \frac{4}{5}$, and $\bar{x} = \bar{y} = 0$ by symmetry;

$I_z = \int_0^{2\pi} \int_0^1 \int_r^1 zr^3\, dz\, dr\, d\theta = \frac{1}{2}\int_0^{2\pi} \int_0^1 (r^3 - r^5)\, dr\, d\theta$

$= \frac{1}{24}\int_0^{2\pi} d\theta = \frac{\pi}{12}$

$R_z = \sqrt{\frac{I_z}{M}} = \sqrt{\frac{1}{3}}$

b. $M = \int_0^{2\pi} \int_0^1 \int_r^1 z^2\, dz\, r\, dr\, d\theta = \frac{\pi}{5}$ from part (a);

$M_{xy} = \int_0^{2\pi} \int_0^1 \int_r^1 z^3\, dz\, r\, dr\, d\theta$

$= \frac{1}{4}\int_0^{2\pi} \int_0^1 (r - r^5)\, dr\, d\theta = \frac{1}{12}\int_0^{2\pi} d\theta = \frac{\pi}{6}$

$\bar{z} = \frac{5}{6}$, and $\bar{x} = \bar{y} = 0$ by symmetry;

$I_z = \int_0^{2\pi} \int_0^1 \int_r^1 z^2 r^3\, dz\, dr\, d\theta$

$= \frac{1}{3}\int_0^{2\pi} \int_0^1 (r^3 - r^6)\, dr\, d\theta = \frac{1}{28}\int_0^{2\pi} d\theta = \frac{\pi}{14}$

$R_z = \sqrt{\frac{I_z}{M}} = \sqrt{\frac{5}{14}}$

66. a. $\nabla f = x\mathbf{i} + y\mathbf{j}$

$D_{\mathbf{u}}f = u_1 x + u_2 y$; the area of the region of integration is $\frac{1}{2}$

average $= 2\int_0^1 \int_0^{1-x} (u_1 x + u_2 y)\, dy\, dx$

$= 2\int_0^1 \left[u_1 x(1 - x) + \frac{1}{2}u_2(1 - x)^2\right] dx$

$= 2\left[u_1\left(\frac{x^2}{2} - \frac{x^3}{3}\right) - \left(\frac{1}{2}u_2\right)\frac{(1 - x)^3}{3}\right]_0^1$

$= 2\left(\frac{1}{6}u_1 + \frac{1}{6}u_2\right) = \frac{1}{3}(u_1 + u_2)$

b. average $= \frac{1}{\text{area}}\int\int_R (u_1 x + u_2 y)\, dA$

$= \frac{u_1}{\text{area}}\int\int_R x\, dA + \frac{u_2}{\text{area}}\int\int_R y\, dA$

$= u_1\left(\frac{M_y}{M}\right) + u_2\left(\frac{M_x}{M}\right) = u_1\bar{x} + u_2\bar{y}$

67. $Q = \int_0^{2\pi} \int_0^R kr^2(1 - \sin\theta)\, dr\, d\theta$

$= \frac{kR^3}{3}\int_0^{2\pi} (1 - \sin\theta)\, d\theta = \frac{kR^3}{3}\left[\theta + \cos\theta\right]_0^{2\pi} = \frac{2\pi kR^3}{3}$

68. For a height h in the bowl the volume of water is

$$V = \int_{-\sqrt{h}}^{\sqrt{h}} \int_{-\sqrt{h-x^2}}^{\sqrt{h-x^2}} \int_{x^2+y^2}^{h} dz\, dy\, dx$$

$$= \int_{-\sqrt{h}}^{\sqrt{h}} \int_{-\sqrt{h-x^2}}^{\sqrt{h-x^2}} (h - x^2 - y^2)\, dy\, dx$$

$$= \int_0^{2\pi} \int_0^{\sqrt{h}} (h - r^2)\, r\, dr\, d\theta$$

$$= \int_0^{2\pi} \left[\frac{hr^2}{2} - \frac{r^4}{4}\right]_0^{\sqrt{h}} d\theta = \int_0^{2\pi} \frac{h^2}{4}\, d\theta = \frac{h^2\pi}{2}$$

Since the top of the bowl has area 10π, then we calibrate the bowl by comparing it to a right circular cylinder whose cross sectional area is 10π from $z = 0$ to $z = 10$. If such a cylinder contains $\frac{h^2\pi}{2}$ cubic inches of water to a depth w then we have $10\pi w = \frac{h^2\pi}{2}$

$w = \frac{h^2}{20}$. So for 1 inch of rain, $w = 1$ and $h = \sqrt{20}$; for 3 inches of rain, $w = 3$ and $h = \sqrt{60}$.

69. The cylinder is given by $x^2 + y^2 = 1$ from $z = 1$ to ∞

$$\int\int\int_D z(r^2 + z^2)^{-5/2}\, dV$$

$$= \int_0^{2\pi} \int_0^1 \int_1^{\infty} \frac{z}{(r^2 + z^2)^{5/2}}\, dz\, r\, dr\, d\theta$$

$$= \lim_{a\to\infty} \int_0^{2\pi} \int_0^1 \int_1^{\infty} \frac{rz}{(r^2 + z^2)^{5/2}}\, dz\, dr\, d\theta$$

$$= \lim_{a\to\infty} \int_0^{2\pi} \int_0^1 \left[\left(-\frac{1}{3}\right)\frac{r}{(r^2 + z^2)^{3/2}}\right]_1^a dr\, d\theta =$$

$$= \lim_{a\to\infty} \int_0^{2\pi} \int_0^1 \left[\left(-\frac{1}{3}\right)\frac{r}{(r^2 + a^2)^{3/2}} + \left(\frac{1}{3}\right)\frac{r}{(r^2 + 1)^{3/2}}\right] dr\, d\theta$$

$$= \lim_{a\to\infty} \int_0^{2\pi} \left[\frac{1}{3}(r^2 + a^2)^{-1/2} - \frac{1}{3}(r^2 + 1)^{-1/2}\right]_0^1 d\theta$$

$$= \lim_{a\to\infty} \int_0^{2\pi} \left[\frac{1}{3}(1 + a^2)^{-1/2} - \frac{1}{3}(2^{-1/2}) - \frac{1}{3}(a^2)^{-1/2} + \frac{1}{3}\right] d\theta$$

$$= \lim_{a\to\infty} 2\pi\left[\frac{1}{3}(1 + a^2)^{-1/2} - \frac{1}{3}\left(\frac{\sqrt{2}}{2}\right) - \frac{1}{3}\left(\frac{1}{a}\right) + \frac{1}{3}\right]$$

$$= 2\pi\left[\frac{1}{3} - \left(\frac{1}{3}\right)\frac{\sqrt{2}}{2}\right].$$

70. $\mathbf{r}(t) = 0 + t(\mathbf{i} + \mathbf{j} + 2\mathbf{k}) = t\mathbf{i} + t\mathbf{j} + 2t\mathbf{k}$, $0 \le t \le 1$

$$\frac{d\mathbf{r}}{dt} = \mathbf{i} + \mathbf{j} + 2\mathbf{k}$$

$\left|\frac{d\mathbf{r}}{dt}\right| = \sqrt{1 + 1 + 4} = \sqrt{6}$; $f(x, y, z) = x^2 - 3y + z^2$

$= (t)^2 + (-3t) + (2t)^2 = 5t^2 - 3t$

$$\int_C f(x, y, z)\, ds = \int_0^1 (5t^2 - 3t)\sqrt{6}\, dt$$

$$= \sqrt{6}\left[\frac{5}{3}t^3 - \frac{3}{2}t^2\right]_0^1 = \frac{\sqrt{6}}{6}$$

71. $\mathbf{r}(t) = (2\cos t)\mathbf{i} + (2\sin t)\mathbf{j} + 3t\mathbf{k}$, $0 \le t \le 2\pi$

$$\frac{d\mathbf{r}}{dt} = -(2\sin t)\mathbf{i} + (2\cos t)\mathbf{j} + 3\mathbf{k}$$

$$\left|\frac{d\mathbf{r}}{dt}\right| = \sqrt{4\sin^2 t + 4\cos^2 t + 9} = \sqrt{13};$$

$f(x, y, z) = x + y + z = 2\cos t + 2\sin t + 3t$

$$\int_C f(x, y, z)\, ds = \int_0^{2\pi} (2\cos t + 2\sin t + 3t)\sqrt{13}\, dt$$

$$= \sqrt{13}\left[2\sin t - 2\cos t + \frac{3}{2}t^2\right]_0^{2\pi}$$

$$= 6\pi^2\sqrt{13}$$

72. Let $M = y$ and $N = y - x$

$\frac{\partial M}{\partial y} = 0$ and $\frac{\partial N}{\partial y} = 1$

$$\int_C y\, dy - (y - x)\, dx = \int\int_R (0 + 1)\, dy\, dx$$

$$= \int_0^{2\pi} \int_0^1 r\, dr\, d\theta = \int_0^{2\pi} \frac{1}{2}\left[r^2\right]_0^1 d\theta$$

$$= \frac{1}{2}\left[\theta\right]_0^{2\pi} = \pi$$

73. $\mathbf{F}_1 = (9\cos^2 t)\mathbf{i} + (16\sin^2 t)\mathbf{j}$,

$$\frac{d\mathbf{r}_1}{dt} = (-3\sin t)\mathbf{i} + (4\cos t)\mathbf{j}$$

$$\mathbf{F}_1 \cdot \frac{d\mathbf{r}_1}{dt} = -27\cos^2 t\sin t + 64\sin^2 t\cos t$$

$$\text{Circ} = \int_0^{\pi} (-27\cos^2 t\sin t + 64\sin^2 t\cos t)\, dt$$

$= \left[27\left(\frac{1}{3}\right)\cos^3 t + 64\left(\frac{1}{3}\right)\sin^3 t\right]_0^{\pi} = -18$; $M = 9\cos^2 t$,

$N = 16\sin^2 t$, $dy = 4\cos t$, $dx = -3\sin t$

$$\text{Flux} = \int_C M\, dy - N\, dx = \int_0^{\pi} (36\cos^3 t - 48\sin^3 t)\, dt$$

$$= \left[36\left(\frac{2}{3}\right)\sin t(1 + \cos^2 t) - 48\left(-\frac{1}{3}\right)\cos t(2 + \sin^2 t)\right]_0^{\pi}$$

$$= -64$$

74. $\mathbf{r} = (2\cos t)\mathbf{i} + (2\sin t)\mathbf{j}$; $0 \le t \le 2\pi$, and

$\mathbf{F} = xy\mathbf{i} + (y - x)\mathbf{j}$

$= (4\sin t\cos t)\mathbf{i} + (2\sin t - 2\cos t)\mathbf{j}$ and $\frac{d\mathbf{r}}{dt}$

$= (-2\sin t)\mathbf{i} + (2\cos t)\mathbf{j}$

$$\mathbf{F} \cdot \frac{d\mathbf{r}}{dt} = -8\sin^2 t\cos t + (4\sin t\cos t - 4\cos^2 t)$$

$$\text{work} = \int_C \mathbf{F} \cdot \frac{d\mathbf{r}}{dt}\, dt$$

$$= \int_0^{2\pi} (-8\sin^2 t\cos t + 4\sin t\cos t - 4\cos^2 t)\, dt$$

$$= \left[-\frac{8}{3}\sin^3 t + 2\sin^2 t - 2\sin t\cos t - 2t\right]_0^{2\pi}$$

75. $\frac{\partial P}{\partial y} = x^2 - \frac{4yz}{(y^2 + z)^2} = \frac{\partial N}{\partial z}, \frac{\partial M}{\partial z} = -\sin(x + z) + 2xy$

$\ne 2xy = \frac{\partial P}{\partial x}, \frac{\partial N}{\partial x} = 2xz = \frac{\partial M}{\partial y}$ so the field is not conservative

no potential function

76. Let $\mathbf{F}(x, y, z) = (yz + 2x)\mathbf{i} + (xz + 2y)\mathbf{j} + (xy + 2z)\mathbf{k}$

$$\frac{\partial P}{\partial y} = x = \frac{\partial N}{\partial z}, \frac{\partial M}{\partial z} = y = \frac{\partial P}{\partial x}, \frac{\partial N}{\partial x} = z = \frac{\partial M}{\partial y}$$

$M dx + N dy + P dz$ is exact; $\frac{\partial f}{\partial x} = yz + 2x$

$f(x, y, z) = xyz + x^2 + g(y, z)$

$\frac{\partial f}{\partial y} = xz + \frac{\partial g}{\partial y} = xz + 2y$

$\frac{\partial g}{\partial y} = 2y$

$g(y, z) = y^2 + h(z)$

$f(x, y, z) = xyz + x^2 + y^2 + h(z)$

$\frac{\partial f}{\partial z} = xy + h'(z) = 2y + 2z$

$h'(z) = 2z$

$h(z) = z^2 + c$

$f(x, y, z) = xyz + x^2 + y^2 + z^2 + c$

$$\int_{(1,1,1)}^{(2,3,4)} (yz + 2x)dx + (xz + 2y)dy + (xy + 2z)dz$$

$= f(2, 3, 4) - f(1, 1, 1)$

$= (24 + 4 + 9 + 16) - (1 + 1 + 1 + 1) = 49$

77. $M = y^2 + 2x, N = x^2 + 2y$

$$\frac{\partial M}{\partial x} = 2, \frac{\partial M}{\partial y} = 2y, \frac{\partial N}{\partial x} = 2x, \frac{\partial N}{\partial y} = 2$$

$$\text{Circ} = \int\int_R (2x - 2y)\, dx\, dy = \int_0^2 \int_0^{2x} (2x - 2)\, dy\, dx$$

$$= \int_0^2 0\, dx = 0$$

$$\text{Flux} = \int\int_R (2 + 2)\, dx\, dy = \int_0^2 \int_0^{2x} 4\, dy\, dx$$

$$= 4\int_0^2 2x\, dx = 16$$

78. $M = yx^2, N = xy^2$

$$\frac{\partial M}{\partial y} = x^2, \frac{\partial N}{\partial x} = y^2$$

$$\oint_C (yx^2\, dx + y^2\, dy) = \int\int_R (y^2 - x^2)\, dx\, dy$$

$$= \int_{-\pi/2}^{\pi/2} \int_0^{\cos x} (y^2 - x^2) dy\, dx$$

$$= \int_{-\pi/2}^{\pi/2} \left(\frac{1}{3}\cos^3 x - x^2 \cos x\right) dx$$

$$= \frac{80 - 9\pi^2}{18} \approx -0.49$$

79. $\mathbf{p} = \mathbf{k}, \nabla f = 2x\mathbf{i} + 2y\mathbf{j} + 2z\mathbf{k}$

$|\nabla f| = \sqrt{4x^2 + 4y^2 + 4z^2} = 2\sqrt{x^2 + y^2 + z^2} = 4$ and

$|\nabla f \cdot \mathbf{p}| = \sqrt{0 + 0 + (2z)^2} = 2z$; $x^2 + y^2 + z^2 = 4$ and

$z = \sqrt{x^2 + y^2}$

$$x^2 + y^2 = 2; \text{ thus, } S = \int\int_R \frac{|\nabla f|}{|\nabla f \cdot \mathbf{p}|}\, dA = \int\int_R \frac{4}{2z}\, dA$$

$$= 2\int\int_R \frac{1}{z}\, dA = 2\int\int_R \frac{1}{\sqrt{4 - (x^2 + y^2)}}\, dA$$

$$= 2\int_0^{2\pi} \int_0^{2} \frac{1}{\sqrt{4 - r^2}}\, r\, dr\, d\theta = 2\int_0^{2\pi} \left[-\sqrt{4 - r^2}\right]_0^2 d\theta$$

$$= 4\int_0^{2\pi} d\theta = 8\pi$$

80. On the faces in the coordinate planes, $f(x, y, z) = 0$ the integral over these faces is 0. On the face $x = 2$, we have

$f(x, y, z) = f(2, y, z) = 2yz$

$\mathbf{p} = \mathbf{i}$ and $\nabla f = \mathbf{i}$

$|\nabla f| = 1$ and $|\nabla f \cdot \mathbf{p}| = 1$

$d\sigma = dy\, dz$

$$\int\int_S f\, d\sigma = \int\int_S 2yz\, d\sigma = \int_0^5 \int_0^3 2yz\, dy\, dz = \int_0^5 9z\, dz$$

$$= \frac{225}{2}$$

On the face $y = 3$, we have $f(x, y, z) = f(x, 3, z) = 3xz$

$\mathbf{p} = \mathbf{j}$ and $\nabla f = \mathbf{j}$

$|\nabla f| = 1$ and $|\nabla f \cdot \mathbf{p}| = 1$

$d\sigma = dx\, dz$

$$\int\int_S f\, d\sigma = \int\int_S 3\, xz\, d\sigma = \int_0^5 \int_0^2 3xz\, dx\, dz$$

$$= \int_0^5 6z\, dz = 75$$

On the face $z = 5$, we have $f(x, y, z) = f(x, y, 5) = 5xy$

$\mathbf{p} = \mathbf{k}$ and $\nabla f = \mathbf{k}$

$|\nabla f| = 1$ and $|\nabla f \cdot \mathbf{p}| = 1$

$d\sigma = dx\, dy$

$$\int\int_S f\, d\sigma = \int\int_S 5xy\, d\sigma = \int_0^3 \int_0^2 5xy\, dx\, dy$$

$$= \int_0^3 10y\, dy = 45$$

Therefore, $\int\int_S f(x, y, z)\, d\sigma = \frac{225}{2} + 75 + 45 = \frac{465}{2}$

81. a. $\mathbf{r}_\phi = 3\cos\phi\cos\theta\,\mathbf{i} + 3\cos\phi\sin\theta\,\mathbf{j} - 3\sin\phi\,\mathbf{k}$,

$\mathbf{r}_\theta = -3\sin\phi\sin\theta\,\mathbf{i} + 3\sin\phi\cos\theta\,\mathbf{j}$

$$\mathbf{r}_\phi \times \mathbf{r}_\theta = \begin{vmatrix} \mathbf{i} & \mathbf{j} & \mathbf{k} \\ 3\cos\phi\cos\theta & 3\cos\phi\sin\theta & -3\sin\phi \\ -3\sin\phi\sin\theta & 3\sin\phi\cos\theta & 0 \end{vmatrix}$$

$$= 9\sin^2\phi\cos\theta\,\mathbf{i} + 9\sin^2\phi\sin\theta\,\mathbf{j} + 9\sin\phi\cos\phi\,\mathbf{k}$$

$|\mathbf{r}_\phi \times \mathbf{r}_\theta| = 9\sin\phi$

$$\text{Surface Area} = \int\int_R |\mathbf{r}_\phi \times \mathbf{r}_\theta|\, d\phi\, d\theta$$
$$= \int_0^{2\pi}\int_{\pi/3}^{2\pi/3} 9\sin\phi\, d\phi\, d\theta$$
$$= 9\int_0^{2\pi} [-\cos\phi]_{\pi/3}^{2\pi/3}\, d\theta$$
$$= 9\int_0^{2\pi} d\theta = 9\cdot 2\pi = 18\pi$$

b. From part (a), $|\mathbf{r}_\phi \times \mathbf{r}_\theta| = 9\sin\phi$

$$\int\int_S (x^2 + y^2 - z^2)\, d\sigma$$
$$= \int_0^{2\pi}\int_{\pi/3}^{2\pi/3} [(3\cos\phi\cos\theta)^2 + (3\cos\phi\sin\theta)^2 - (3\sin\phi)^2](9\sin\phi)\, d\phi\, d\theta$$
$$= 81\int_0^{2\pi}\int_{\pi/3}^{2\pi/3} (2\cos^2\phi - 1)\sin\phi\, d\phi\, d\theta$$
$$= 81\int_0^{2\pi}\left(-\frac{5}{6}\right) d\theta = -135\pi$$

82. a. Let $w^2 + \frac{z^2}{c^2} = 1$ where $w = \cos\phi$ and $\frac{z}{c} = \sin\phi$

$$\frac{x^2}{a^2} + \frac{y^2}{b^2} = \cos^2\phi$$
$$\frac{x}{a} = \cos\phi\cos\theta \text{ and } \frac{y}{b} = \cos\phi\sin\theta$$

$x = a\cos\theta\cos\phi$, $y = b\sin\theta\cos\phi$, and $z = c\sin\phi$

$\mathbf{r}(\theta, \phi) = (a\cos\theta\cos\phi)\mathbf{i} + (b\sin\theta\cos\phi)\mathbf{j} + (c\sin\phi)\mathbf{k}$

b. $\mathbf{r}_\theta = (-a\sin\theta\cos\phi)\mathbf{i} + (b\cos\theta\cos\phi)\mathbf{j}$ and
$\mathbf{r}_\phi = (-a\cos\theta\sin\phi)\mathbf{i} - (b\sin\theta\sin\phi)\mathbf{j} + (c\cos\phi)\mathbf{k}$

$$\mathbf{r}_\theta \times \mathbf{r}_\phi = \begin{vmatrix} \mathbf{i} & \mathbf{j} & \mathbf{k} \\ -a\sin\theta\cos\phi & b\cos\theta\cos\phi & 0 \\ -a\cos\theta\sin\phi & -b\sin\theta\sin\phi & c\cos\phi \end{vmatrix}$$
$$= (bc\cos\theta\cos^2\phi)\mathbf{i} + (ac\sin\theta\cos^2\phi)\mathbf{j} + (ab\cos\phi\sin\phi\sin^2\phi + ab\cos\phi\cos^2\theta\sin\phi)\mathbf{k}$$

$$|\mathbf{r}_\theta \times \mathbf{r}_\phi| = \sqrt{a^2b^2\cos^2\phi\sin^2\phi + c^2\cos^4\phi(b^2\cos^2\theta + a^2\sin^2\theta)}$$
$$\int_0^{2\pi}\int_0^{\pi} \sqrt{a^2b^2\cos^2\phi\sin^2\phi + c^2\cos^4\phi(b^2\cos^2\theta + a^2\sin^2\theta)}\, d\phi\, d\theta$$

83. $\mathbf{F} = yz^2\mathbf{i} + xz^2\mathbf{j} + 2xyz\mathbf{k}$ and $\mathbf{n} = \frac{x\mathbf{i} + y\mathbf{j} + z\mathbf{k}}{\sqrt{x^2 + y^2 + z^2}}$

$= \frac{x\mathbf{i} + y\mathbf{j} + z\mathbf{k}}{R}$, so $\mathbf{F}$ is parallel to $\mathbf{n}$ when $yz^2 = \frac{cx}{R}$, $xz^2 = \frac{cy}{R}$, and $2xyz = \frac{cz}{R}$

$$\frac{yz^2}{x} = \frac{xz^2}{y} = 2xy$$
$$y^2 = x^2$$
$$y = \pm x \text{ and } z^2 = \pm\frac{c}{R} = 2x^2$$

$z = \pm\sqrt{2}x$. Also, $x^2 + y^2 + z^2 = R^2$

$$x^2 + x^2 + 2x^2 = R^2$$
$$4x^2 = R^2$$

$x = \pm\frac{R}{2}$. Thus the points are: $\left(\frac{R}{2}, \frac{R}{2}, \frac{\sqrt{2}R}{2}\right)$, $\left(\frac{R}{2}, \frac{R}{2}, -\frac{\sqrt{2}R}{2}\right)$, $\left(-\frac{R}{2}, -\frac{R}{2}, \frac{\sqrt{2}R}{2}\right)$, $\left(-\frac{R}{2}, -\frac{R}{2}, -\frac{\sqrt{2}R}{2}\right)$, $\left(\frac{R}{2}, -\frac{R}{2}, \frac{\sqrt{2}R}{2}\right)$, $\left(\frac{R}{2}, -\frac{R}{2}, -\frac{\sqrt{2}R}{2}\right)$, $\left(-\frac{R}{2}, \frac{R}{2}, \frac{\sqrt{2}R}{2}\right)$, $\left(-\frac{R}{2}, \frac{R}{2}, -\frac{\sqrt{2}R}{2}\right)$

84. Set up the coordinate system so that $(a, b, c) = (0, R, 0)$

$\delta(x, y, z) = \sqrt{x^2 + (y - R)^2 + z^2}$
$= \sqrt{x^2 + y^2 + z^2 - 2Ry + R^2} = \sqrt{2R^2 - 2Ry}$;

let $f(x, y, z) = x^2 + y^2 + z^2 = R^2$ and $\mathbf{p} = \mathbf{i}$

$\nabla f = 2x\mathbf{i} + 2y\mathbf{j} + 2z\mathbf{k}$

$|\nabla f| = 2\sqrt{x^2 + y^2 + z^2} = 2R$

$$d\sigma = \frac{|\nabla f|}{|\nabla f\cdot\mathbf{i}|}\, dz\, dy = \frac{2R}{2x}\, dz\, dy$$

$$\text{Mass} = \int\int_S \delta(x, y, z)\, d\sigma$$
$$= \int\int_{R_{yz}} \sqrt{2R^2 - 2Ry}\left(\frac{R}{x}\right) dz\, dy$$
$$= R\int\int_{R_{yz}} \frac{\sqrt{2R^2 - 2Ry}}{\sqrt{R^2 - y^2 - z^2}}\, dz\, dy$$
$$= 4R\int_{-R}^{R}\int_0^{\sqrt{R^2 - y^2}} \frac{\sqrt{2R^2 - 2Ry}}{\sqrt{R^2 - y^2 - z^2}}\, dz\, dy = \frac{16\pi R^3}{3}$$

(we used a CAS integrator)

85. $\mathbf{r}(r, \theta) = (r\cos\theta)\mathbf{i} + (r\sin\theta)\mathbf{j} + \theta\mathbf{k}$, $0 \le r \le 1$, $0 \le \theta \le 2\pi$

$$\mathbf{r}_r \times \mathbf{r}_\theta = \begin{vmatrix} \mathbf{i} & \mathbf{j} & \mathbf{k} \\ \cos\theta & \sin\theta & 0 \\ -r\sin\theta & r\cos\theta & 1 \end{vmatrix}$$
$$= (\sin\theta)\mathbf{i} - (\cos\theta)\mathbf{j} + r\mathbf{k}$$

$|\mathbf{r}_r \times \mathbf{r}_\theta| = \sqrt{1 + r^2}$; $\delta = 2\sqrt{x^2 + y^2}$
$= 2\sqrt{r^2\cos^2\theta + r^2\sin^2\theta} = 2r$

$$\text{Mass} = \int\int_S \delta(x, y, z)\, d\sigma = \int_0^{2\pi}\int_0^1 2r\sqrt{1 + r^2}\, dr\, d\theta$$
$$= \int_0^{2\pi}\left[\frac{2}{3}(1 + r^2)^{3/2}\right]_0^1 d\theta$$
$$= \int_0^{2\pi}\frac{2}{3}(2\sqrt{2} - 1)\, d\theta = \frac{4\pi}{3}(2\sqrt{2} - 1)$$

86. $M = x^2 + 4xy$ and $N = -6y$

$$\frac{\partial M}{\partial x} = 2x + 4y \text{ and } \frac{\partial N}{\partial x} = -6$$

$$\text{Flux} = \int_0^b \int_0^a (2x + 4y - 6)\,dx\,dy$$
$$= \int_0^b (a^2 + 4ay - 6a)\,dy = a^2b + 2ab^2 - 6ab.$$

We want to minimize $f(a, b) = a^2b + 2ab^2 - 6ab$
$= ab(a + 2b - 6)$. Thus $f_a(a, b) = 2ab + 2b^2 - 6b = 0$
and $f_b(a, b) = a^2 + 4ab - 6a = 0$
$b(2a + 2b - 6) = 0$
$b = 0$ or $b = -a + 3$. Now $b = 0$
$a^2 - 6a = 0$
$a = 0$ or $a = 6$
$(0, 0)$ and $(6, 0)$ are critical points. On the other hand,
$b = -a + 3$
$a^2 + 4a(-a + 3) - 6a = 0$
$-3a^2 + 6a = 0$
$a = 0$ or $a = 2$
$(0, 3)$ and $(2, 1)$ are also critical points. The flux at $(0, 0) = 0$, the flux at $(6, 0) = 0$, the flux at $(0, 3) = 0$ and the flux at $(2, 1) = -4$. Thus, the rectangular region defined by: $0 \le x \le 2$, $0 \le y \le 1$ has the least total outward flux with value -4.

87. Let the plane be given by $z = ax + by$ and let $f(x, y, z) = x^2 + y^2 + z^2 = 4$. Let C denote the circle of intersection of the plane with the sphere. By Stokes's Theorem,

$\oint_C \mathbf{F} \cdot d\mathbf{r} = \iint_S \nabla \times \mathbf{F} \cdot \mathbf{n}\,d\sigma$, where $\mathbf{n}$ is a unit normal to the plane. Let $r(x, y) = x\mathbf{i} + y\mathbf{j} + (ax + by)\mathbf{k}$ be a parametrization of the surface.

$$\text{Then } \mathbf{r}_x \times \mathbf{r}_y = \begin{vmatrix} \mathbf{i} & \mathbf{j} & \mathbf{k} \\ 1 & 0 & a \\ 0 & 1 & b \end{vmatrix} = -a\mathbf{i} - b\mathbf{j} + \mathbf{k}$$

$d\sigma = |\mathbf{r}_x \times \mathbf{r}_y|\,dx\,dy = \sqrt{a^2 + b^2 + 1}\,dx\,dy$. Also,

$$\nabla \times \mathbf{F} = \begin{vmatrix} \mathbf{i} & \mathbf{j} & \mathbf{k} \\ \frac{\partial}{\partial x} & \frac{\partial}{\partial y} & \frac{\partial}{\partial z} \\ z & x & y \end{vmatrix} = \mathbf{i} + \mathbf{j} + \mathbf{k} \text{ and}$$

$$\mathbf{n} = \frac{a\mathbf{i} + b\mathbf{j} - \mathbf{k}}{\sqrt{a^2 + b^2 + 1}}$$

$$\iint_S \nabla \times \mathbf{F} \cdot \mathbf{n}\,d\sigma$$
$$= \iint_{R_{xy}} \frac{a + b + 1}{\sqrt{a^2 + b^2 + 1}} \sqrt{a^2 + b^2 + 1}\,dx\,dy$$
$$= \iint_{R_{xy}} (a + b - 1)\,dx\,dy = (a + b - 1)\iint_{R_{xy}} dx\,dy.$$

Now $x^2 + y^2 + (ax + by)^2 = 4$

$$\left(\frac{a^2 + 1}{4}\right)x^2 + \left(\frac{b^2 + 1}{4}\right)y^2 + \left(\frac{ab}{2}\right)xy = 1$$

the region R_{xy} is the interior of the ellipse $Ax^2 + Bxy + Cy^2 = 1$ in the xy-plane, where $A = \frac{a^2 + 1}{4}$, $B = \frac{ab}{2}$, and $C = \frac{b^2 + 1}{4}$. By Exercise 47 in Section 9.3, the area of the ellipse is $\frac{2\pi}{\sqrt{4AC - B^2}} = \frac{4\pi}{\sqrt{a^2 + b^2 + 1}}$

$\oint_C \mathbf{F} \cdot d\mathbf{r} = h(a, b) = \frac{4\pi(a + b - 1)}{\sqrt{a^2 + b^2 + 1}}$. thus we optimize

$$H(a, b) = \frac{(a + b - 1)^2}{a^2 + b^2 + 1}:$$

$$\frac{\partial H}{\partial a} = \frac{2(a + b - 1)(b^2 + 1 + a - ab)}{a^2 + b^2 + 1} = 0 \text{ and}$$

$$\frac{\partial H}{\partial b} = \frac{2(a + b - 1)(a^2 + 1 + b - ab)}{a^2 + b^2 + 1} = 0$$

$a + b - 1 = 0$, or $b^2 + 1 + a - ab = 0$ and
$a^2 + 1 + b - ab = 0$
$a + b - 1 = 0$, or $a^2 - b^2 + (b - a) = 0$
$a + b - 1 = 0$, or $(a - b)(a + b - 1) = 0$
$a + b - 1 = 0$ or $a = b$. The critical values $a + b - 1 = 0$ give a saddle. If $a = b$, then $0 = b^2 + 1 + a - ab$
$a^2 + 1 + a - a^2 = 0$
$a = -1$
$b = -1$. Thus, the point $(a, b) = (-1, -1)$ gives a local extremum for $\oint_C \mathbf{F} \cdot d\mathbf{r}$

$z = -x - y$
$x + y + z = 0$ is the desired plane. Note: Since $h(-1, -1)$ is negative, the circulation about $\mathbf{n}$ is clockwise, so $-\mathbf{n}$ is the correct pointing normal for the counterclockwise circulation. Thus $\iint_S \nabla \times \mathbf{F} \cdot (-\mathbf{n})\,d\sigma$ actually gives the maximum circulation. You may wish to obtain 3D or contour plots for the surface $H(a, b)$.

Appendix A5, Section 4

1.
$$r\cos\left(\theta - \frac{\pi}{6}\right) = 5$$
$$r\left(\cos\theta\cos\frac{\pi}{6} + \sin\theta\sin\frac{\pi}{6}\right) = 5$$
$$\frac{\sqrt{3}}{2}r\cos\theta + \frac{1}{2}r\sin\theta = 5$$
$$\frac{\sqrt{3}}{2}x + \frac{1}{2}y = 5$$
$$\sqrt{3}x + y = 10$$
$$y = -\sqrt{3}x + 10$$

3.

$$\begin{aligned} r\cos\left(\theta - \frac{4\pi}{3}\right) &= 3 \\ r\left(\cos\theta\cos\frac{4\pi}{3} + \sin\theta\sin\frac{4\pi}{3}\right) &= 3 \\ -\frac{1}{2}r\cos\theta - \frac{\sqrt{3}}{2}r\sin\theta &= 3 \\ -\frac{1}{2}x - \frac{\sqrt{3}}{2}y &= 3 \\ -\frac{\sqrt{3}}{2}y &= \frac{1}{2}x + 3 \\ y &= -\frac{\sqrt{3}}{3}x - 2\sqrt{3} \end{aligned}$$

5.

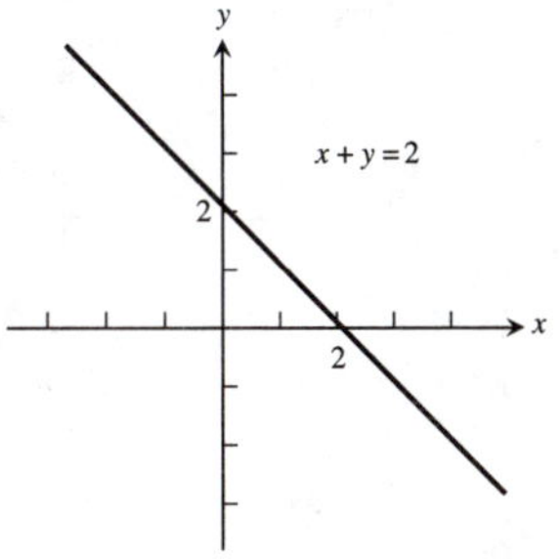

$$\begin{aligned} r\cos\left(\theta - \frac{\pi}{4}\right) &= \sqrt{2} \\ r\left(\cos\theta\cos\frac{\pi}{4} + \sin\theta\sin\frac{\pi}{4}\right) &= \sqrt{2} \\ \frac{\sqrt{2}}{2}r\cos\theta + \frac{\sqrt{2}}{2}r\sin\theta &= \sqrt{2} \end{aligned}$$

7.

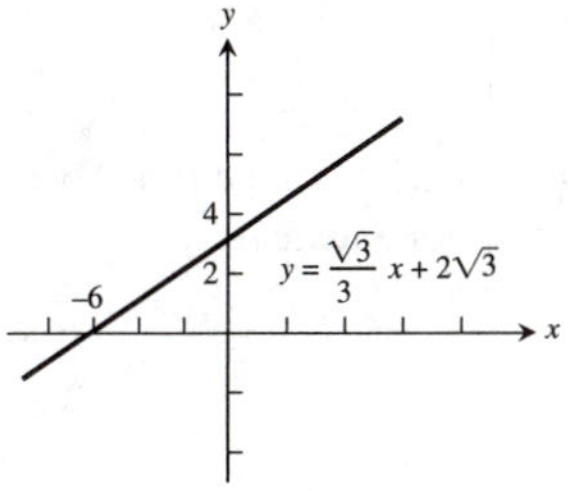

$$\begin{aligned} r\cos\left(\theta - \frac{2\pi}{3}\right) &= 3 \\ r\left(\cos\theta\cos\frac{2\pi}{3} + \sin\theta\sin\frac{2\pi}{3}\right) &= 3 \\ -\frac{1}{2}r\cos\theta + \frac{\sqrt{3}}{2}r\sin\theta &= 3 \\ -x + \sqrt{3}y &= 6 \end{aligned}$$

9.

$$\begin{aligned} \sqrt{2}x + \sqrt{2}y &= 6 \\ \frac{\sqrt{2}}{2}x + \frac{\sqrt{2}}{2}y &= 3 \\ \frac{\sqrt{2}}{2}r\cos\theta + \frac{\sqrt{2}}{2}r\sin\theta &= 3 \\ r\left(\cos\theta\cos\frac{\pi}{4} + \sin\theta\sin\frac{\pi}{4}\right) &= 3 \\ r\cos\left(\theta - \frac{\pi}{4}\right) &= 3 \end{aligned}$$

11.

$$\begin{aligned} y &= -5 \\ 0\cdot x - 1\cdot y &= 5 \\ 0\cdot r\cos\theta - 1\cdot r\sin\theta &= 5 \\ r\left(\cos\theta\cos\frac{\pi}{2} - \sin\theta\sin\frac{\pi}{2}\right) &= 5 \\ r\cos\left(\theta + \frac{\pi}{2}\right) &= 5 \end{aligned}$$

13. $r = 2a\cos\theta$
$r = 2(4)\cos\theta$
$r = 8\cos\theta$

15. $r = 2a\sin\theta$
$r = 2(\sqrt{2})\sin\theta$
$r = 2\sqrt{2}\sin\theta$

17.

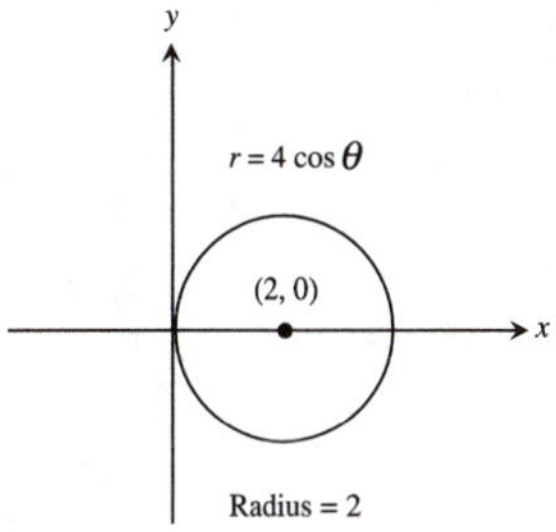

19.

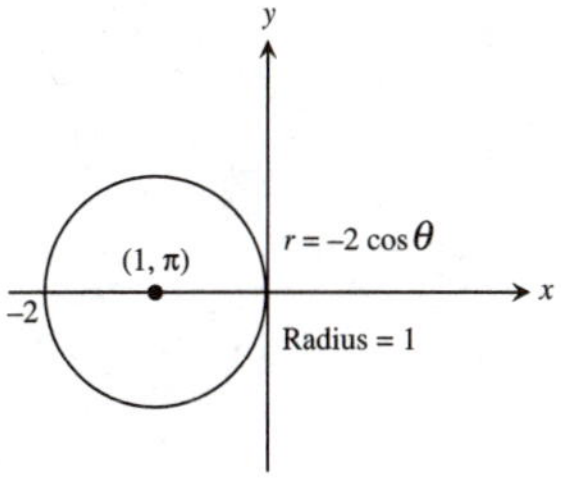

21. $(x - 6)^2 + y^2 = 6^2$

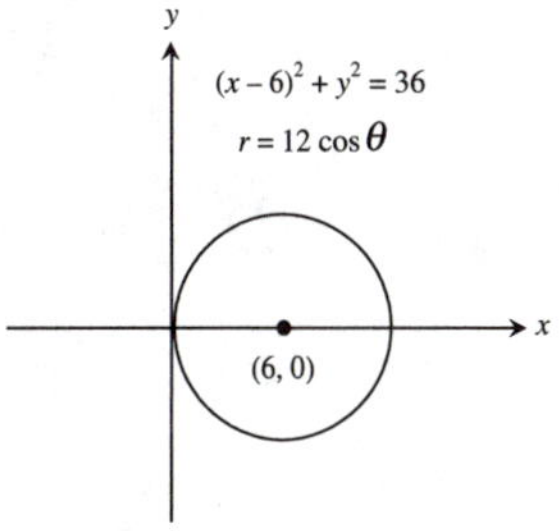

$r = 2(6)\cos\theta$
$r = 12\cos\theta$

23. $x^2 + 2x + y^2 = 0$
$x^2 + 2x + 1 + y^2 = 1$
$(x + 1)^2 + y^2 = 1$

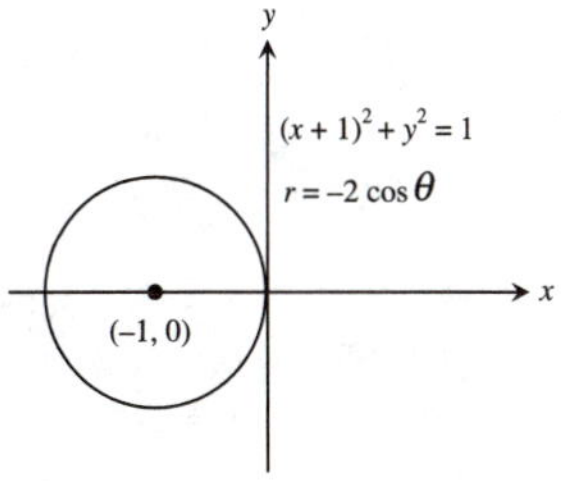

$r = -2(1)\cos\theta$
$r = -2\cos\theta$

25. $r = \dfrac{ek}{1 + e\cos\theta} = \dfrac{1 \cdot 2}{1 + 1 \cdot \cos\theta}$

$r = \dfrac{2}{1 + \cos\theta}$

27. $r = \dfrac{5 \cdot 6}{1 - 5 \cdot \sin\theta}$

$r = \dfrac{30}{1 - 5\sin\theta}$

29. $r = \dfrac{\frac{1}{2} \cdot 1}{1 + \frac{1}{2} \cdot \cos\theta} = \dfrac{\frac{1}{2}}{1 + \frac{1}{2}\cos\theta}$

$r = \dfrac{1}{2 + \cos\theta}$

31. $r = \dfrac{\frac{1}{5} \cdot 10}{1 - \frac{1}{5} \cdot \sin\theta} = \dfrac{2}{1 - \frac{1}{5}\sin\theta}$

$r = \dfrac{10}{5 - \sin\theta}$

33.

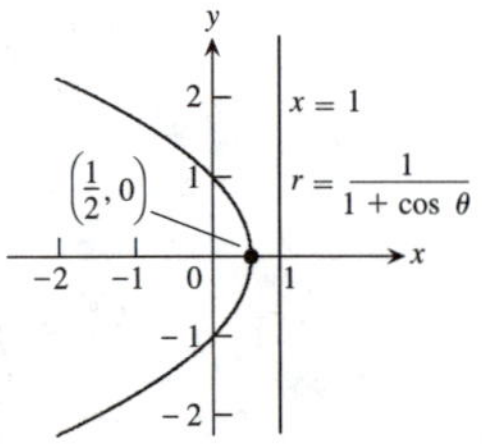

35.

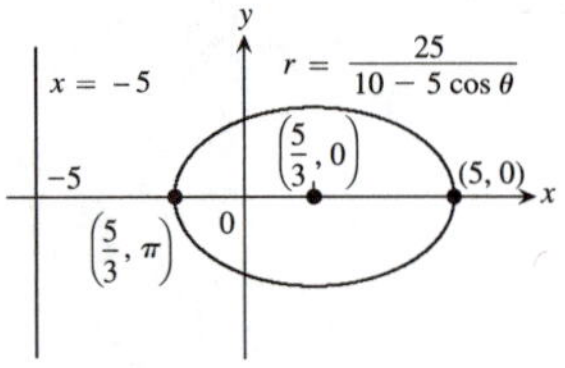

37.

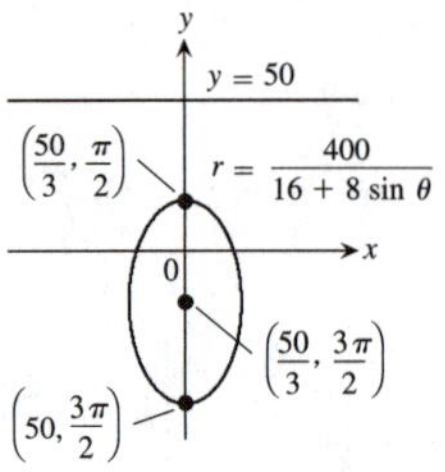

39.

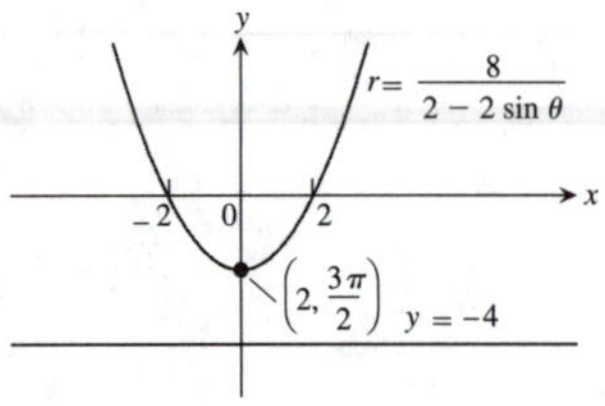

41. a.

$$r = 4\sin\theta \qquad r = \sqrt{3}\sec\theta$$
$$r^2 = 4r\sin\theta \qquad r\cos\theta = \sqrt{3}$$
$$x^2 + y^2 = 4y \qquad x = \sqrt{3}$$

b.

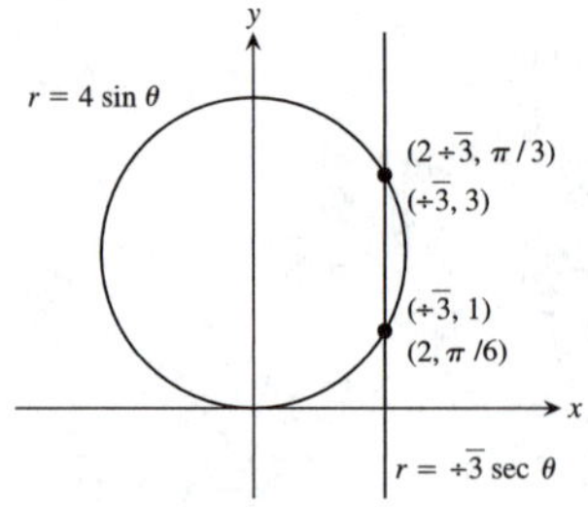

$$(\sqrt{3})^2 + y^2 = 4y$$
$$y^2 - 4y + 3 = 0$$
$$(y - 1)(y - 3) = 0$$
$$y = 1, 3$$

43. For a parabola, $e = 1$.

$r\cos\theta = x = 4$ so $k = 4$

$r = \dfrac{ke}{1 + e\cos\theta}$

$r = \dfrac{4}{1 + \cos\theta}$

45.

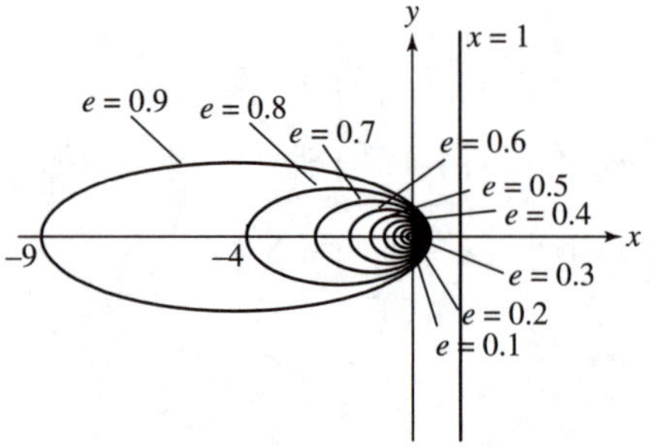

This is a sequence of ellipses, each with the right focus at the origin and directrix $x = 1$. As e increases, the ellipses stretch further to the left.

47.

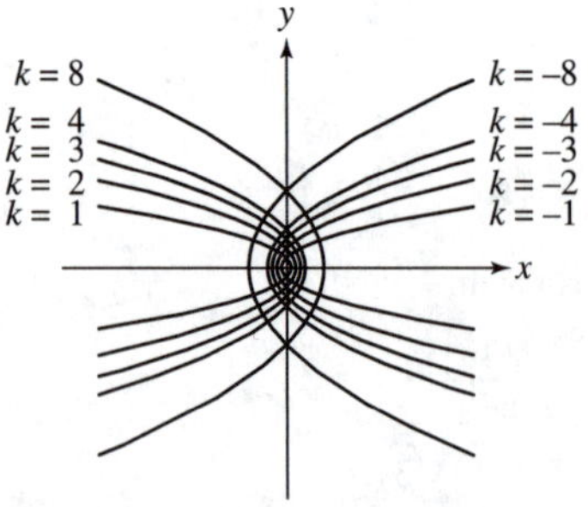

This is a sequence of parabolas, each with focus at the origin and vertex at $\left(\frac{k}{2}, 0\right)$. For negative values of k, the parabolas open right; for positive values, they open left. As $|k|$ increases, the parabolas open wider.

49.

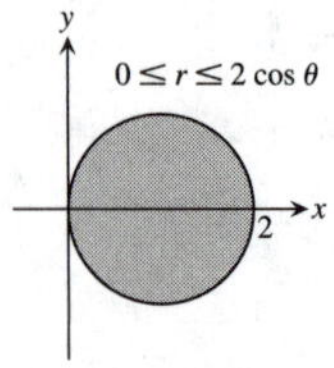

51. a. $r = \dfrac{a(1 - e^2)}{1 + e\cos\theta}$

The point on the ellipse closest to the focus is when $\theta = 0$.

$$r = \frac{a(1 - e^2)}{1 + e} = \frac{a(1 - 3)(1 + e)}{1 + e} = a(1 - e)$$

The point on the ellipse furthest from the focus is when $\theta = \pi$.

$$r = \frac{a(1 - e)^2}{1 - e} = \frac{a(1 - e)(1 + e)}{1 - e} = a(1 + e)$$

b.

Planet	Smallest Distance	Greatest Distance
Mercury	0.3871(0.7944) = 0.3075 *AU*	0.3871(1.2056) = 0.4667 *AU*
Venus	0.7233(0.9932) = 0.7184 *AU*	0.7233(1.0068) = 0.7282 *AU*
Earth	(1)(0.9833) = 0.9833 *AU*	(1)(1.0167) = 1.0167 *AU*
Mars	1.524(0.9066) = 1.382 *AU*	1.524(1.0934) = 1.6663 *AU*
Jupiter	5.203(0.9516) = 4.9512 *AU*	5.203(1.0484) = 5.4548 *AU*
Saturn	9.539(0.9457) = 9.0210 *AU*	9.359(1.0543) = 10.057 *AU*
Uranus	19.18(0.954) = 18.2977 *AU*	19.18(1.0460) = 20.0623 *AU*
Neptune	30.06(0.9918) = 29.8135 *AU*	30.06(1.0082) = 30.3065 *AU*
Pluto	39.44(0.7519) = 29.655 *AU*	39.44(1.2481) = 49.2250 *AU*

53. a. Refer to Figure A5.33: $r_{max} = a + ea$ and $r_{min} = a - ea$. $\dfrac{r_{max} - r_{min}}{r_{max} + r_{min}} = \dfrac{(a + ea) - (a - ea)}{(a + ea) + (a - ea)} = \dfrac{2ea}{2a} = e$

b. The distance from the center to one focus is c given by the relationship $e = \dfrac{c}{a}$. From the method illustrated in Figure A5.7, we have $2a = 10$ or $a = 5$. Therefore $\dfrac{c}{a} = \dfrac{c}{5} = 0.2$ so $c = 1$. The pins should be placed $2c$ or 2 inches apart.

Appendix A8

1. $\begin{vmatrix} 2 & 3 & 1 \\ 4 & 5 & 2 \\ 1 & 2 & 3 \end{vmatrix}$ Expand according to cofactors of the first row.

$2\begin{vmatrix} 5 & 2 \\ 2 & 3 \end{vmatrix} - 3\begin{vmatrix} 4 & 2 \\ 1 & 3 \end{vmatrix} + 1\begin{vmatrix} 4 & 5 \\ 1 & 2 \end{vmatrix} = 2(15 - 4) - 3(12 - 2)$
$+ 1(8 - 5) = -5$

3. $\begin{vmatrix} 1 & 2 & 3 & 4 \\ 0 & 1 & 2 & 3 \\ 0 & 0 & 2 & 1 \\ 0 & 0 & 3 & 2 \end{vmatrix}$ Expand according to cofactors of the first column.

$1\begin{vmatrix} 1 & 2 & 3 \\ 0 & 2 & 1 \\ 0 & 3 & 2 \end{vmatrix} - 0 + 0 - 0 = 1\left(1\begin{vmatrix} 2 & 1 \\ 3 & 2 \end{vmatrix} - 0 + 0\right)$
$= 1 \cdot 1(4 - 3) = 1$

5. $\begin{vmatrix} 2 & -1 & 2 \\ 1 & 0 & 3 \\ 0 & 2 & 1 \end{vmatrix}$

a. $0\begin{vmatrix} -1 & 2 \\ 0 & 3 \end{vmatrix} - 2\begin{vmatrix} 2 & 2 \\ 1 & 3 \end{vmatrix} + 1\begin{vmatrix} 2 & -1 \\ 1 & 0 \end{vmatrix} = 0 - 2(6 - 2)$
$+ 1(0 + 1) = -7$

b. $-(-1)\begin{vmatrix} 1 & 3 \\ 0 & 1 \end{vmatrix} + 0\begin{vmatrix} 2 & 2 \\ 0 & 1 \end{vmatrix} - 2\begin{vmatrix} 2 & 2 \\ 1 & 3 \end{vmatrix}$
$= 1(1 - 0) + 0 - 2(6 - 2) = -7$

7. $\begin{vmatrix} 1 & 1 & 0 & 0 \\ 0 & 0 & -2 & 1 \\ 0 & -1 & 0 & 7 \\ 3 & 0 & 2 & 1 \end{vmatrix}$

a. $0\begin{vmatrix} 1 & 0 & 0 \\ 0 & -2 & 1 \\ 0 & 2 & 1 \end{vmatrix} - (-1)\begin{vmatrix} 1 & 0 & 0 \\ 0 & -2 & 1 \\ 3 & 2 & 1 \end{vmatrix} + 0\begin{vmatrix} 1 & 1 & 0 \\ 0 & 0 & 1 \\ 3 & 0 & 1 \end{vmatrix} - 7\begin{vmatrix} 1 & 1 & 0 \\ 0 & 0 & -2 \\ 3 & 0 & 2 \end{vmatrix}$

$= 0 + 1\left(1\begin{vmatrix} -2 & 1 \\ 2 & 1 \end{vmatrix} - 0 + 0\right) + 0$

$- 7\left(-1\begin{vmatrix} 0 & -2 \\ 3 & 2 \end{vmatrix} + 0 - 0\right) = 1(-2 - 2) + 7(0 + 6) = 38$

b. $-1\begin{vmatrix} 0 & -2 & 1 \\ 0 & 0 & 7 \\ 3 & 2 & 1 \end{vmatrix} + 0\begin{vmatrix} 1 & 0 & 0 \\ 0 & 0 & 7 \\ 3 & 2 & 1 \end{vmatrix} - (-1)\begin{vmatrix} 1 & 1 & 0 \\ 0 & -2 & 1 \\ 3 & 2 & 1 \end{vmatrix} + 0\begin{vmatrix} 1 & 0 & 0 \\ 0 & -2 & 1 \\ 0 & 0 & 7 \end{vmatrix}$

$= -1\left(0 - 0 + 3\begin{vmatrix} -2 & 1 \\ 0 & 7 \end{vmatrix}\right) + 0 + 1\left(1\begin{vmatrix} -2 & 1 \\ 2 & 1 \end{vmatrix} - 0 + 0\right)$
$+ 0 = -3(-14 - 0) + 1(-2 - 2) = 38$

9. $D = \begin{vmatrix} 1 & 8 \\ 3 & -1 \end{vmatrix} = -1 - 24 = -25$

$$x = \frac{\begin{vmatrix} 4 & 8 \\ -13 & -1 \end{vmatrix}}{-25} = \frac{-4 + 104}{-25} = -4$$

$$y = \frac{\begin{vmatrix} 1 & 4 \\ 3 & -13 \end{vmatrix}}{-25} = \frac{-13 - 12}{-25} = 1$$

11. $D = \begin{vmatrix} 4 & -3 \\ 3 & -2 \end{vmatrix} = -8 + 9 = 1$

$$x = \frac{\begin{vmatrix} 6 & -3 \\ 5 & -2 \end{vmatrix}}{1} = \frac{-12 + 15}{1} = 3$$

$$y = \frac{\begin{vmatrix} 4 & 6 \\ 3 & 5 \end{vmatrix}}{1} = \frac{20 - 18}{1} = 2$$

13. $D = \begin{vmatrix} 2 & 1 & -1 \\ 1 & -1 & 1 \\ 2 & 2 & 1 \end{vmatrix} = (-2) + 2 + (-2) - 2 - 4 - 1 = -9$

$$x = \frac{\begin{vmatrix} 2 & 1 & -1 \\ 7 & -1 & 1 \\ 4 & 2 & 1 \end{vmatrix}}{-9} = \frac{(-2) + 4 + (-14) - 4 - 4 - 7}{-9} = 3$$

$$y = \frac{\begin{vmatrix} 2 & 2 & -1 \\ 1 & 7 & 1 \\ 2 & 4 & 1 \end{vmatrix}}{-9} = \frac{14 + 4 + (-4) - (-14) - 8 - 2}{-9} = -2$$

$$z = \frac{\begin{vmatrix} 2 & 1 & 2 \\ 1 & -1 & 7 \\ 2 & 2 & 4 \end{vmatrix}}{-9} = \frac{(-8) + 14 + 4 - (-4) - 28 - 4}{-9} = 2$$

15. $D = \begin{vmatrix} 1 & 0 & -1 \\ 0 & 2 & -2 \\ 2 & 0 & 1 \end{vmatrix} = 2 + 0 + 0 - (-4) - 0 - 0 = 6$

$$x = \frac{\begin{vmatrix} 3 & 0 & -1 \\ 2 & 2 & -2 \\ 3 & 0 & 1 \end{vmatrix}}{6} = \frac{6 + 0 + 0 - (-6) - 0 - 0}{6} = 2$$

$$y = \frac{\begin{vmatrix} 1 & 3 & -1 \\ 0 & 2 & -2 \\ 2 & 3 & 1 \end{vmatrix}}{6} = \frac{2 + (-12) + 0 - (-4) - (-6) - 0}{6} = 0$$

$$z = \frac{\begin{vmatrix} 1 & 0 & 3 \\ 0 & 2 & 2 \\ 2 & 0 & 3 \end{vmatrix}}{6} = \frac{6 + 0 + 0 - 12 - 0 - 0}{6} = -1$$

17. $D = \begin{vmatrix} 2 & h \\ 1 & 3 \end{vmatrix} = 6 - h$ and $D = 0$ if $h = 6$

$$y = \frac{\begin{vmatrix} 2 & 8 \\ 1 & k \end{vmatrix}}{6 - h} = \frac{2k - 8}{6 - h} \text{ and } 2k - 8 = 0 \text{ if } k = 4$$

The system will have (a) infinitely many solutions if $h = 6$, $k = 4$ and (b) no solutions if $h = 6$, $k \neq 4$.

19. Given $au + bv + cw = 0$, differentiate to obtain $au' + bv' + cw' = 0$. Since the functions are twice differentiable, differentiate again to obtain $au'' + bv'' + cw'' = 0$. We have the system of three equations in the three unknowns a, b, and c.

$$\begin{aligned} au + bv + cw &= 0 \\ au' + bv' + cw' &= 0 \\ au'' + bv'' + cw'' &= 0 \end{aligned}$$

Since a, b, and c are not all zero, then we must have

$$\begin{vmatrix} u & v & w \\ u' & v' & w' \\ u'' & v'' & w'' \end{vmatrix} = 0.$$